Autodesk Revit 2016 BIM Management: Template and Family Creation

ASCENT – Center for Technical Knowledge®

SDC Publications
P.O. Box 1334
Mission KS 66222
913-262-2664
www.SDCpublications.com
Publisher: Stephen Schroff

ISBN-13: 978-1-58503-976-0
ISBN-10: 1-58503-976-4

Printed and bound in the United States of America.

Preface .... **vii**

In this Guide .... **ix**

Practice Files .... **xiii**

Feedback .... **xv**

Customizing the Interface .... **xvii**

Chapter 1: Creating Custom Templates **1-1**

1.1 Preparing Project Templates **1-2**
Managing Settings 1-4
Families in Templates 1-4
Views and Sheets 1-5
Setting Default Template Files 1-7

1.2 Customizing Annotation Types **1-8**
Creating Text Types 1-8
Creating Dimension Types 1-10
Creating System Tags 1-12
Loading Tags in a Template 1-14

Practice 1a Prepare Templates (All Disciplines) **1-16**

1.3 Creating Title Blocks **1-19**
Adding Labels 1-20
Adding Revision Schedules 1-22
Adding Sheets to Project Templates 1-24

Practice 1b Create Title Blocks (All Disciplines) **1-26**

1.4 View Templates **1-33**
Applying View Templates 1-34
Creating View Templates 1-38
Visibility/ Graphic Overrides Filters 1-40

Practice 1c View Templates for Architectural and Structural Projects **1-43**

Practice 1d View Templates for MEP Projects 1-46

Chapter Review Questions 1-50

Command Summary 1-52

Chapter 2: Schedules 2-1

2.1 Introduction to Schedules 2-2
Importing and Exporting Schedules 2-3

2.2 Creating Building Component Schedules 2-5
Schedule Properties 2-11
Filtering Elements from Schedules 2-12

Practice 2a Create Schedules for Architectural Projects 2-14

Practice 2b Create Schedules for MEP Projects 2-20

Practice 2c Create Schedules for Structural Projects 2-24

2.3 Modifying Schedule Appearance 2-29
Adding Image Fields and Images 2-36

Practice 2d Modify Schedules for Architectural Projects 2-39

Practice 2e Modify Schedules for MEP Projects 2-43

Practice 2f Modify Schedules for Structural Projects 2-47

2.4 Creating Key Schedules 2-50

2.5 Advanced Schedule Options 2-54
Creating Project Parameters 2-54
Creating Fields from Formulas 2-58
Conditional Formatting 2-61
Embedded Schedules 2-62

2.6 Creating Material Takeoff Schedules 2-64

Practice 2g Create Complex Schedules for Architectural Projects 2-67

Practice 2h Create Complex Schedules for MEP Projects 2-74

Practice 2i Create Complex Schedules for Structural Projects 2-79

Chapter Review Questions 2-86

Command Summary 2-88

Chapter 3: Custom System Families 3-1

3.1 Creating Wall, Roof, Floor, and Ceiling Types 3-2

Practice 3a Create Compound Wall Types 3-7

3.2 Vertically Compound Walls 3-11

Practice 3b Create Vertically Compound Walls 3-16

3.3 Stacked and Embedded Walls **3-20**

Practice 3c Create Stacked and Embedded Walls **3-23**

3.4 Creating MEP System Families **3-25**

Chapter Review Questions **3-32**

Command Summary **3-34**

Chapter 4: Component Family Concepts **4-1**

4.1 Creating Component Families **4-2**
Preparing to Create Families 4-4
Saving Custom Family Files 4-5

4.2 Creating the Parametric Framework **4-6**
Placing Reference Planes and Lines 4-7
Adding Dimensions and Labels 4-8
Flexing Geometry 4-10
Creating and Modifying Parameters 4-11

Practice 4a Set up a Bookcase Family **4-15**

Practice 4b Set up a Heat Pump Family **4-20**

Practice 4c Set up a Structural Column Family **4-24**

4.3 Creating Family Elements **4-29**
Creating 3D Elements 4-29
Extrusions 4-30
Blends 4-31
Revolves 4-32
Sweeps 4-33
Swept Blends 4-34
Aligning and Locking 4-35

Practice 4d Create Family Geometry for the Bookcase **4-36**

Practice 4e Create Family Geometry for the Heat Pump **4-43**

Practice 4f Create Family Geometry for the Structural Column **4-50**

4.4 Creating Family Types **4-57**
Working with Families in Projects 4-59

Practice 4g Create Family Types for the Bookcase **4-60**

Practice 4h Create Family Types for the Heat Pump **4-62**

Practice 4i Create Family Types for the Structural Column **4-64**

Chapter Review Questions **4-67**

Command Summary **4-69**

Chapter 5: Advanced Family Techniques 5-1

5.1 Additional Tools for Families 5-2
Adding Controls 5-2
Setting Room Calculation Points 5-3
Adding Connectors 5-4
Adding Openings 5-5
Adding Components 5-6

Practice 5a Add a Component to the Bookcase 5-8

Practice 5b Add a Component and Connectors to the Heat Pump . 5-16

Practice 5c Modify the Structural Column 5-25

5.2 Visibility Display Settings 5-31
Adding Lines 5-32
Creating Masking Regions 5-33

Practice 5d Modify the Visibility of Elements in the Bookcase 5-34

Practice 5e Modify the Visibility of Elements in the Heat Pump 5-40

Practice 5f Modify the Visibility of Elements in the Structural Column 5-47

Chapter Review Questions 5-52

Command Summary 5-54

Chapter 6: Additional Family Types 6-1

6.1 Creating In-Place Families 6-2

6.2 Creating Profiles 6-4

Practice 6a In-Place Families and Profiles: Door Opening 6-5

Practice 6b In-Place Families and Profiles: Concrete Corbeling 6-8

Practice 6c Use a Profile to Create a Structural Floor Type 6-13

6.3 Creating Annotation Families 6-18

Practice 6d Create Annotation Families: Arrow Symbol (All Disciplines) 6-20

6.4 Working with Project and Shared Parameters 6-22
Project Parameters 6-22
Shared Parameters 6-23

Practice 6e Work with Shared Parameters in Architectural Projects 6-28

Practice 6f Work with Shared Parameters in MEP Projects 6-37

Practice 6g Work with Shared Parameters in Structural Projects 6-45

Chapter Review Questions 6-54

Command Summary 6-56

Chapter 7: Creating Architectural Specific Families 7-1

7.1 Creating Custom Doors and Windows 7-2

Practice 7a Create Custom Doors 7-3

Practice 7b Create Custom Windows 7-10

7.2 Creating Angled Cornices and Copings 7-15
Creating Fascias 7-15

Practice 7c Create Angled Cornices and Copings 7-18

7.3 Creating Custom Railings 7-22

7.4 Families for Railings, Balusters, and Panels 7-26
Creating Rails 7-27
Creating Baluster, Post, and Panel Families 7-30
Adding Custom Posts 7-31

Practice 7d Create Custom Railings 7-33

Chapter Review Questions 7-42

Command Summary 7-45

Chapter 8: Creating MEP Specific Families 8-1

Practice 8a Upgrade an Architectural Plumbing Fixture to MEP 8-2

Practice 8b Create a Custom Lighting Fixture Family 8-6

Practice 8c Create an Data Device with Annotation Parameters 8-11

Practice 8d Create a Pipe Fitting Flange (Advanced) 8-18

Chapter Review Questions 8-34

Chapter 9: Creating Structural Specific Families 9-1

Practice 9a Parametric Gusset Plate 9-2

Practice 9b Column Stiffeners (In-Place Family) 9-9

Practice 9c In-Place Slab Depression 9-14

Practice 9d Built-Up Column 9-19

Practice 9e Tapered Concrete Column 9-26

Practice 9f Truss Family 9-32

Practice 9g Precast Hollow Core Slab 9-39

Practice 9h Tapered Moment Frame 9-47

Chapter Review Questions 9-59

Appendix A: Additional Management Tools A-1

A.1 General Settings A-2
Specifying Units A-2
Snap Settings A-4
Temporary Dimension Settings A-5
Setting Up Arrowheads A-6

A.2 Creating Object Styles A-7
Line Color A-9
Line Patterns A-10
Line Styles A-11

A.3 Creating Fill Patterns A-12

A.4 Creating Materials A-16
Working with Assets A-23

A.5 Settings for Mechanical and Electrical Projects A-24
Mechanical Settings A-24
Electrical Settings A-30

A.6 Settings for Structural Projects A-35

A.7 Sheet List Schedules A-39

A.8 Basic User Interface Customization A-42
Customizing Shortcuts A-42
Customizing Double-click Options A-44
Customizing the Browser Organization A-45

Command Summary A-47

Appendix B: Autodesk Revit Architecture Certification Exam Objectives B-1

Appendix C: Autodesk Revit MEP Certification Exam Objectives C-1

Appendix D: Autodesk Revit Structure Certification Exam Objectives D-1

Index Index-1

Building Information Modeling (BIM) is an approach to the entire building life cycle. Autodesk® Revit® Architecture, Autodesk® Revit® MEP, and Autodesk® Revit® Structure are powerful BIM programs that support the ability to coordinate, update, and share design data with team members throughout the design construction and management phases of a building's life. A key component in managing the BIM process is to establish a company foundation for different types of projects by creating standard templates and custom elements. Having this in place makes the process of any new project flow smoothly and efficiently.

The objective of the *Autodesk® Revit® 2016 BIM Management: Template and Family Creation* training guide is to enable students who have worked with the software to expand their knowledge in setting up office standards with templates that include annotation styles, preset views, sheets, and schedules, as well as creating custom element types and families.

This training guide can be used with any of the Autodesk Revit software programs (i.e., Architecture, MEP, or Structure), or a combination of them. The guide contains practices that are specific to each discipline.

Topics Covered

- Create custom templates with annotation styles, title blocks, and custom element types.
- Create schedules, including material takeoff schedules with formula.
- Create custom wall, roof, and floor types as well as MEP system families.
- Set up a component family file with a parametric framework.
- Create family geometry.
- Create family types.
- Create specific families, including in-place families, profiles, annotations, and parameters.

The training guide also contains discipline-specific practices for families, including: doors, windows, railings, pipe fittings, light fixtures, gusset plates, and built-up columns.

Note on Software Setup

This training guide assumes a standard installation of the software using the default preferences during installation. Lectures and practices use the standard software templates and default options for the Content Libraries.

Students and Educators can Access Free Autodesk Software and Resources

Autodesk challenges you to get started with free educational licenses for professional software and creativity apps used by millions of architects, engineers, designers, and hobbyists today. Bring Autodesk software into your classroom, studio, or workshop to learn, teach, and explore real-world design challenges the way professionals do.

Get started today - register at the Autodesk Education Community and download one of the many Autodesk software applications available.

Visit www.autodesk.com/joinedu/

Note: Free products are subject to the terms and conditions of the end-user license and services agreement that accompanies the software. The software is for personal use for education purposes and is not intended for classroom or lab use.

In this Guide

The following images highlight some of the features that can be found in this Training Guide.

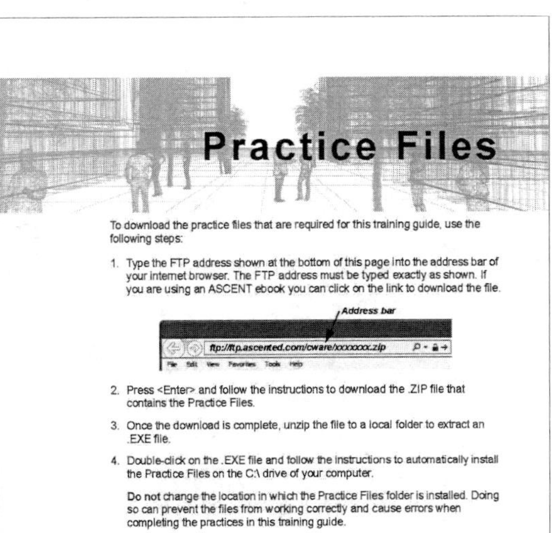

Practice Files

To download the practice files that are required for this training guide, use the following steps:

1. Type the FTP address shown at the bottom of this page into the address bar of your internet browser. The FTP address must be typed exactly as shown. If you are using an ASCENT ebook you can click on the link to download the file.
2. Press <Enter> and follow the instructions to download the .ZIP file that contains the Practice Files.
3. Once the download is complete, unzip the file to a local folder to extract an .EXE file.
4. Double-click on the .EXE file and follow the instructions to automatically install the Practice Files on the C:\ drive of your computer.

   Do not change the location in which the Practice Files folder is installed. Doing so can prevent the files from working correctly and cause errors when completing the practices in this training guide.

ftp://ftp.ascented.com/cware/xxxxxxxx.zip

FTP link for practice files

Practice Files

The Practice Files page tells you how to download and install the practice files that are provided with this training guide.

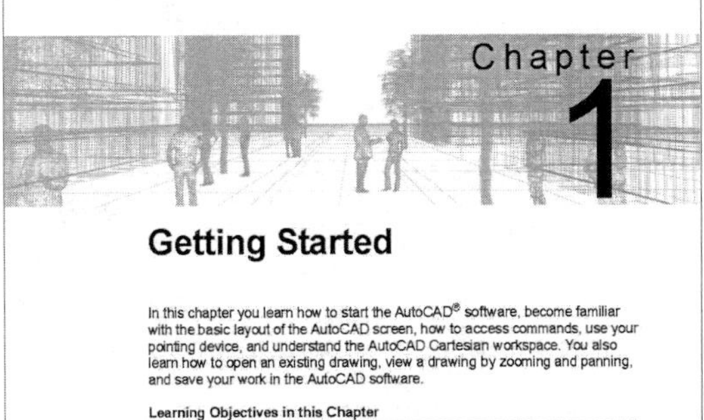

Chapter 1

Getting Started

In this chapter you learn how to start the AutoCAD® software, become familiar with the basic layout of the AutoCAD screen, how to access commands, use your pointing device, and understand the AutoCAD Cartesian workspace. You also learn how to open an existing drawing, view a drawing by zooming and panning, and save your work in the AutoCAD software.

Learning Objectives in this Chapter

- Launch the AutoCAD software and complete a basic initial setup of the drawing environment.
- Identify the basic layout and features of AutoCAD interface including the Ribbon, Drawing Window, and Application Menu.
- Locate commands and launch them using the Ribbon, shortcut menus, Application Menu, and Quick Access Toolbar.
- Locate points in the AutoCAD Cartesian workspace.
- Open and close existing drawings and navigate to file locations.
- Move around a drawing using the mouse, the **Zoom** and **Pan** commands, and the Navigation Bar.
- Save drawings in various formats and set the automatic save options using the **Save** commands.

Learning Objectives for the chapter

Chapters

Each chapter begins with a brief introduction and a list of the chapter's Learning Objectives.

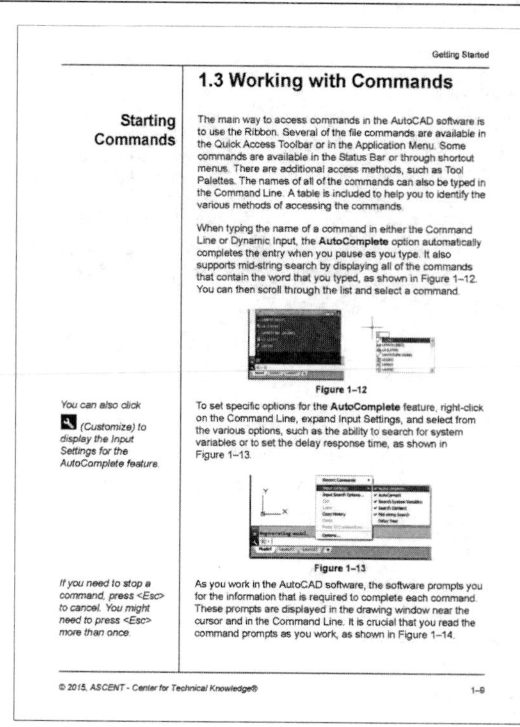
Getting Started

1.3 Working with Commands

Starting Commands

The main way to access commands in the AutoCAD software is to use the Ribbon. Several of the file commands are available in the Quick Access Toolbar or in the Application Menu. Some commands are available in the Status Bar or through shortcut menus. There are additional access methods, such as Tool Palettes. The names of all of the commands can also be typed in the Command Line. A table is included to help you to identify the various methods of accessing the commands.

When typing the name of a command in either the Command Line or Dynamic Input, the **AutoComplete** option automatically completes the entry when you pause as you type. It also supports mid-string search by displaying all of the commands that contain the word that you typed, as shown in Figure 1–12. You can then scroll through the list and select a command.

Figure 1–12

You can also click (Customize) to display the Input Settings for the AutoComplete feature.

To set specific options for the **AutoComplete** feature, right-click on the Command Line, expand Input Settings, and select from the various options, such as the ability to search for system variables or to set the delay response time, as shown in Figure 1–13.

Figure 1–13

If you need to stop a command, press <Esc> to cancel. You might need to press <Esc> more than once.

As you work in the AutoCAD software, the software prompts you for the information that is required to complete each command. These prompts are displayed in the drawing window near the cursor and in the Command Line. It is crucial that you read the command prompts as you work, as shown in Figure 1–14.

© 2015, ASCENT - Center for Technical Knowledge® 1–9

Side notes

Side notes are hints or additional information for the current topic.

Instructional Content

Each chapter is split into a series of sections of instructional content on specific topics. These lectures include the descriptions, step-by-step procedures, figures, hints, and information you need to achieve the chapter's Learning Objectives.

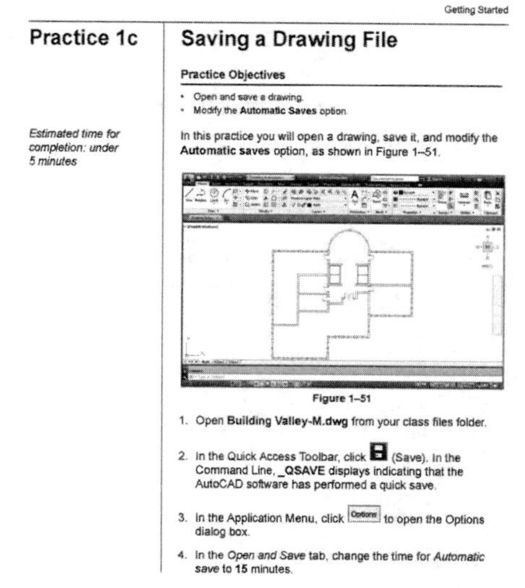
Getting Started

Practice 1c

Saving a Drawing File

Practice Objectives

- Open and save a drawing.
- Modify the **Automatic Saves** option.

Estimated time for completion: under 5 minutes

In this practice you will open a drawing, save it, and modify the **Automatic saves** option, as shown in Figure 1–51.

Figure 1–51

1. Open **Building Valley-M.dwg** from your class files folder.
2. In the Quick Access Toolbar, click (Save). In the Command Line, **_QSAVE** displays indicating that the AutoCAD software has performed a quick save.
3. In the Application Menu, click Options to open the Options dialog box.
4. In the *Open and Save* tab, change the time for *Automatic save* to **15** minutes.

Practice Objectives

Practices

Practices enable you to use the software to perform a hands-on review of a topic.

Some practices require you to use prepared practice files, which can be downloaded from the link found on the Practice Files page.

Getting Started

Chapter Review Questions

1. How do you switch from the drawing window to the text window?
   a. Use the icons in the Status Bar.
   b. Press <Tab>.
   c. Press <F2>.
   d. Press the <Spacebar>.
2. How can you cancel a command using the keyboard?
   a. Press <F2>.
   b. Press <Esc>.
   c. Press <Ctrl>.
   d. Press <Delete>.
3. What is the quickest way to repeat a command?
   a. Press <Esc>.
   b. Press <F2>.
   c. Press <Enter>.
   d. Press <Ctrl>.
4. To display a specific Ribbon panel, you can right-click on the Ribbon and select the required panel in the shortcut menu.
   a. True
   b. False
5. How are points specified in the AutoCAD Cartesian workspace?
   a. X value x Y value

Chapter Review Questions

Chapter review questions, located at the end of each chapter, enable you to review the key concepts and learning objectives of the chapter.

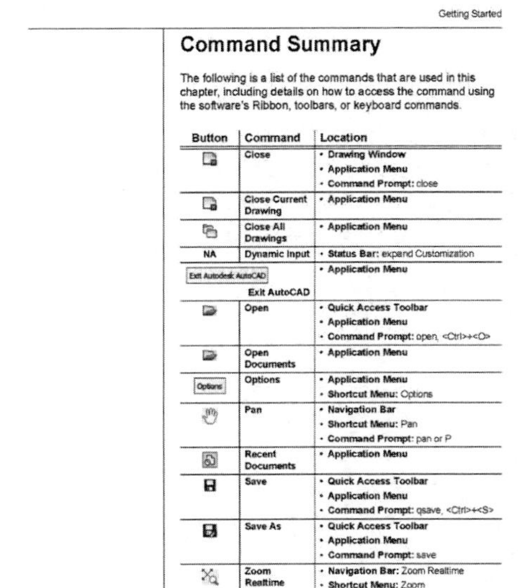

Getting Started

Command Summary

The following is a list of the commands that are used in this chapter, including details on how to access the command using the software's Ribbon, toolbars, or keyboard commands.

Button	Command	Location
	Close	• **Drawing Window** • **Application Menu** • **Command Prompt:** close
	Close Current Drawing	• **Application Menu**
	Close All Drawings	• **Application Menu**
NA	Dynamic Input	• **Status Bar:** expand Customization
Exit Autodesk AutoCAD	Exit AutoCAD	• **Application Menu**
	Open	• **Quick Access Toolbar** • **Application Menu** • **Command Prompt:** open, <Ctrl>+<O>
	Open Documents	• **Application Menu**
Options	Options	• **Application Menu** • **Shortcut Menu:** Options
	Pan	• **Navigation Bar** • **Shortcut Menu:** Pan • **Command Prompt:** pan or P
	Recent Documents	• **Application Menu**
	Save	• **Quick Access Toolbar** • **Application Menu** • **Command Prompt:** qsave, <Ctrl>+<S>
	Save As	• **Quick Access Toolbar** • **Application Menu** • **Command Prompt:** save
	Zoom Realtime	• **Navigation Bar:** Zoom Realtime • **Shortcut Menu:** Zoom

Command Summary

The Command Summary is located at the end of each chapter. It contains a list of the software commands that are used throughout the chapter, and provides information on where the command is found in the software.

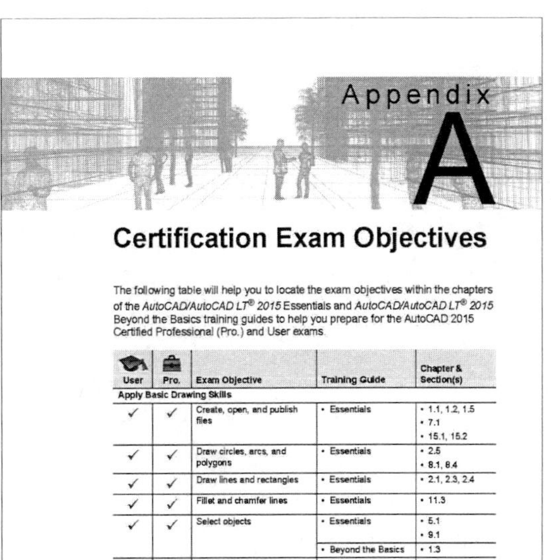

Appendix A

Certification Exam Objectives

The following table will help you to locate the exam objectives within the chapters of the *AutoCAD/AutoCAD LT® 2015* Essentials and *AutoCAD/AutoCAD LT® 2015* Beyond the Basics training guides to help you prepare for the AutoCAD 2015 Certified Professional (Pro.) and User exams.

User	Pro.	Exam Objective	Training Guide	Chapter & Section(s)
Apply Basic Drawing Skills				
✓	✓	Create, open, and publish files	• Essentials	• 1.1, 1.2, 1.5 • 7.1 • 15.1, 15.2
✓	✓	Draw circles, arcs, and polygons	• Essentials	• 2.5 • 8.1, 8.4
✓	✓	Draw lines and rectangles	• Essentials	• 2.1, 2.3, 2.4
✓	✓	Fillet and chamfer lines	• Essentials	• 11.3
✓	✓	Select objects	• Essentials	• 5.1 • 9.1
			• Beyond the Basics	• 1.3
✓	✓	Use coordinate systems	• Essentials	• 1.4
			• Beyond the Basics	• 2.1

Autodesk Certification Exam Appendix

This appendix includes a list of the topics and objectives for the Autodesk Certification exams, and the chapter and section in which the relevant content can be found.

Icons in this Training Guide

The following icons are used to help you quickly and easily find helpful information.

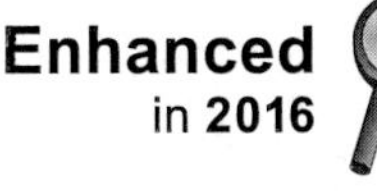	Indicates items that are new in the Autodesk Revit 2016 software.
Enhanced in **2016**	Indicates items that have been enhanced in the Autodesk Revit 2016 software.

Practice Files

To download the practice files that are required for this training guide, type the following in the address bar of your web browser:

SDCpublications.com/downloads/978-1-58503-976-0

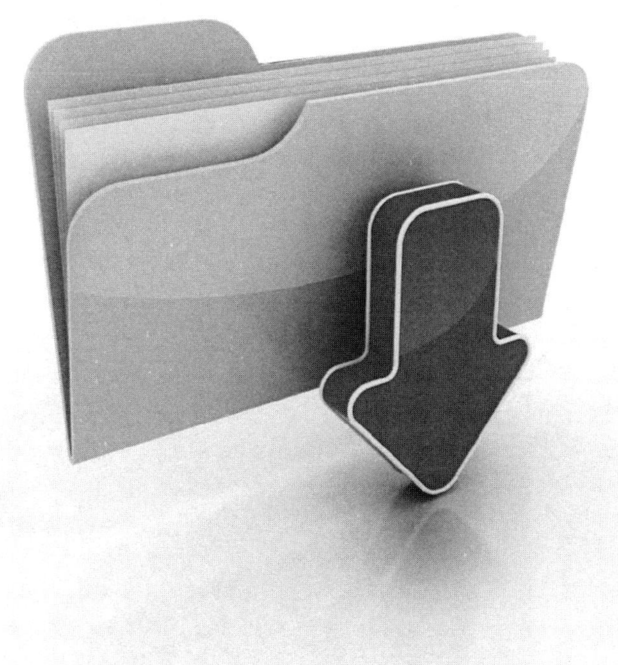

Feedback

As a part of our efforts to ensure the ongoing quality and effectiveness of our products, we invite you to submit feedback regarding this training guide. Please complete our feedback survey, or contact us at feedback@ASCENTed.com.

To access and complete the survey, type the following URL into the address bar of your Internet browser, or scan the QR code using your smartphone or tablet.

http://www.ASCENTed.com/feedback

Customizing the Interface

The Autodesk® Revit® software has three disciplines: Architecture, Structure, and MEP (Mechanical, Electrical, and Plumbing, which is also know as Systems). When using the Autodesk Building Design Suite, all of the tools for these disciplines are installed in one copy of the software. By default, all of the tools, templates, and sample files are available, as shown in Figure 1. Most users only need access to their specific set of tools and the interface can be customized to suit those needs.

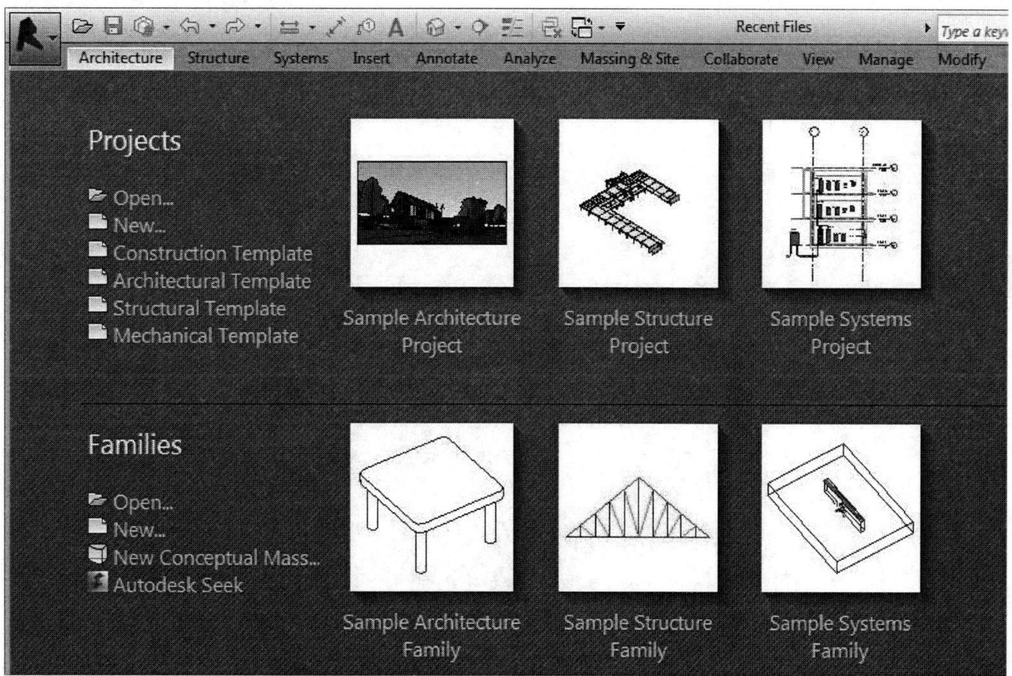

Figure 1

- The following steps describe how to customize the suite-based software with special reference to the layout of the discipline-specific software.
- This training guide uses the all-discipline interface.

How To: Set the User Interface

1. In the upper left corner of the screen, expand  (Application Menu) and click **Options**.
2. In the Options dialog box, in the left pane, select **User Interface**.

   In the *Configure* area, under *Tools and analyses* (as shown in Figure 2), clear all of options that you do not want to use.

You are not deleting these tools, just removing them from the current user interface.

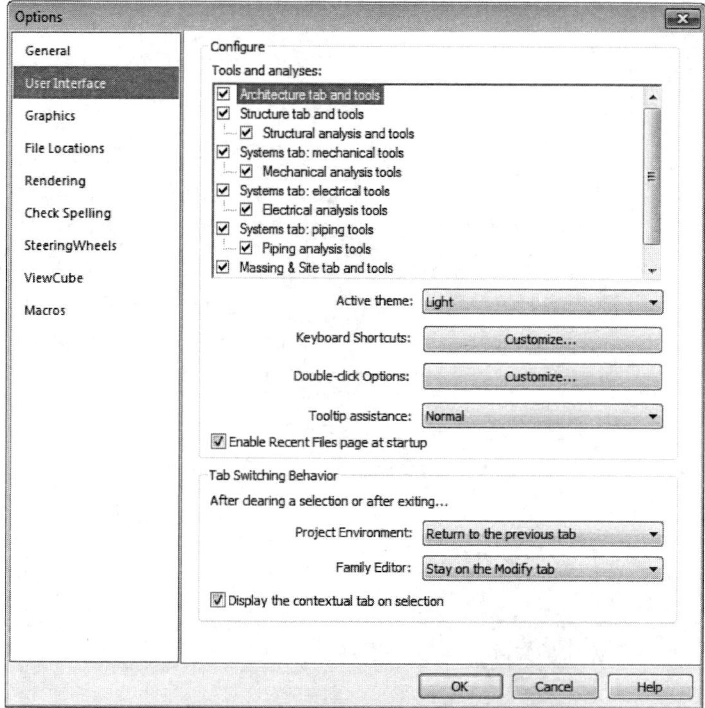

Figure 2

- To match the Autodesk Revit Architecture interface, select only the following options:
  - *Architecture* tab and tools
  - *Structure* tab and tools (but not Structural analysis and tools)
  - *Massing and Site* tab and tools
  - Energy analysis and tools

- To match the Autodesk Revit Structure interface, select only the following options:
  - *Architecture* tab and tools
  - *Structure* tab and tools including Structural analysis and tools
  - *Massing and Site* tab and tools

- To match the Autodesk Revit MEP interface, clear only the following option:
  - *Structure* tab and tools

How To: Set Template File Locations

All default templates are found in the following location: C:\ProgramData> Autodesk>RVT 2016> Templates>[units].

1. In the Options dialog box, in the left pane, select **File Locations**.
2. In the right pane (as shown in Figure 3), select and order the templates that you want to display. Typically, these are set up by the company.

Name	Path
Construction Tem...	C:\ProgramData\Autodesk\RVT 2016\Tem...
Architectural Temp...	C:\ProgramData\Autodesk\RVT 2016\Tem...
Structural Template	C:\ProgramData\Autodesk\RVT 2016\Tem...
Mechanical Templ...	C:\ProgramData\Autodesk\RVT 2016\Tem...

Figure 3

- To match the Autodesk Revit Architecture interface, select **Architectural Template** and move it to the top of the list. Remove the **Structural Template** and **Mechanical Template**.

- To match the Autodesk Revit Structure interface, select **Structural Template** and move it to the top of the list. Remove the **Architectural Template**, **Construction Template**, and **Mechanical Template**.

- To match the Autodesk Revit MEP interface, remove the **Construction Template, Architectural Template**, and **Structural Template**. Add the **Electrical Template** (**Electrical-Default.rte**) and **Systems Template** (**Systems-Default.rte**).

Setting Tab Locations

You might also want to move the tabs to a different order. To do so, select the tab, hold <Ctrl>, and drag the tab to the new location.

- To match the  Autodesk Revit Architecture interface you are not required to modify the tab locations.
- To match the  Autodesk Revit Structure interface, select the *Structure* tab and drag it to the front of the tabs.
- To match the  Autodesk Revit MEP interface, select the *Systems* tab and drag it to the front of the tabs.

Creating Custom Templates

Custom templates can save you time and provide consistency in applying office standards when creating similar projects. Templates can include items such as levels, views, sheets, schedules, and annotation types for text, dimensions, and tags. A typical family that is added to templates is a custom title block, which helps you to ensure that sheets are created with the appropriate information.

Learning Objectives in this Chapter

- Create project templates.
- Create standard text and dimension types for use in your projects.
- Modify callout, elevation, and section tags and specify which tags are loaded in a template.
- Create title blocks including detail lines, text, labels, symbols, regions, and revision schedules.
- Apply and set view templates to views.
- Create view templates.
- Set up Visibility/Graphic override filters for various categories of elements.

1.1 Preparing Project Templates

A project template is a file that contains information that can be used over and over to create new projects. The goal is to save time by using standards, enabling you to concentrate on the design. Some items in a project template include:

- Levels
- Project-based settings (e.g., Project Units, Object Styles and discipline-specific settings, etc.).
- Annotation Types
- Sheets with associated title blocks
- Views and View Templates
- Schedules and Legends
- System Families
- Component Families

You can have several project templates for different types of projects or building types, such as residential, commercial, and industrial.

If you do a lot of work for a specific client (e.g., a school system), you can also create a template specifically for their projects with associated title blocks and other information.

Defining Levels in a project template is helpful. They could be just a few basic levels for a residential project (as shown in Figure 1–1), or 50 levels for a high-rise.

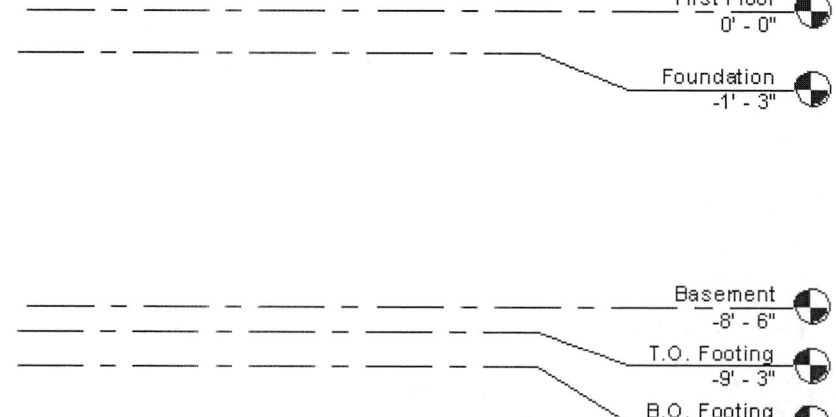

Figure 1–1

How To: Create a Project Template File

Project template files have the extension .RTE.

1. In the Application Menu, expand (New) and click (Project).
2. In the New Project dialog box, in the *Template File* area, select a template file to build from or select **None** to use a blank project file.
3. In the *Create new* area, select **Project template**, as shown in Figure 1–2.

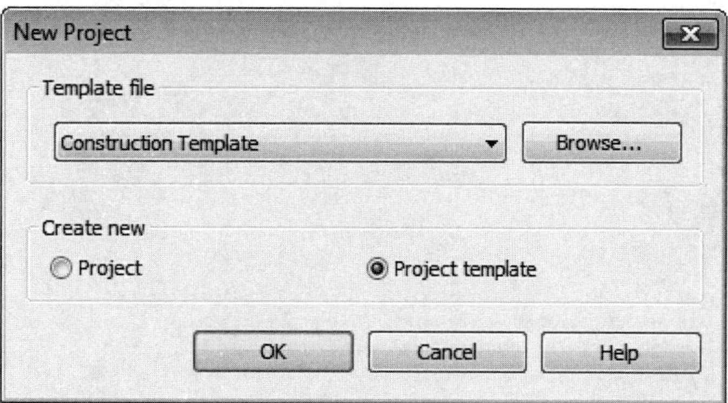

Figure 1–2

4. Click **OK**.
5. Add settings, families, views, etc., as needed to the new file.
6. Save the project template file.

- To save time, use an existing project or template that includes some of the basics you need rather than starting from scratch.

- If you select <None> in the Template file list, you are prompted to specify the initial unit system for the project: **Imperial** or **Metric**, as shown in Figure 1–3.

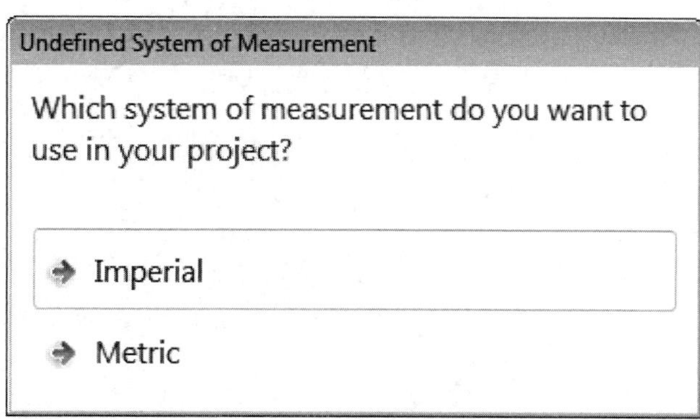

Figure 1–3

- If you need to modify the units type **UN** to quickly open the Project Units dialog box.

Managing Settings

Most of the settings stored in a template file are found in the *Manage* tab>Settings panel as shown in Figure 1–4. These settings include Units, Snaps, Temporary Dimensions, Object Styles (Lineweights, Line color, and Line patterns), Line Styles, Materials, Fill Patterns, etc.

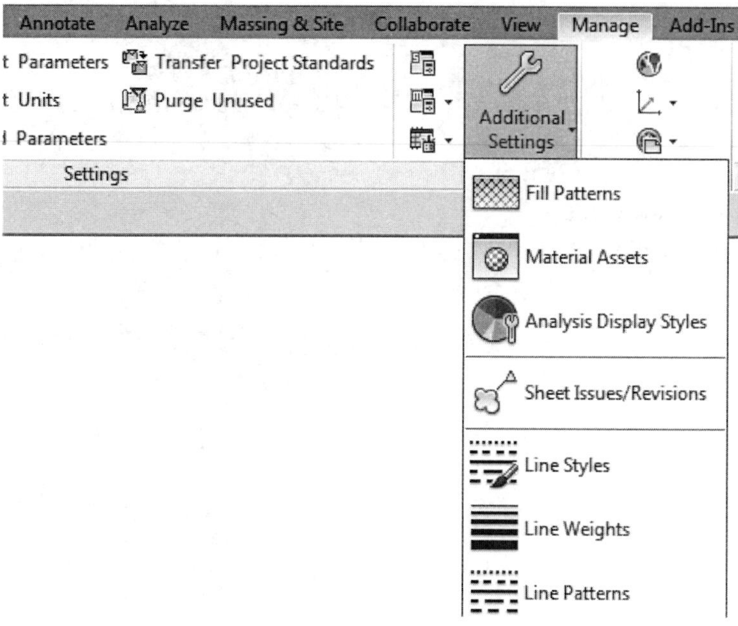

Figure 1–4

- Specific Structural Settings, MEP Settings, and Panel Schedule Templates are also included in this grouping.

Families in Templates

Walls, Wall Foundations, Floors, Slabs, Ceilings, and Roofs as well as Duct, Pipe, Cable Tray and Conduit types are created by duplicating and modifying an existing type as shown in Figure 1–5. These are known as System Families, because they are created in the system rather than in a separate family file.

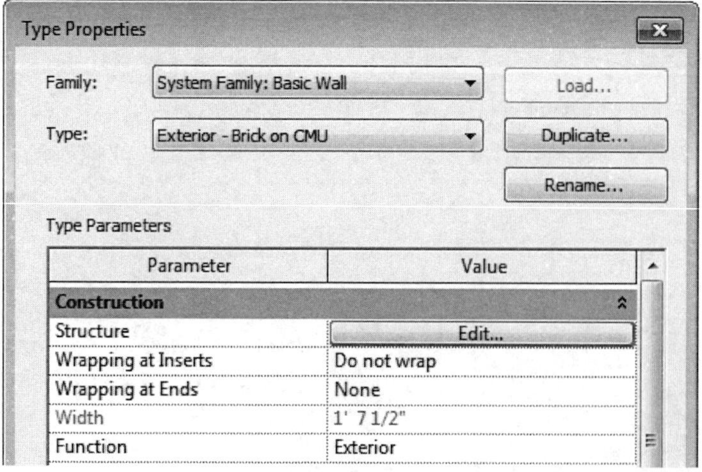

Figure 1–5

- Annotation Styles for text and dimensions are also system families and are modified in Type Properties.
- System Families can also be added when a project is in process, but it helps to have company standards created in the template.
- Component Families, such as furniture, trees, beams, columns, mechanical equipment, and electrical devices can be loaded directly in a template file if there are types and sizes that are used frequently. Otherwise, they can be loaded from a library as the project progresses.

Views and Sheets

You can create standard sheets and place empty views on the sheets. They are filled in as you proceed through a project. Views can be a few basic floor and/or ceiling plans for a residential project, as shown in Figure 1–6, or a more complex mix of Mechanical, Electrical, and Plumbing categories as shown in Figure 1–7.

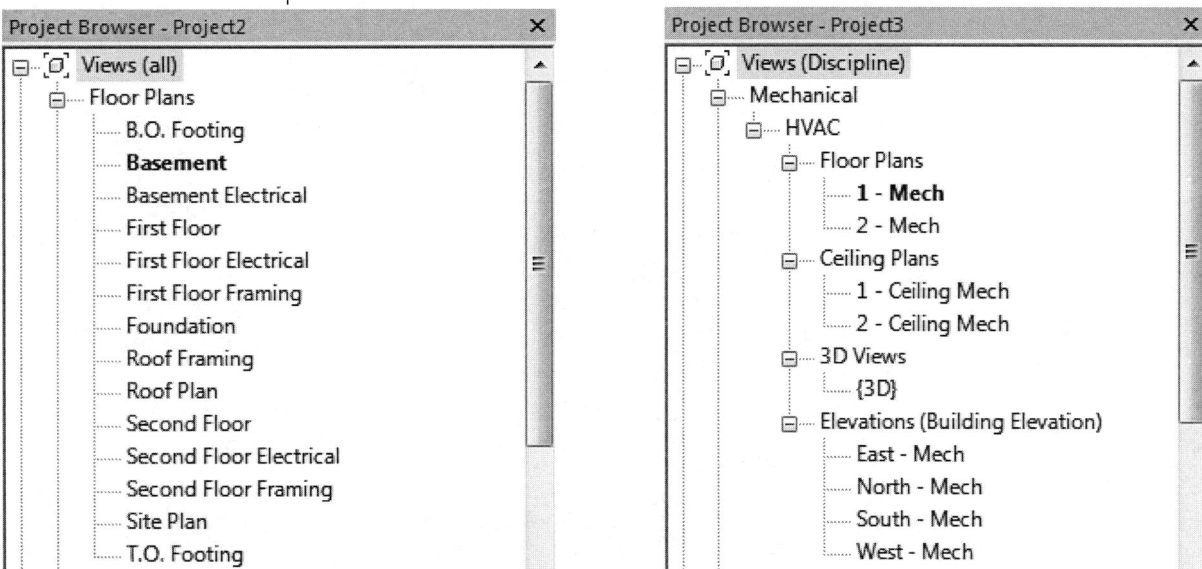

Figure 1–6 **Figure 1–7**

- You can create as many sheets as are typically used in a project or you can create a sheet schedule and populate it with sheet names that can be used as placeholder sheets.

Presetting a Starting View

When you create a project template or project, it can help to specify a starting view. This can be any of the standard views, such as plan, elevation, 3D view, or one that is specifically created. This is often a Drafting View (as shown in Figure 1–8), or a Legend View with information about the project or the cover sheet for the project.

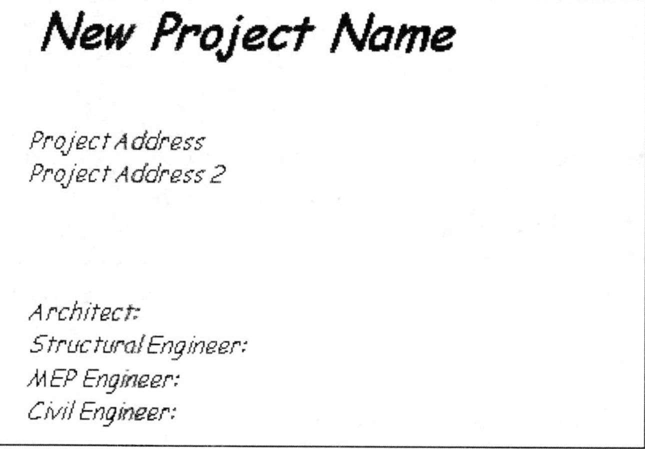

Figure 1–8

How To: Set a Starting View

1. Set up the view or sheet that you want to use as the starting view.
2. In the *Manage* tab>Manage Project panel, click (Starting View).
3. In the Starting View dialog box, select the view or sheet that you want to use as shown in Figure 1–9.

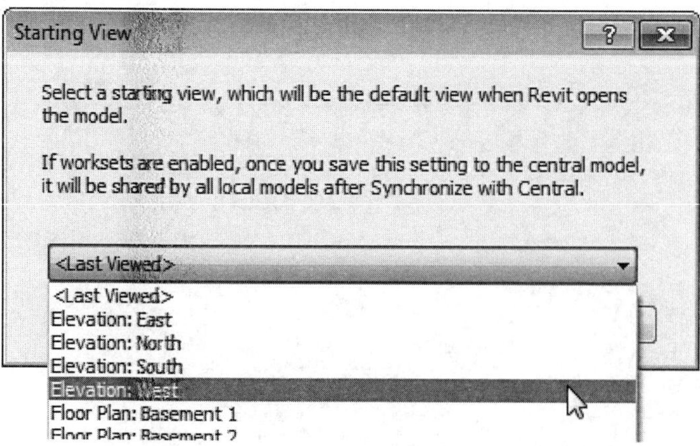

Figure 1–9

4. Click **OK**.

Setting Default Template Files

If your company uses several different templates, you can create a list that displays in the New Project dialog box, as shown in Figure 1–10. The top five templates also display on the Recent Files page.

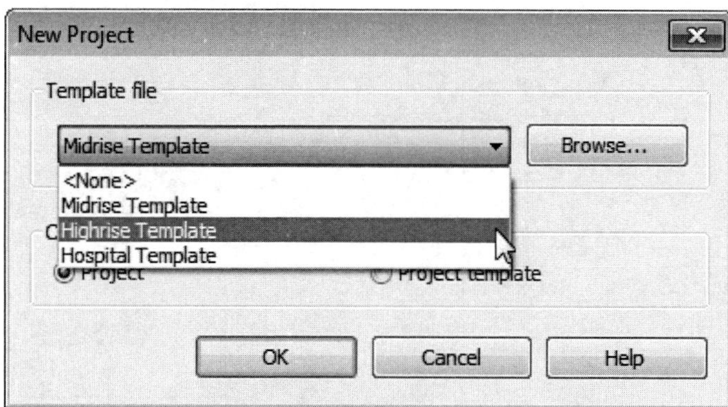

Figure 1–10

- To set the project template file list, in the Application Menu, click **Options**. In the Options dialog box, in the left pane, select **File Locations**, as shown in Figure 1–11. Specify the *Name* and *Path* for each template. Use (Add Value) to select additional templates.

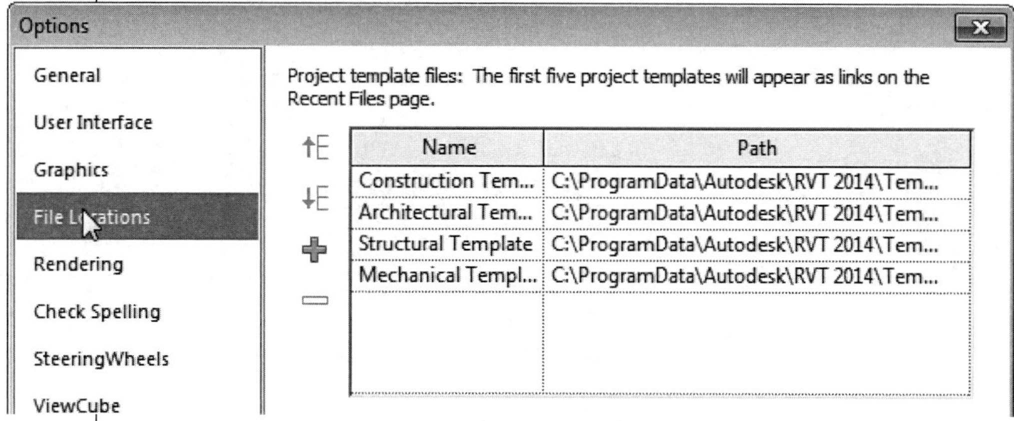

Figure 1–11

- Use (Move Rows Up) and (Move Rows Down) to order the templates. Move the templates that are used most often to the top.
- If you do not require a template anymore, select it and click (Remove Value).

1.2 Customizing Annotation Types

You can customize Annotation types in your project template file including dimensions, text types, and tags, as shown in Figure 1–12.

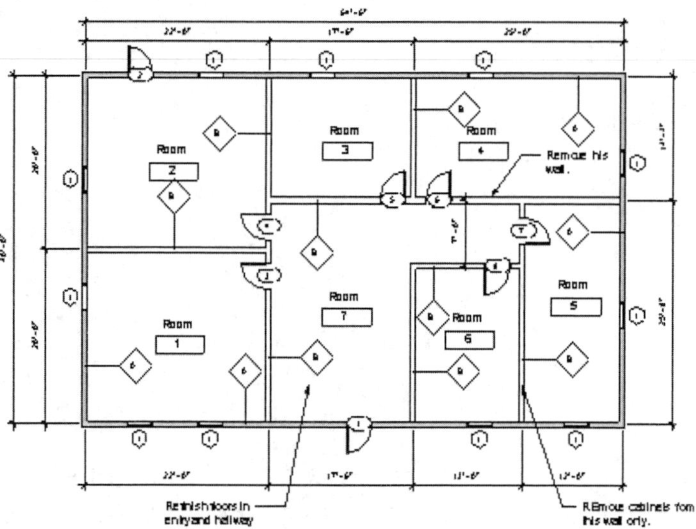

Figure 1–12

Text, Dimensions, and Arrowheads are all system families. This means they have a standard set of parameters, which you can modify and save as a type. Callout, Section, and Elevation tags can be modified within the Autodesk Revit software. Most other tags are created using families.

Creating Text Types

Text types are used to standardize text formatting (such as the font, text height, etc.), as shown in Figure 1–13. They can be created for to both annotative text and Model Text.

A FANCY FONT AT 1/4"

A HAND LETTERING FONT AT 1/8"

A HAND LETTERING FONT AT 3/32"

Figure 1–13

- The **Text** command places text at the height you need for the final plot. The view scale controls the height of the standard text in the views.

- The **Model Text** command places text that is typically used on the building, such as the signage shown in Figure 1–14. Text types for Model Text should be the full height of the text and are not affected by the view scale.

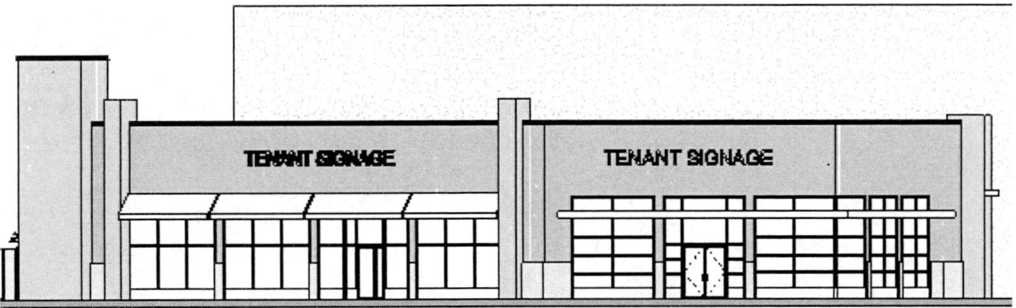

Figure 1–14

How To: Create a Text Type

1. Start the **Text** command. In Properties, click (Edit Type) or in the *Annotate* tab>Text panel title, click (Text Types). For **Model Text** you first have to place an instance, select it, and modify the type properties.
2. In the Type Properties dialog box, click **Duplicate...**.
3. Type a new name and click **OK**.The new type is activated.
4. Modify the parameters as needed for the new type, as shown for annotation text in Figure 1–15.

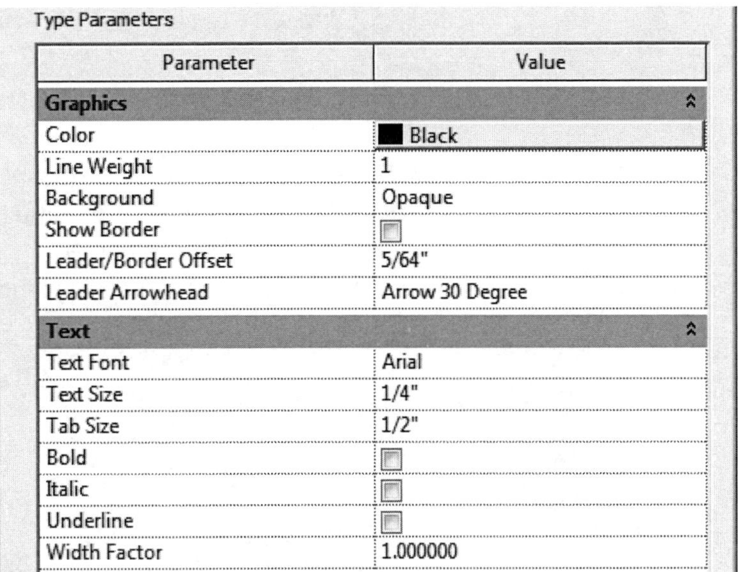

Type Parameters

Parameter	Value
Graphics	
Color	Black
Line Weight	1
Background	Opaque
Show Border	☐
Leader/Border Offset	5/64"
Leader Arrowhead	Arrow 30 Degree
Text	
Text Font	Arial
Text Size	1/4"
Tab Size	1/2"
Bold	☐
Italic	☐
Underline	☐
Width Factor	1.000000

Figure 1–15

5. Click **OK** to finish.

Creating Dimension Types

Dimensions are one of the more complex system families in terms of the number of parameters you can modify. You can create types for each dimension method.

How To: Create Dimension Types

1. In the *Annotate* tab>Dimension panel, expand the Dimension panel title (as shown in Figure 1–16), and click (wrench icon) next to the dimension type you want to create.

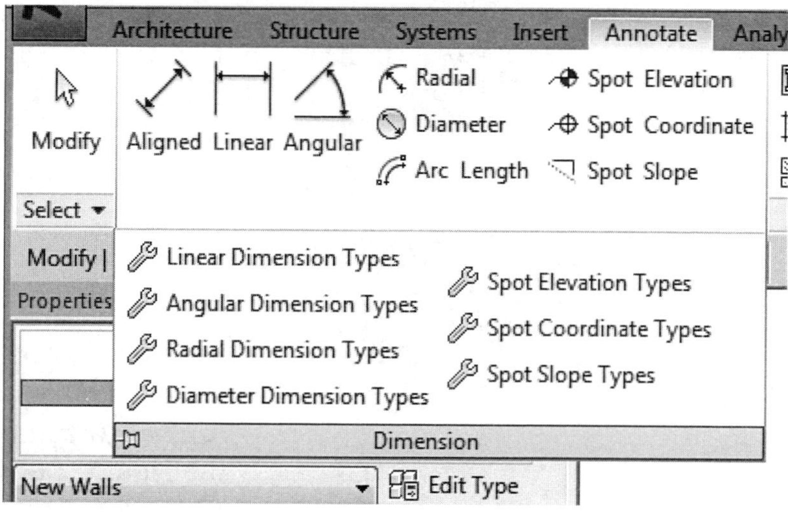

Figure 1–16

2. In the Type Properties dialog box, click **Duplicate...**.
3. Type a new name and click **OK**. The new type is activated.
4. Modify the parameters as needed for the new type.
5. Click **OK** when you are finished.

Dimension Type Options

The dimension type parameters include the *Graphics* of the dimension (such as **Tick Mark** and **Line Weight**), as shown in Figure 1–17, for Linear dimensions, and the *Text* formatting (scroll down in the Type Parameter dialog box to view).

Values for parameters (such as text size, witness line extension, etc.) are the actual plot size for these elements. The view scale controls how large they are in the specific view.

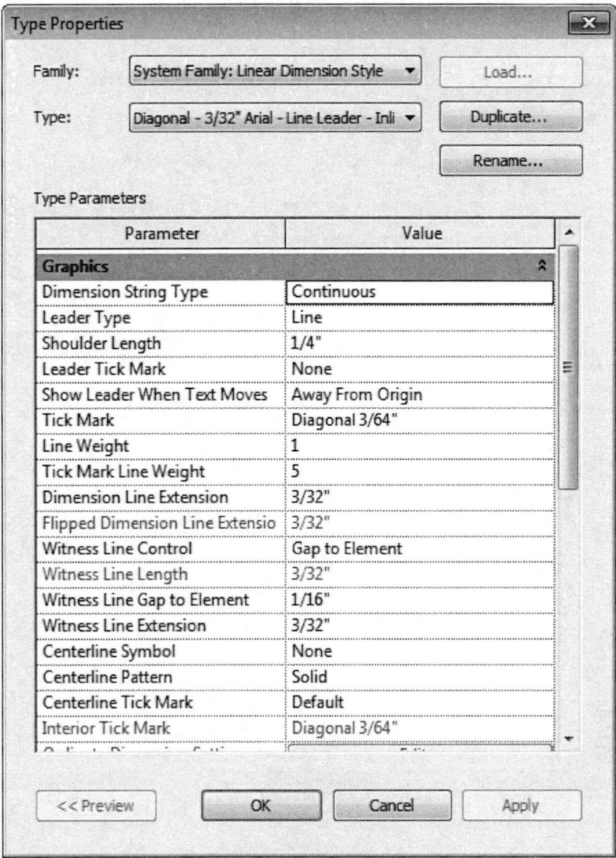

Figure 1–17

- For Linear dimensions, you can specify a *Leader Type, Shoulder Length, Leader Tick Mark, and Show Leader When Text Moves* that is used when the text is pulled away from the dimension string. You can also specify the text inserted for **Equality Text** (the default is still EQ).

- You can specify a *Text Background* option. If you set the value to **opaque**, it automatically masks any elements behind the text. If it is set to **transparent**, anything the text overlaps is still visible.

- If you are dimensioning doors and windows by their widths rather than their centers, you can also have the opening height displayed with the dimension. Select the **Show Opening Height** option.

Creating System Tags

Callout, Elevation, and Section tags can be modified to suit an office standard. Using these tags is frequently a two-step process. Set up the tag itself and then associate it with a related command.

How To: Create Callout, Elevation, and Section Tags

1. In the *Manage* tab>Settings panel, expand (Additional Settings) and click (Callout Tags), (Elevation Tags), or (Section Tags).
2. In the Type Properties dialog box, click **Duplicate....**
3. Type a new name and click **OK**. The new type is activated in Type Properties.
4. Modify the parameters as needed for the new type. In the example of an Elevation Tag shown in Figure 1–18, a new type was created and a new Elevation Mark was selected.

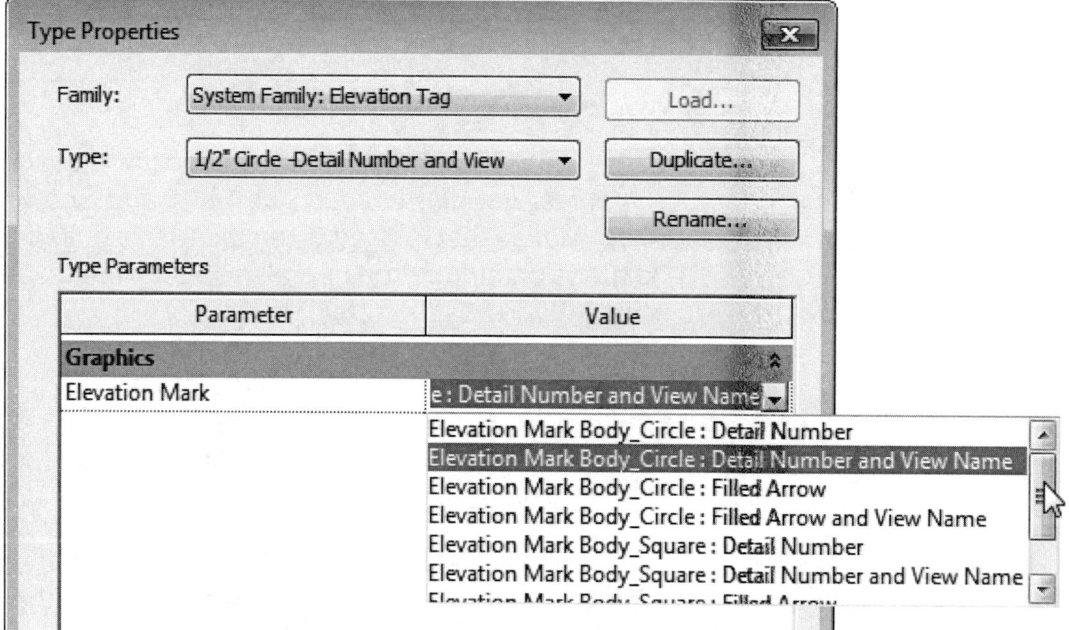

Figure 1–18

5. Click **OK** when finished.
6. Click (Callout), (Elevation), or (Section) as required.
7. In Properties, click (Edit Type).
8. In the Type Properties dialog box, duplicate and name the new view type.

9. Specify the related tag(s) values. When you select the value click (Browse). This opens the Type Properties for the tag in which you can select a different type, as shown in Figure 1–19. (You can also create tag types in this dialog box.)

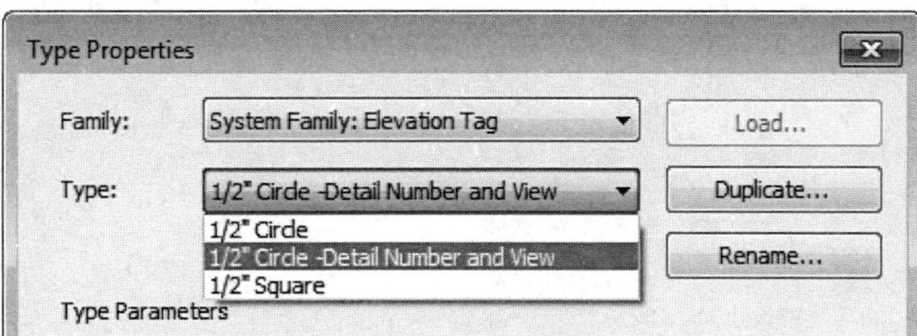

Figure 1–19

10. Click **OK** to apply the tag type.
11. Modify the other parameters as needed and click **OK** to close the Type Properties dialog box and continue using the command.

Type Specific Tag Parameters

The **Callout Tags** parameters specify a *Callout Head* and the *Corner Radius* of the callout box, as shown in Figure 1–20.

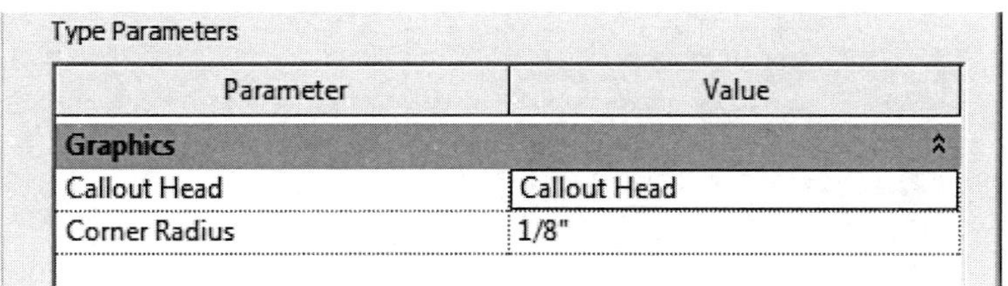

Figure 1–20

The **Elevation Tags** parameter is the *Elevation Mark*. You can select from a variety of types that come with the Autodesk Revit software, as shown in Figure 1–21. For example, you might want to set the exterior elevation mark to display a square body and the detail number on the sheet where it is placed.

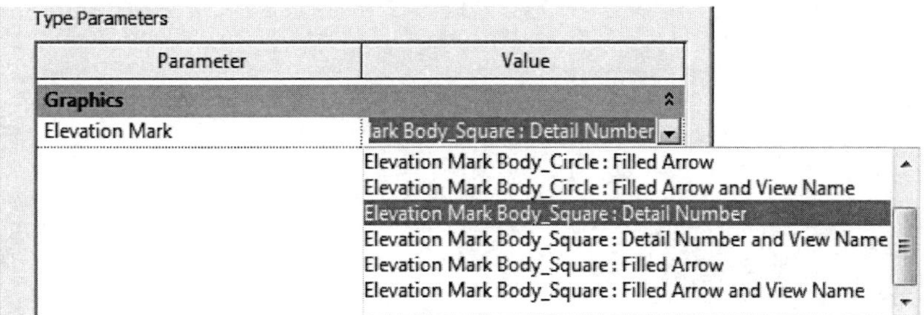

Figure 1–21

The **Section Tags** parameters include both the *Section Head* and *Section Tail*, as well as the *Broken Section Display Style*, as shown in Figure 1–22.

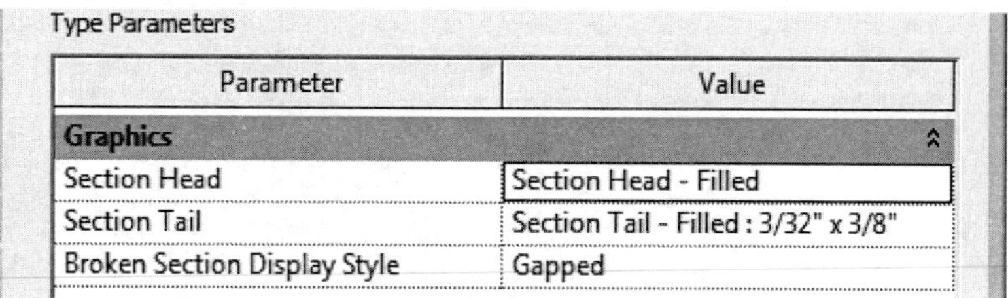

Type Parameters

Parameter	Value
Graphics	
Section Head	Section Head - Filled
Section Tail	Section Tail - Filled : 3/32" x 3/8"
Broken Section Display Style	Gapped

Figure 1–22

Loading Tags in a Template

You can load tags, such as door, window, and wall tags (as shown in Figure 1–23), from the libraries that are included with the software in the *Annotations* folder, or from a specified network drive, into a project template. Having the specific tags required for your projects increases adherence to company standards.

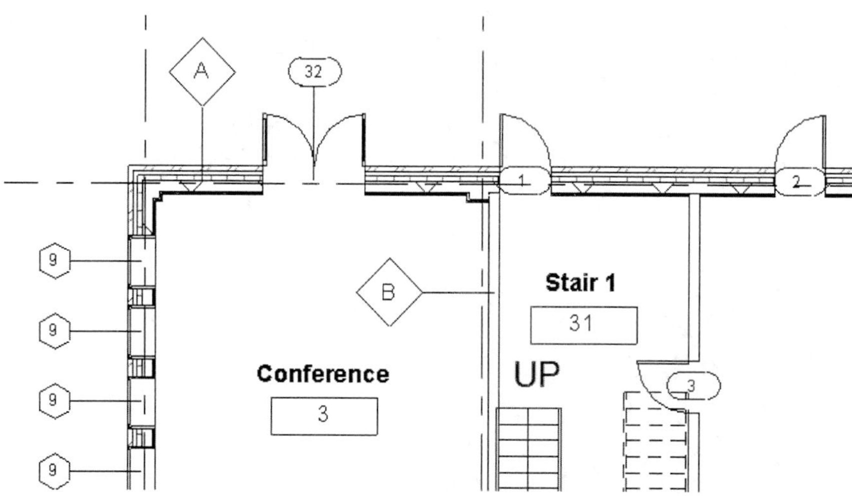

Figure 1–23

Tags are created as separate family files (RFA) and stored in the assigned library (frequently on a network drive). In the Loaded Tags and Symbols dialog box (shown in Figure 1–24), you can easily load the tags you need from the Library into the project template Therefore, when you start the **Tag** command, the one you need is available.

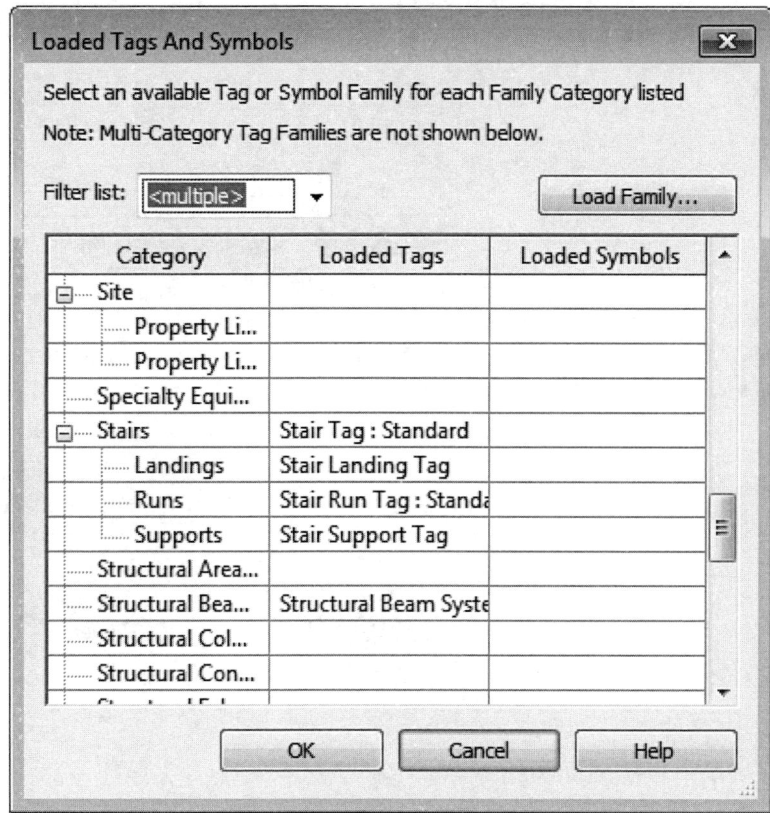

Figure 1–24

- Some elements, such as Floors and Structural Fabric Reinforcement, have both tags and symbols. They can both be specified in this dialog box.

How To: Specify Loaded Tags

1. In the *Annotate* tab>Tag panel, expand the panel title and click (Loaded Tags).
2. In the Tags dialog box, Use the *Filter* field to limit the types of tags you are looking for by discipline.
3. Click **Load Family...**.
4. In the Load Family dialog box, navigate to the appropriate library and folder.
5. Select the required tags and click **Open**. Hold <Ctrl> or <Shift> to select multiple tags.
6. When you have loaded all of the tags that you typically need for a project, click **OK**.

Practice 1a Prepare Templates (All Disciplines)

Estimated time for completion: 15 minutes

Practice Objectives

- Create a new project template file and specify the units and levels.
- Set up text and dimension types and load tags.

In this practice you will create a new project template file, modify the units, snaps, temporary dimensions, and add several new levels. You will create several text types, add a dimension style that uses arrows (as shown in Figure 1–25), and load typical tags into the project template file.

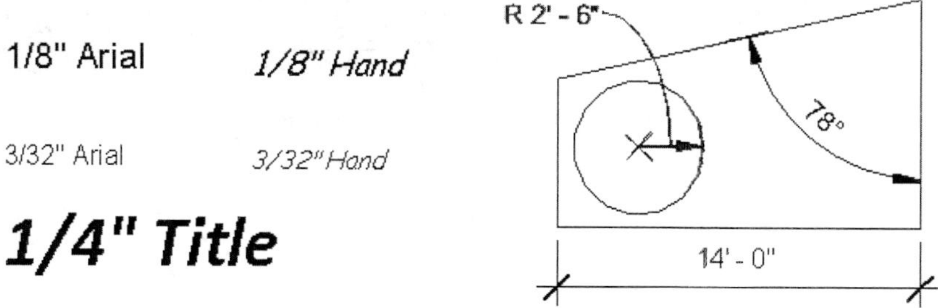

Figure 1–25

Task 1 - Establish a project template file.

1. In the Application Menu, expand (New) and click (Project).
2. In the New Project dialog box, select an existing template file similar to your primary discipline. In the *Create New* area, select **Project template** and click **OK**.
3. In the Quick Access Toolbar, click (Save) and save the template in the practice files folder as **Midrise-Template.rte**.
4. In the *Manage* tab>Settings panel, click (Project Units) or type **UN**.

5. In the Project Units dialog box, set the *Length* formats as follows:

Length:	Units	Feet and Fractional Inches
	Rounding	To the nearest 1/16"
	Suppress 0 feet	Check
Angle:	Rounding	0 decimal places

6. Click **OK** to close the Units dialog box.
7. Open an elevation view.
8. Change Level 2 to **16'-0"**. Add three more levels above the current levels at **12'-0"** apart. Add two levels below Level 1 and set them **10'-0"** apart. Rename them as **Basement 1** and **Basement 2**, as shown in Figure 1–26.

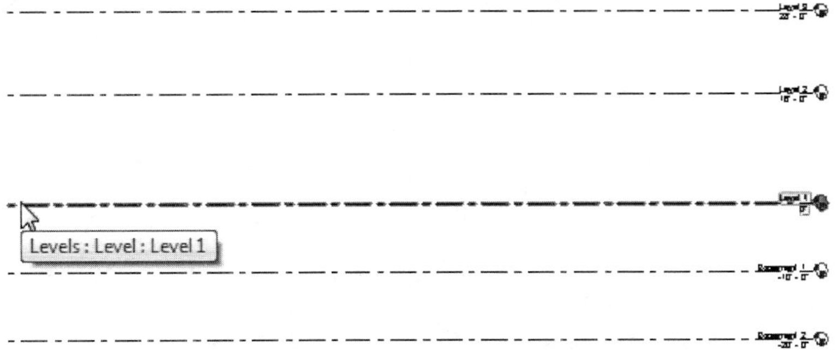

Figure 1–26

- If you draw the levels using (Level), you can select the **Make Plan View** option in the Options Bar to create the associated views.
- If you copy the existing levels, you need to create the plan views and ceiling plan views. In the *View* tab>Create panel, expand (Plan Views) and click (Floor Plan), (Structural Plan), and/or (Reflected Ceiling Plan).
- **MEP only:** Rename above ground levels and corresponding views to match the other levels in the HVAC sub-discipline (3 - Mech, 4 - Mech, etc) and expand the ??? node and select all of the new ceiling plans. In Properties, change the *Sub-Discipline* to **HVAC**.

9. Switch back to the original plan view and save the project template.

Task 2 - Preset annotation types.

1. Create the following text types:

1/8" Arial	Use the default, but set the *Height* to **1/8"**.
1/8" Hand	Use a hand-lettering *Font*, such as **Comic Sans MS**, and select *Italic*.
3/32" Hand	Same as above with a *Height* of **3/32"**.
1/4" Title	Use the font of your choice, in bold.
1" Title	Duplicate **1/4" Title** and change the *Height* to **1"**

2. Create or modify a Linear Dimension type and change the Units Format *Rounding* to the **nearest 1/8"**. Modify the angular dimension type to use arrowheads. Modify other parameters if required.

3. Load the following tags from the Revit Library into the discipline-specific project template:

Architecture:

Folder	Tag Name
Architectural	Casework, furniture, and furniture systems
Civil	Parking and planting

MEP:

Folder	Tag Name
Fire Protection	Sprinkler
Electrical	Lighting fixture and panel name

Structure:

Folder	Tag Name
Architectural	Wall and Floor
Structural	Select any different Structural tags you might want to use in place of those already loaded.

4. Save the project template.

1.3 Creating Title Blocks

Title blocks contain information about the company and consultants designing the project, project information, and sheet-specific information, as shown in Figure 1–27. This information might include but is not limited to the following: the project name, address, number, sheet number, revisions, and other parameters. Some of these parameters never change, some are project-specific, and some are sheet-specific.

Labels and Revision Schedules are specifically used in title blocks.

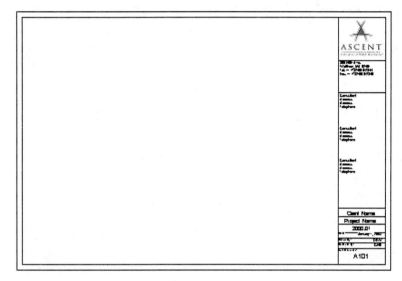

Figure 1–27

- Create the title block by sketching detail lines and adding text, symbols, and regions, as well as image files for company logos. The variable information is stored in labels.

How To: Create a Title Block

You can select from several preset sizes or create a custom size.

1. In the Application Menu, expand (New) and click (Title Block).
2. In the New Title Block - Select Template File dialog box, select a template file size from the list and click **Open**. A new family file opens and the Family tools display in the Ribbon, as shown in Figure 1–28.

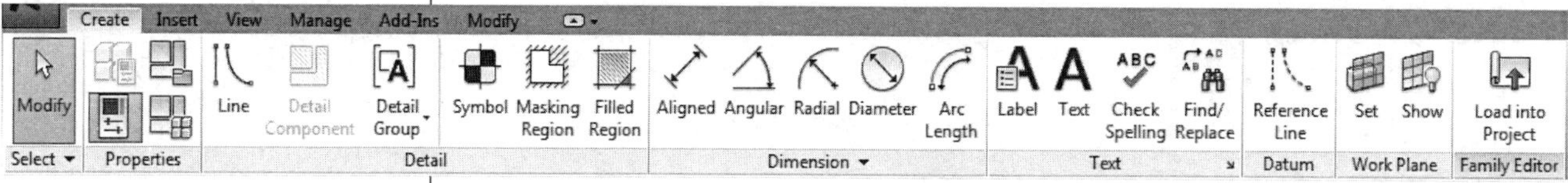

Figure 1–28

   - If you select a template with a standard size, a rectangle of that size displays in the view.
   - If you select **New Size**, a rectangle with dimensions displays. Edit the dimensions to modify the size.
3. Add lines, filled regions, symbols, text, labels, etc., as needed.
4. Save the file and close it.

- You can use dimensions to help place the elements in the title block family, they are not displayed when the title block is inserted.

Adding Labels

Labels are not just text but elements that are assigned to specific parameters and can be added to title blocks or tags. They can change without modifying the rest of the elements. For example, you would use annotation text for the words **Drawn By:** and a label for the initials of the person who did the work (by default displaying DRW in Figure 1–29), because that varies from sheet to sheet.

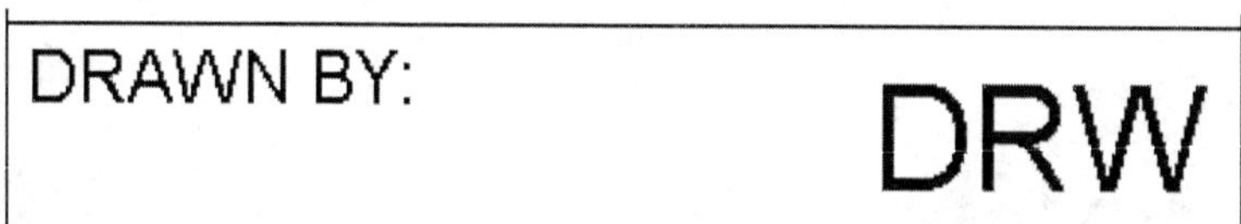

Figure 1–29

- The title block template comes with one text type and one label type already defined. You can create additional types in Properties by duplicating types. The **Text** and **Label** parameters are similar, but you must create separate types for each of them.

How To: Create a Label

1. In the Family Editor, in the *Create* tab>Text panel, click (Label).
2. In the *Modify | Place Label* tab>Format panel, specify the alignments: **Left**, **Center**, **Right**, **Top**, **Center Middle**, or **Bottom**, as shown in Figure 1–30.

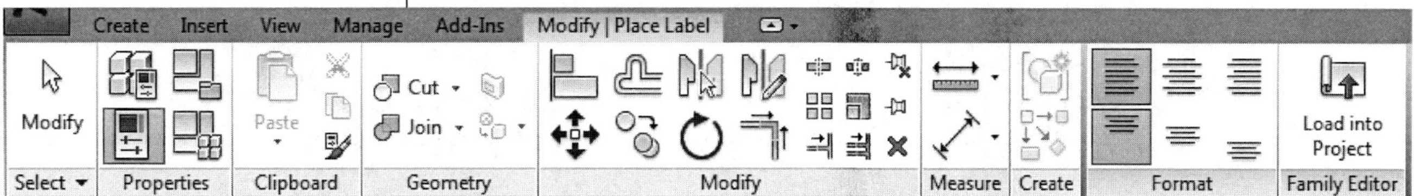

Figure 1–30

3. Click in the view window to place the label, as shown in Figure 1–31.

Figure 1–31

4. In the Edit Label dialog box shown in Figure 1–32, select a label in the *Category Parameters* list and double-click or click (Add parameter(s) to label). You can select more than one by holding down <Ctrl> or <Shift>.

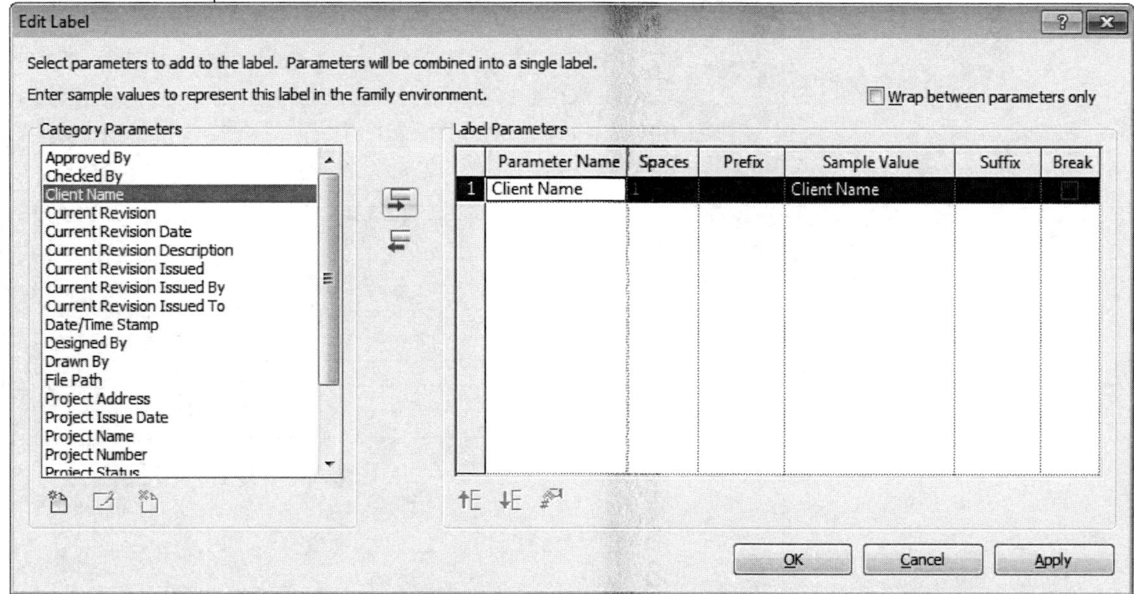

Figure 1–32

5. Enter the *Sample Value* and specify any other options as needed.
   - If you are using several parameters in one label, select **Wrap between parameters only** and **Break** (in column) options to separate them while still permitting a word wrap.
   - Click (Add Parameter) to create a new parameter for the project.
   - Click (Move parameter up) and (Move parameter down) to reorder multiple parameters.
   - If you select a numerical parameter, click (Edit parameter's units format) to change, if necessary.
6. Click **OK** when you have finished editing the label.
7. While still placing the label, rotate or stretch it as needed (as shown in Figure 1–33), or select a point for an additional label.

Figure 1–33

- You can also rotate or stretch a label once it has been placed in the title block.

Adding Revision Schedules

A table of revisions included in a project and/or sheet is typically added to a company title block, as shown in Figure 1–34. In the Autodesk Revit software, you can create a Revision Schedule that is then linked to the Revision Table in the project.

No.	Description	Date

Figure 1–34

How To: Add Revision Schedules to Title Blocks

1. In the Family Editor, in the *View* tab>Create panel, click (Revision Schedule).
2. In the Revision Properties dialog box, select the fields you want to use. Several are already selected for you, as shown in Figure 1–35.

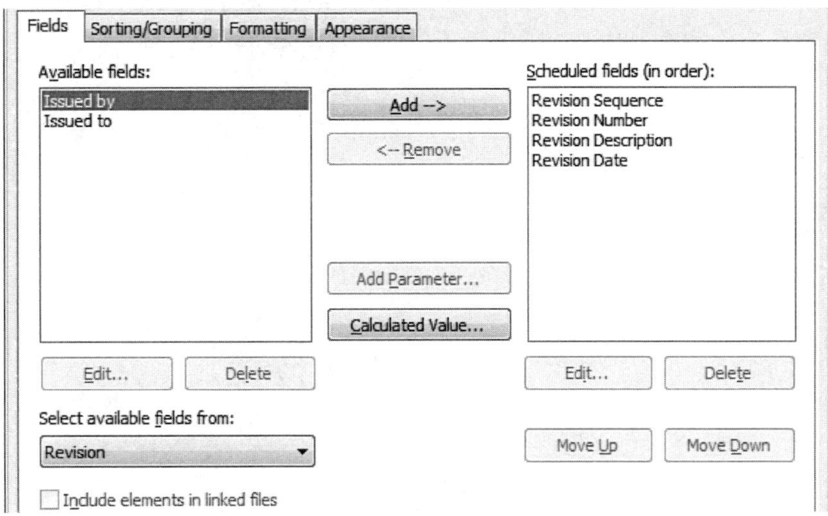

Figure 1–35

3. Modify the options in the *Sorting/Grouping* and *Formatting* tabs as needed.
4. In the *Appearance* tab, select how you want to build the schedule: from the **Top-down** or **Bottom-up**. You can also set the *Height* to **Variable** or **User-Defined**.
5. Click **OK**. The schedule view displays.
6. In the Project Browser, open the Sheet view (it has no name).
7. Drag and drop the schedule onto the sheet and modify the controls, as shown in Figure 1–36.

If the height is set to ***User-Defined****, an additional control is displayed at the bottom of the schedule. Use it to set the height of the schedule.*

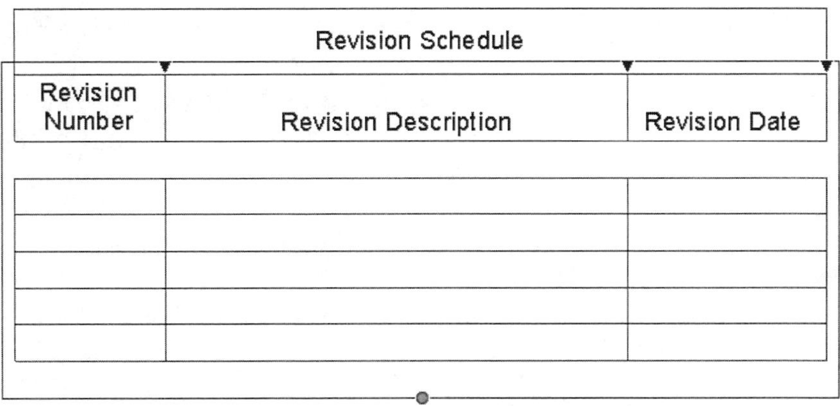

Figure 1–36

- In the Options Bar, you can change the *Rotation on the Sheet* to **None**, **90° Clockwise**, or **90° Counterclockwise**.

Adding Sheets to Project Templates

You can set up project templates using the custom title block in two ways. Add all or most of the sheets typically needed for a project, or create a Sheet List Schedule that can be used to automatically create sheets when they are needed.

In most cases, it is best to create sheets that are always used in projects in the template and even put typical views on them, as shown in Figure 1–37. Note that the view is empty in the template, but as you draw elements in the project they automatically display in the view and on the sheet.

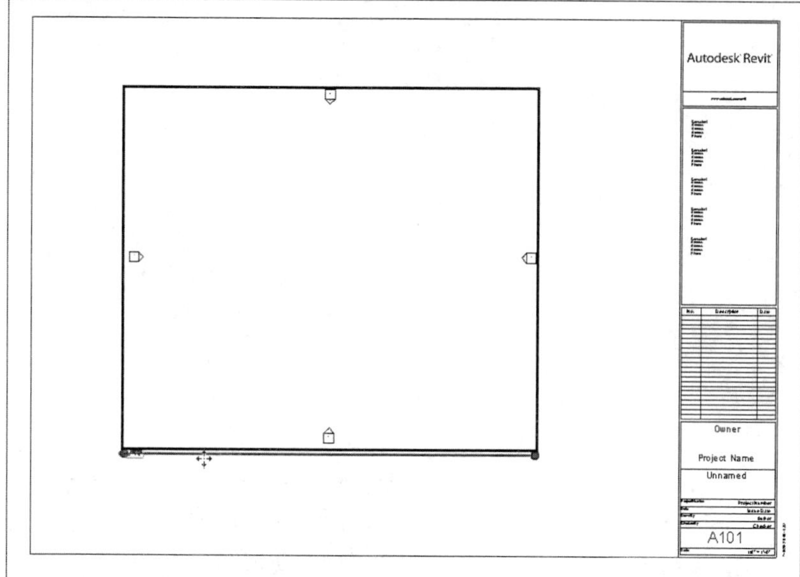

Figure 1–37

How To: Load a Title Block into a Project Template

1. Open the project template.
2. Return to the custom title block family. In the Family Editor panel, click (Load into Project).
3. In the Load into Projects dialog box, select the project you want the title block loaded into and click **OK**. If only one project is open, it is loaded into the project automatically.
4. The title block is now available when you create a sheet.

- If the title block family is not open, select *Insert* tab>Load from Library panel and click (Load Family) to access it, or click **Load...** in the New Sheet dialog box.

- When you save a title block, it should be on the network where everyone has access to it. That way, it is not deleted if someone reinstalls the Autodesk Revit software. Set the default location for the family template files in the Options dialog box, in the *File Locations* tab, as shown in Figure 1–38.

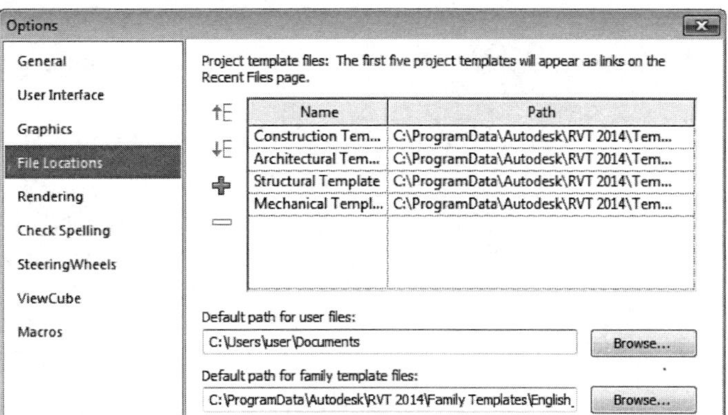

Figure 1–38

Practice 1b Create Title Blocks (All Disciplines)

Practice Objectives

- Draw a custom title block including detail lines, text, labels, and a revision table.
- Set up sheets in the template using the new title block.

Estimated time for completion: 15 minutes

In this practice you will create a new title block by adding lines, text, labels, logo, and a Revision Schedule similar to Figure 1–39. You will then load it into a project template file and create several standard sheets.

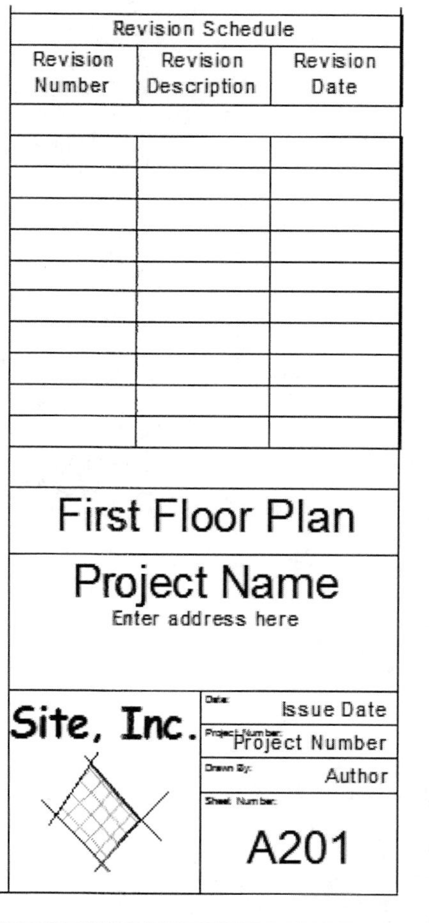

Figure 1–39

Task 1 - Create a title block.

1. In the Application Menu, expand (New) and click (Title Block).

2. In the New Title Block - Select Template File dialog box, select **D - 36 x24.rft** and click **Open**.

3. In the *Create* tab>Detail panel, click (Line) and create lines on the inside of the existing rectangle **1/4"** away from the top, bottom, and right sides. Draw a line **1"** away on the left margin. Trim the lines as needed.

4. Draw lines in the lower right corner of the title block, as shown in Figure 1–40.

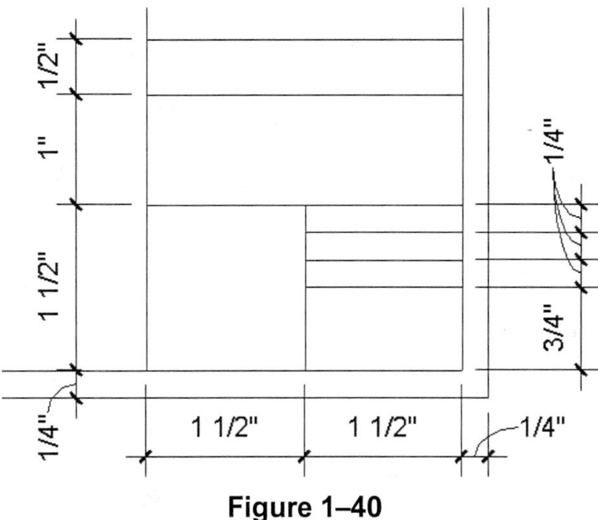

Figure 1–40

5. In the *Create* tab>Text panel, click **A** (Text).

6. In Properties, click (Edit Type) and create the following Text types:

Type Name	Font	Size	Bold	Background
Arial 1/16"	Arial	1/16"	No	Transparent
Title	Comic SansMS	1"	Yes	Transparent
Logo	Your choice	1/4"	Yes	Transparent

7. In the *Create* tab>Text panel, click (Label).

8. In Properties, click (Edit Type) and create the following Label types:

Type Name	Font	Size	Bold	Background
Arial 1/8"	Arial	1/8"	No	Transparent
Arial 1/4"	Arial	1/4"	No	Transparent

9. Create a graphic and company name of your own design in the lower left box, for example Jepson+Johnson Architects. as shown in Figure 1–41. Using the following steps, add text, labels, and graphics to the title block.

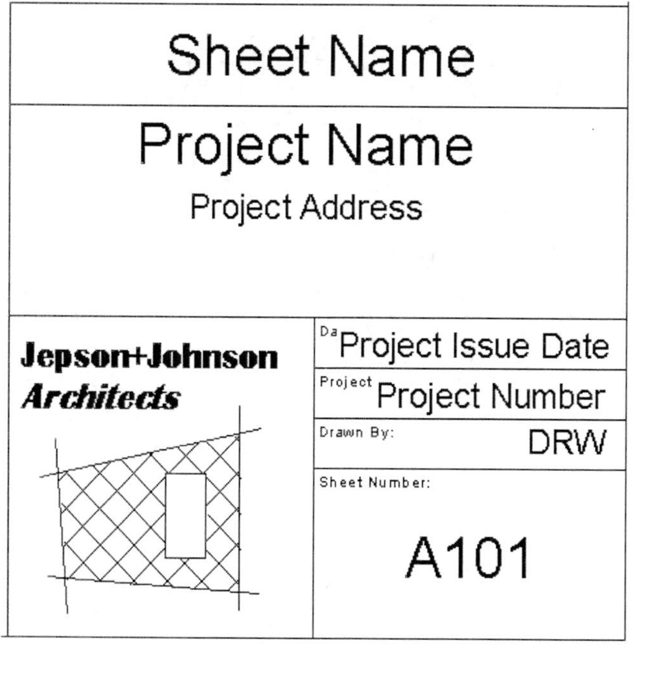

Sheet Name
Project Name
Project Address
Jepson+Johnson
Architects
Da
Project Issue Date
Project
Project Number
Drawn By:
DRW
Sheet Number:
A101

Figure 1–41

- Use (Text) with the *text type* **Arial 1/16"** to add text in the lower right spaces for the *Date*, *Project Number*, *Drawn By*, and *Sheet Number*.
- Use (Label) with the *label type* **Arial 1/4"** and (Align Center) justification to add the *Sheet Name*, *Project Name*, and *Sheet Number*. Move and stretch the labels to fit in the title block.
- Using the *label type* **Arial 1/8"** and (Align Center) justification, add the *Project Address* below the *Project Name*.

- Using the *label type* **Arial 1/8"** and (Align Right) justification, add the *Project Issue Date*, *Project Number*, and *Drawn By*.
- Add the **Logo** text in the box on the lower left side, shown as Site, Inc. in the title block. Draw lines and add a filled region for a graphic logo as needed.

10. Save the title block in the practice files folder as **TBLK-D.rfa**.

Task 2 - Add a Revision Schedule.

1. In the title block Family Editor, in the *View* tab>Create panel, click (Revision Schedule).
2. In the Revision Properties dialog box, *Fields* tab, set up the fields as shown in Figure 1–42.

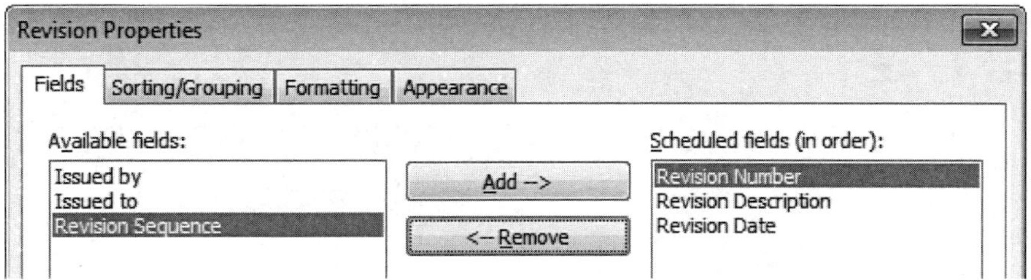

Figure 1–42

3. Accept the defaults for the *Sorting/Grouping* and *Formatting* tabs.
4. Select the *Appearance* tab and change the *Height* to **User defined**.
5. Click **OK**.
6. In the Project Browser, expand *Views (all)* to display the *Schedules* and *Sheets (all)* areas, as shown in Figure 1–43. In the *Sheets (all)* area, open the **-** view.

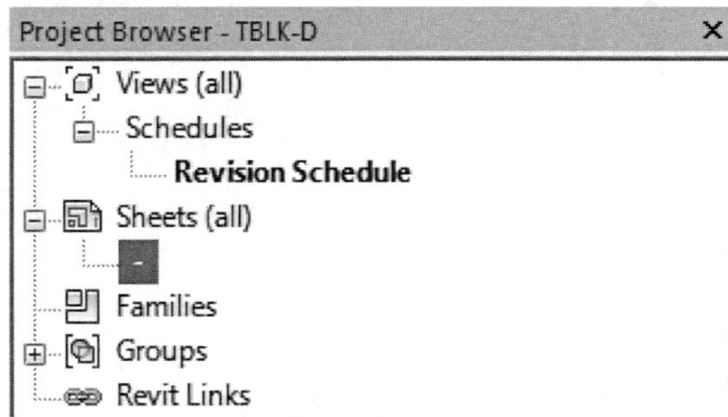

Figure 1–43

7. Drag and drop the Revision Schedule onto the sheet.

8. Move it above the sheet name and resize it to display several lines, as shown in Figure 1–44.

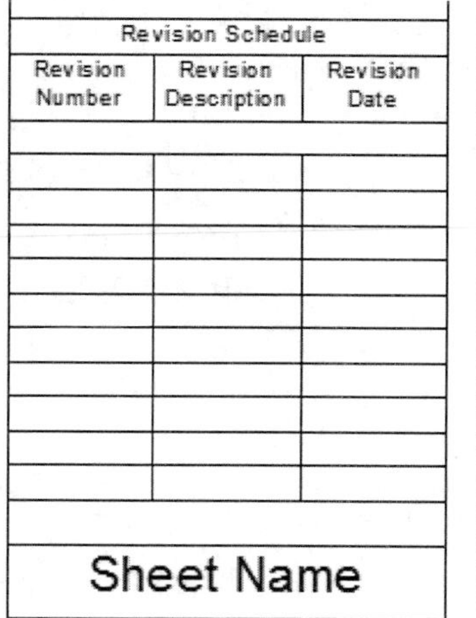

Figure 1–44

9. Save and close the title block.

Task 3 - Set up sheets in a project template using the new title block.

1. Open the project template **Midrise-Template.rte** that you created in the previous practice. If you did not complete the previous practice, open the template found in the practice files folder:
   - *.../Architectural/***Midrise-Template-Architectural.rte**
   - *.../MEP/***Midrise-Template-MEP.rte**
   - *.../Structural/***Midrise-Template-Structural.rte**

2. In the *View* tab>Sheet Composition panel, click (Sheet).

3. In the New Sheet dialog box, click **Load...**.

4. In the Load Family dialog box, navigate to the practice files folder, select **TBLK-D.rfa** (that you just created), and click **Open**. If you did not complete the previous task, open the title block found in the corresponding practice files folder:
   - *.../Architectural/***TBLK-D-Architectural.rfa**
   - *.../MEP/***TBLK-D-MEP.rfa**
   - *.../Structural/***TBLK-D-Structural.rfa**

5. Select the title block that you just loaded and click **OK**.

6. In the Project Browser, select the sheet and rename it as **CS000 – Cover Sheet**.

7. Using the **Title** text type, add placeholder text for the project name and address, as shown in Figure 1–45.

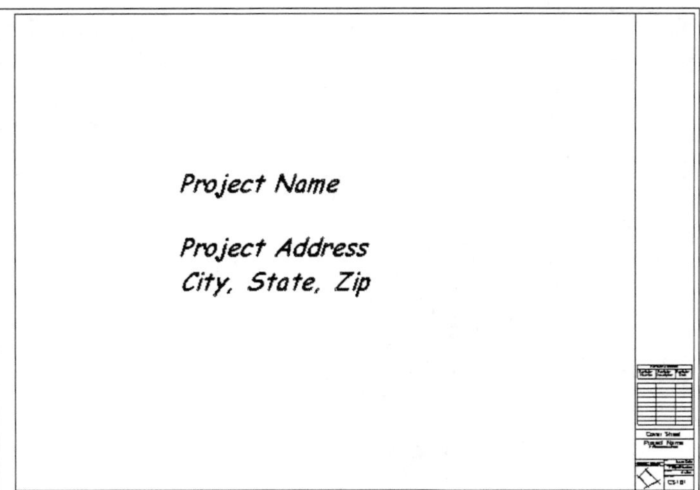

Figure 1–45

8. In the *Manage* tab>Manage Project panel, click (Starting View).

9. In the Starting View dialog box, select **Sheet: CS000 - Cover Sheet** as shown in Figure 1–46.

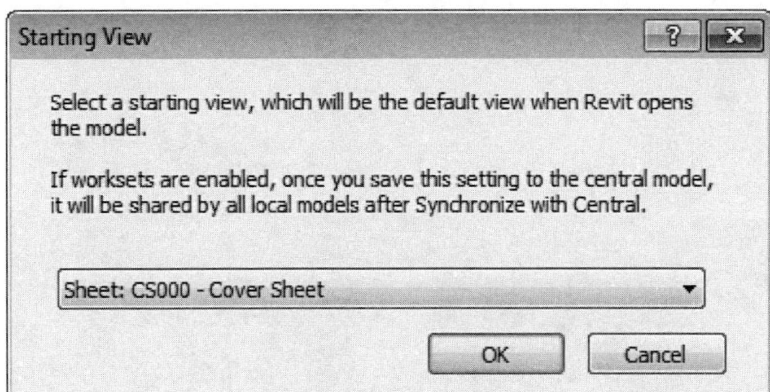

Figure 1–46

10. Create another sheet with the new title block and name it according to your discipline:
    - **A201 – First Floor Plan**
    - **M201 - First Floor Plan Mechanical**
    - **E201 - First Floor Plan Electrical**
    - **P201 - First Floor Plan Plumbing**
    - **S201 - First Floor Plan Structural**

No elements are on the view, but it acts as a placeholder on the sheet. Elements are displayed as they are drawn.

11. Open this sheet view. Drag the associated **Level 1** floor plan view onto the sheet, as shown in Figure 1–47.

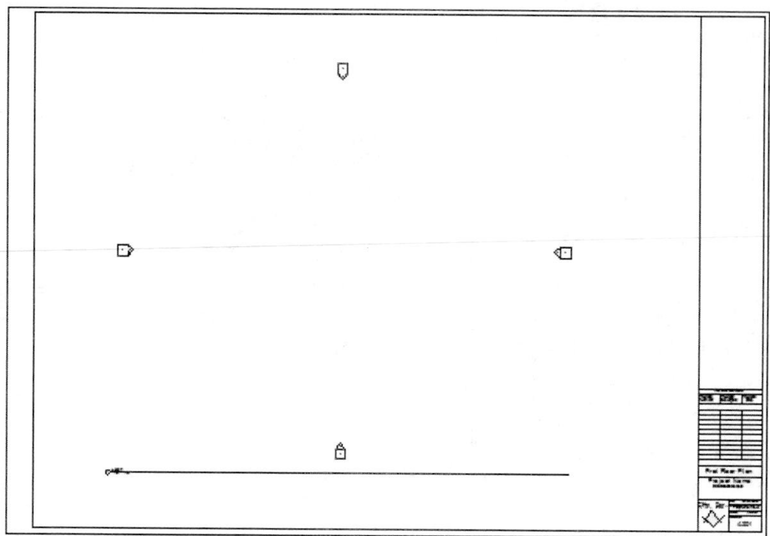

Figure 1–47

1.4 View Templates

View templates enable you to specify all of the view properties and visibility options for a view by selecting another view or a view template as a base. For example, you can create a furniture plan view template that sets the scale, detail level, and visibility of objects so that everything except the furniture displays in halftone, as shown in Figure 1–48.

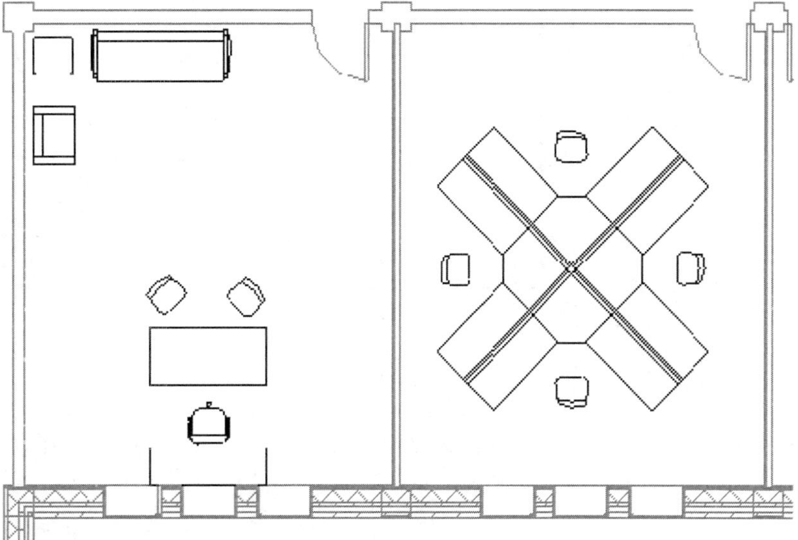

Figure 1–48

View templates can be applied to specific views when you create a project template file and as you work in a project. You can also specify a view template in view properties that restricts you from making changes, such as the Detail Level shown in Figure 1–49. This is useful for views that have been set for printing.

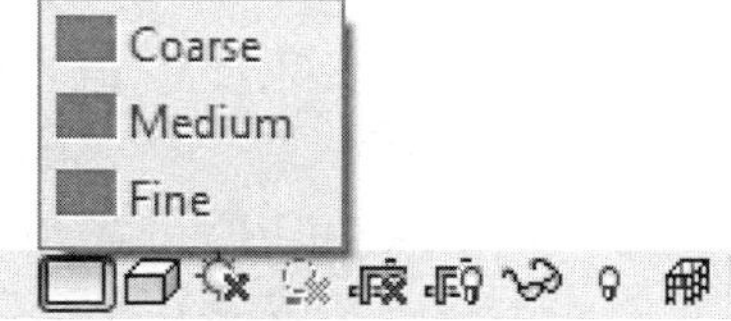

Figure 1–49

Applying View Templates

There are two typical ways to use view templates. The first is while you are working to display different view parameters in the same view. For example, you might want to display the spaces in a section and use a view template to switch to the mechanical layout in the same section, as shown in Figure 1–50. You can continue to make changes to the views even after the view template has been applied. The other way is to preset a view using a default view template so you cannot make changes to it.

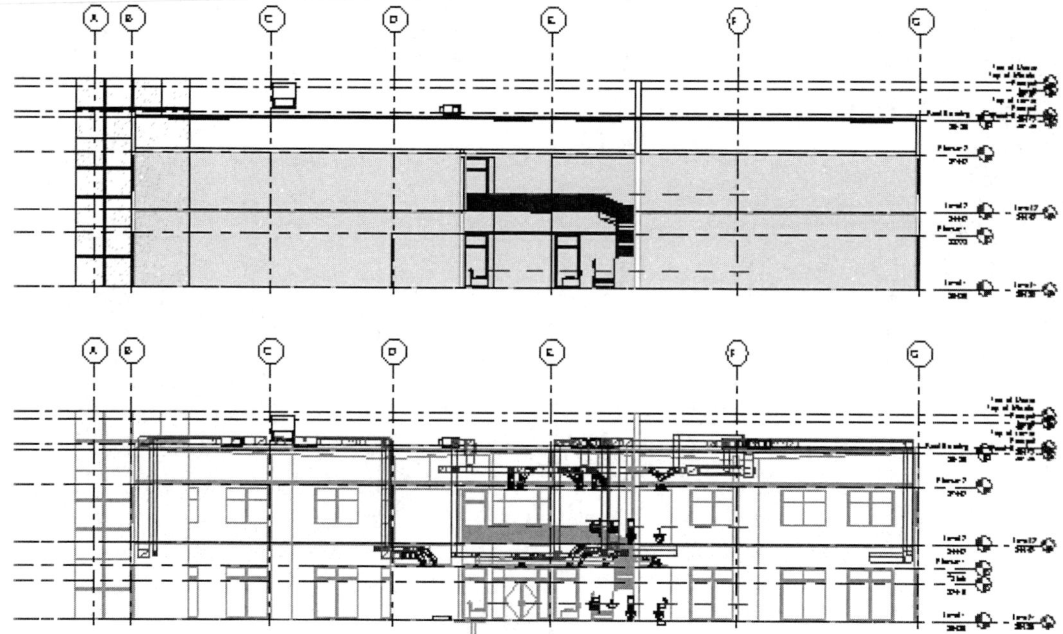

Figure 1–50

How To: Apply a Temporary View Template Override

1. Select one or more views in the Project Browser. (Hold down <Ctrl> or <Shift> to select multiple views.)
2. Right-click in the Project Browser and select **Apply Template Properties**, or on the *View* tab>Graphics panel, expand (View Template) and click (Apply Template Properties to Current View).
3. In the Apply View Template dialog box, select the view template you want to use from the *Names* list, as shown in Figure 1–51. Several standard view templates are available when you create a project using one of the Autodesk Revit project templates.

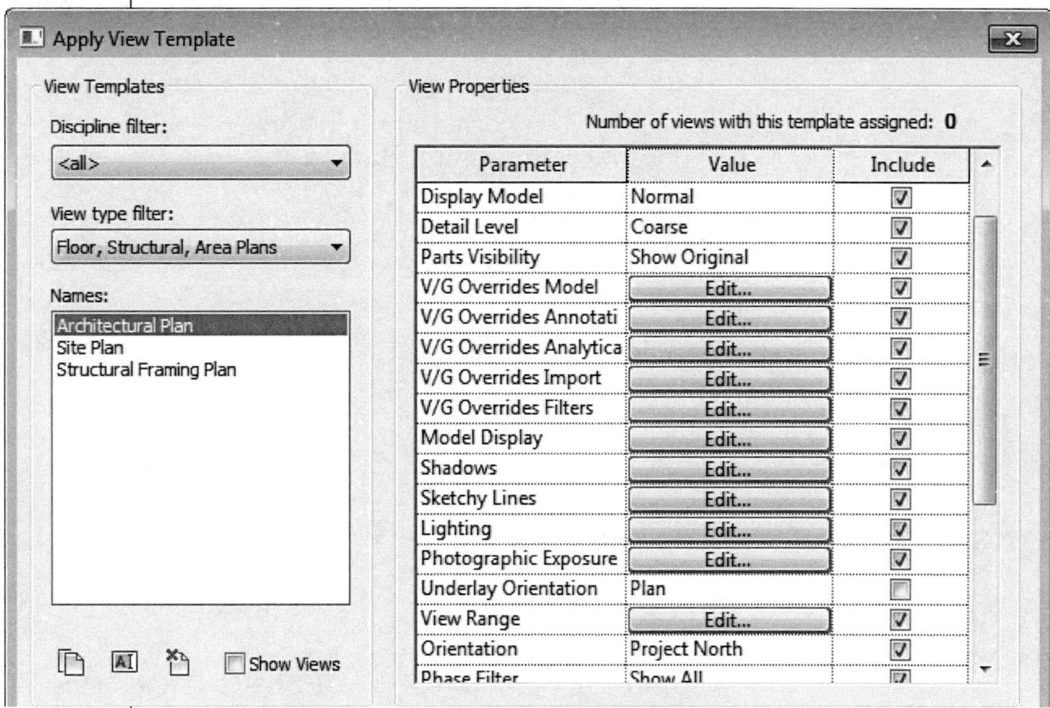

Figure 1–51

4. If you do not want the entire view template to be applied to the selected view, clear the check mark in the *Include* column for the parameters you do not want to include. You can also create overrides to the view template by changing parameter values.
5. Click **OK** to finish.

- To limit the number of view templates that display, filter the list by selecting an option in the Discipline filter and View type filter drop-down lists, as shown in Figure 1–52.

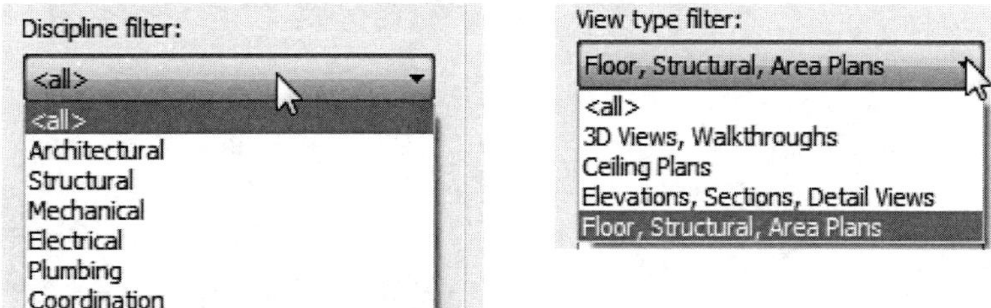

Figure 1–52

- You can apply view templates to any view as many times as required, unless it is set as the default template in the view properties.

- If you want to start a new view template based on an existing view, select **Show Views**. The *Names* list expands to include all of the related views in the project.

Hint: Temporary View Properties

As you are working, it can be helpful to temporarily override the view. You can do this by selecting a view template from the View Control Bar, as shown in Figure 1–53

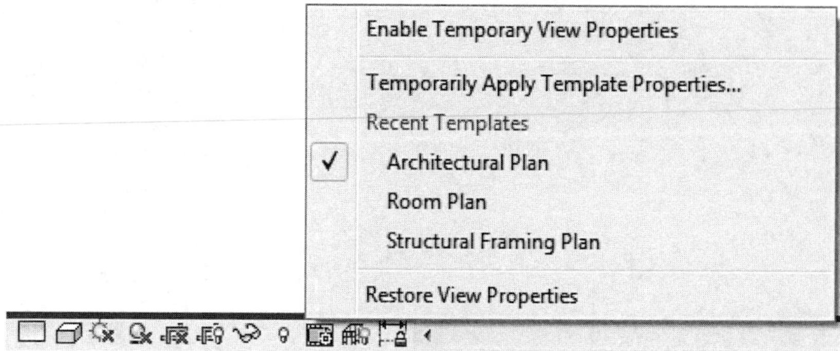

Figure 1–53

1. In the View Control Panel, expand (Temporary View Properties) and select **Enable Temporary View Properties**.
2. Expand expand (Temporary View Properties) again and select **Temporarily Apply Template Properties**. Alternatively, if you have already used the process, you can select from a list of view templates, as shown in Figure 1–53.
3. In the dialog box, select the view template you want to apply and click **OK**.
4. When you are finished, expand (Temporary View Properties) and select **Restore View Properties**.

How To: Set the Default View Template for a View

1. Select one or more views in the Project Browser.
2. In Properties, in the *Identity Data* section, click the button next to *View Template*, as shown in Figure 1–54. The name on the button varies according to any template that has already been applied.

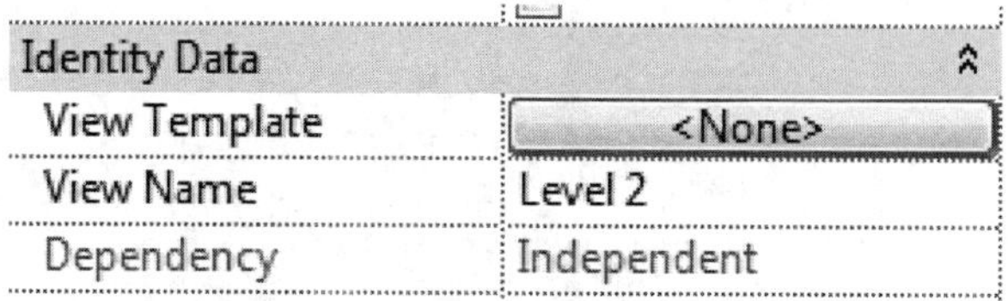

Figure 1–54

3. In the Apply View Template dialog box, select the appropriate template and click **OK**.
4. The view is now locked using the parameters set in the view template.

- View Templates can also be set by the type properties of a view. For example, in a view's Type Properties, duplicate a view type, such as Space Plan shown in Figure 1–55. Apply the appropriate template and select **New views are dependent on template**.

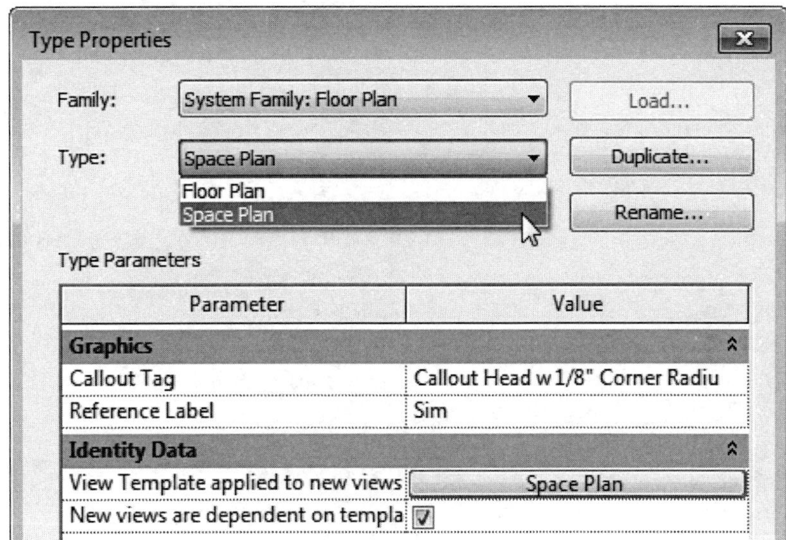

Figure 1–55

- When you use the View type, the view modifies to match the view template and is also placed in its own group in the Project Browser, as shown in Figure 1–56.

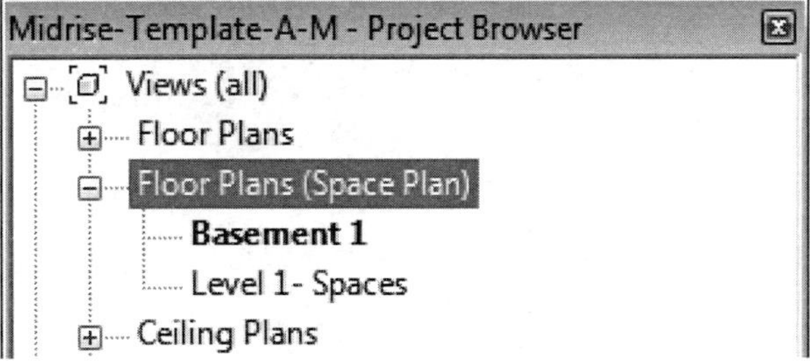

Figure 1–56

Creating View Templates

Many view templates come with the Autodesk Revit software, but you might need to create your own custom view types that are best applied by view template. You can create a view template from the view settings in an existing view or through the View Template dialog box.

How To: Create a View Template from an Existing View

1. Set up a view the way you want it with *Scale*, *Detail Level*, *Visibility Graphic Overrides*, and other *View Settings*.
2. In the Project Browser, right-click on the view and select **Create View Template from View** or, on the *View* tab> Graphics panel, expand (View Template) and click (Create Template from Current View).
3. In the New View Template dialog box, type a name for the view template.
4. Click **OK**.
5. In the View Template dialog box, make other adjustments as needed and click **OK**.

How To: Create a View Template

1. In the *View* tab>Graphics panel, expand (View Template) and click (Manage View Templates).
2. In the Apply View Template dialog box, in the *Names* list, select a view similar to the one you want to create and click (Duplicate).
3. In the New View Template dialog box, type a new name for the view template and click **OK**.

4. Select the new view template as shown in Figure 1–57, and modify the View Properties parameter values as needed.

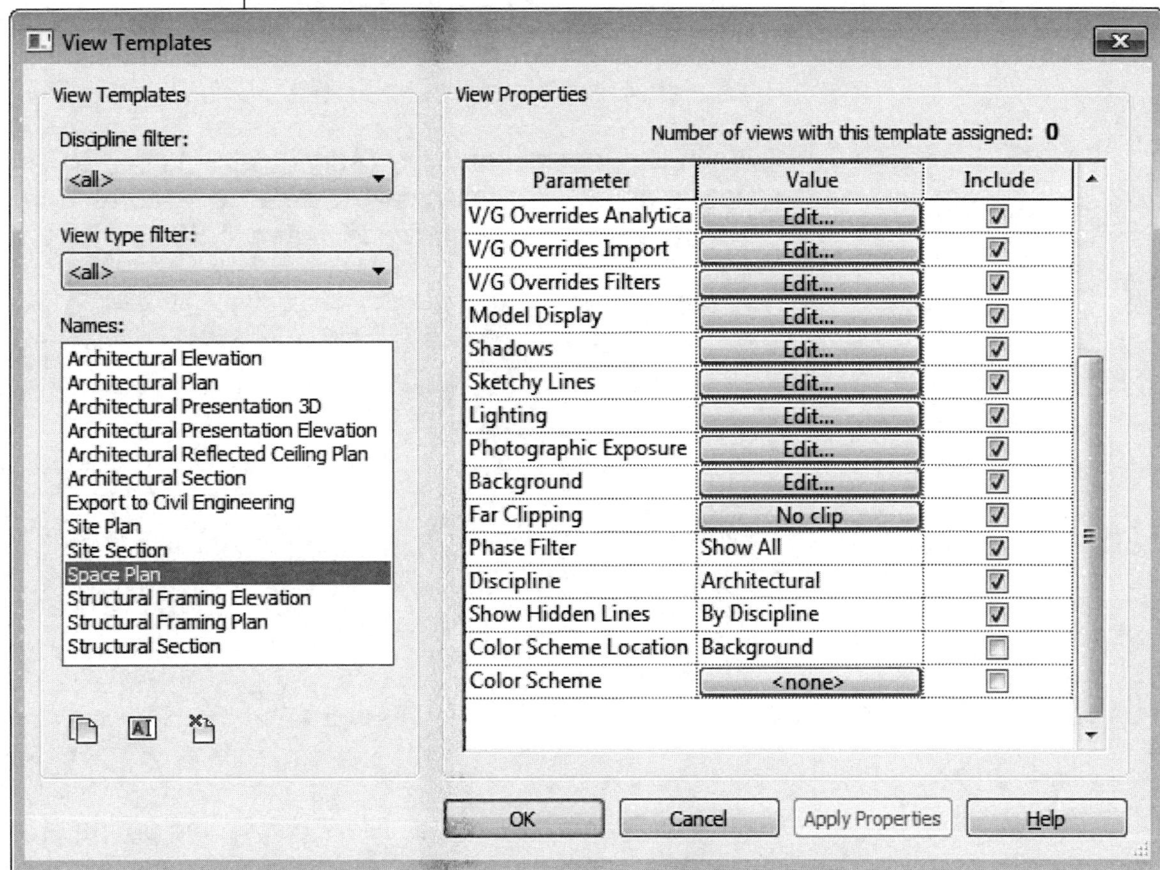

Figure 1–57

5. Click **OK** to finish.

- Not all view types have the same view parameters. For example, 3D views have options for Rendering and Perspectives that plans do not have. When you apply a view template across view types, only the options that are mutual are updated.
- The only parameter saved to a view template for schedule views is the appearance of the schedule.

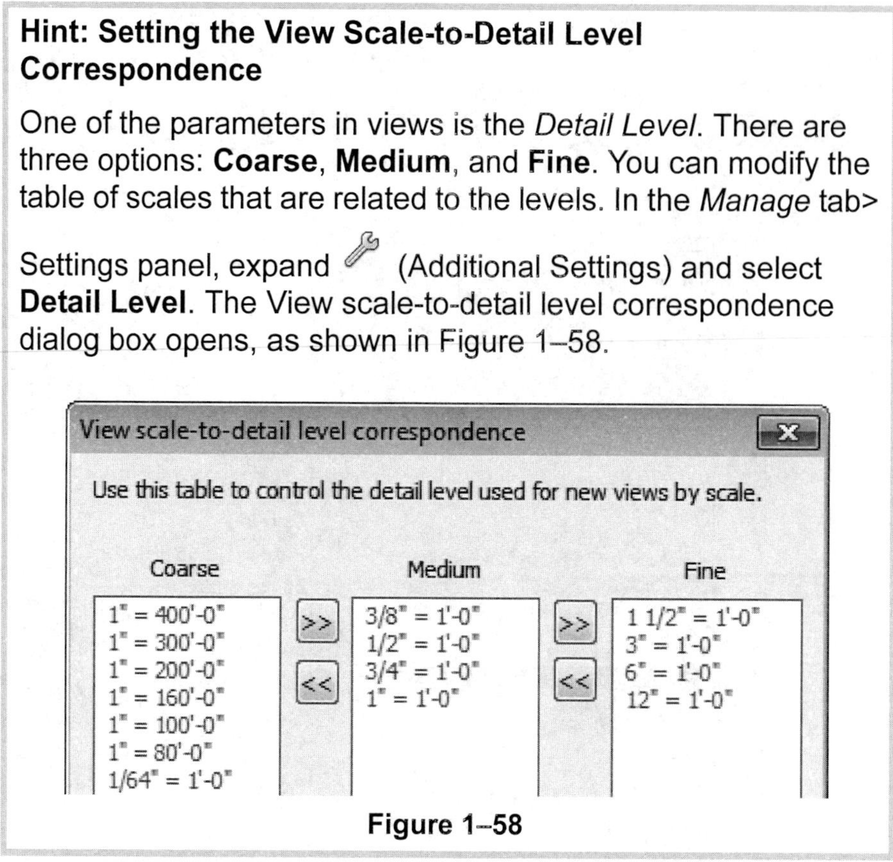

Hint: Setting the View Scale-to-Detail Level Correspondence

One of the parameters in views is the *Detail Level*. There are three options: **Coarse**, **Medium**, and **Fine**. You can modify the table of scales that are related to the levels. In the *Manage* tab>Settings panel, expand (Additional Settings) and select **Detail Level**. The View scale-to-detail level correspondence dialog box opens, as shown in Figure 1–58.

Figure 1–58

Visibility/Graphic Overrides Filters

Visibility/Graphic Overrides filters can speed up the process of specifying categories that are modified in View Templates. You might want to display certain elements in halftone, or with a different color or lineweight in specific views. For example, a fire evacuation plan might have walls with different fire ratings that display with thicker lineweights, as shown in Figure 1–59. By creating these filters and then applying them in a view template, you can reuse them without having to recreate them each time.

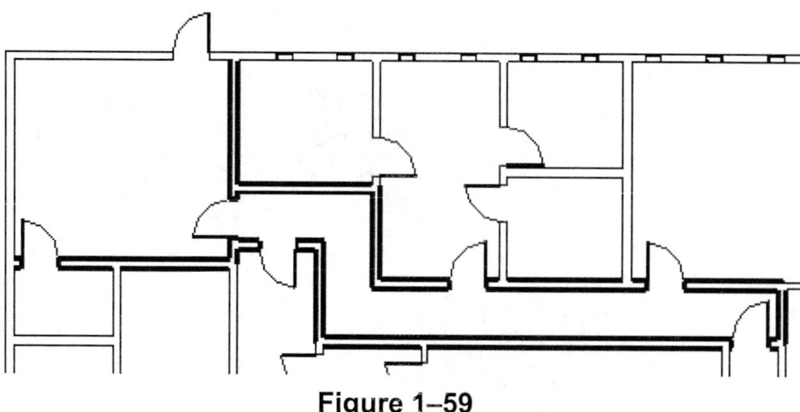

Figure 1–59

How To: Create Filters

If you are using the non-discipline specific version of the Autodesk Revit software, an additional Filters dialog box opens in which you can create new filters, edit, rename, or delete them without having to open the full Filters dialog box.

1. In the *View* tab>Graphics panel, click (Filters). The Filters dialog box opens as shown in Figure 1–60.

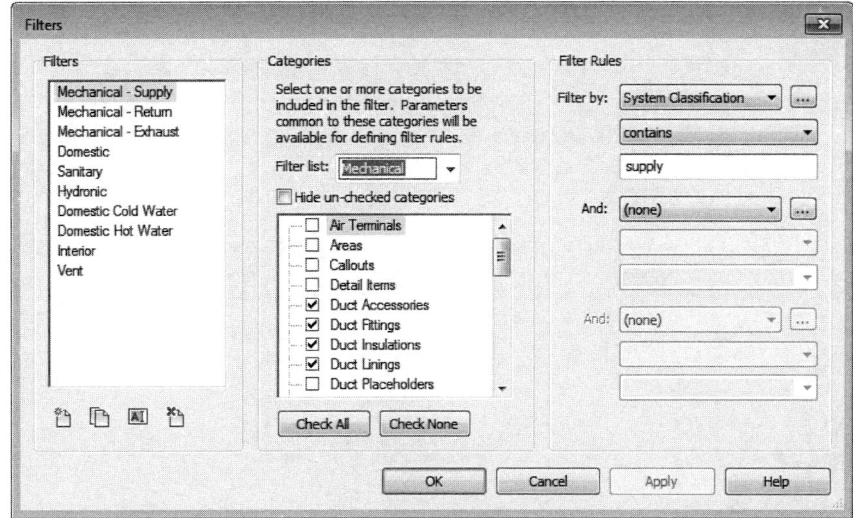

Figure 1–60

2. In the *Filters* area, click (New) or (Duplicate). Name the new filter and click **OK**.
3. In the *Categories* area, filter the list by discipline and then select the categories to include in the filter. Use **Check All** and **Check None** to help select the categories.
4. In the *Filter Rules* area, select what you what to *Filter by*, the filter operator(s), and the value for the filter. If more than one category is selected, the *Filter by* list is limited to parameters shared by the categories you selected. You can create more than one filter rule. They are applied in order. Set *Filter by* to **None** if you do not want to filter by parameters. Filter operators are shown in Figure 1–61.

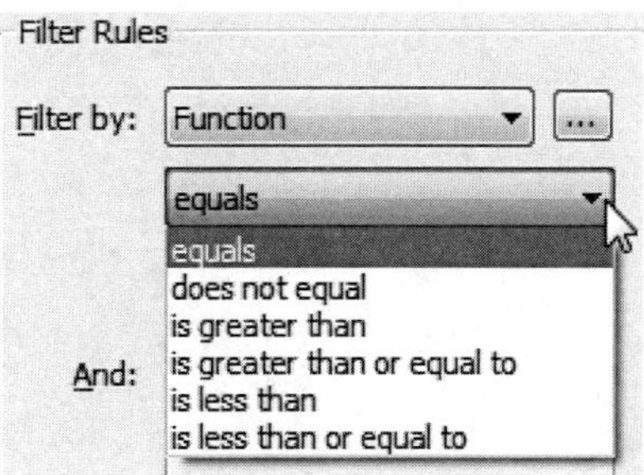

Figure 1–61

5. Click **Apply** to save the changes and remain in the dialog box, or click **OK** to finish.

How To: Apply Visibility/Graphic Overrides Filters

1. Type **VG** or **VV** or in the *View* tab>Graphics panel, click (Visibility/Graphics) to open the Visibility/Graphic Overrides dialog box. Select the *Filters* tab, as shown in Figure 1–62.

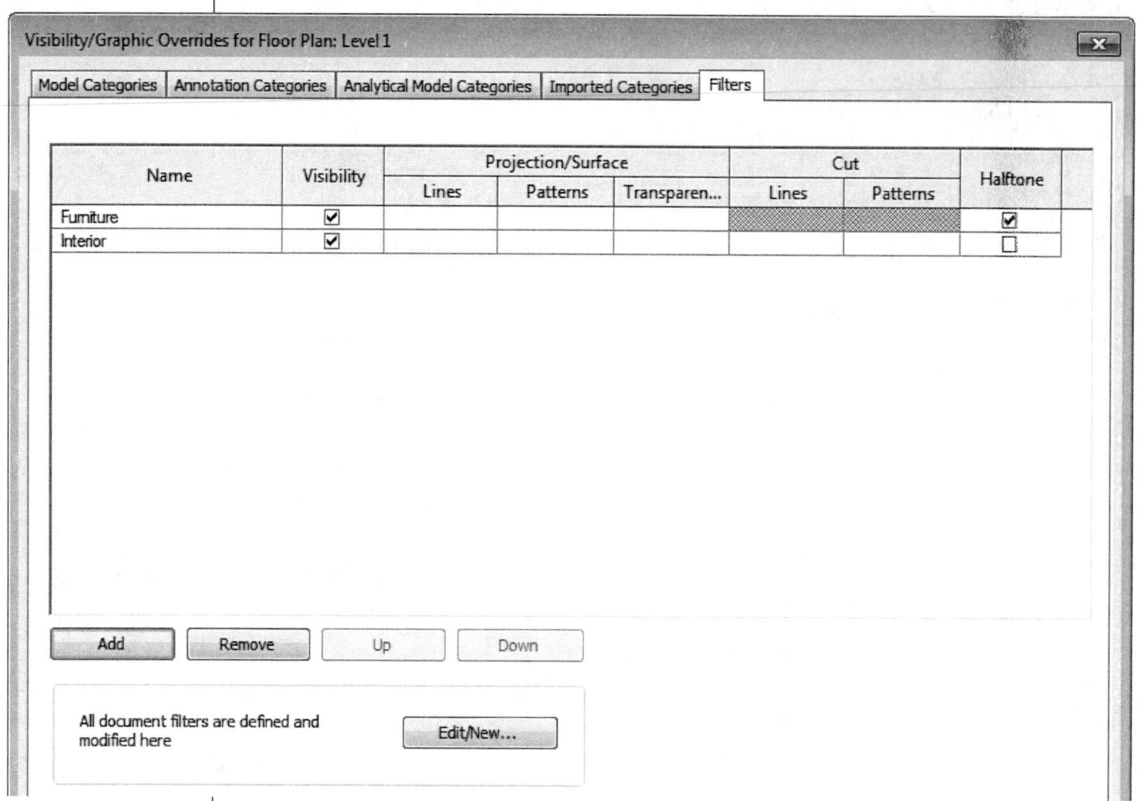

Figure 1–62

2. Click **Add** to add a filter to the list.
3. In the Add Filters dialog box, select the filter(s) you want to add and click **OK**.
   - If the filter you want is not defined, click **Edit/New...** to open the Filters dialog box, where you can define a new filter or edit an existing one.
4. In the Visibility/Graphic Overrides dialog box, assign the overrides you want for the filter. For example, you might want items to be **Halftone**, as shown in Figure 1–63.

Name	Visibility	Projection/Surface			Cut		Halftone
		Lines	Patterns	Transparen...	Lines	Patterns	
Furniture	☑						☑
Interior	☑						☐

Figure 1–63

5. Click **OK**.

Practice 1c View Templates for Architectural and Structural Projects

Practice Objectives

- Create visibility filters.
- Create a view template and apply it to several views.

Estimated time for completion: 15 minutes

In this practice you will create filters that you will use in a structural view template, and apply the view template to several structural plan views, as shown in Figure 1–64. When the filter is working correctly you can add the same information to a project template file.

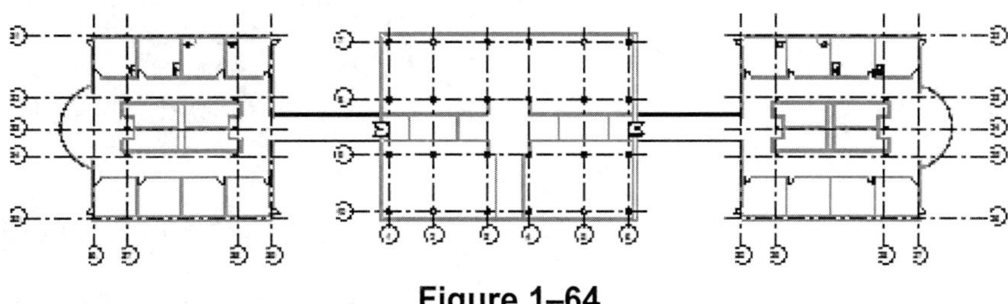

Figure 1–64

Task 1 - Create filters.

1. Open the project found in the practice files folder:
   - *.../Architectural/***Office-Architectural.rvt**
   - *.../Structural/***Office-Structural.rvt**
2. In the *View* tab>Graphics panel, click (Filters).
3. In the Filters dialog box, click (New), name the new filter **All But Structural**, and click **OK**.
4. In the *Categories* area, set the *Filter list* to **Architecture** and **Structure**.
5. Click **Check All**, and clear **Columns**, all categories starting with **Analytical**, **Structural**, and **Walls**. Click **Apply**.
6. Click (New). Name the new filter **Non-Structural Walls** and click **OK**.
7. In the *Categories* area, click **Check None** and select **Walls**.

In the non-discipline specific Autodesk Revit software, an additional Filters dialog box opens. Click ***New...*** *and name the new filter. Select* ***Define Criteria*** *and click* ***OK****. This opens the full Filters dialog box.*

8. In the *Filter Rules* area, set *Filter by* to **Structural Usage**, **equals**, and **Non-bearing**, as shown in Figure 1–65.

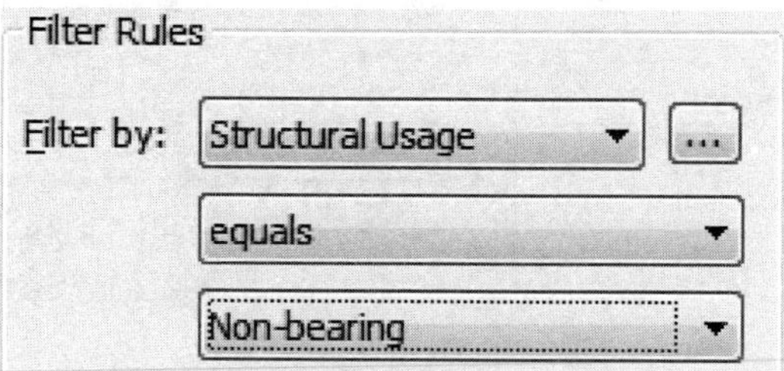

Figure 1–65

9. Click **OK** to close the Filters dialog box.

Task 2 - Create a View Template.

1. In the Project Browser, in the Floor Plans area, right-click on the **Level 1** view and select **Create View Template From View**.
2. Name the new view template **Structural Plan** and click **OK**.
3. In the View Templates dialog box, set the *View Scale* to **1/16"=1'-0"** and the *Detail Level* to **Coarse**.
4. Next to **V/G Override Filters**, click **Edit...** The Visibility/Graphic Overrides dialog box opens with the *Filter* tab selected.
5. Click **Add**.
6. In the Add Filters dialog box, select the two filters you just created and click **OK**.
7. Place a check mark in the *Halftone* column for both filters. Continue working in the Visibility/Graphics Overrides dialog box.
8. In the *Annotation Categories* tab, turn off everything except Grids and all Structural annotations and tags.
9. In the *Model Categories* tab, turn off items, such as **Casework**, **Furniture**, **Furniture Systems** and **Plumbing**, and **Mechanical equipment**.
10. Click **OK** to close the Visibility/Graphic Overrides dialog box, and click **OK** to complete the view template.

Task 3 - Apply a View Template to several views.

1. In the *Floor Plans* area, use *Duplicate with Detailing* on the **Level 1**, **Level 2**, **Level 3**, and **Level 4** views. Rename them as **Level 1 - Structural**, **Level 2 - Structural**, **Level 3 - Structural**, and **Level 4 - Structural**.
2. Select the new Structural views.
3. Right-click and select **Apply View Template**.
4. In the Apply View Template dialog box, select the **Structural Plan** view template you just created and click **OK** to apply it.
5. Look at the different views and compare them to the standard floor plan views. Finish with the **Level 1** view.
6. Save the project.

Practice 1d View Templates for MEP Projects

Estimated time for completion: 15 minutes

Practice Objectives

- Create visibility filters.
- Create a view template and apply it to several views.

In this practice you will create velocity based filters for ductwork that you will use in a mechanical view template, and apply the view template to a duplicated plan view, as shown in Figure 1–66. When the filter is working correctly, you can add the same information to a project template file.

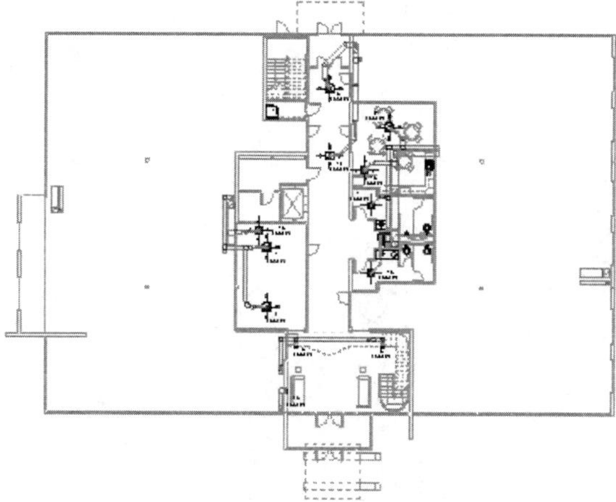

Figure 1–66

Task 1 - Create filters.

1. In the ...*MEP*\ practice files folder, open **Office-MEP.rvt**.
2. In the *View* tab>Graphics panel, click (Filters).
3. If you are using the non-discipline specific version of the Autodesk Revit software, an additional Filters dialog box opens. Click **New...**, name the new filter, select **Define Criteria**, and click **OK**. This opens the full Filters dialog box.
4. If you are using the Autodesk Revit MEP software, in the Filters dialog box, click (New), name the new filter **Mechanical - Low Velocity**, and click **OK**.
5. In the *Categories* area, set the *Filter list* to **Mechanical**. Select **Ducts** and **Flex Ducts**.

6. In the *Filter Rules* area, set *Filter by* to **Velocity**, **is less than or equal to**, and **900 FPM**, as shown in Figure 1–67.

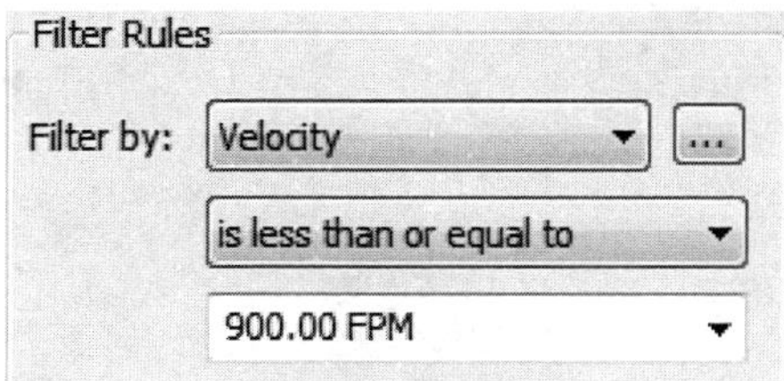

Figure 1–67

7. Click **Apply** and remain in the Filters dialog box.

8. Select the new **Mechanical-Low Velocity** filter and click (Duplicate). Right-click on the duplicated filter and click (Rename). In the Rename dialog box type **Mechanical-Medium/High Velocity** and click **OK**.

9. In the *Filter Rules* area, set *Filter by* to **Velocity**, **is greater than**, and **1900 FPM**, as shown in Figure 1–68.

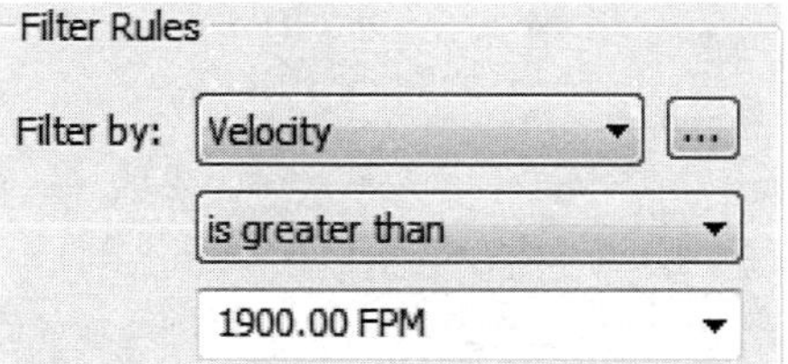

Figure 1–68

10. Click **OK** twice to close the Filters dialog boxes.

Task 2 - Create a View Template.

1. In the Project Browser, in the *Mechanical>Floor Plans* area, right-click on the **1 - Mech** view and select **Create View Template From View**.

2. Name the new view template **Velocity Duct Plan** and click **OK**.

3. Next to **V/G Override Filters**, click **Edit...** The Visibility/Graphic Overrides dialog box opens with the *Filter* tab selected.

4. Click **Add**.

5. In the Add Filters dialog box, select the two filters you just created and click **OK**.

6. For the **Mechanical-Low Velocity** filter, in the *Projection/Surface>Lines* column, select **Override**.

7. In the Line Graphics dialog box, change *Weight* to **6** and the *Color* to an orange, as shown in Figure 1–69. Click **OK**.

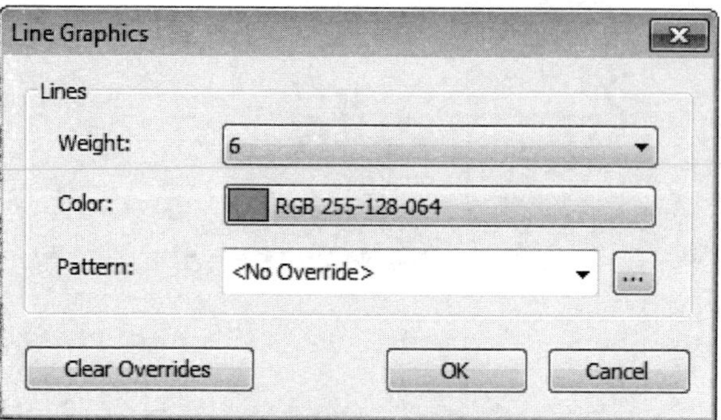

Figure 1–69

8. Repeat for the **Mechanical-Medium/High Velocity** filter and set the *Weight* to **6** and the *Color* to a green. Click **OK**.

9. In the Visibility/Graphics Overrides dialog box, move the new filters to the top of the list. For **Mechanical-Exhaust**, in the *Visibility* column, clear the checkbox as shown in Figure 1–70.

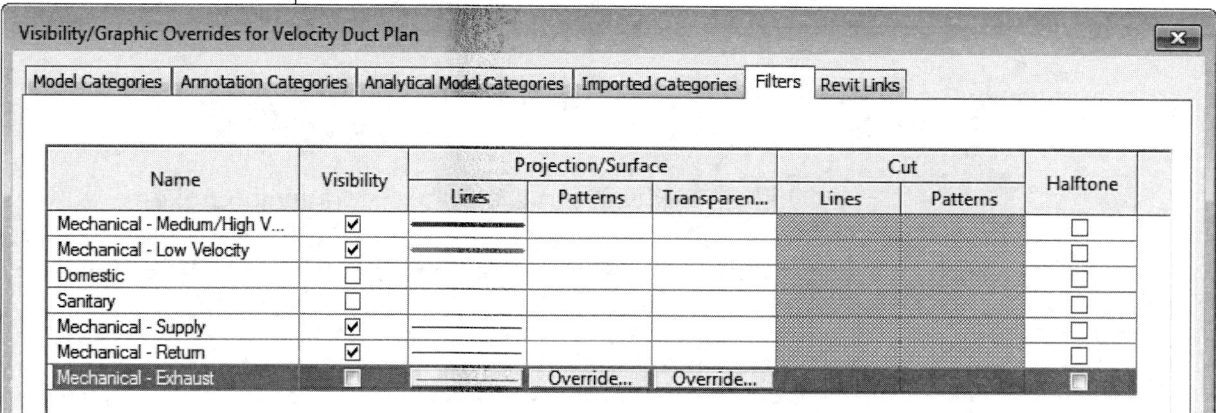

Figure 1–70

10. Click **OK** to close the Visibility/Graphic Overrides dialog box, and click **OK** to complete the view template.

Task 3 - Apply a View Template the views.

1. In the Project Browser, use *Duplicate with Detailing* to duplicate the **1 - Mech** view. Rename it **1 - Mech - Velocity.**
2. Right-click and select **Apply View Template**.
3. In the Apply View Template dialog box, select the **Velocity Duct Plan** view template you just created and click **OK** to apply it.
4. Compare the new floor plan view with the existing **1 - Mech** view.
5. Duplicate the **2 - Mech** view, rename it as **2 - Mech - Velocity**, and apply the same view template to it.
6. Finish with the **1 - Mech** view.
7. Save the project.

Chapter Review Questions

1. Which of the following items are NOT set in a template file?
    a. Units
    b. Annotation Types
    c. Title blocks
    d. Keyboard Shortcuts
2. What is a label?
    a. A text type used in title blocks.
    b. A dimension with text instead of numbers.
    c. A type of text with variable information.
3. When you want to create new Text Types, such as those shown in Figure 1–71, you need to copy an existing one.

    A FANCY FONT AT 1/4"

    A HAND LETTERING FONT AT 1/8"

    A HAND LETTERING FONT AT 5/32"

    Figure 1–71

    a. True
    b. False
4. Which of the following enables you to assign a view template consistently to a view so that no changes can be made to the view parameters?
    a. In the View Status Bar, lock the view.
    b. In Properties, select a View Template.
    c. In the Project Browser, right-click on the view and select **Lock**.
    d. In the Apply View Template dialog box, select a template.

5. Which of the following parameters are included in a View Template (as shown in Figure 1–72)? (Select all that apply.)

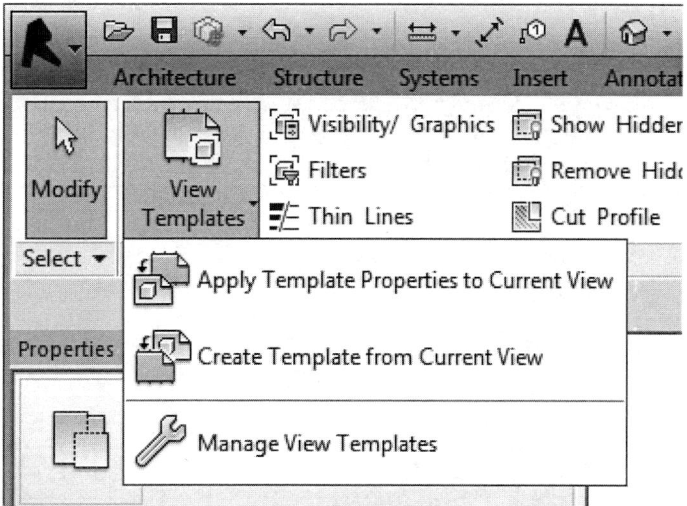

Figure 1–72

a. View Scale
b. V/G Overrides
c. Project Units
d. Detail Level

Command Summary

Button	Command	Location
Templates		
	Callout Tags	• **Ribbon:** *Manage* tab>Settings panel> expand Additional Settings
	Elevation Tags	• **Ribbon:** *Manage* tab>Settings panel> expand Additional Settings
	Floor Plan	• **Ribbon:** *View* tab>Create panel> expand Plan Views
	Loaded Tags	• **Ribbon:** *Annotate* tab>Tag panel> expand the panel title
	Project Units	• **Ribbon:** *Manage* tab>Settings panel • **Shortcut:** UN
	Section Tags	• **Ribbon:** *Manage* tab>Settings panel> expand Additional Settings
Annotation		
	Dimension Types	• **Ribbon:** *Annotate* tab>Dimensions panel>expand the panel title
	Text	• **Family Editor** • **Ribbon:** *Create* tab>Text panel
Title Blocks		
	Label	• **Family Editor** • **Ribbon:** *Create* tab>Text panel
	New Title Block	• **Application Menu:** expand New> Title block
	Revision Schedule	• **Family Editor** • **Ribbon:** *View* tab>Create panel
View Templates		
	ApplyTemplate Properties to Current View	• **Ribbon:** *View* tab>Graphics panel> expand View Templates • **Project Browser:** (*right-click on a view*) **Apply Template Properties...**
	Create View Template From View	• **Ribbon:** *View* tab>Graphics panel> expand View Templates • **Project Browser:** (*right-click on a view*)
	Manage View Templates	• **Ribbon:** *View* tab>Graphics panel> expand View Templates
	Temporary View Properties	• View Control Bar

Schedules

Schedules are a critical component of BIM projects. They gather information about the model that is used in construction documents, and can also be used to test your model (for example, to verify the amount of air flow in a space). Several types of schedules can be created, including: building component schedules, key schedules, and material takeoff schedules. These schedules are all created using the same basic methods.

Learning Objectives in this Chapter

- Schedule building components.
- Customize schedule tables using filters, sorting/grouping, formatting, and appearance.
- Modify the appearance and content of individual schedule cells.
- Create a Key Style schedule by defining keys for elements with similar characteristics.
- Create additional parameters including Calculated Value Fields and Percentage Fields.
- Apply conditional formatting to fields in a schedule.
- Embed a schedule into an existing schedule (Autodesk® Revit® MEP software only).
- Create material takeoff schedules.

2.1 Introduction to Schedules

Schedules can be created at any point in a project or included in template files. Each schedule is a separate view and can be inserted on a sheet, as shown in Figure 2–1.

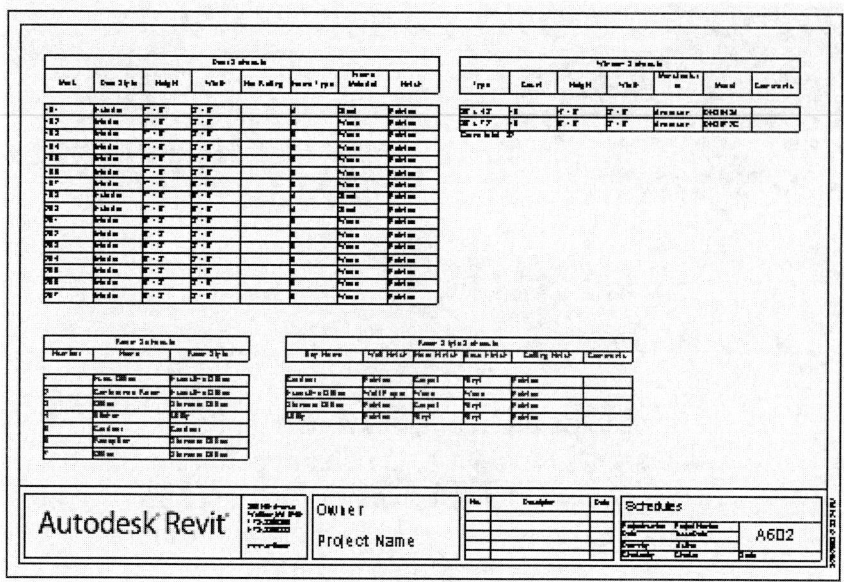

Figure 2–1

You can create the following types of schedules:

(Schedule/ Quantities)	Generates schedules using parameters available at category level. There are two types of Schedule/ Quantities schedules: **Building Component** and **Key**.
(Material Takeoff)	Generates schedules from information about materials in a project and can be used to do quantity takeoffs.
(Sheet List)	Generates schedule from sheet information including who has designed and checked the sheet.
(Note Block)	Generates schedules that keep count of all of the symbols used in the project.
(View List)	Generates schedules that keep count of all of the views in the project, including their scale, phase, and use on sheets.
(Graphical Column Schedule)	(Autodesk Revit Structure only) Generates graphical schedules displaying heights of all columns in a project compared to levels.

Importing and Exporting Schedules

Schedules are views and can be copied into your project from other projects. Only the formatting information is copied; the information about individually scheduled items is not included. That information is automatically gathered from the elements in the current project. You can also export the schedule information to be used in spreadsheets.

How To: Import Schedules

1. In the *Insert* tab>Import panel, expand (Insert From File) and click (Insert Views From File).
2. In the Open dialog box, locate the project file containing the schedule that you want to use.
3. Select the schedules that you want to import, as shown in Figure 2–2.

*If the referenced project contains many types of views, change Views: to **Show schedules and reports only**.*

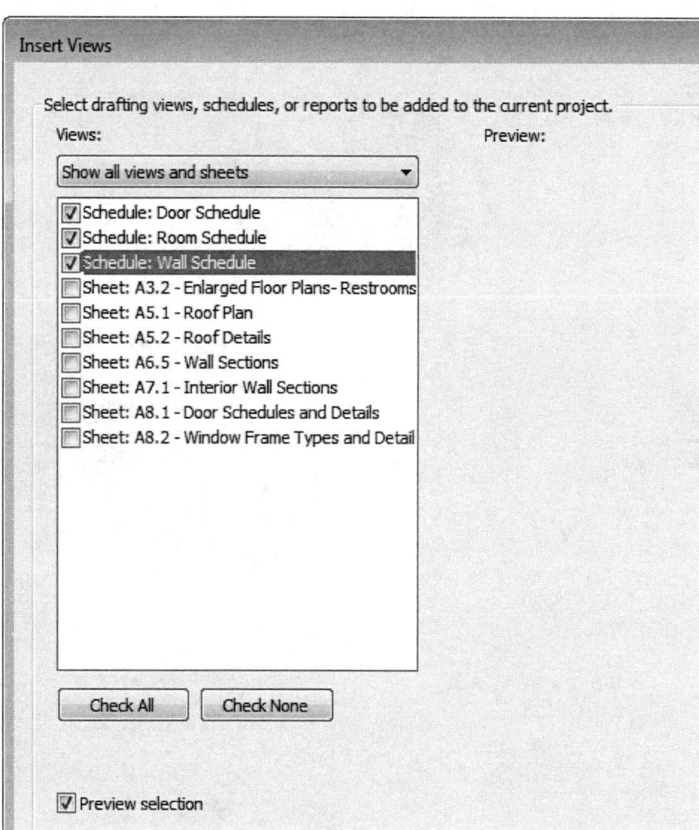

Figure 2–2

4. Click **OK**.

How To: Export Schedule Information

1. Switch to the schedule view that you want to export.
2. In the Application Menu, expand (Export), expand (Reports), and click (Schedule).
3. In the Export Schedule dialog box, select a location and name for the text file and click **Save**.
4. In the Export Schedule dialog box, set the options in the *Schedule appearance* and *Output options* areas that best suit your spreadsheet software, as shown in Figure 2–3.

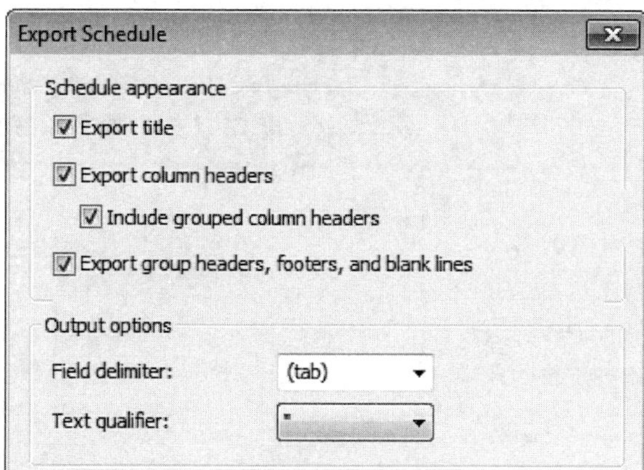

Figure 2–3

5. Click **OK**. A new text file is created that you can open in a spreadsheet, as shown in Figure 2–4.

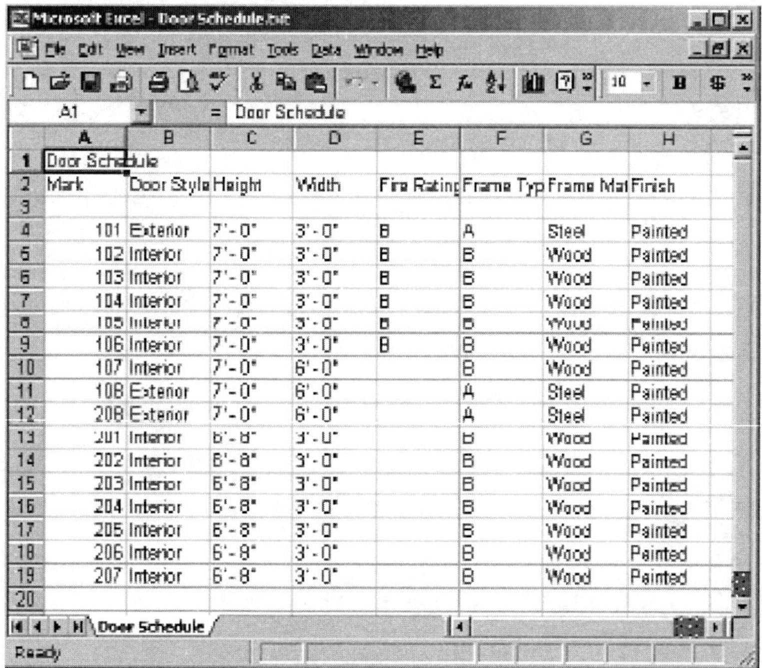

A	B	C	D	E	F	G	H
Door Schedule							
Mark	Door Style	Height	Width	Fire Ratin	Frame Typ	Frame Mat	Finish
101	Exterior	7' - 0"	3' - 0"	B	A	Steel	Painted
102	Interior	7' - 0"	3' - 0"	B	B	Wood	Painted
103	Interior	7' - 0"	3' - 0"	B	B	Wood	Painted
104	Interior	7' - 0"	3' - 0"	B	B	Wood	Painted
105	Interior	7' - 0"	3' - 0"	B	B	Wood	Painted
106	Interior	7' - 0"	3' - 0"	B	B	Wood	Painted
107	Interior	7' - 0"	6' - 0"		B	Wood	Painted
108	Exterior	7' - 0"	6' - 0"		A	Steel	Painted
208	Exterior	7' - 0"	6' - 0"		A	Steel	Painted
201	Interior	6' - 8"	3' - 0"		B	Wood	Painted
202	Interior	6' - 8"	3' - 0"		B	Wood	Painted
203	Interior	6' - 8"	3' - 0"		B	Wood	Painted
204	Interior	6' - 8"	3' - 0"		B	Wood	Painted
205	Interior	6' - 8"	3' - 0"		B	Wood	Painted
206	Interior	6' - 8"	3' - 0"		B	Wood	Painted
207	Interior	6' - 8"	3' - 0"		B	Wood	Painted

Figure 2–4

2.2 Creating Building Component Schedules

A Building Component schedule is a table view of the type and instance parameters of a specific element. You can specify the parameters (fields) you want to include in the schedule. All of the parameters found in the type of element you are scheduling are available to use. For example, a door schedule (as shown in Figure 2–5) can include instance parameters that are automatically filled in (such as the **Height** and **Width**) and type parameters that might need to have the information assigned in the schedule or element type (such as the **Fire Rating** and **Frame**).

<Door Schedule>

A	B	C	D	E	F	G	H
Mark	Height	Width	Fire Rating	Frame Type	Frame Material	Finish	Function
2	7' - 0"	10' - 0"	A	A	Steel	Brushed	Exterior
3	6' - 8"	3' - 0"	B	B	Wood	Paint	Interior
4	6' - 8"	3' - 0"	B	B	Wood	Paint	Interior
5	6' - 8"	3' - 0"	B	B	Wood	Paint	Interior
6	6' - 8"	3' - 0"	B	B	Wood	Paint	Interior

Figure 2–5

How To: Create a Building Component Schedule

1. In the *View* tab>Create panel, expand (Schedules) and click (Schedule/Quantities) or in the Project Browser, right-click on the Schedule/Quantities node and select **New Schedule/Quantities**.

2. In the New Schedule dialog box, select the type of schedule you want to create (e.g., Doors) from the *Category* list, as shown in Figure 2–6.

In the Filter list drop-down list, you can specify the discipline(s) to show only the categories that you want to display.

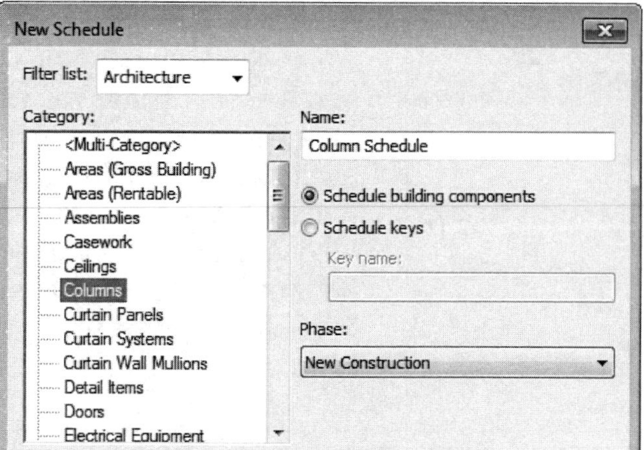

Figure 2–6

3. Type a new *Name*, if the default does not suit.
4. Select **Schedule building components.**
5. Specify the *Phase* as required.
6. Click **OK**.
7. Fill out the information in the Schedule Properties dialog box. This includes the information in the *Fields*, *Filter*, *Sorting/Grouping*, *Formatting*, and *Appearance* tabs.
8. Once you have entering the schedule properties, click **OK**. A schedule report is created in its own view.

Schedule Properties – Fields Tab

In the *Fields* tab, you can select from a list of available fields and organize them in the order in which you want them to display in the schedule, as shown in Figure 2–7.

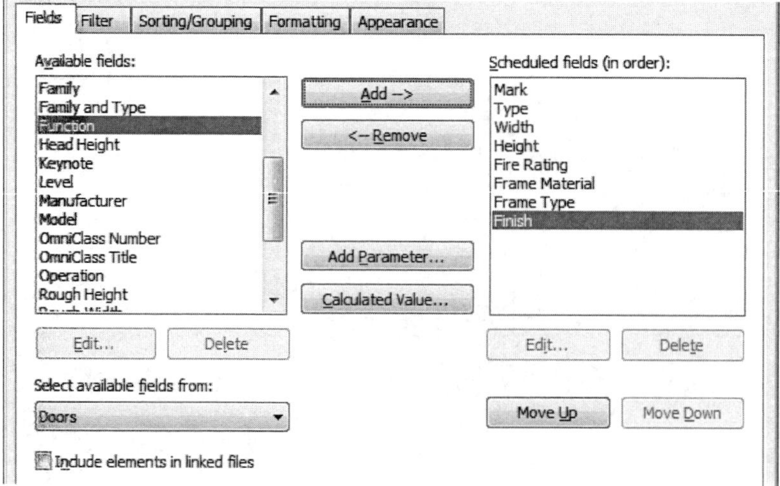

Figure 2–7

How To: Fill out the Fields Tab

You can also double-click on a field to move it from the Available fields to the Scheduled fields list.

1. In the *Available fields* list, select one or more fields you want to add to the schedule and click **Add -->**. The field(s) are placed in the *Scheduled fields (in order)* list.
2. Continue adding fields as required. If you add one you did not want to use, select it in the *Scheduled fields* list and click **<-- Remove** to move it back to the *Available fields* list.
3. Use **Move Up** and **Move Down** to change the order of the scheduled fields.

Other Fields Tab Options

Select available fields from	Enables you to select additional category fields for the specified schedule. The available list of additional fields depends on the original category of the schedule. Typically, they include room information.
Include elements in linked files	Includes elements that are in files linked to the current project, so that their elements can be included in the schedule.
Add Parameter...	Adds a new field according to your specification. New fields can be placed by instance or by type.
Calculated Value...	Enables you to create a field that uses a formula based on other fields.
Edit...	Enables you to edit custom fields. This is grayed out if you select a standard field.
Delete...	Deletes selected custom fields. This is grayed out if you select a standard field.

Schedule Properties – Filter Tab

In the *Filter* tab, you can set up filters so that only elements meeting specific criteria are included in the schedule. For example, you might only want to show information for one level, as shown in Figure 2–8. You can create filters for up to eight values. All values must be satisfied for the elements to display.

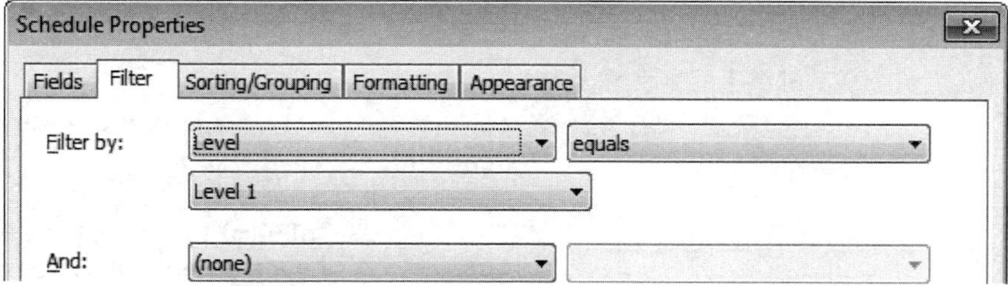

Figure 2–8

- The parameter you want to use as a filter must be included in the schedule. You can hide the parameter once you have completed the schedule, if required.

Filter by	Specifies the field to filter. Not all fields are available to be filtered.
Condition	Specifies the condition that must be met. This includes options such as **equal**, **not equal**, **greater than**, and **less than**.
Value	Specifies the value of the element to be filtered. You can select from a drop-down list of appropriate values. For example, if you set ***Filter By*** to **Level**, it displays the list of levels in the project.

Schedule Properties – Sorting/Grouping Tab

In the *Sorting/Grouping* tab, you can set how you want the information to be sorted, as shown in Figure 2–9. For example, you can sort by **Mark** (number) and then **Type**.

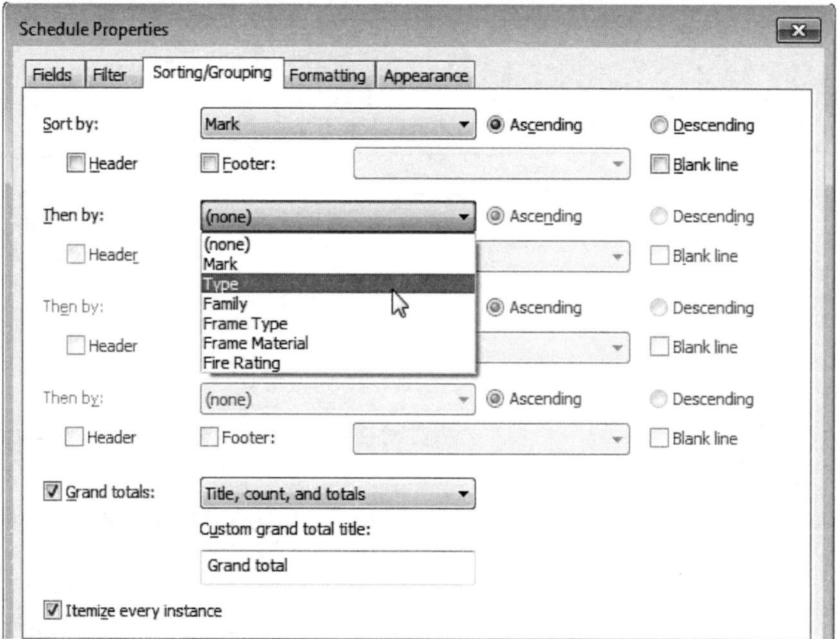

Figure 2–9

Sort by	Enables you to select the field(s) you want to sort by. You can select up to four levels of sorting.
Ascending/ Descending	Sorts fields in **Ascending** or **Descending** order.
Header/ Footer	Enables you to group similar information and separate it by a **Header** with a title and/or a **Footer** with quantity information.

Blank line	Adds a blank line between groups.
Grand totals	Selects which totals to display for the entire schedule. You can specify a name to display in the schedule for the Grand total.
Itemize every instance	If selected, displays each instance of the element in the schedule. If not selected, displays only one instance of each type, as shown in Figure 2–10.

<Window Schedule>

A	B	C	D	E	F	G
Type	Count	Height	Width	Manufacturer	Model	Comments
36 x 36	6	3' 0"	3' - 0"	Anderson	FX3636	
36" x 48"	7	4' - 0"	3' - 0"	Anderson	FX3648	
Grand total: 13						

Figure 2–10

Schedule Properties – Formatting Tab

In the *Formatting* tab, you can control how the headers of each field display, as shown in Figure 2–11.

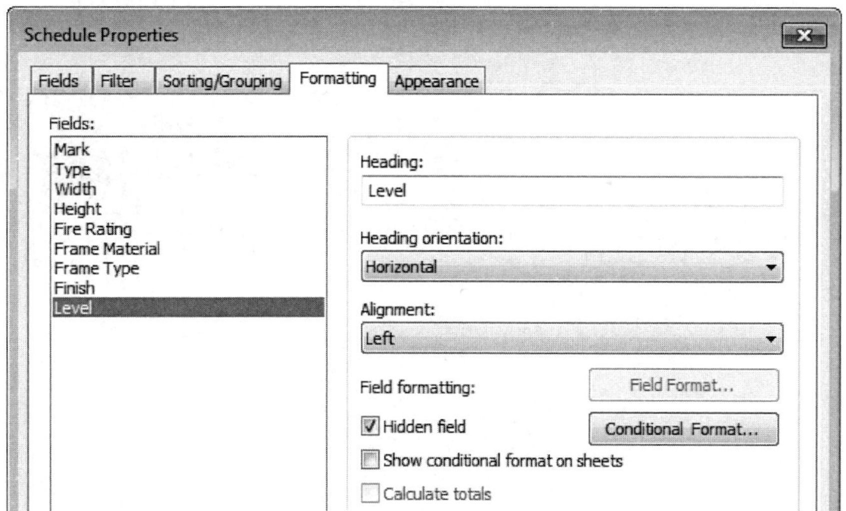

Figure 2–11

Fields	Enables you to select the field for which you want to modify the formatting.
Heading	Enables you to change the heading of the field if you want it to be different from the field name. For example, you might want to replace **Mark** (a generic name) with the more specific **Door Number** in a door schedule.

Heading orientation	Enables you to set the heading on sheets to **Horizontal** or **Vertical**. This does not impact the schedule view.
Alignment	Aligns the text in rows under the heading to be **Left**, **Right**, or **Center**.
Field Format...	Sets the units format for the length, area, volume, angle, or number field. By default, this is set to use the project settings.
Conditional Format...	Sets up the schedule to display visual feedback based on the conditions listed.
Hidden field	Enables you to hide a field. For example, you might want to use a field for sorting purposes, but not have it display in the schedule. You can also modify this option in the schedule view later.
Show conditional format on sheets	Select if you want the color code set up in the Conditional Format dialog box to display on sheets.
Calculate totals	Displays the subtotals of numerical columns in a group.

Schedule Properties – Appearance Tab

In the *Appearance* tab, you can set the text style and grid options for a schedule, as shown in Figure 2–12.

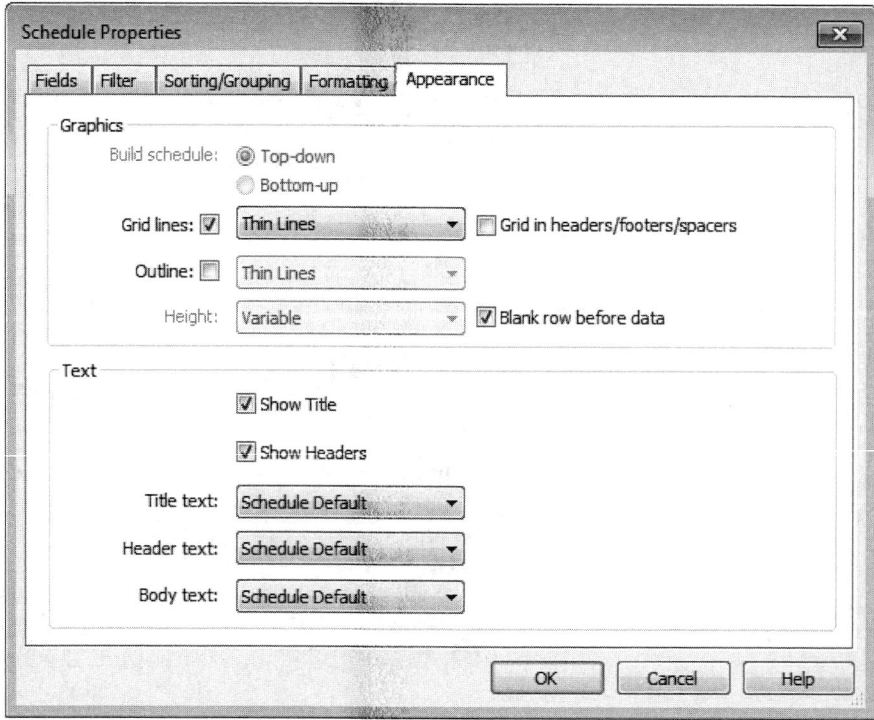

Figure 2–12

Grid lines	Displays lines between each instance listed and around the outside of the schedule. Select the style of lines from the drop-down list; this controls all lines for the schedule, unless modified.
Grid in headers/ footers/spacers	Extends the vertical grid lines between the columns.
Outline	Specify a different line type for the outline of the schedule.
Blank row before data	Select this option if you want a blank row to be displayed before the data begins in the schedule.
Show Title/Show Headers	Select these options to include the text in the schedule.
Title text/Header text/Body Text	Select the text style for the title, header, and body text.

Schedule Properties

Schedule views have properties including the *View Name*, *Phases* and methods of returning to the Schedule Properties dialog box as shown in Figure 2–13. In the *Other* area, select the button next to the tab that you want to open in the Schedule Properties dialog box. In the dialog box, you can switch from tab to tab and make any required changes to the overall schedule.

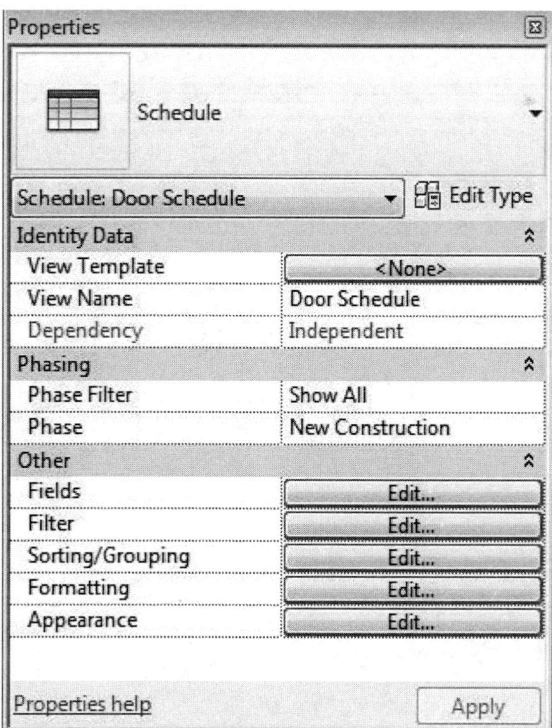

Figure 2–13

Filtering Elements from Schedules

When you create schedules based on a category you might need to filter out some of the element types in that category. For example, in the Autodesk Revit software, doors (and windows) in curtain walls are automatically added to a door schedule, as shown at the top in Figure 2–14, but are typically estimated as part of the curtain wall rather than as a separate door. To remove them from the schedule, as shown at the bottom in Figure 2–14, assign a parameter that identifies them and then use that parameter to filter them out of the schedule.

Door Schedule- 1st						
	Door Size			Frame		
Door Type	Width	Height	Thickness	Frame Type	Frame Material	Head Detail
	8' - 3 1/2"	9' - 4 1/4"				
	3' - 0"	7' - 0"	0' - 2"	A	Aluminum	
	3' - 0"	7' - 0"	0' - 2"	B	Aluminum	
	3' - 0"	6' - 8"	0' - 2"	C	Aluminum	

Door Schedule- 1st						
	Door Size			Frame		
Door Type	Width	Height	Thickness	Frame Type	Frame Material	Head Detail
	3' - 0"	7' - 0"	0' - 2"	A	Aluminum	
	3' - 0"	7' - 0"	0' - 2"	B	Aluminum	
	3' - 0"	6' - 8"	0' - 2"	C	Aluminum	
	3' - 0"	6' - 8"	0' - 2"	C	Aluminum	

Figure 2–14

- This type of filtering can be used for any schedule in any discipline.

How To: Filter Elements in a Schedule

Create a type specifically for this if you are using one that is also used elsewhere.

1. Select an element (such as a door used in curtain walls) and modify the Type Parameters. Add a value to one of the parameters that you are not otherwise using in your schedule. For example, set *Construction Type* to **CW**, as shown Figure 2–15

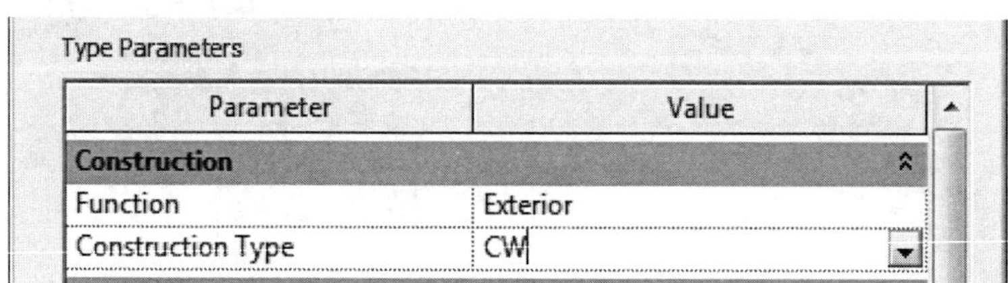

Figure 2–15

2. Create a schedule and include the field, such as *Construction Type.*

3. Modify the *Filter* of the schedule so the parameter does not equal the specified value. In the example shown in Figure 2–16, *Construction Type* **does not equal CW**. Any types that match this filter are excluded from the schedule.

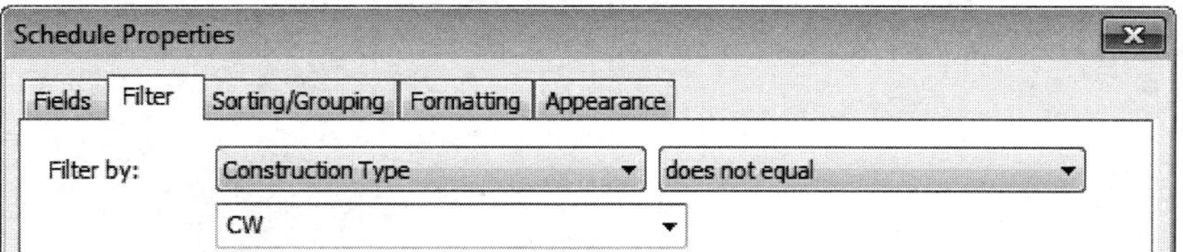

Figure 2–16

4. In the final schedule, the elements display with the specified value. Right-click on the column header and select **Hide Column(s)**, as shown in Figure 2–17. It is just used as a filter and not part of the final schedule.

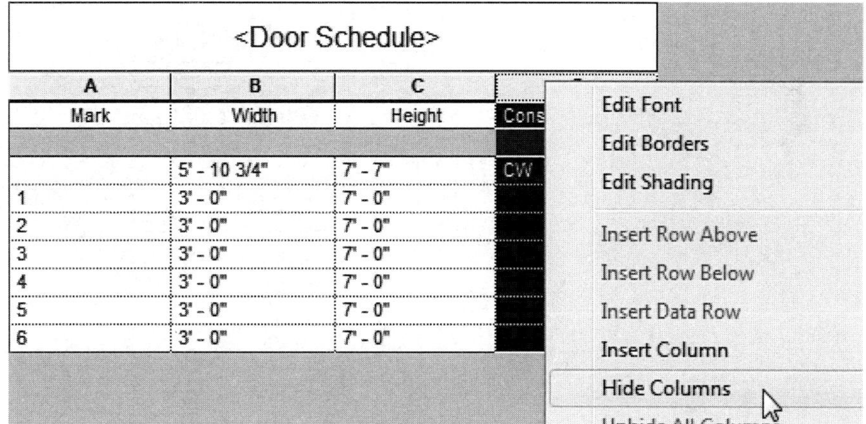

Figure 2–17

Practice 2a Create Schedules for Architectural Projects

Practice Objectives

- Create building component schedules.
- Apply filters.
- Enter information.
- Place schedules on a sheet.

Estimated time for completion: 15 minutes

n this practice you will create Door and Window schedules, filter out curtain wall doors, and add information to cells in the door schedule. You will also place the schedules on a sheet, as shown in Figure 2–18.

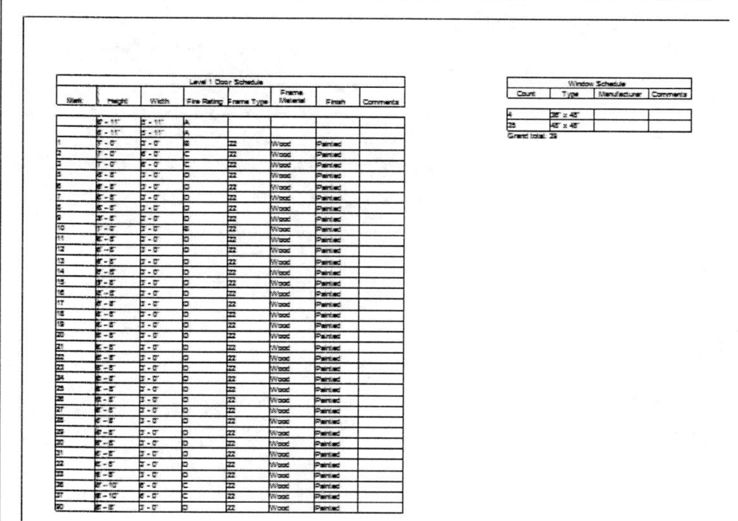

Figure 2–18

Task 1 - Create a Door schedule.

1. In the ...*Architectural* folder open **Clark-Hall-Schedules-Create-A.rvt**.

2. In the *View* tab>Create panel, expand (Schedules) and click (Schedule/Quantities).

3. In the New Schedule dialog box, set the *Filter List* to **Architecture** only to limit the number of categories and then set the *Category* to **Doors.**

4. Name the schedule **Door Schedule-Level 1**. Click **OK**.

5. In the Schedule Properties dialog box, in the *Fields* tab, add the following fields, as shown in Figure 2–19:
   - *Level*
   - *Mark*
   - *Height*
   - *Width*
   - *Fire Rating*
   - *Frame Type*
   - *Frame Material*
   - *Finish*
   - *Comments*
   - *Construction Type*

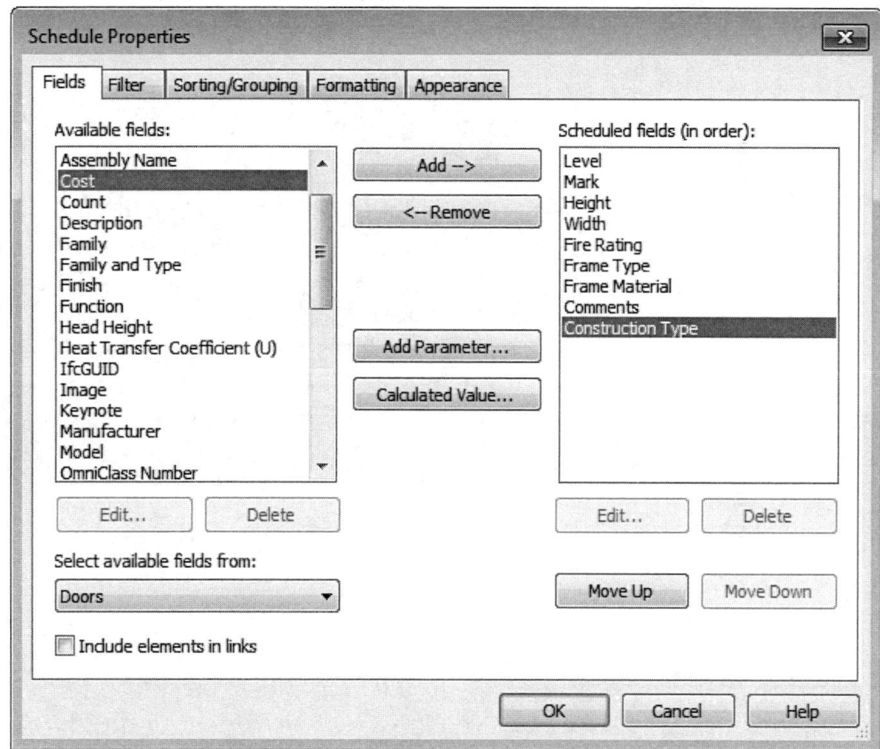

Figure 2–19

6. Click **OK** to create the schedule. Note that Level 1, Level 2, and Level 3 doors are included.

7. In Properties, in the *Other* category, next to **Filter**, click **Edit...**. The Schedule Properties dialog box opens with the *Filter* tab selected. Set *Filter by* to **Level - equals - Level 1**.

8. In the dialog box, in the *Sorting/Grouping* tab, sort the doors by **Mark**, and click **OK** to update the schedule.

9. Save the project.

Task 2 - Remove the Curtain Wall Doors from the schedule.

1. Continue working in the new **Door Schedule-Level 1** view, The doors in the curtain walls do not display with a **Mark**, as shown in Figure 2–20. Additionally, they are not included in door schedules and need to be removed.

A	B	C	D
Level	Mark	Height	Width
Level 1		6' - 11"	5' - 11"
Level 1		6' - 11"	5' - 11"
Level 1	1	7' - 0"	3' - 0"
Level 1	2	7' - 0"	6' - 0"
Level 1	3	7' - 0"	6' - 0"
Level 1	4	6' - 8"	3' - 0"
Level 1	5	6' - 8"	3' - 0"

Figure 2–20

2. In Properties, in the *Other* category, next to **Filter**, click **Edit...**.

3. In the Schedule Properties dialog box, in the *Filter* tab, add the additional Filter: **Construction Type - does not equal - CW** as shown in Figure 2–21, and click **OK**.

*You need to type **CW** because the option does not yet exist in the project.*

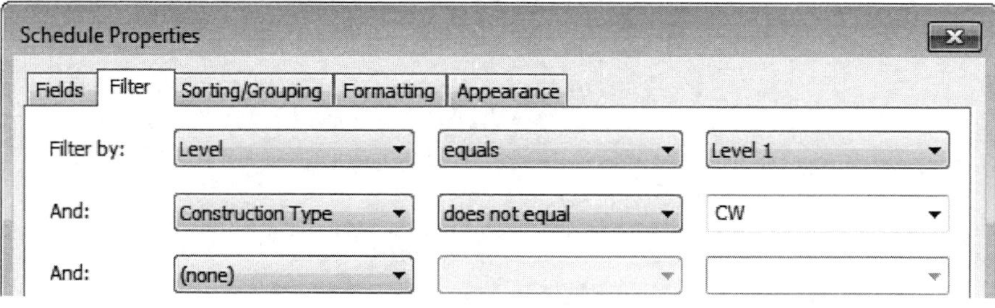

Figure 2–21

4. In the Schedule, next to one of the curtain wall doors, in the *Construction Type* column, type **CW**. An alert displays prompting you that this is a type property. Click **OK** and the curtain wall doors are no longer included in the schedule, as shown in Figure 2–22.

A	B	C	D
Level	Mark	Height	Width
Level 1	1	7' - 0"	3' - 0"
Level 1	2	7' - 0"	6' - 0"
Level 1	3	7' - 0"	6' - 0"
Level 1	4	6' - 8"	3' - 0"

Figure 2–22

5. Right-click on the Construction Type column header and select **Hide Columns**.

6. Save the project

Task 3 - Fill in additional information in the Door Schedule

1. Continue working in the **Door Schedule-Level 1** view,

2. In the *Fire Rating* column, type a letter for one of the door types. Because the door type controls the **Fire Rating** parameter, an alert box opens, as shown in Figure 2–23. Click **OK**.

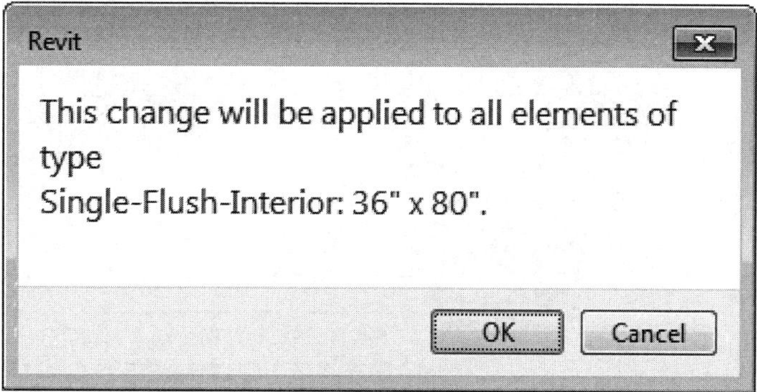

Figure 2–23

3. Repeat with the other doors until they all have a fire rating.

4. Open the **Floor Plans: Level 1** view and select one of the interior single doors.

5. Right-click and select **Select All Instances>In Entire Project** in Properties. Set the following parameters:

   - **Frame Type:** 22
   - **Frame Material:** Wood
   - **Finish:** Stained

6. Switch back to the schedule to verify that it has updated with the new information.

7. In the schedule view, set the rest of the doors to the following parameters:

   - **Frame Type:** 21
   - **Frame Material:** Aluminum
   - **Finish:** Brushed

8. Right-click on the *Level* header and select **Hide Columns**. Repeat the procedure with the *Construction Type* header. This removes the columns from the schedule as shown in Figure 2–24, while still enabling you to use the fields as filters.

<Door Schedule-Level 1>

A	B	C	D	E	F	G	H
Mark	Height	Width	Fire Rating	Frame Type	Frame Material	Finish	Comments
1	7' - 0"	3' - 0"	A	21	Aluminum	Brushed	
2	7' - 0"	6' - 0"	A	21	Aluminum	Brushed	
3	7' - 0"	6' - 0"	A	21	Aluminum	Brushed	
4	6' - 8"	3' - 0"	B	22	Wood	Stained	
5	6' - 8"	3' - 0"	B	22	Wood	Stained	
6	6' - 8"	3' - 0"	B	22	Wood	Stained	
7	6' - 8"	3' - 0"	B	22	Wood	Stained	
8	6' - 8"	3' - 0"	B	22	Wood	Stained	

Figure 2–24

9. Save the project.

Task 4 - Create a Window schedule.

1. Create a new window schedule with the following fields: **Count**, **Type**, **Manufacturer**, and **Comments**.
2. In the *Sorting/Grouping* tab, sort the windows by **Type**. Toggle on **Grand totals** and toggle off **Itemize every instance**.
3. Click **OK** to create the schedule. It lists the total count for a single type (only one type is used in the project), as shown in Figure 2–25.

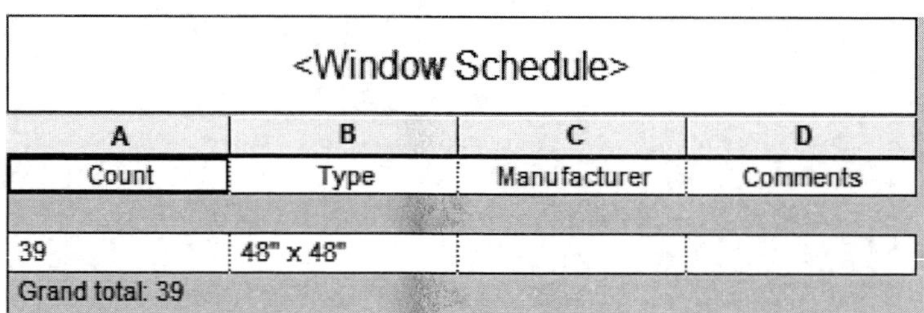

<Window Schedule>

A	B	C	D
Count	Type	Manufacturer	Comments
39	48" x 48"		
Grand total: 39			

Figure 2–25

4. In the **Floor Plans: Level 1** view, select the three windows in the Conference Room in the office wing. In Properties change the type to **Fixed with Trim: 36" x 72"**.

5. Switch back to the schedule view. A change should display in the schedule that reflects the new window type, as shown in Figure 2–26.

<Window Schedule>			
A	B	C	D
Count	Type	Manufacturer	Comments
3	36" x 72"		
36	48" x 48"		
Grand total: 39			

Figure 2–26

6. Open the **A601- Schedules** sheet. Drag and drop the schedules to this sheet. Modify the width of each field as needed.

7. Save the project.

Practice 2b Create Schedules for MEP Projects

Practice Objectives

- Create building component schedules.
- Place the schedules on a sheet.

Estimated time for completion: 15 minutes

In this practice you will create schedules for two types of typical MEP schedules: a system schedule and a fixture schedule. In this case, they will be for duct systems and plumbing fixtures. If you have time, you will also create schedules for electrical circuits and lighting fixtures. You will then place the schedules on a sheet, as shown in Figure 2–27.

Duct Systems		
System Name	Comments	Flow
Exhaust Air		
Mechanical Exhaust Air 1		2000 CFM
Mechanical Exhaust Air 7		420 CFM
Mechanical Exhaust Air 9		200 CFM
Return Air		
Mechanical Return Air 1		0 CFM
Mechanical Return Air 2		0 CFM
Mechanical Return Air 3		205 CFM
Mechanical Return Air 4		500 CFM
Mechanical Return Air 5		500 CFM
Mechanical Return Air 6		500 CFM
Mechanical Return Air 10		0 CFM
Mechanical Return Air 14		205 CFM
Supply Air		
Mechanical Supply Air 1		4000 CFM
Mechanical Supply Air 2		2400 CFM
Mechanical Supply Air 3		2900 CFM
Mechanical Supply Air 4		3500 CFM
Mechanical Supply Air 8		0 CFM
Mechanical Supply Air 9		2900 CFM
Mechanical Supply Air 10		3900 CFM
Mechanical Supply Air 15		3090 CFM
Mechanical Supply Air 19		2400 CFM

Figure 2–27

Task 1 - Create a Duct System schedule.

1. In the ...*MEP* practice files folder, open **Freel Office-Create.rvt**.
2. In the *View* tab>Create panel, expand (Schedules) and click (Schedule/Quantities).

3. In the New Schedule dialog box, set the *Filter list* to **Mechanical** and the *Category* to **Duct Systems** and name it **Duct Systems**, as shown in Figure 2–28.

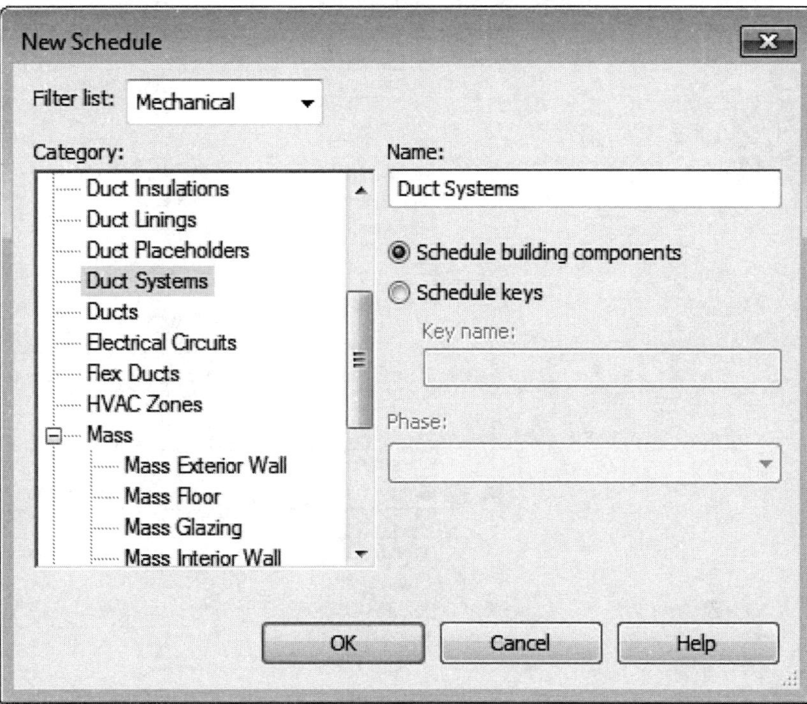

Figure 2–28

4. Click **OK**.

5. In the Schedule Properties dialog box, in the *Fields* tab, add the following fields:

   - *System Classification*
   - *System Name*
   - *Comments*
   - *Flow*

6. In the *Sorting/Grouping* tab, sort by **System Classification**. Select **Ascending** and **Header**.

7. Click **OK** to create the schedule. Stretch the columns so that you can read all of the information, as shown in Figure 2–29.

<Duct System Schedule>			
A	B	C	D
System Classification	System Name	Comments	Flow
Exhaust Air			
Exhaust Air	Mechanical Exhaust Air 1		2000 CFM
Exhaust Air	Mechanical Exhaust Air 7		420 CFM
Exhaust Air	Mechanical Exhaust Air 9		200 CFM
Return Air			
Return Air	Mechanical Return Air 1		0 CFM
Return Air	Mechanical Return Air 2		0 CFM
Return Air	Mechanical Return Air 3		205 CFM
Return Air	Mechanical Return Air 4		500 CFM
Return Air	Mechanical Return Air 5		500 CFM
Return Air	Mechanical Return Air 6		500 CFM
Return Air	Mechanical Return Air 10		0 CFM
Return Air	Mechanical Return Air 14		205 CFM
Supply Air			
Supply Air	Mechanical Supply Air 1		4000 CFM
Supply Air	Mechanical Supply Air 2		2400 CFM
Supply Air	Mechanical Supply Air 3		2900 CFM
Supply Air	Mechanical Supply Air 4		3500 CFM
Supply Air	Mechanical Supply Air 8		0 CFM
Supply Air	Mechanical Supply Air 9		2900 CFM
Supply Air	Mechanical Supply Air 10		3900 CFM
Supply Air	Mechanical Supply Air 15		3090 CFM
Supply Air	Mechanical Supply Air 19		2400 CFM

Figure 2–29

8. Right-click on the *System Classification* column and select **Hide Column**. (The System Classification is specified by the Header).

9. Open the **M601- Mechanical Schedules** sheet. Drag and drop the duct system schedule to this sheet. Modify the width of each field as needed.

10. Save the project.

Task 2 - Create a Plumbing Fixture schedule.

1. In the *View* tab>Create panel, expand (Schedules) and click (Schedule/Quantities).

2. In the New Schedule dialog box, set the *Filter list* to **Piping**, set *Category* to **Plumbing Fixtures**, and name it **Plumbing Fixtures**. Click **OK**.

3. In the Schedule Properties dialog box, in the *Fields* tab, add the following fields:
   - *Family*
   - *Type*
   - *Manufacturer*
   - *Count*
4. In the *Sorting/Grouping* tab, sort by Family, and clear **Itemize every instance**.
5. Click **OK** to create the schedule. Stretch the columns such that all the words are visible, as shown in Figure 2–30.

Plumbing Fixture Schedule			
Family	Type	Manufacturer	Count
Drinking Fountain-Hi-Lo-3D	Drinking Fountain-Hi-Lo-3D		2
Shower Base-Round Corner	36" x 36"		1
Sink Kitchen-Double	42" x 21"		3
Sink Vanity-Round	19" x 19"		8
Toilet-Commercial-Wall-3D	15" Seat Height		6
Urinal-Wall-3D	Urinal-Wall-3D		2

Figure 2–30

6. Open the **P601- Plumbing Schedules** sheet. Drag and drop the schedule to this sheet. Modify the width of each field as needed.
7. Save the project.

Task 3 - Create electrical-related schedules and a sheet.

1. If you have time, create an electrical circuit system schedule similar to the duct system schedule and a lighting fixture schedule similar to the plumbing fixture schedule. Place these schedules on the E601 - Electrical and Lighting Schedules sheet.
2. Save the project.

Practice 2c Create Schedules for Structural Projects

Practice Objectives

- Create a structural schedule that includes all types of structural elements.
- Reorganize the schedule content by sorting and grouping.
- Place the schedule on a sheet.

Estimated time for completion: 15 minutes

In this practice you will create a schedule displaying structural elements by category, You will then place it on a sheet and modify it to fit, as shown in Figure 2–31.

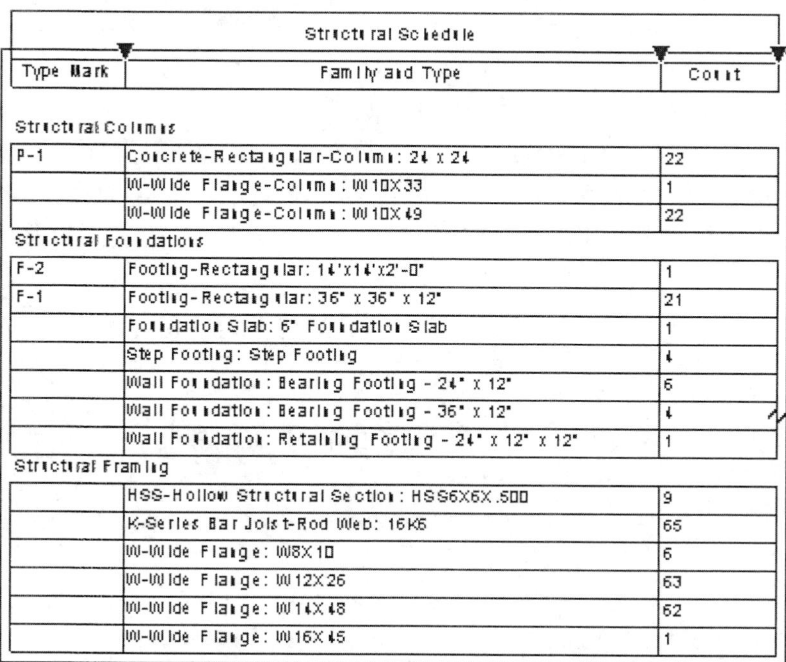

Structural Schedule		
Type Mark	Family and Type	Count
Structural Columns		
P-1	Concrete-Rectangular-Column: 24 x 24	22
	W-Wide Flange-Column: W10X33	1
	W-Wide Flange-Column: W10X49	22
Structural Foundations		
F-2	Footing-Rectangular: 14'x14'x2'-0"	1
F-1	Footing-Rectangular: 36" x 36" x 12"	21
	Foundation Slab: 6" Foundation Slab	1
	Step Footing: Step Footing	4
	Wall Foundation: Bearing Footing - 24" x 12"	6
	Wall Foundation: Bearing Footing - 36" x 12"	4
	Wall Foundation: Retaining Footing - 24" x 12" x 12"	1
Structural Framing		
	HSS-Hollow Structural Section: HSS6X6X.500	9
	K-Series Bar Joist-Rod Web: 16K5	65
	W-Wide Flange: W8X10	6
	W-Wide Flange: W12X26	63
	W-Wide Flange: W14X48	62
	W-Wide Flange: W16X45	1

Figure 2–31

Task 1 - Create a structural schedule.

1. In the ...*Structural* practice files folder, open **Axiom-Building-Create.rvt**.
2. In the *View* tab>Create panel, expand (Schedules) and click (Schedule/Quantities).
3. In the New Schedule dialog box, set *Filter list* to **Structure**, set *Category* to **<Multi-Category>**, and name it **Structural Schedule**. Click **OK**.

4. In the Schedule Properties dialog box, in the *Fields* tab, add the following fields:
   - *Type Mark*
   - *Family and Type*
   - *Count*
   - *Category*
5. Use **Move Up** and **Move Down** to change the order of the scheduled fields if needed so they display as shown in Figure 2–32.

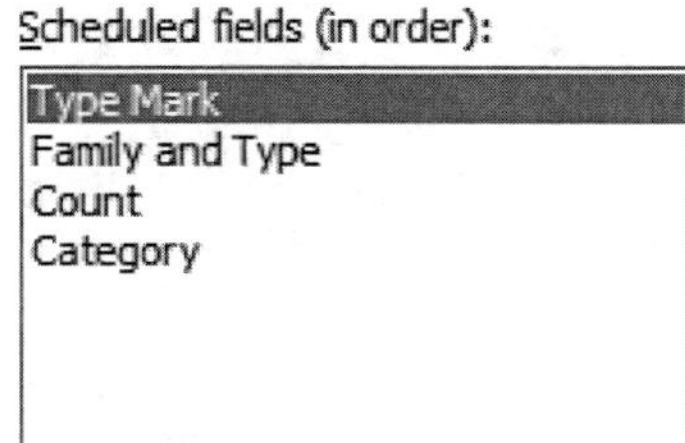

Figure 2–32

6. Click **OK**. The new schedule displays with all of the structural elements in the project in a long list, as shown in part in Figure 2–33.

<Structural Schedule>

A	B	C	D
Type Mark	Family and Type	Count	Category
	W-Wide Flange-Colum	1	Structural Columns
	W-Wide Flange-Colum	1	Structural Columns
	W-Wide Flange-Colum	1	Structural Columns
	W-Wide Flange-Colum	1	Structural Columns
	W-Wide Flange-Colum	1	Structural Columns
	W-Wide Flange-Colum	1	Structural Columns
	W-Wide Flange-Colum	1	Structural Columns
	W-Wide Flange-Colum	1	Structural Columns
	W-Wide Flange-Colum	1	Structural Columns
	W-Wide Flange-Colum	1	Structural Columns
	W-Wide Flange-Colum	1	Structural Columns
	W-Wide Flange-Colum	1	Structural Columns
	W-Wide Flange-Colum	1	Structural Columns
	W-Wide Flange-Colum	1	Structural Columns
	W-Wide Flange-Colum	1	Structural Columns
	W-Wide Flange-Colum	1	Structural Columns
	W-Wide Flange-Colum	1	Structural Columns
	W-Wide Flange-Colum	1	Structural Columns
	W-Wide Flange-Colum	1	Structural Columns
	Wall Foundation: Beari	1	Structural Foundati
	Wall Foundation: Beari	1	Structural Foundati
	Wall Foundation: Beari	1	Structural Foundati

Figure 2–33

7. Save the project.

Task 2 - Add filtering, sorting, and grouping to the schedule.

1. In Properties, in the *Other* category, next to **Sorting/ Grouping**, click **Edit....** The Schedule Properties dialog box opens with the *Sorting/Grouping* tab selected.

2. In the *Sorting/Grouping* tab, sort by **Category**, and select **Header**. Then sort by **Family and Type** with no header. Clear **Itemize every instance**, as shown in Figure 2–34.

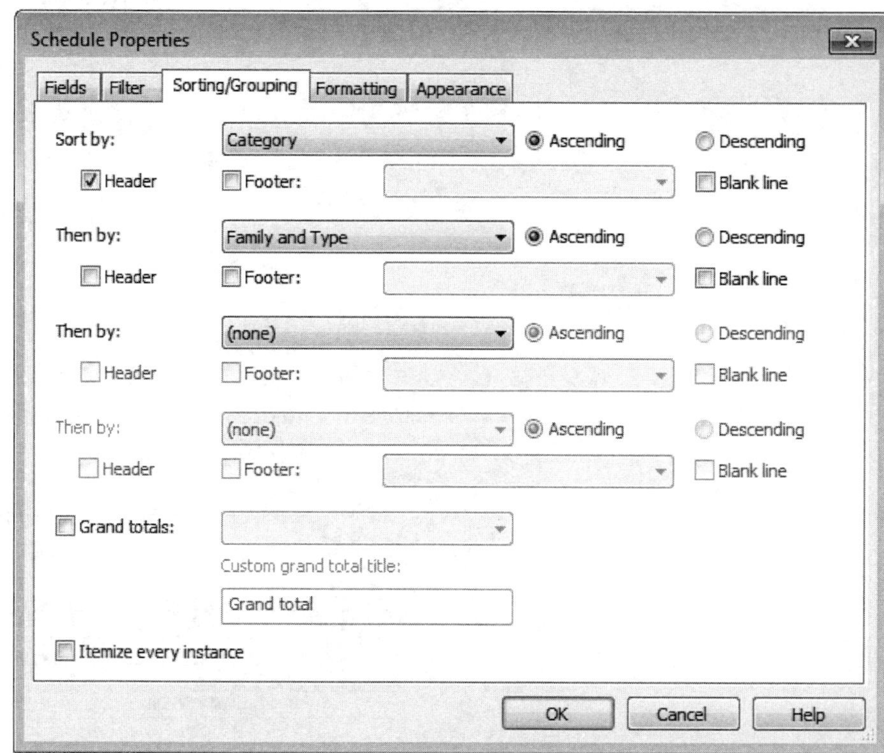

Figure 2–34

3. In the Schedule Properties dialog box, in the *Filter* tab, add the Filter: **Category Type does not equal Structural Rebar**, as shown in Figure 2–35.

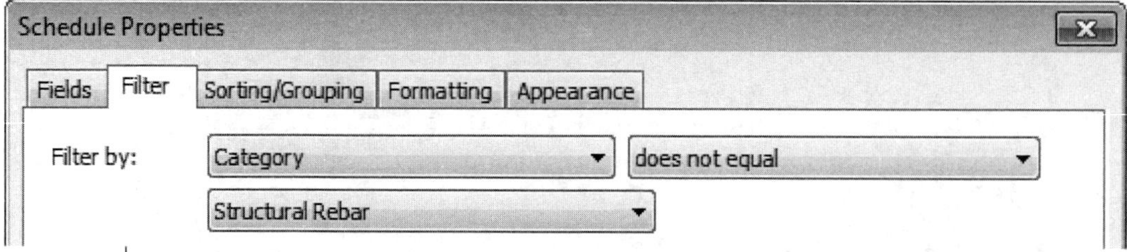

Figure 2–35

4. Click **OK** to create the schedule.

5. Resize the columns to see the information. Right-click on the *Category* column and select **Hide Columns**. The schedule displays as shown in Figure 2–36.

<Structural Schedule>		
A	B	C
Type Mark	Family and Type	Count
Structural Columns		
P-1	Concrete-Rectangular-Column: 24 x 24	22
	W-Wide Flange-Column: W10X33	1
	W-Wide Flange-Column: W10X49	22
Structural Foundations		
F-2	Footing-Rectangular: 14'x14'x2'-0"	1
F-1	Footing-Rectangular: 36" x 36" x 12"	21
	Foundation Slab: 6" Foundation Slab	1
	Step Footing: Step Footing	4
	Wall Foundation: Bearing Footing - 24" x 12"	6
	Wall Foundation: Bearing Footing - 36" x 12"	4
	Wall Foundation: Retaining Footing - 24" x 12" x 12"	1
Structural Framing		
	HSS-Hollow Structural Section: HSS6X6X.500	9
	K-Series Bar Joist-Rod Web: 16K6	65
	W-Wide Flange: W8X10	6
	W-Wide Flange: W12X26	63
	W-Wide Flange: W14X48	62
	W-Wide Flange: W16X45	1

Figure 2–36

6. Open the **S601- Schedules** sheet. Drag and drop the schedule to this sheet. Modify the width of each field as needed.

7. Save the project.

Task 3 - Optional Schedules

1. If you have time, create schedules that incorporate the information shown in Figure 2–37 and Figure 2–38.

<Structural Foundation Schedule>

A	B	C	D	E
Type Mark	**Count**	**Type**	**Width**	**Length**
F-1	21	36" x 36" x 12"	3' - 0"	3' - 0"
F-2	1	14'x14'x2'-0"	14' - 0"	14' - 0"
Grand total: 22				

Figure 2–37

- Hint: Filter by **Type Mark begins with "F"** so that only the individual footing foundations display and sort by Type Mark.

<Structural Foundation Column Schedule>

A	B	C	D	E
Type Mark	Count	Type	Top Level	Base Level
P-1	22	24 x 24	Level 1	T.O. Footing

Figure 2–38

- Hint: Create a Structural Column Schedule, change the name and Filter by **Type Mark begins with "P"** so that only the foundation columns display.

2. Save the project.

2.3 Modifying Schedule Appearance

While you assign global appearance settings in the Schedule Properties dialog box, you can also modify the appearance of each cell in a schedule and add other rows and columns. Title and Heading cells can include both text and parameter values and have different text formatting, as shown in Figure 2–39. You can also add images to title cells and to individual rows in a schedule.

<Window Schedule>						
<Toronto Towers>					Project No.	<1234-5678>
A	B	C	D	E	F	G
Type	Count	Height	Width	Manufacturer	Model	Comments
36 x 36	72	3' - 0"	3' - 0"	Anderson	FX3636	
36" x 48"	84	4' - 0"	3' - 0"	Anderson	FX3648	
Grand total: 156						

Figure 2–39

- These modifications work with all types of schedules.
- Most of these modifications can be accessed in the *Modify Schedule/Quantities* tab (as shown in Figure 2–40), which is available when you are in a schedule view.

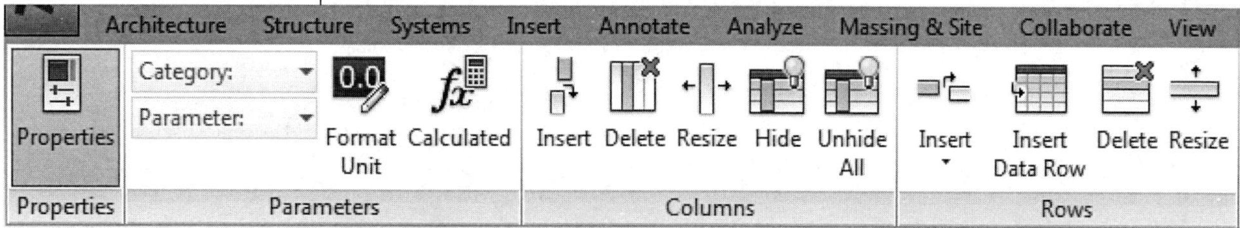

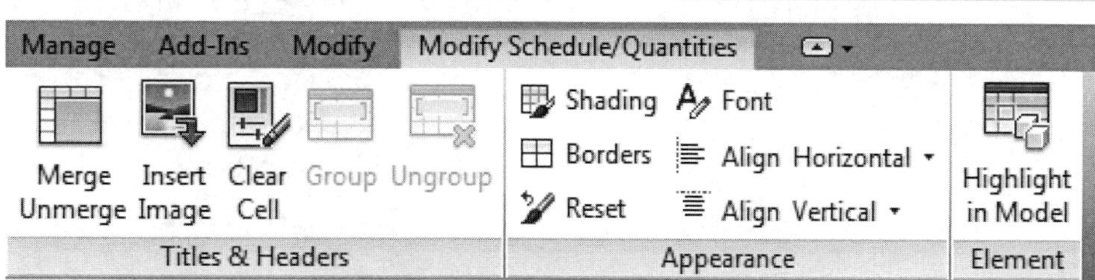

Figure 2–40

- Many of these options are also found in the right-click menu.

Parameter Values

Parameter values and options (as shown in Figure 2–41) can be set up by cell. In title cells you can use the **Schedule** and **Project Information** parameters. If you select a header cell the parameters match the type of schedule in which you are working. If you change this you are changing the field as you did in the Schedule Properties dialog box, in the *Field* tab. In number-based data cells you can modify the **Format Unit** or add **Calculated** values (formulas).

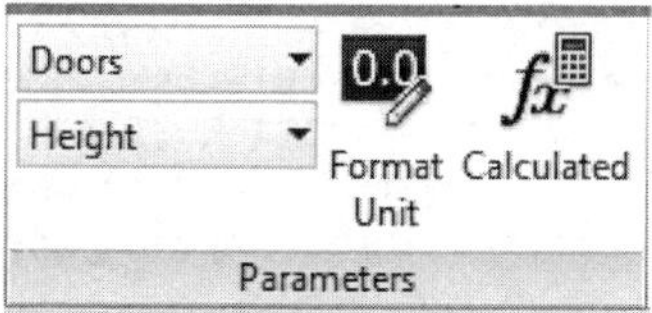

Figure 2–41

- Parameter values in titles are displayed in the schedule view with **<>** at either end of the value. This does not display on the sheet view.

How To: Add Parameters to Schedule Cells

1. Select the cell in which you want to place the parameter.
2. In the Parameters panel, expand the top drop-down list (as shown for Title options in Figure 2–42), and select the type of parameter that you want to use.

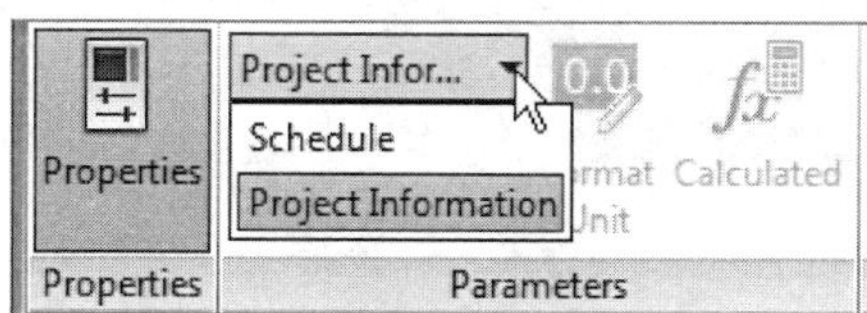

Figure 2–42

3. Expand the lower drop-down list and select the parameter that you want to use as shown in Figure 2–43.
   - The list of parameters for Project Information is shown in Figure 2–43. Schedule parameters include **Phase**, **Phase Filter**, and **View Name**.

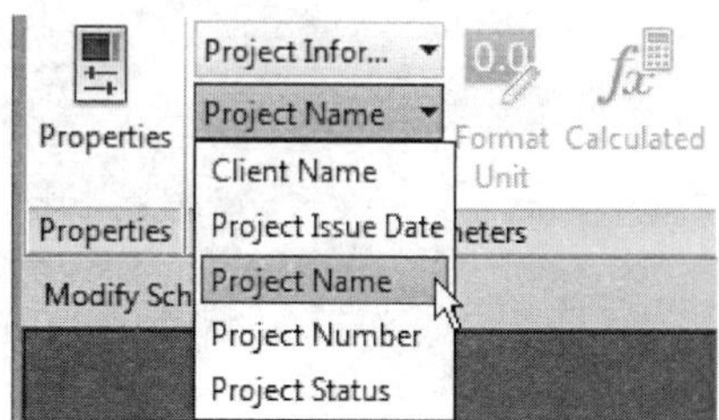

Figure 2–43

- If you select a data cell you can modify the units of the cell by clicking (Format Unit) or create a formula in a cell by clicking (Calculated).

Columns and Rows

The options for Columns and Rows (as shown in Figure 2–44), vary according to the type of cell selected. You can **Insert**, **Delete**, and **Resize** title columns and rows. If you select a header column you can insert, delete, resize, and hide the column.

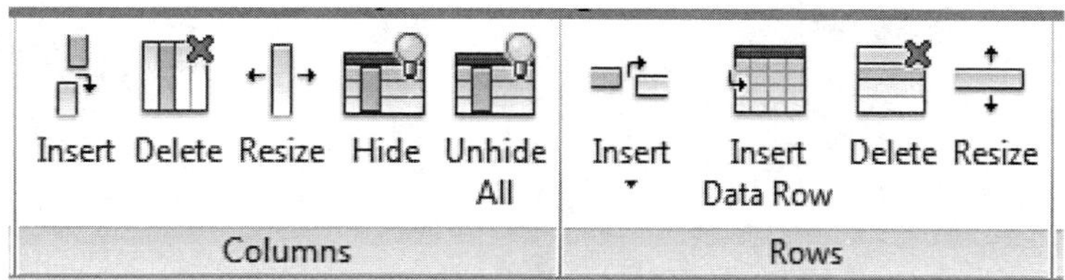

Figure 2–44

- Be careful about making changes in the Ribbon when you are in a data cell because any changes you make apply to the entire row or column.

How To: Hide a Column

1. Select any cell in the column that you want to hide.
2. In the *Modify Schedule/Quantities* tab>Columns panel, click (Hide). Alternatively, if you have selected the header cell, right-click and select **Hide Columns**.
3. The column is hidden.

- If you have hidden any columns and no longer want them hidden, you can click (Unhide All) or right-click and select **Unhide All Columns** at any time.
- Hiding columns is typically used when you want to sort or filter and do not want them to display in the final schedule.

- You can also hide columns when you are creating the schedule. In the Schedule Properties dialog box, in the *Formatting* tab, select the Field you want to hide and in the *Field formating* area, select **Hidden field** as shown in Figure 2–45.

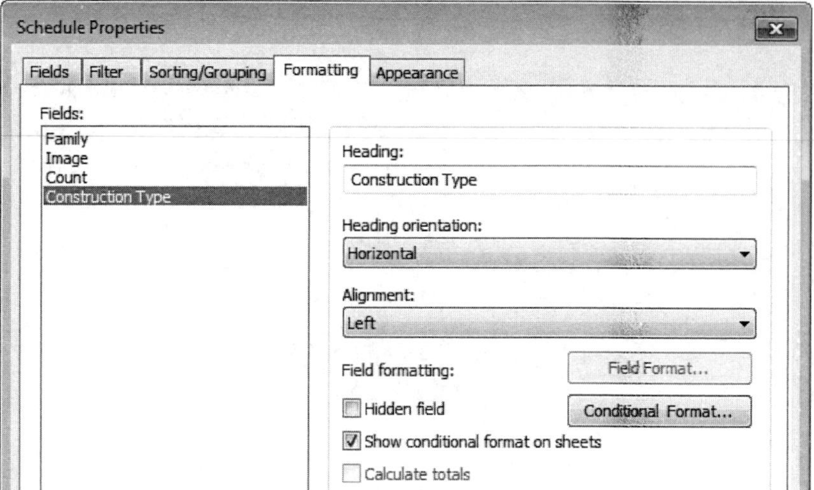

Figure 2–45

How To: Resize a Row or Column

1. Select a cell. If the cell is a title cell you can resize both the row and column
2. In the *Modify Schedule/Quantities* tab>Columns panel, click (Resize) or in the Rows panel, click (Resize).
3. In the Resize Column (or Row) dialog box (as shown in Figure 2–46), type a size and click **OK**.

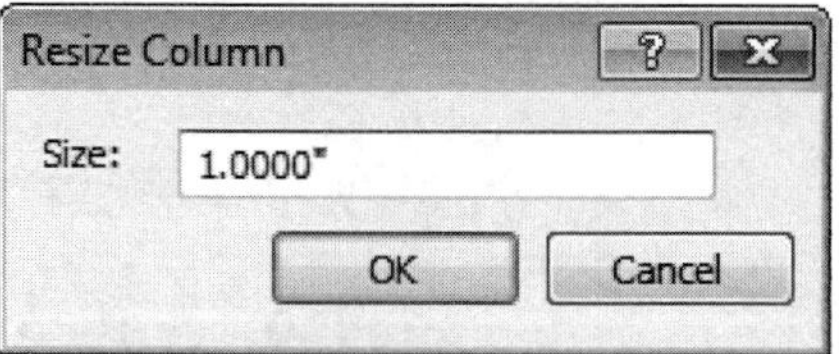

Figure 2–46

4. The column (or row) is resized in the schedule view and on the sheet.

Titles and Headers

The more complex and extensive a schedule, the more it can help to modify the title and header cells, using the commands shown in Figure 2–47. The title cells can be merged or unmerged. You can also insert images and delete the contents of title cells. Header cells can be grouped and ungrouped.

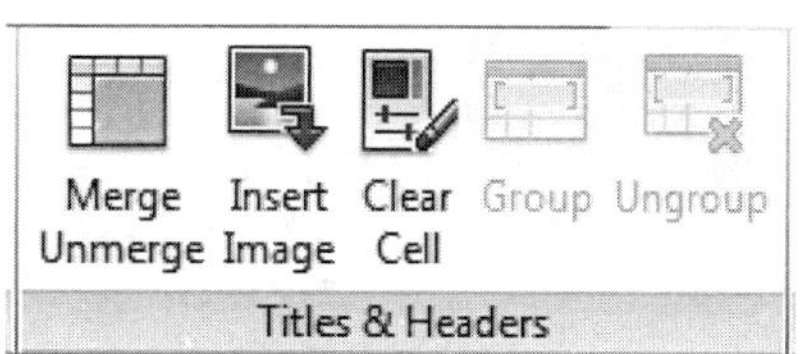

Figure 2–47

How To: Add and Merge Title Cells

1. Click inside a title cell.
2. In the *Modify Schedule/Quantities* tab>Rows panel, expand (Insert) and click (Above Selected) or (Below Selected). When inserting below the primary cell it typically comes in matching the number of header cells as shown in Figure 2–48.

<Door Schedule>						
A	B	C	D	E	F	G
Mark	Family and Type	Height	Width	Fire Rating	Frame Material	Frame Type

Before

<Door Schedule>						
A	B	C	D	E	F	G
Mark	Family and Type	Height	Width	Fire Rating	Frame Material	Frame Type

After

Figure 2–48

3. Pick and drag across the cells that you want to merge.
4. Click (Merge Unmerge).

5. The cells are now merged together as shown in Figure 2–49.

<Door Schedule>						
A	B	C	D	E	F	G
Mark	Family and Type	Height	Width	Fire Rating	Frame Material	Frame Type

Figure 2–49

6. Add a parameter, text, or image in the newly merged cell.

- If you no longer want the cells to be merged, select the cell and click (Merge Unmerge).
- To insert an image file (such as a .BMP, .JPG, or .PNG) in a title cell, select the title cell and click (Insert Image) and select the image from the list.
- Use (Clear Cell) to delete the text, image, or parameter used in that cell (Titles only).

How To: Setting up Sub-headers

1. In the Schedule view, drag the cursor over the headers you want to group.
2. In the *Modify Schedule/Quantities* tab>Titles & Headers panel, click (Group).
3. A new header area is created above the selected columns. Enter a new name as shown in Figure 2–50.

A	B	C	D	E
		Window Dimensions		
Window Type	Count	Height	Width	Sill Height
36 x 36	72	3' - 0"	3' - 0"	3' - 0"
36" x 48"	84	4' - 0"	3' - 0"	3' - 0"
Grand total: 156				

Figure 2–50

- If you no longer need the grouped header cell, select the grouped cell and click (Ungroup).

Appearance

The overall appearance of a schedule, including the text types used in the various cells, is set up in the Schedule Properties. You can also modify individual title and header cells using the options in the Appearance panel as shown in Figure 2–51.

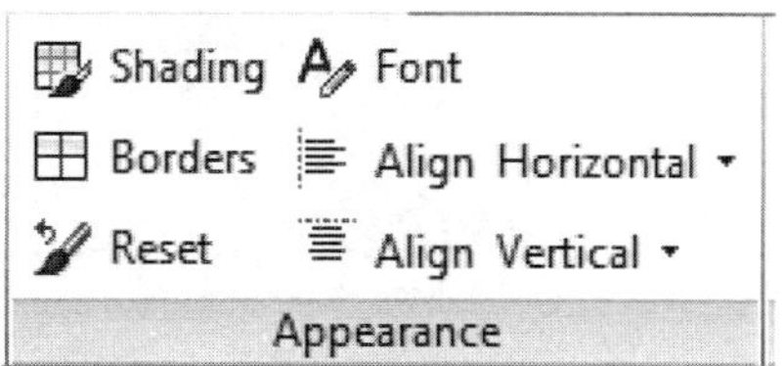

Figure 2–51

(Shading)	Opens the Color dialog box in which you can select a background color for the selected cells.
(Borders)	Opens the Edit Borders dialog box in which you can specify the line style and locations of borders around the selected cells.
(Reset)	Returns the formatting of selected cells to the original specified in the schedule properties.
(Font)	Opens the Edit Font dialog box in which you can specify the **Font**, **Font size**, **Font Style** (**Bold**, **Italic**, or **Underline**) and **Font Color** for the selected cells.
(Align Horizontal)	Aligns the text in the selected cells to the **Left**, **Center**, or **Right** justification.
(Align Vertical)	Aligns the text in the selected cells to the **Top**, **Middle**, or **Bottom** justification.

Adding Image Fields and Images

In addition to adding images in the title cells of a schedule you can assign images to elements and have them display in the schedule on a sheet, as shown in Figure 2–52.

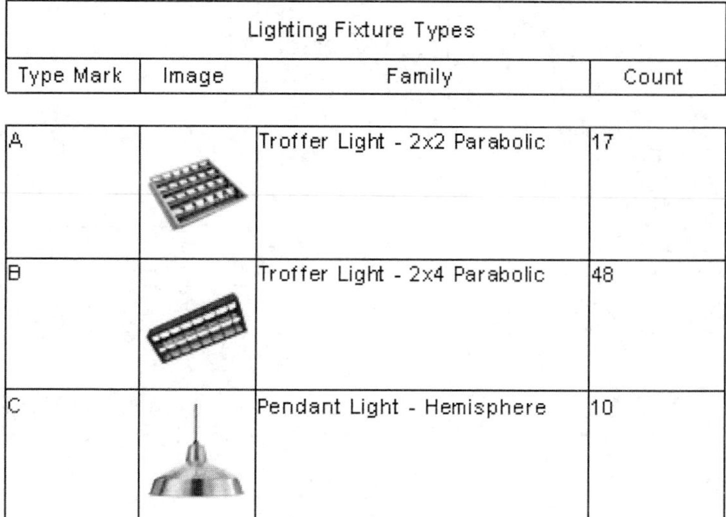

Lighting Fixture Types			
Type Mark	Image	Family	Count
A		Troffer Light - 2x2 Parabolic	17
B		Troffer Light - 2x4 Parabolic	48
C		Pendant Light - Hemisphere	10

Figure 2–52

- Images can be set by instance or by type.

How To: Add Images to Schedules

1. In the schedule view, open the Schedule Properties dialog box in the *Fields* tab.
2. Add the *Image* (or *Type Image*) field to the scheduled fields, as shown in Figure 2–53.

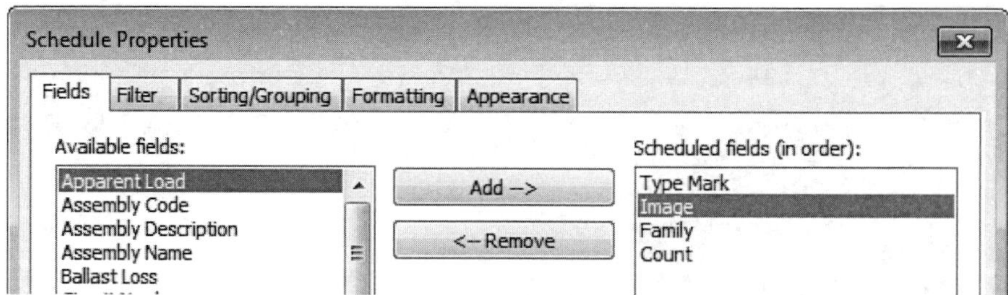

Figure 2–53

3. Complete the rest of your schedule design and click **OK**.

4. In the schedule view, the *Image* (or *Type Image*) column displays as part of the schedules, as shown in Figure 2–54.

<Lighting Fixture Types>			
A	B	C	D
Type Mark	Image	Family	Count
A	<None>	Troffer Light - 2x2 Parabolic	17
B	<None>	Troffer Light - 2x4 Parabolic	48
C	<None>	Pendant Light - Hemisphere	10

Figure 2–54

5. Assign the appropriate image file to the instances or types.
    - Type images for component families (such as light fixtures) are assigned directly in the family file and are read-only in the project. Type images for system families (such as walls) are assigned in the Type Properties.
    - Instance images are assigned in Properties for each instance of the element.
6. To display the images, place the schedule on a sheet.

How To: Assign Image Files

1. Navigate to the appropriate place at which to assign the Image or Type Image. (Type Images for component families are assigned using Type Properties in the family file, not the project.)
2. Next to the **Image** or **Type Image** parameter, click (Browse), as shown in Figure 2–55.

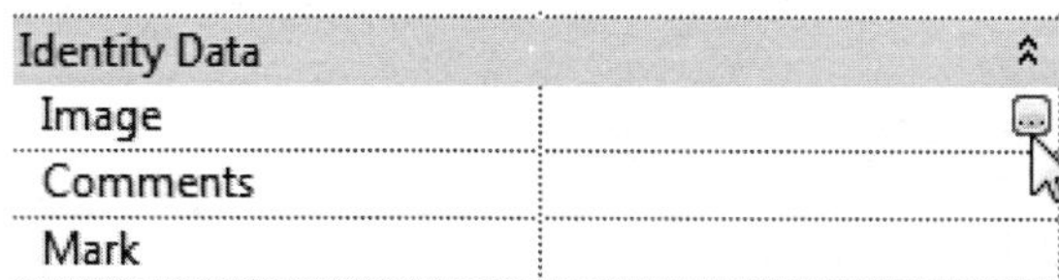

Figure 2–55

3. In the Manage Images dialog box, select the image that you want to use, as shown in Figure 2–56

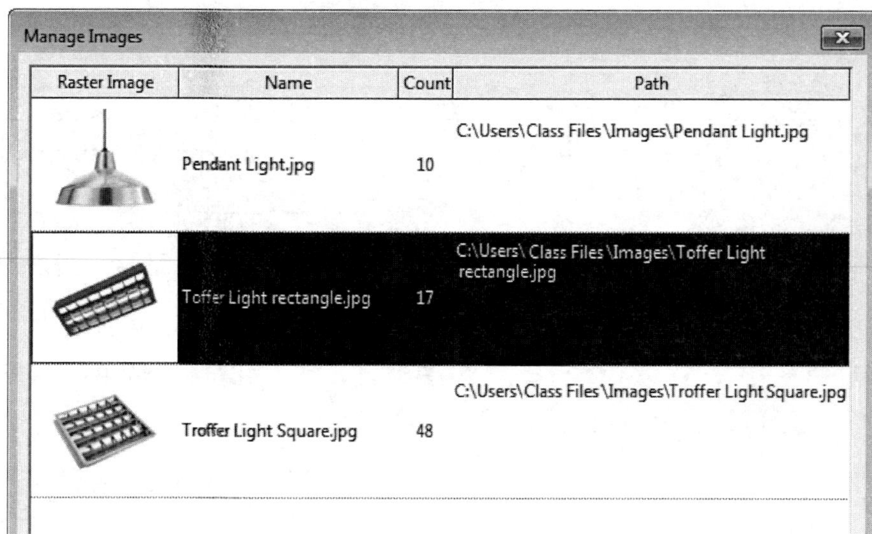

Figure 2–56

4. Click **OK**

- To add images to the Manage Image dialog box, click **Add...** and navigate to the location in which the images are stored.
- You can also delete and reload images as you would reload linked files.
- Image file types include BMP, JPG, JPEG, PNG, and TIF.

Practice 2d Modify Schedules for Architectural Projects

Estimated time for completion: 15 minutes

Practice Objectives

- Modify the look of the schedule using parameters, and by merging and grouping cells.
- Add images to a schedule.

In this practice you will modify the appearance of a schedule by adding parameters and text to the title and by merging group headers, as shown in Figure 2–57 (on a sheet). You will also add an image field (column) and assign images to a schedule.

<table>
<tr><th colspan="8">Door Schedule - Level 1</th></tr>
<tr><td colspan="4">Clark Hall Performing Arts Building</td><td colspan="4">Project Number: 1234.56</td></tr>
<tr><td rowspan="2">Mark</td><td colspan="2">Door Size</td><td rowspan="2">Fire Rating</td><td colspan="3">Frame Information</td><td rowspan="2">Comments</td></tr>
<tr><td>Height</td><td>Width</td><td>Frame Type</td><td>Frame Material</td><td>Finish</td></tr>
</table>

Figure 2–57

Task 1 - Modify the appearance of the schedule.

1. In the *...\Architectural* folder, open **Clark-Hall-Schedules-Modify.rvt**.
2. Open the **Door Schedule-Level 1** view.
3. Select the top title row of the schedule.
4. In the *Modify Schedule/Quantities* tab>Row panel, expand (Insert) and click (Below Selected).
5. Drag the cursor over approximately half of the new heading columns on the left and, in the Titles & Headers panel, click (Merge Unmerge). The cells are merged as shown in Figure 2–58.

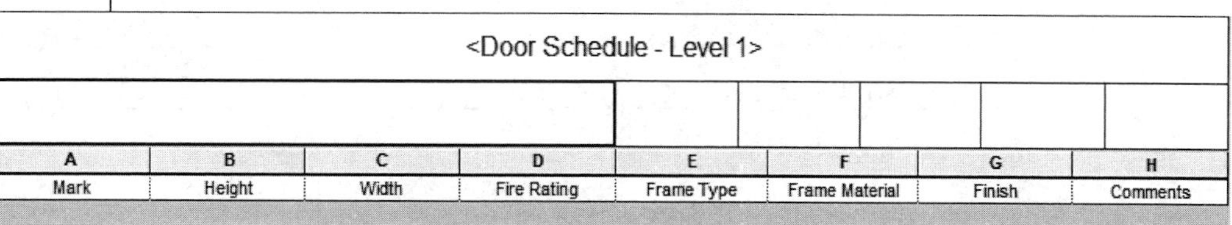

Figure 2–58

6. Select the newly merged cell. In the Parameters panel, expand Category and select **Project Information**. Expand Parameter and select **Project Name**. The name of the project displays in brackets indicating that it is a parameter rather than text, as shown in Figure 2–59.

7. Merge the other cells and add text (**Project Number**) and another parameter (**Project Number**), as shown in Figure 2–59.

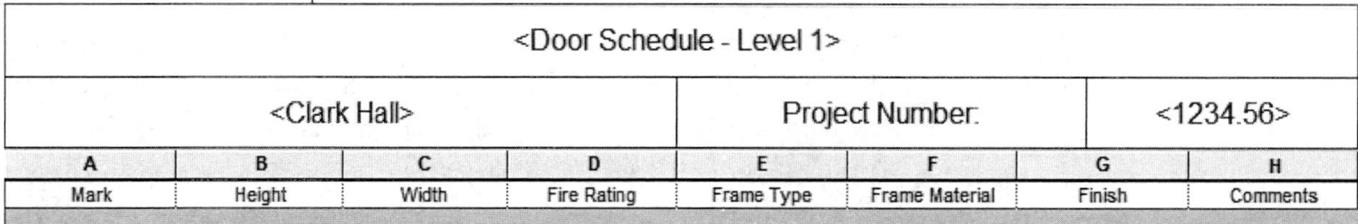

<Door Schedule - Level 1>							
<Clark Hall>				Project Number:		<1234.56>	
A	B	C	D	E	F	G	H
Mark	Height	Width	Fire Rating	Frame Type	Frame Material	Finish	Comments

Figure 2–59

8. In the *Manage* tab>Settings Panel, click (Project Information).

9. In the Project Properties dialog box, update the *Project Name* to **Clark Hall Performing Arts Building**. Click **OK**.

10. The Project Name updates, as shown in Figure 2–60.

<Door Schedule - Level 1>							
<Clark Hall Performing Arts Building>				Project Number:		<1234.56>	
A	B	C	D	E	F	G	H
Mark	Height	Width	Fire Rating	Frame Type	Frame Material	Finish	Comments

Figure 2–60

11. Drag the cursor over the *Height* and *Width* header cells.

12. In the Titles & Headers panel, click (Group).

13. Select the new cell and type **Door Size**.

14. Repeat for Frame Type, Frame Material, and Finish and type **Frame Information** as shown in Figure 2–61.

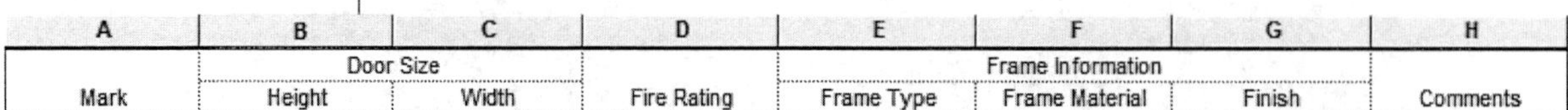

A	B	C	D	E	F	G	H
	Door Size			Frame Information			
Mark	Height	Width	Fire Rating	Frame Type	Frame Material	Finish	Comments

Figure 2–61

15. Save the project.

Task 2 - Modify the Appearance of the Title

1. Select the title cell and, in the *Schedules/Quantities* tab> Appearance panel, click (Font).
2. In the Edit Font dialog box, change the *Font size* to **1/4"** and set the *Font style* to **Bold**.
3. Modify the other title and header cells as needed. In the example shown in Figure 2–62, in addition to changing the font appearance, the Project Number text and parameter have been aligned horizontally to be closer to each other and the border between the two of them has been removed.

<Door Schedule - Level 1>							
<Clark Hall Performing Arts Building>				Project Number: <1234.56>			
A	B	C	D	E	F	G	H
	Door Size			Frame Information			
Mark	Height	Width	Fire Rating	Frame Type	Frame Material	Finish	Comments

Figure 2–62

4. Open the **A601-Schedules** view to display the images and adjust it as needed.
5. Save the project.

Task 3 - Add Images to a schedule.

1. Open the **Window Schedule** view.
2. In the Schedule Properties dialog box, in the *Fields* tab add the **Type Mark** and **Image** parameters to the Scheduled Fields above the **Type** parameter. Move the **Count** parameter to the bottom of the list (as shown in Figure 2–63), and click **OK**.

In this situation the Type Image is only accessible by opening and adjusting the family file. Therefore, you can use the instance Image parameter.

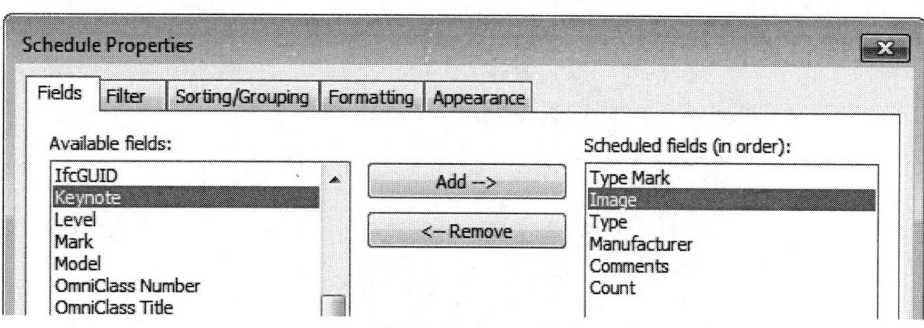

Figure 2–63

3. In the Window Schedule view, select the Image cell for the first type and click ....
4. In the Manage Images dialog box, click **Add...**
5. In the Import Image dialog box, navigate to the *...\Architectural>Images* folder, select the two window images, and click **OK**.
6. In the Manage Images dialog box, select the image that matches the window row in which you are working.
7. Repeat the process with the other window image. The schedule should look similar to Figure 2–64.

<Window Schedule>

A	B	C	D	E	F
Type Mark	Image	Type	Manufacturer	Comments	Count
18	Window Rectangle.png	36" x 72"			3
19	Window Square.png	48" x 48"			36
Grand total: 39					

Figure 2–64

8. Open the **Sheets (all): A601 - Schedules** view. The images should display as shown in Figure 2–65

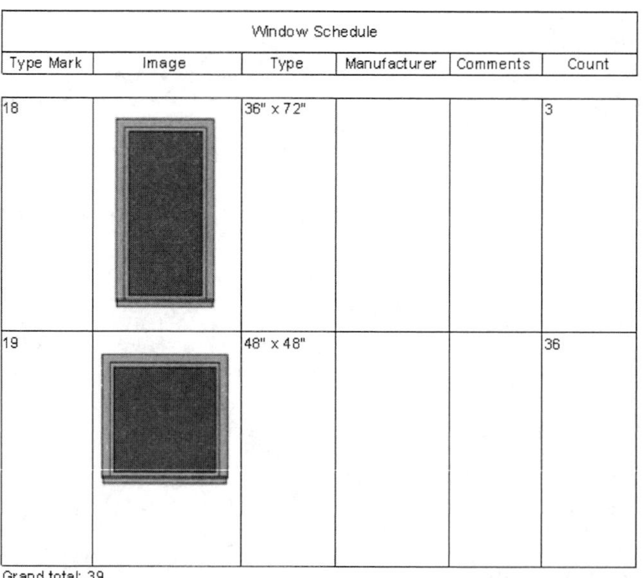
Window Schedule

Type Mark	Image	Type	Manufacturer	Comments	Count
18		36" x 72"			3
19		48" x 48"			36

Grand total: 39

Figure 2–65

9. If you have time, modify the schedule appearance to match the Door Schedule.
10. Save the project.

Practice 2e

Modify Schedules for MEP Projects

Practice Objectives

- Modify the look of the schedule using parameters and by merging and grouping cells.
- Add images to a schedule.

Estimated time for completion: 15 minutes

In this practice you will modify the appearance of a schedule by adding parameters and text to the title and by merging headers. You will also add an image field (column) and assign images to a schedule, as shown on a sheet in Figure 2–66.

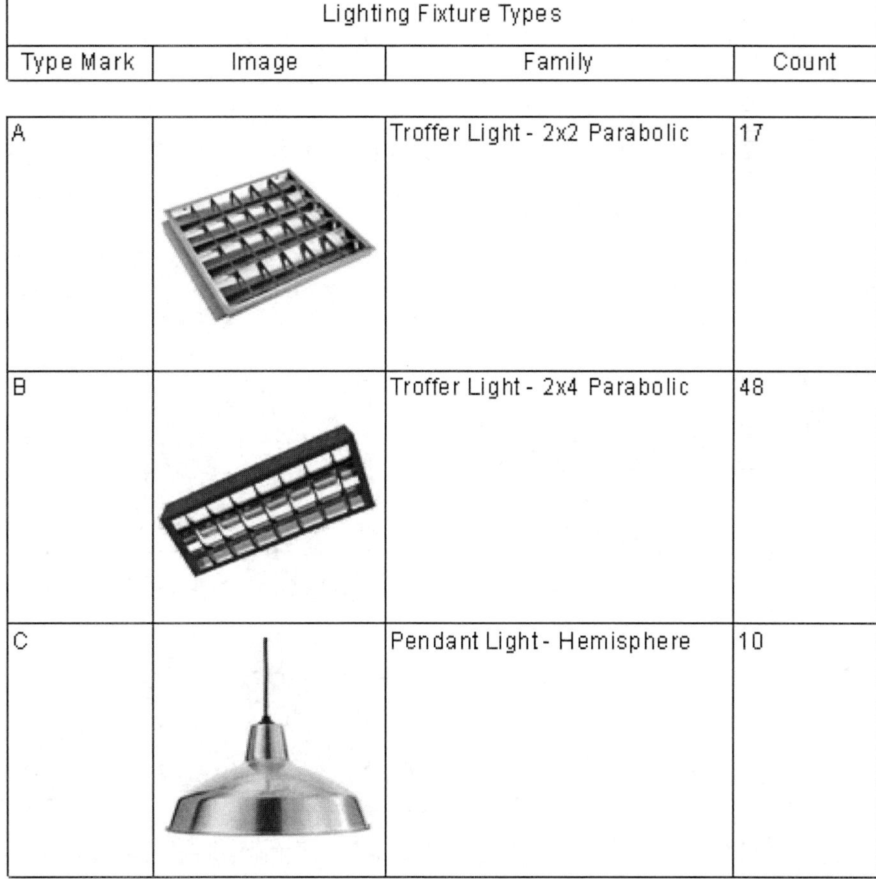

Lighting Fixture Types			
Type Mark	Image	Family	Count
A		Troffer Light - 2x2 Parabolic	17
B		Troffer Light - 2x4 Parabolic	48
C		Pendant Light - Hemisphere	10

Figure 2–66

Task 1 - Modify the appearance of the schedule.

1. In the ...*MEP* practice files folder, open **Freel-Office-Modify.rvt**.
2. Open the **Plumbing Fixtures** schedule view.

3. Select the top title row of the schedule.
4. In the *Modify Schedule/Quantities* tab>Row panel, expand (Insert) and click (Below Selected).
5. Drag the cursor over half of the new heading columns on the left. In the Titles & Headers panel, click (Merge Unmerge). The cells are merged, as shown in Figure 2–67.

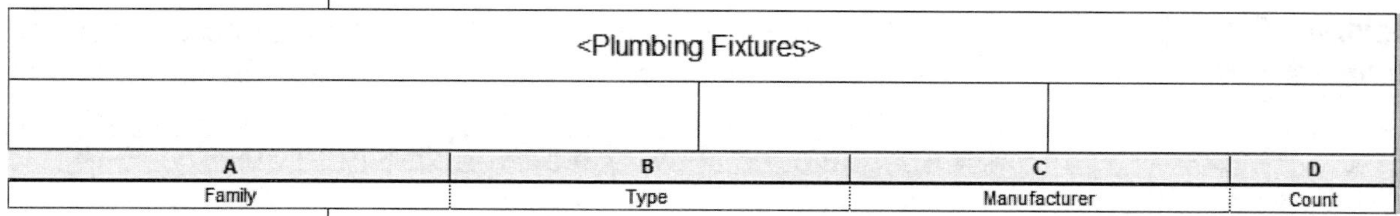

Figure 2–67

6. Select the newly merged cell. In the Parameters panel, expand Category and select **Project Information**. Then expand Parameter and select **Project Name**. The name of the project displays in brackets, indicating that it is a parameter
7. In the other cells and add text (**Project Number:**) and another parameter (**Project Number**), as shown in Figure 2–68.

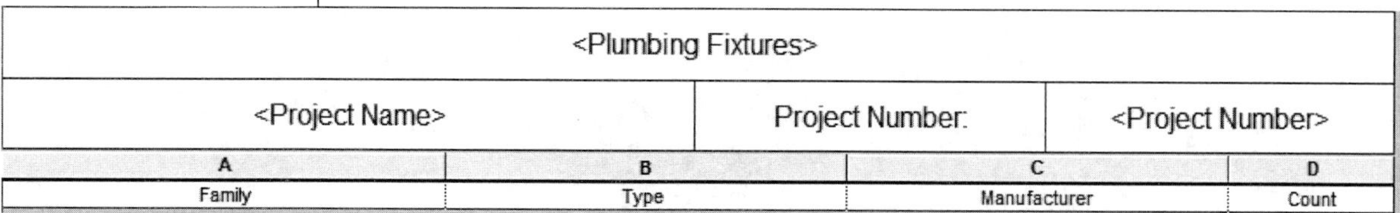

Figure 2–68

8. In the *Manage* tab>Settings Panel, click (Project Information).
9. In the Project Properties dialog box, update *Project Name* to **Freel Office Building** and *Project Number* to **1234.56.** Click **OK**. The parameters update as shown in Figure 2–69.

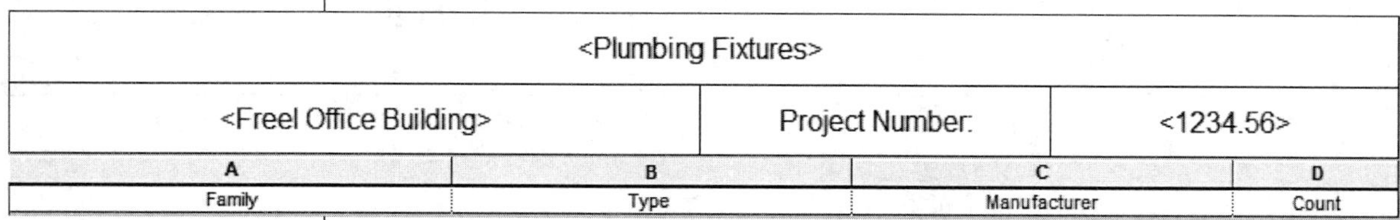

Figure 2–69

10. Save the project.

Task 2 - Modify the Appearance of the Title

1. Select the title cell and, in the Schedules/Quantities tab> Appearance panel click (Font).
2. In the Edit Font dialog box change the *Font size* to **1/4"** and set the *Font style* to **Bold**.
3. Modify the other title and header cells as desired. In the example shown in Figure 2–70, besides changing the font appearance, the Project Number text and parameter have been aligned horizontally so they are closer to each other and the border between the two of them has been removed.

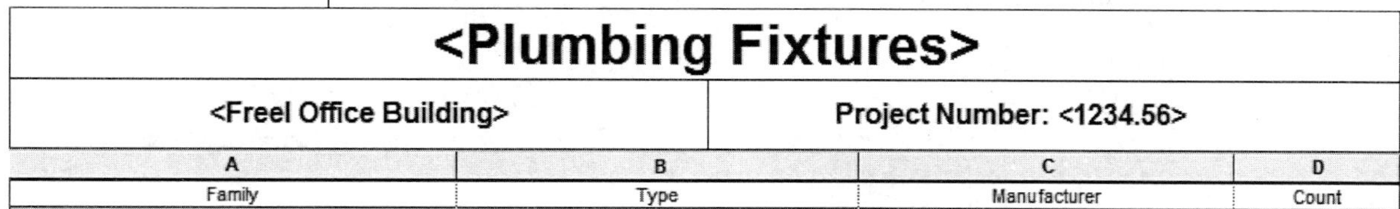

Figure 2–70

4. Save the project.

Task 3 - Add Images to a schedule.

1. Open the **Lighting Fixture Types** view.
2. In the Schedule Properties dialog box, in the *Fields* tab, add the **Image** parameter to the Scheduled Fields after the **Type** parameter as shown in Figure 2–71, and click **OK**.

In this situation the Type Image is only accessible by opening and adjusting the family file. Therefore, you can use the instance Image parameter.

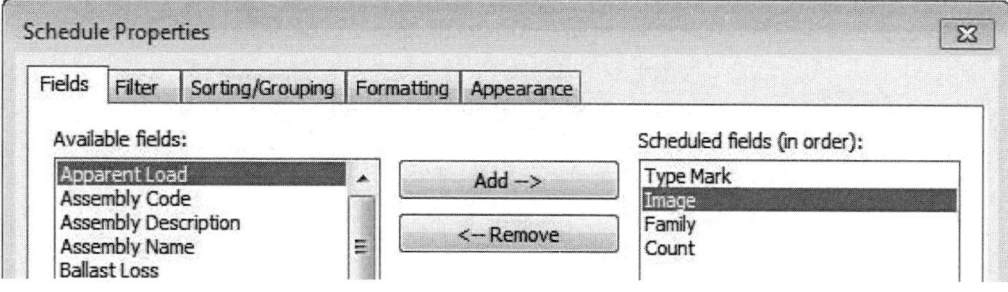

Figure 2–71

3. In the Lighting Fixture Types view, select the Image cell for the first type and click .
4. In the Manage Images dialog box, click **Add...**
5. In the Import Image dialog box, navigate to the ...*MEP>Images* folder. select the three light images. and click **OK**.

6. In the Manage Images dialog box, select the image that matches the row in which you are working.

7. Repeat the process with the other lighting fixture images. The schedule should look similar to the one shown in Figure 2–72.

<Lighting Fixture Types>			
A	B	C	D
Type Mark	Image	Family	Count
A	Troffer Light Square.jpg	Troffer Light - 2x2 Parabolic	17
B	Toffer Light rectangle.jpg	Troffer Light - 2x4 Parabolic	48
C	Pendant Light.jpg	Pendant Light - Hemisphere	10

Figure 2–72

8. Open the **Sheets (all): E601 - Electrical and Lighting Schedules** view. The images should display as shown in Figure 2–73

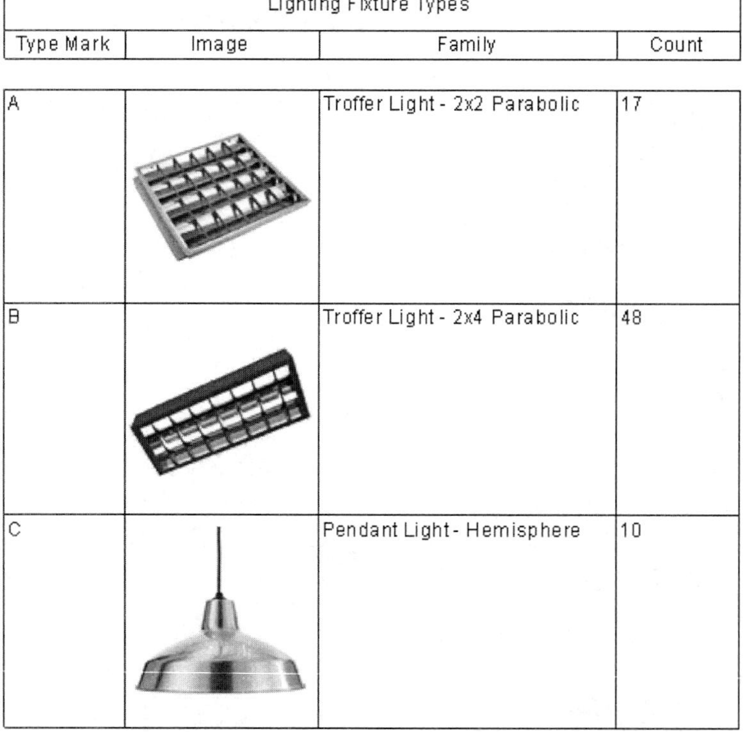

Lighting Fixture Types			
Type Mark	Image	Family	Count
A		Troffer Light - 2x2 Parabolic	17
B		Troffer Light - 2x4 Parabolic	48
C		Pendant Light - Hemisphere	10

Figure 2–73

9. If you have time, modify the schedule appearance to match the other schedule.

10. Save the project.

Practice 2f

Modify Schedules for Structural Projects

Estimated time for completion: 15 minutes

Practice Objective

- Modify the look of the schedule using parameters, and merging and grouping cells.

In this practice you will modify the appearance of a schedule by adding parameters and text to the title and by merging cells, as shown in Figure 2–74.

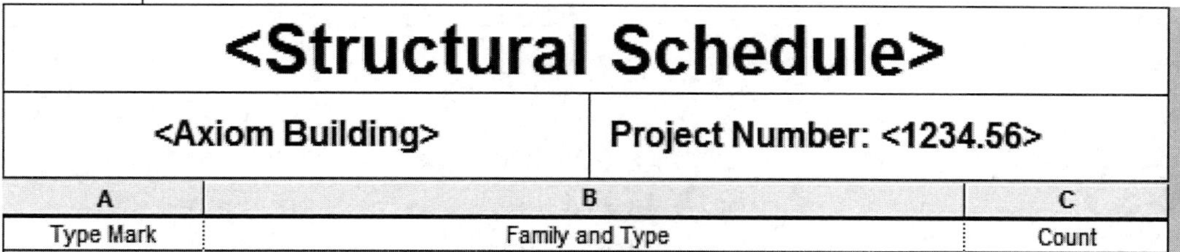

Figure 2–74

Task 1 - Modify the appearance of the schedule.

1. In the ...*Structural* folder, open **Axiom-Building-Modify.rvt**.
2. Open the **Structural Schedule** view.
3. Select the top title row of the schedule.
4. In the *Modify Schedule/Quantities* tab>Row panel, expand (Insert) and click (Below Selected).
5. If there are more than three new cells, drag the cursor over half of the new heading columns on the left and, in the Titles & Headers panel, click (Merge Unmerge). The cells are merged as shown in Figure 2–75.

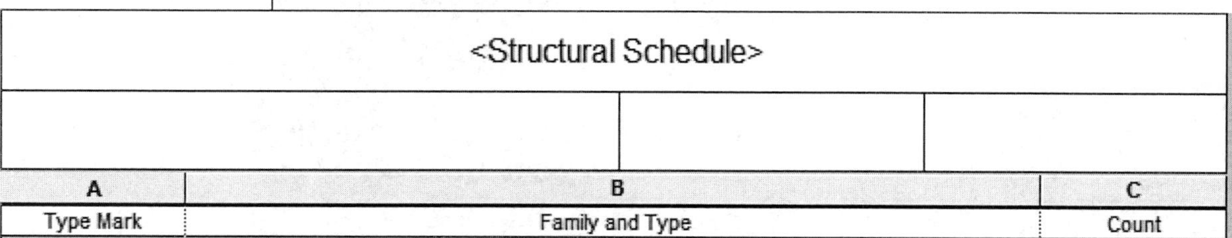

Figure 2–75

6. Select the newly merged cell. In the Parameters panel, expand Category and select **Project Information**. Then expand Parameter and select **Project Name**. The name of the project displays in brackets indicating that it is a parameter rather than text, as shown in Figure 2–76.

7. Add text (**Project Number:**) and another parameter (**Project Number**) as shown in Figure 2–76.

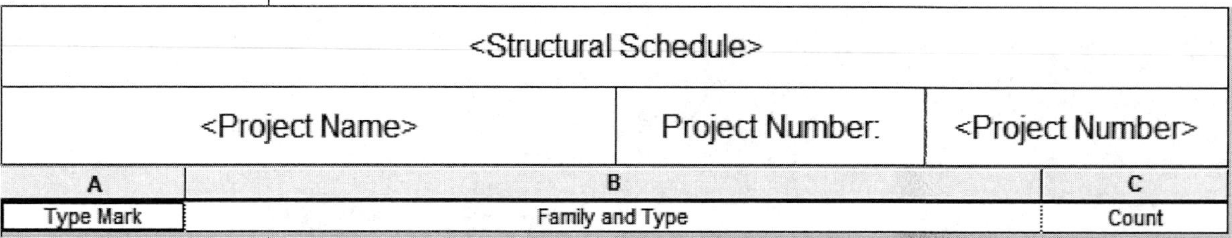

Figure 2–76

8. In the *Manage* tab>Settings Panel, click (Project Information).

9. In the Project Properties dialog box, update *Project Name* to **Axiom Building** and *Project Number* to **1234.56**. Click **OK**. The project name and number update as shown in Figure 2–77.

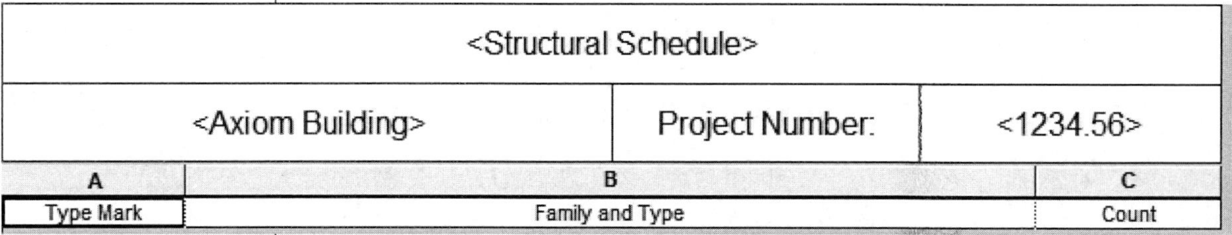

Figure 2–77

10. Save the project.

Task 2 - Modify the appearance of the title.

1. Select the title cell. In the *Schedules/Quantities* tab> Appearance panel, click (Font).

2. In the Edit Font dialog box, change the *Font size* to **1/4"** and set the *Font style* to **Bold**.

3. Modify the other title and header cells as needed. In the example shown in Figure 2–78, in addition to changing the font appearance, the Project Number text and parameter have been aligned horizontally to be closer to each other and the border between them has been removed.

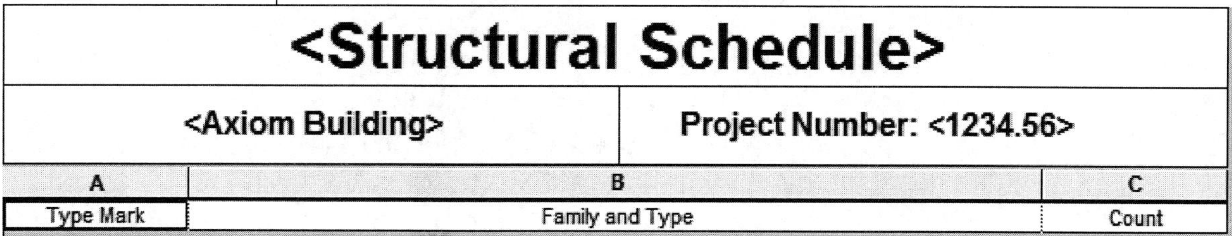

Figure 2–78

4. Modify the other two schedules to match the appearance.
5. Save the project.

2.4 Creating Key Schedules

A *key schedule* lists type information that is not automatically included in an element, such as the finish or frame type of a door. By grouping this information into keys, you can enter it more quickly.

Schedule key styles are useful when you have a large number of elements that need to have the same finish information, as shown in Figure 2–79. Fields in key styles are typically controlled by instance rather than by type. Grouping styles together makes it easier to apply this information to large numbers of instances.

<Room Style Schedule>					
A	B	C	D	E	F
Key Name	Wall Finish	Floor Finish	Base Finish	Ceiling Finish	Comments
Corridor	Painted	Carpet	Vinyl	Painted	
Executive Office	Wall Paper	Wood	Wood	ACT	
Standard Office	Painted	Carpet	Vinyl	ACT	
Utility	Painted	Vinyl	Vinyl	ACT	

Figure 2–79

How To: Create a Schedule Key Style

1. In the *View* tab>Create panel, expand (Schedules) and click (Schedule/Quantities).
2. In the New Schedule dialog box, select a category.
3. Select the **Schedule keys** option and type a *Name* for the key, as shown in Figure 2–80. This adds an **Instance** parameter in the Properties dialog box for the selected category and adds a field to the schedule properties.

Use the Filter list to limit the number of categories with which you can work.

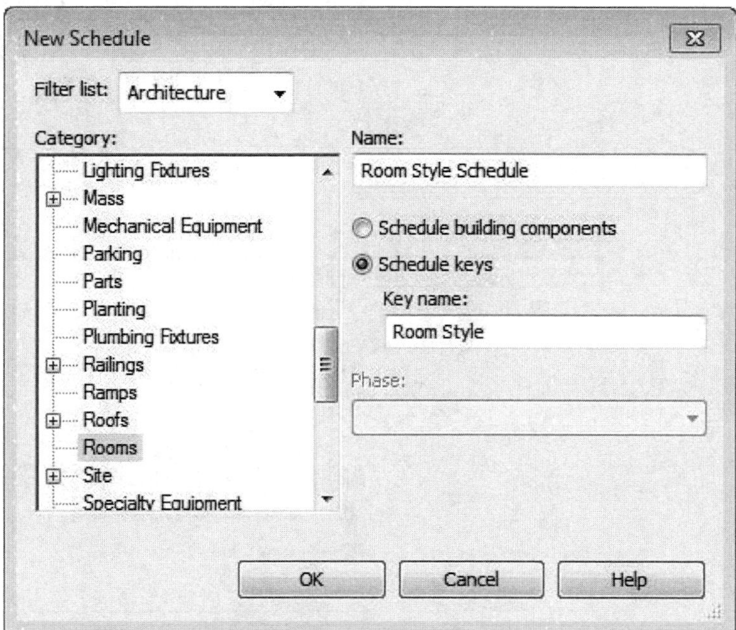

Figure 2–80

4. Click **OK**.
5. In the Schedule Properties dialog box, the *Fields* tab displays the new **Key Name** that has already been added to the *Scheduled fields*, as shown in Figure 2–81.
6. Add other fields from the list of *Available fields*, as shown in Figure 2–81.

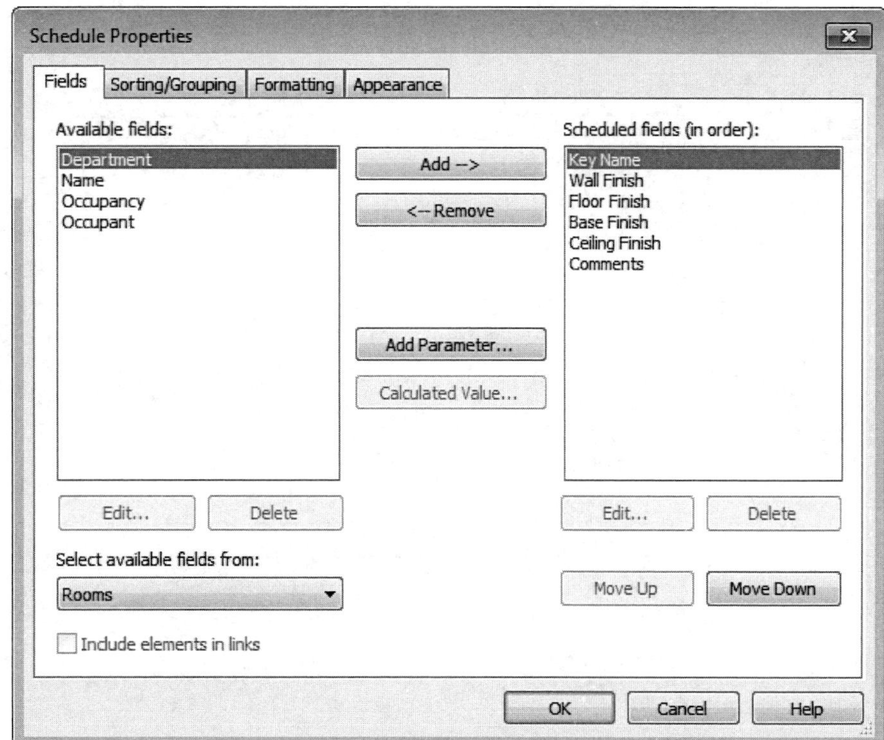

Figure 2–81

7. Fill out the rest of the tabs as needed and click **OK**. The new schedule opens with only the fields displayed, as shown in Figure 2–82.

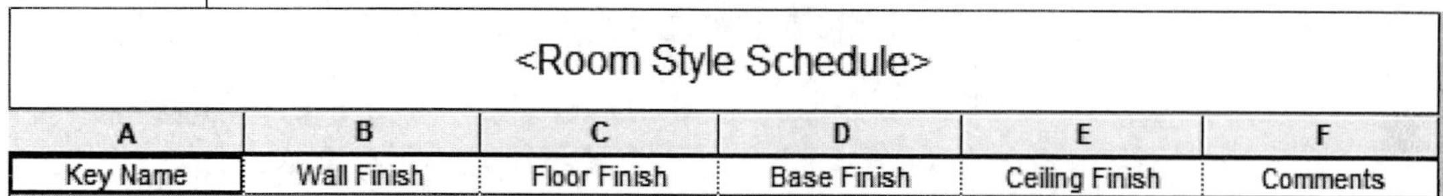

<Room Style Schedule>					
A	B	C	D	E	F
Key Name	Wall Finish	Floor Finish	Base Finish	Ceiling Finish	Comments

Figure 2–82

Enhanced in **2016**

8. In the *Modify | Schedule/Quantities* tab>Rows panel, click (Insert Data Row) or right-click on the schedule and select **Insert Data Row**. A new row with the default label **1** is displayed, as shown in Figure 2–83. Type a new name for the row.

<Room Style Schedule>					
A	B	C	D	E	F
Key Name	Wall Finish	Floor Finish	Base Finish	Ceiling Finish	Comments
1					

Figure 2–83

9. Continue adding rows and names as shown in Figure 2–84.

<Room Style Schedule>					
A	B	C	D	E	F
Key Name	Wall Finish	Floor Finish	Base Finish	Ceiling Finish	Comments
Corridor					
Executive Office					
Standard Office					
Utility					

Figure 2–84

10. Fill in the values for the rows as shown in Figure 2–85.

<Room Style Schedule>					
A	B	C	D	E	F
Key Name	Wall Finish	Floor Finish	Base Finish	Ceiling Finish	Comments
Corridor	Painted	Carpet	Vinyl	Painted	
Executive Office	Wall Paper	Wood	Wood	ACT	
Standard Office	Painted	Carpet	Vinyl	ACT	
Utility	Painted	Vinyl	Vinyl	ACT	

Figure 2–85

11. To assign the key style, add a column to the schedule (such as the *Room Style* column in a Room Schedule shown in Figure 2–86).

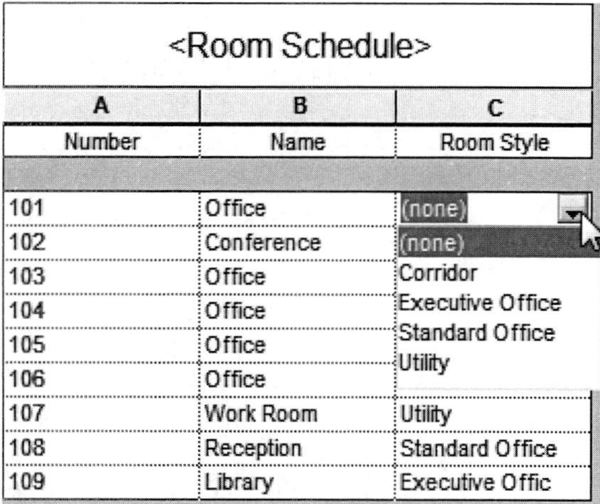

Figure 2–86

You can also modify the properties of various instances in the project as shown in Figure 2–87.

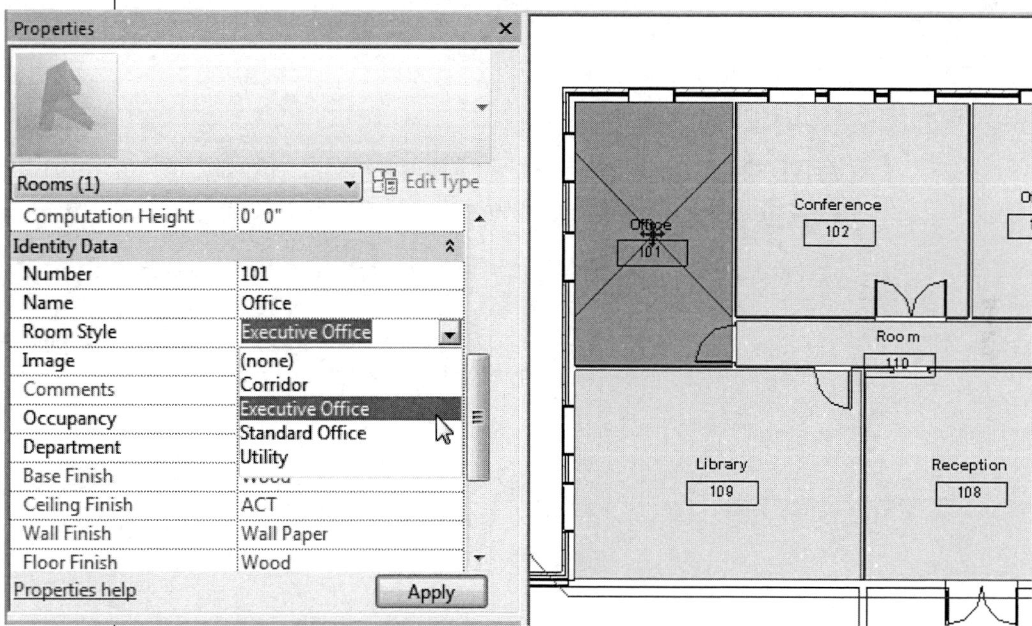

Figure 2–87

2.5 Advanced Schedule Options

When you are working in a variety of schedules, you might want to include additional parameters, create fields that include formulas, as shown for a rounded area in Figure 2–88. You can also add conditional formatting that adds a background color to cells that meet the condition. For MEP projects you can also create nested schedules.

<Room Schedule>			
A	B	C	D
Number	Name	Area	Floor Area
101	Office	280.49 SF	280 SF
102	Conference	337.38 SF	340 SF
103	Office	199.63 SF	200 SF
104	Office	199.63 SF	200 SF
105	Office	193.17 SF	190 SF
106	Office	243.22 SF	240 SF
107	Work Room	154.77 SF	150 SF
108	Reception	327.48 SF	330 SF
109	Library	404.78 SF	400 SF

Figure 2–88

Creating Project Parameters

Two types of custom parameters can be used in schedules: project parameters and shared parameters.

- Project parameters can be used in schedules, but not in tags.
- Shared parameters can be used in schedules and in tags. They can also be used in other projects and exported to a database.
- Both project and shared parameters can be created wherever parameter options exist. They are also available in the *Manage* tab>Settings panel.

How To: Create a Project Parameter

1. In the Schedule Properties dialog box, in the *Fields* tab, click **Add Parameter...** (or other Project Parameter access).
2. In the Parameter Properties dialog box, select **Project parameter**, as shown in Figure 2–89.

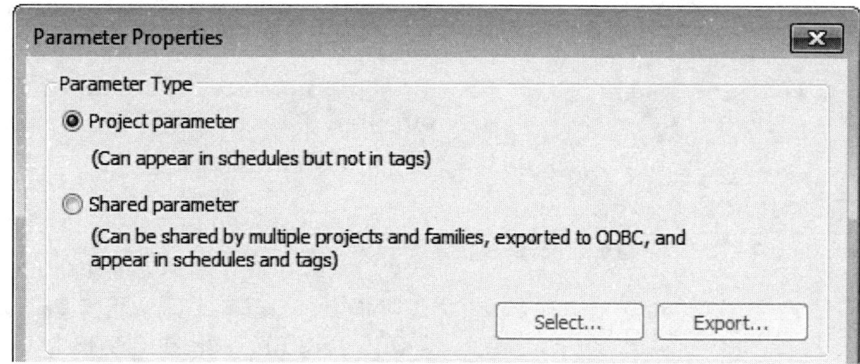

Figure 2–89

3. In the *Parameter Data* area shown in Figure 2–90, Set the value as an **Instance** or a **Type**. and then specify a new *Name* for the parameter and select the *Discipline*, *Group*, and *Type of Parameter* in the drop-down lists. You can also add a custom tooltip.

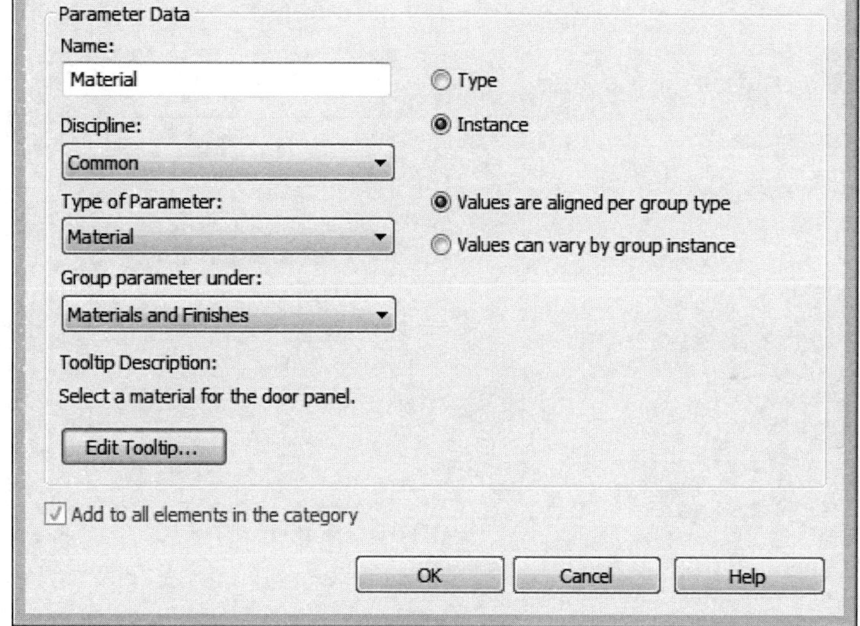

Figure 2–90

- **Instance parameters:** Set for each instance of the associated element. For example, if you create a **Material** parameter when you are creating a furniture schedule, you can assign a different material to each instance of the component.
- **Type parameters:** Modify all instances of a specific type. For example, if you create a **Material** parameter for a furniture component, changing the material for one component changes all of them.

- **Values are aligned per group type:** (Instance Parameters only.) If an element that includes this parameter is part of multiple groups, the value remains the same for all of the groups. Changing the value in one group changes the value in all of the groups.

- Values can vary by group instance: (Instance Parameters only.) If an element that includes this parameter is part of multiple groups, the value can vary according to the group. Changing the value in one group changes the value for all other instances of that group, but does not change the value for instances of other groups that also include the element.

- The *Discipline* can be set to **Common**, **Structural**, **HVAC, Electrical**, **Piping**, or **Energy**.

- The *Type of Parameter* specifies how the parameter value is stored, The options available for Discipline impact the types of parameters in the list, as shown in Figure 2–91.

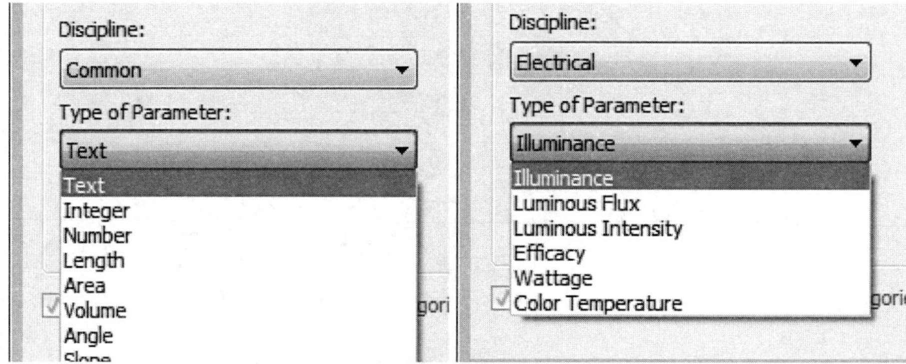

Figure 2–91

Common Parameter Types

Text	A text parameter can host any type of information that consists of both numbers and letters. Examples could be the fabric color or part number of the element.
Integer	Any value that is always represented by a whole number. Examples could be the number of chairs that fit around a conference table or the number of shelves in a bookcase.
Number	Any numeric value: whole, decimal, or fraction. You can use formulas with **Number** parameters.
Length	When you label dimensions, you create **Length** parameters. You can use formulas with **Length** parameters.
Area	Establishes the area of an element. It is numeric and can have formulas applied to it.

Volume	Establishes the volume of an element. It is numeric and can have formulas applied to it.
Angle	Establishes the angle of an element. It is numeric and can have formulas applied to it.
Slope	Displays a slope, as set up in the Field Format.
Currency	Displays the amount in dollars or other currencies, as set up in the Field Format.
Mass Density	Displays the mass per unit volume of material.
URL	Specifies a link to a web site.
Material	Provides a place to assign a material from the list of materials that are set up in the project.
Image	Creates a parameter in which you can add a raster image connected to the family.
Yes/No	Used with instance parameters where you need a **Yes** or **No** answer to a question listed in the name. The default is **Yes**.
<Family Type>	Opens a dialog box where you can select from the list of family types, such as doors, furniture, or tags.

The Group parameter under drop-down list lists groups that categorize parameters in the Properties and Family Types dialog boxes, as shown in Figure 2–92.

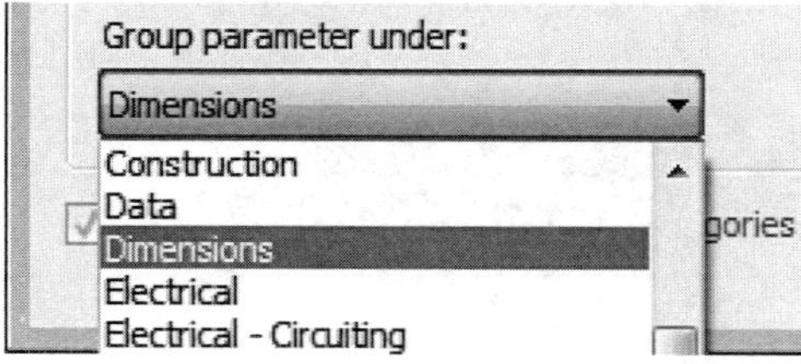

Figure 2–92

To add information to display when you hover the cursor over the new parameter in Properties, click **Edit Tooltip...**, as shown in Figure 2–93.

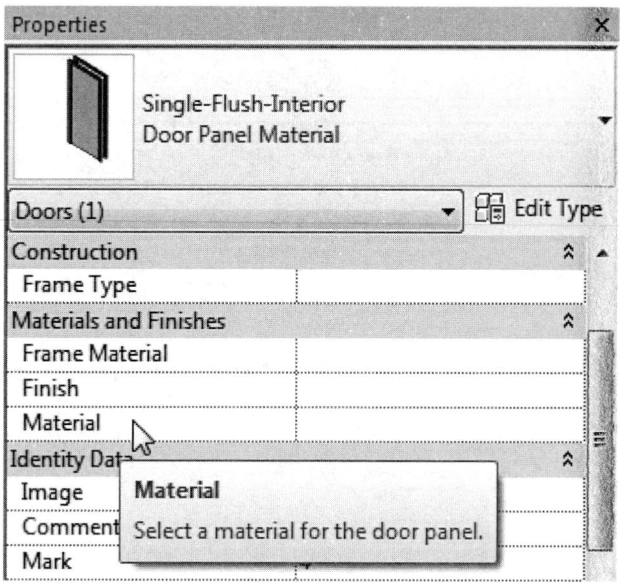

Figure 2–93

Creating Fields from Formulas

In a variety of schedules, you need to gather numerical information and apply formulas to it for a new parameter. For example, you can include cost and quantity information in a Material Takeoff, as shown in the schedule in Figure 2–94.

<Brick - Material Takeoff>					
A	B	C	D	E	F
Material: Name	Material: Area	Material: Manufacturer	Material: Cost	Total Cost	Total Units of Brick
Masonry - Brick	22160 SF	XYZ Masonry	3.00	$66480	155123
Masonry - Brick: 14	22160 SF			$66480	155123

Figure 2–94

- You can create calculated value fields to add formulas to a schedule, or create a percentage field using values in the schedule.
- Gather the information you need to use to create formulas, including the exact name of any fields, before you start the process of creating the calculated value.
- Formula outcomes can be rounded off.

How To: Create a Calculated Value Field

A formula has been added for this figure.

1. In the Schedule Properties dialog box, in the *Fields* tab, click **Calculated Value....** The Calculated Value dialog box opens, as shown in Figure 2–95.

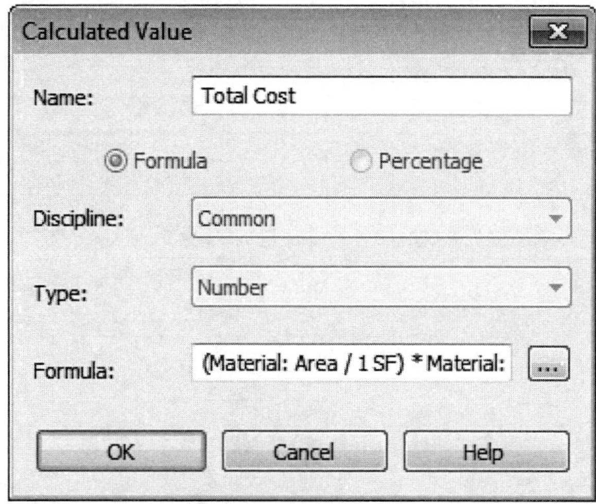

Figure 2–95

Formulas can only reference fields that are included in the schedule.

2. Type a *Name* for the new field.
3. Specify **Formula** as the type of field.
4. Select the *Discipline* and *Type* as needed.
5. Type the formula you want to use in the *Formula* field. (The full formula in is (Material: Area / 1 SF) * Material: Cost).) To ensure you get the exact field name, click (Browse) to select from a list of fields, as shown in Figure 2–96.

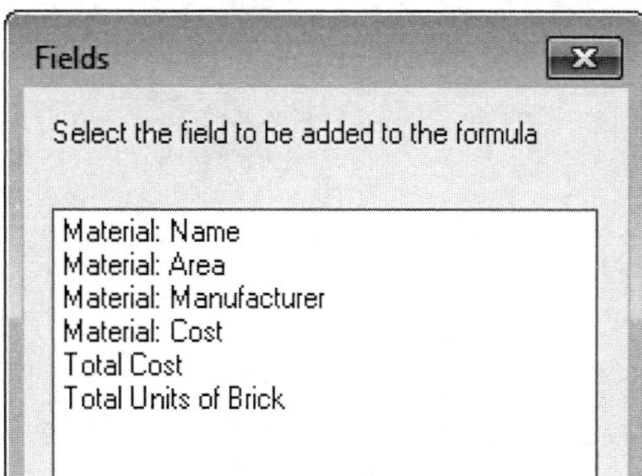

Figure 2–96

6. Click **OK**.

Percentage Fields

Percentage fields can be used with area schedules, as shown in Figure 2–97.

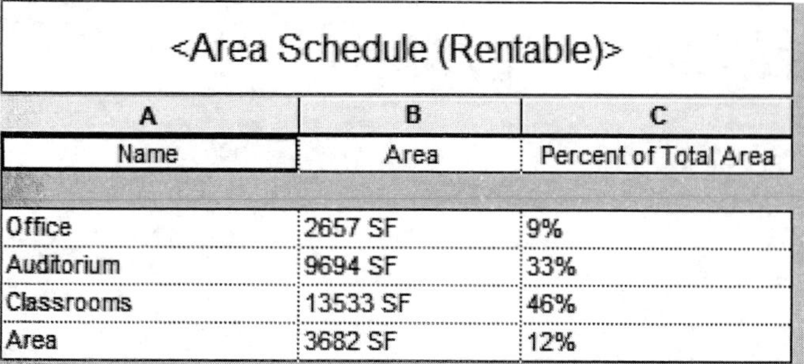

<Area Schedule (Rentable)>

A	B	C
Name	Area	Percent of Total Area
Office	2657 SF	9%
Auditorium	9694 SF	33%
Classrooms	13533 SF	46%
Area	3682 SF	12%

Figure 2–97

How To: Create a Percentage Field

1. In the Schedule Properties dialog box, in the *Fields* tab, click **Calculated Value....** The Calculated Value dialog box opens, as shown in Figure 2–98.

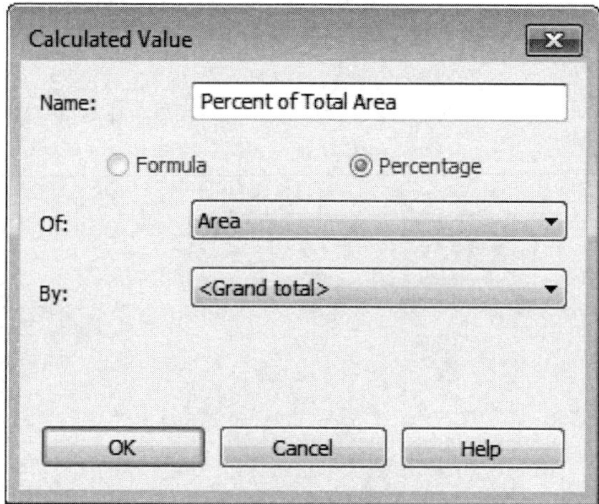

Figure 2–98

2. Type a *Name* for the new field.
3. Specify **Percentage** as the type of field.
4. Select the field you want to take a percentage *Of.*
5. Select the field you want to take a percentage *By.*
    - The default option is **<Grand total>**, which calculates the percentage based on the total of the entire schedule.
    - If you have created groups in the *Sorting/Grouping* tab, they are also available. For example, if Fields are sorted by **Level**.
6. Click **OK**.

Conditional Formatting

Conditional formatting is typically used for testing and does not carry over when placed on a sheet.

Conditional formatting can be applied to any field where you need to display some specific information that can easily be seen at a glance. For example, in Figure 2–99, the *Area* field is highlighted according to a conditional test for any size over 10000 SF.

Area Schedule Over 10000 SF			
A	B	C	D
Number	Name	Area Type	Area
1	Office	Building Common Area	2657 SF
2	Auditorium	Building Common Area	9694 SF
3	Classrooms	Building Common Area	13533 SF
4	Area	Building Common Area	3682 SF

Figure 2–99

How To: Add Conditional Formatting in a schedule

1. Open the schedule you want to work in and select the *Formatting* tab.
2. Select the field to which you want to apply the conditional formatting.
3. Click **Conditional Format...**.
4. In the Conditional Formatting dialog box, select the *Field*, *Test*, and specify a *Value*, as shown in Figure 2–100.

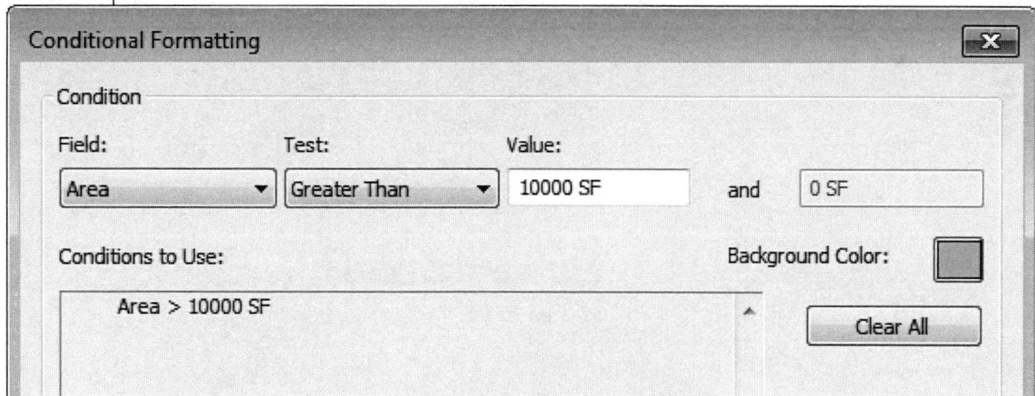

Figure 2–100

5. Click the **Background Color** button and select a color to display in the schedule when the conditions are met.
6. You can select additional fields, test, and values to further refine the conditional statements.
7. When you are finished, click **OK** to close the dialog box and continue working on the Schedule Properties as needed.

Embedded Schedules

When working with Autodesk Revit MEP schedules that need to refer to spaces or rooms, you can use embedded schedules. For example, in Figure 2–101, a space schedule that is displaying the Space **Number** and **Name**, also has an embedded schedule displaying information about the air terminals in each space.

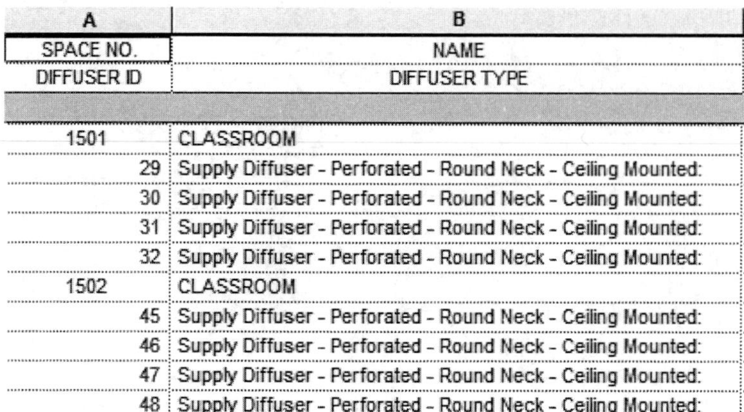

A	B
SPACE NO.	NAME
DIFFUSER ID	DIFFUSER TYPE
1501	CLASSROOM
29	Supply Diffuser - Perforated - Round Neck - Ceiling Mounted:
30	Supply Diffuser - Perforated - Round Neck - Ceiling Mounted:
31	Supply Diffuser - Perforated - Round Neck - Ceiling Mounted:
32	Supply Diffuser - Perforated - Round Neck - Ceiling Mounted:
1502	CLASSROOM
45	Supply Diffuser - Perforated - Round Neck - Ceiling Mounted:
46	Supply Diffuser - Perforated - Round Neck - Ceiling Mounted:
47	Supply Diffuser - Perforated - Round Neck - Ceiling Mounted:
48	Supply Diffuser - Perforated - Round Neck - Ceiling Mounted:

Figure 2–101

How To: Add an embedded schedule

1. Create the base Space or Room schedule.
2. In the Schedule Properties dialog box, select the *Embedded Schedule* tab.
3. Select **Embedded Schedule** and then select a category type from the Category list as shown in Figure 2–102.

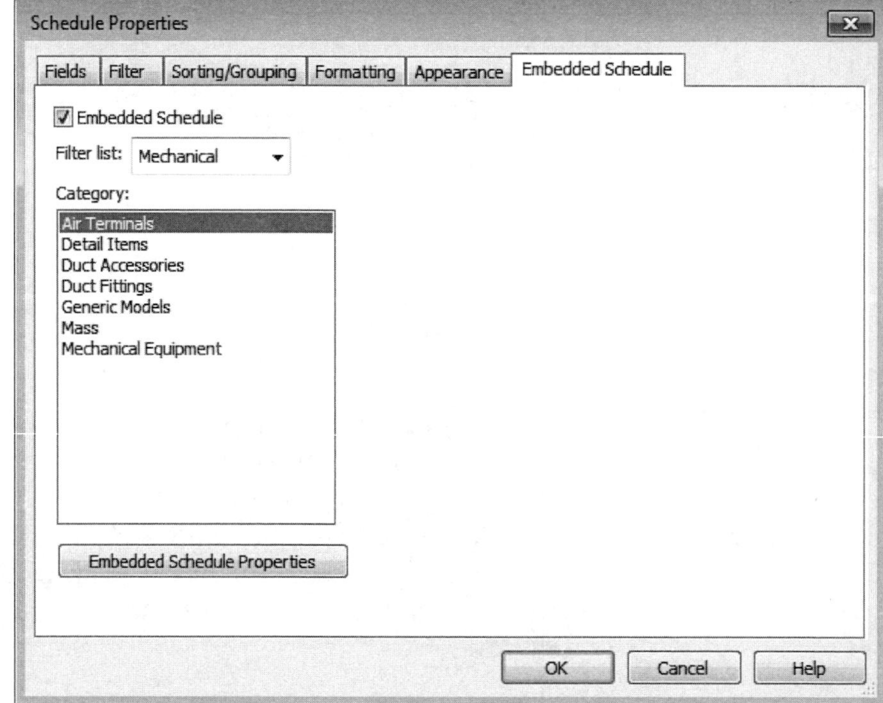

Figure 2–102

4. Click **Embedded Schedule Properties**.
5. In the Schedule Properties dialog box, in the *Field* tab, add the necessary fields. The fields display in order directly under the base schedule columns, as shown in Figure 2–103.

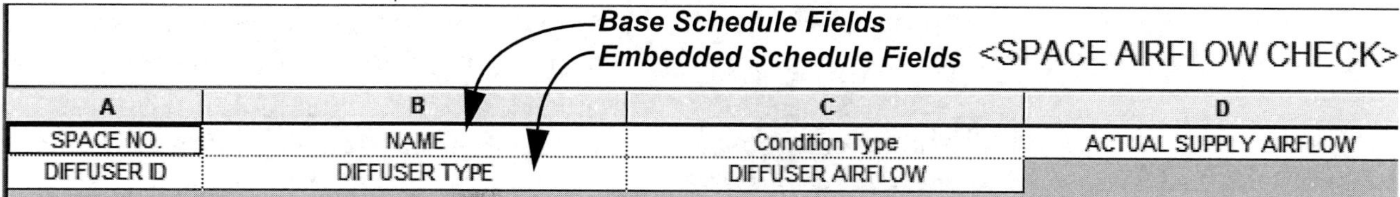

Base Schedule Fields
Embedded Schedule Fields
<SPACE AIRFLOW CHECK>
A
B
C
D
SPACE NO.
NAME
Condition Type
ACTUAL SUPPLY AIRFLOW
DIFFUSER ID
DIFFUSER TYPE
DIFFUSER AIRFLOW

Figure 2–103

6. Add any filters, sorting, and formatting as required.
7. Click **OK** until the schedule displays.

2.6 Creating Material Takeoff Schedules

Schedules are powerful features that interact with all parts of an Autodesk Revit project. A special schedule, called a Material Takeoff, gathers information about the materials in a project (such as the area and volume) and user-designated properties (such as the cost or manufacturer). You can also create formulas or percentages based on material properties to include in a schedule using the **Calculated value** option, as shown in Figure 2–104. Custom parameters are also permitted in Material Takeoffs.

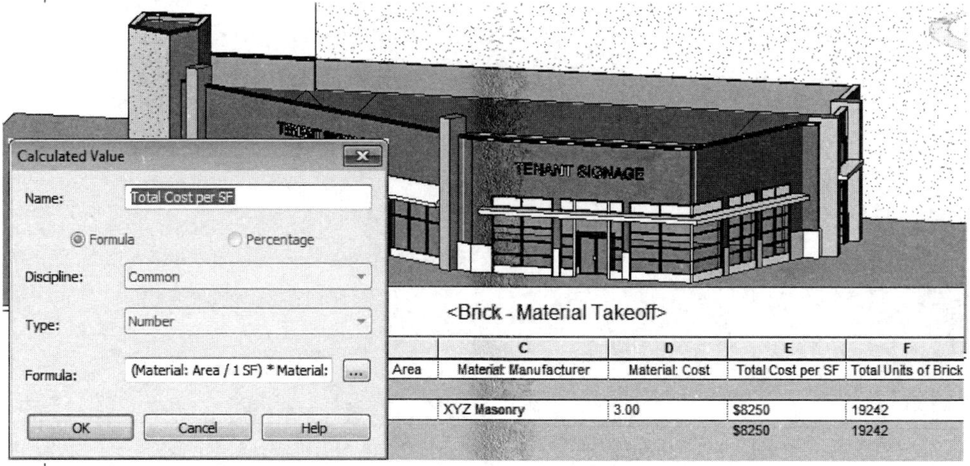

Figure 2–104

- Creating material takeoff schedules follows the same procedure as when creating building component schedules, except that it uses a different command. In the *View* tab> Create panel, expand (Schedules) and click (Material Takeoff) or in the Project Browser, right-click on the Schedule/Quantities node and select **New Material Takeoff**.
- The options for fields include all of the material parameters, as shown in Figure 2–105.

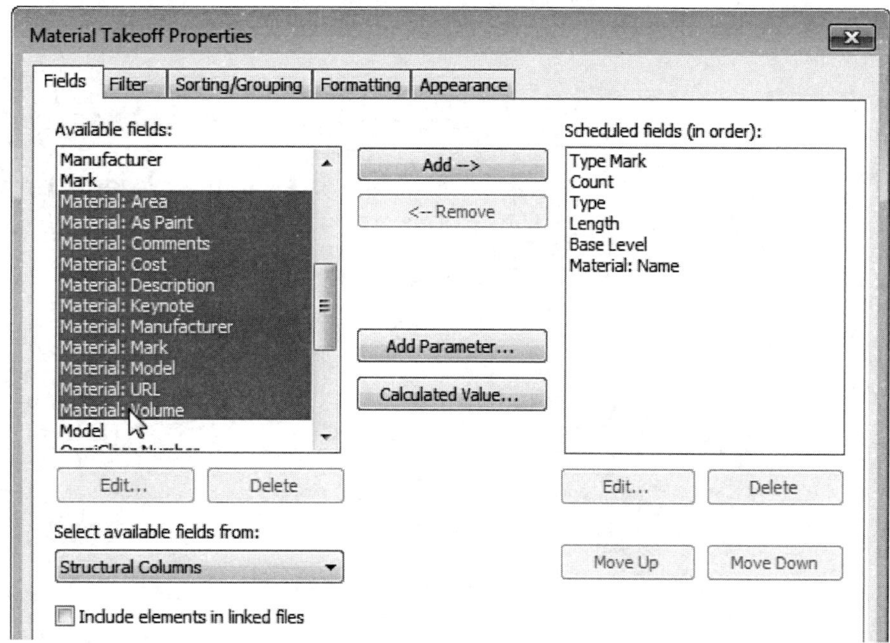

Figure 2–105

- When creating a Material Takeoff Schedule, it can help to modify the unit settings for the various fields. In the Formatting tab, select a field and click **Field Format...**. In the Format dialog box, clear the **Use project settings** option and modify the options, as shown in Figure 2–106. This is only available for numerical fields.

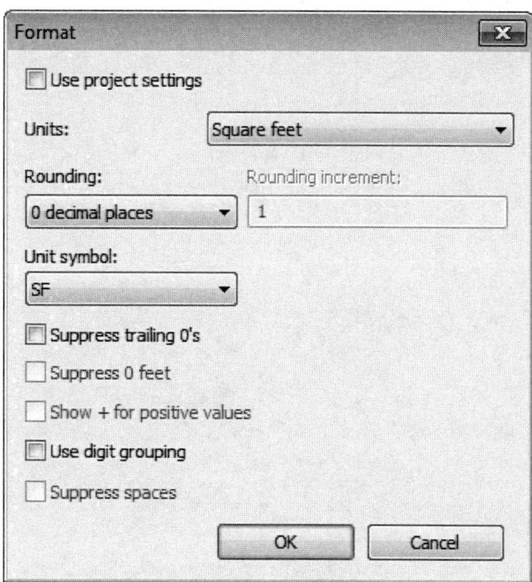

Figure 2–106

- A numerical value can be assigned a currency. Zeros can be suppressed. Spaces can be suppressed and large numbers can include commas as separators.

Hint: Material Properties

Most material properties used in material takeoff schedules can be set by material. In the *Manage* tab>Settings panel, click (Materials). In the Material Browser, select the material that you want to modify. In the *Identity* tab, you can add information, such as **Manufacturer** and **Cost** as shown in Figure 2–107.

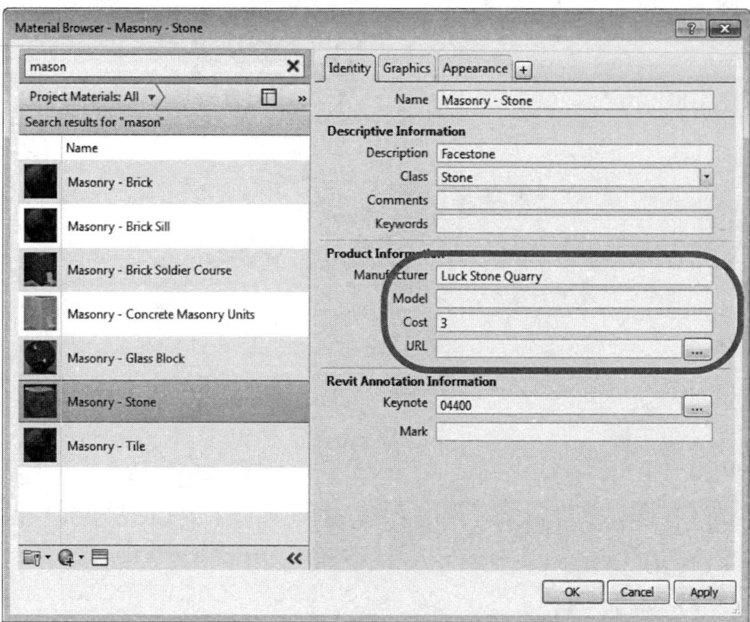

Figure 2–107

Practice 2g Create Complex Schedules for Architectural Projects

Practice Objective

- Create material takeoff schedules.
- (Optional) Create a percentage-based calculated value field for use in a schedule.

Estimated time for completion: 15 minutes

In this practice you will create a material takeoff schedule that includes several calculated value formulas. You will then add it to a sheet, as shown in Figure 2–108, and test it and other material takeoff schedules by modifying the wall type and material properties. As an optional task, you will create a percentage-based field for an area schedule.

Brick - Material Takeoff					
Material: Name	Material: Area	Material: Manufacturer	Material: Cost	Total Cost per SF	Total Units of Brick
Masonry - Brick	2749 SF	XYZ Masonry	3.00	$8250	19242
Masonry - Brick: 11	2749 SF			$8250	19242

Figure 2–108

Task 1 - Create a material takeoff schedule.

1. In the ...*Architecture* folder, open **Mall-Addition.rvt** .
2. In the *View* tab>Create panel, expand (Schedules) and click (Material Takeoff).
3. In the New Material Takeoff dialog box, in the *Filter* list, select **Architecture**, and in the *Category* list, select **Walls**. Name the schedule **Brick – Material Takeoff** and select **New Construction** from the *Phase* list.
4. Click **OK**.

5. In the *Fields* tab of the Material Takeoff Properties dialog box, add the fields **Material: Name**, **Material: Area**, **Material: Manufacturer**, and **Material: Cost**, as shown in Figure 2–109.

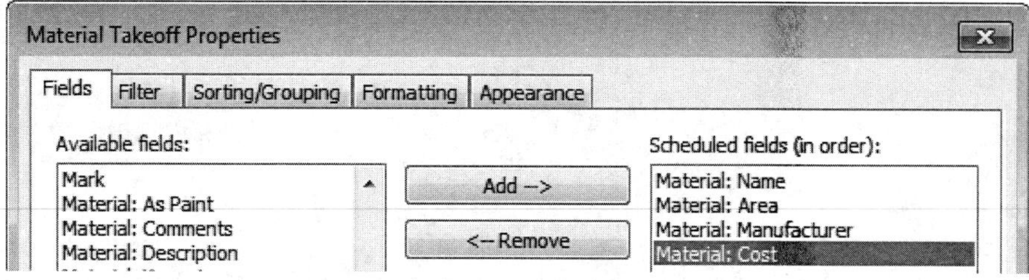

Figure 2–109

Area and Cost are two different types of parameters. Therefore, in this practice, you need to divide the ***Material: Area*** *parameter by 1 to get the correct result.*

6. Click **Calculated Value...** and set the following parameters, as shown in Figure 2–110. (Make sure you account for the spaces in the field names when typing the formulas. There should be a space after the colon for the software to recognize **Material: Area** as a field.)

   - **Name:** Total Cost per SF
   - **Discipline: Common**
   - **Type:** Number
   - **Formula:** (Material: Area / 1SF) * Material: Cost

You can click (Browse) to select from a list of fields as you are creating the formula to prevent you from misspelling parameter names.

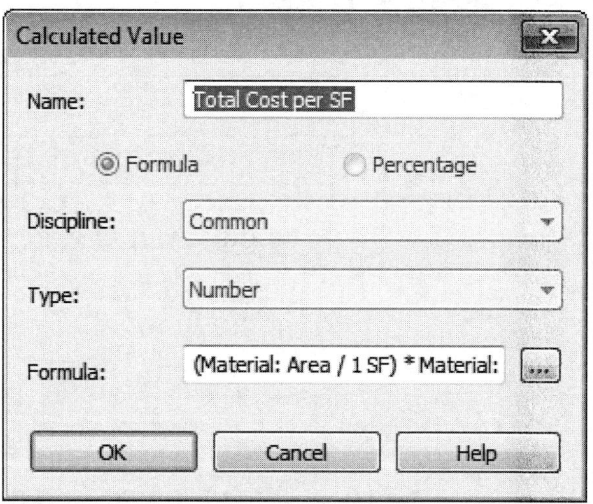

Figure 2–110

7. Create another calculated value fields using the following parameters. (There are seven standard bricks in a square foot. The type and size of bricks being used might vary and that can affect the number of bricks.)

   - **Name:** Total Units of Brick
   - **Discipline:** Common
   - **Type:** Number
   - **Formula:** (Material: Area / 1) *7

8. In the *Filter* tab, filter the schedule using **Material: Name equals Masonry - Brick**, as shown in Figure 2–111.

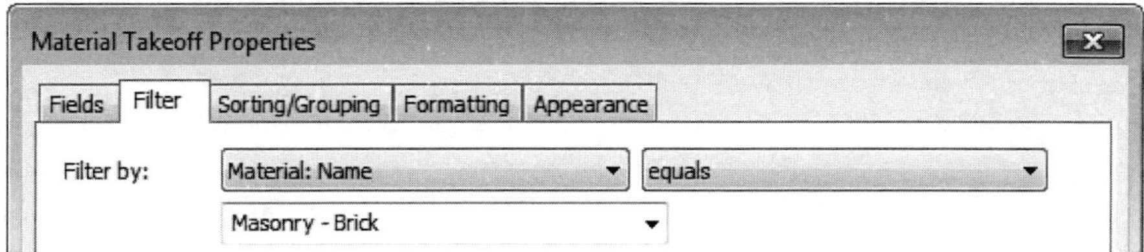

Figure 2–111

9. In the *Sorting/Grouping* tab, sort by **Material: Name** and select **Footer**, as shown in Figure 2–112.

Figure 2–112

10. In the *Formatting* tab, in the *Fields* list, select **Material: Area, Total Cost per SF**, and **Total Units of Brick.** In the *Field formatting* area, select the **Calculate totals** option, as shown in Figure 2–113.

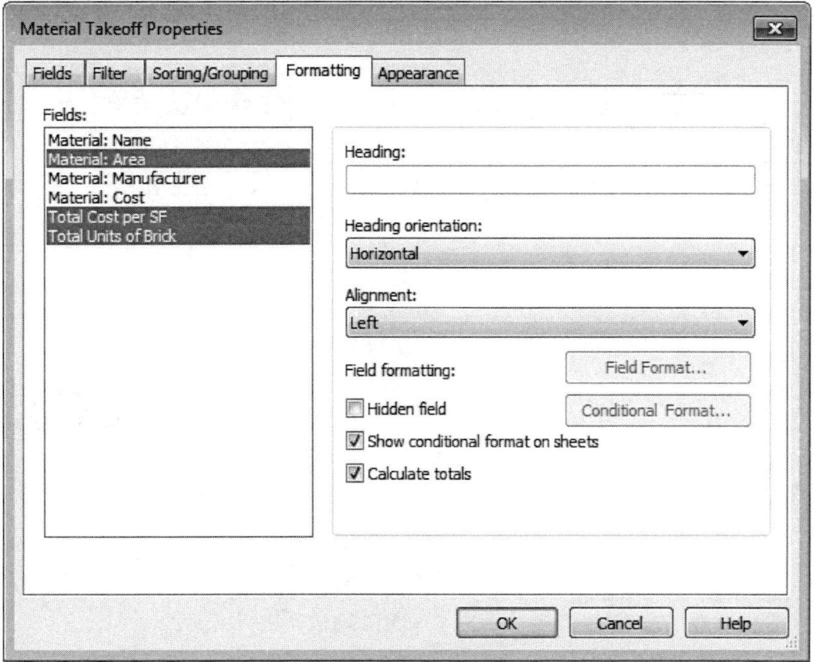

Figure 2–113

11. In the *Appearance* tab, modify the grid lines or text fonts as needed and click **OK**.

Task 2 - Modify the material takeoff schedule.

1. The units for the *Total Cost per SF* and *Total Units of Brick* fields have six decimal places, as shown in Figure 2–114. This is more than required.

<Brick - Material Takeoff>

A	B	C	D	E	F
Material: Name	Material: Area	Material: Manufact	Material: Cost	Total Cost per SF	Total Units of Brick
Masonry - Brick	192 SF		4.00	769.675167	1346.931543
Masonry - Brick	756 SF		4.00	3023.630924	5291.354117

Figure 2–114

2. In Properties, next to *Formatting*, click **Edit...**. The Material Takeoff Properties dialog box opens with the *Formatting* tab selected.

3. Select the *Total Cost per SF* field and click **Field Format...**.

4. In the Format dialog box, clear the **Use default settings** option. Set the *Units* to **Currency**, *Rounding* to **To the nearest 10** and *Unit symbol* to **$**, as shown in Figure 2–115. Click **OK**.

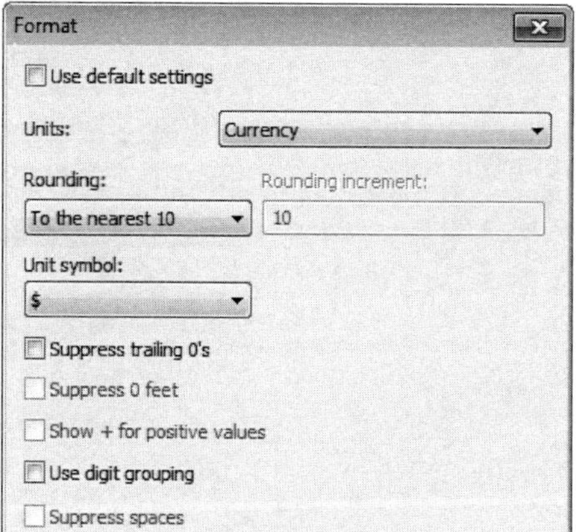

Figure 2–115

5. Select the *Total Units of Brick* field and click **Field Format...**.

6. In the Format dialog box, clear the **Use default settings** option. Set the *Units* to **Fixed** and *Rounding* to **0 decimal places**. Click **OK**.

7. Click **OK** to close the dialog box.

8. In the schedule view, add a manufacturer's name and change the *Material: Cost* of the Brick to **3.00**. The Total Cost updates, but the Total Units of Brick remains the same.

Task 3 - Test the material takeoff schedule.

1. Open the **A102 – Plan and Schedules** sheet.
2. Zoom in on the existing schedules to one side.
3. Add the new **Brick – Material Takeoff** schedule to the sheet. Use grips to modify the size of the schedule as needed.
4. This schedule is different from other schedules because it includes all the instances of Brick. In the Project Browser, select the **Brick - Material Takeoff** schedule (you do not have to open it to change the properties).
5. In Properties, in the *Other* area, next to Sorting/Grouping, click **Edit...**.
6. In the Material Takeoff Properties dialog box, in the *Sorting/Grouping* tab, clear the **Itemize every instance** option. The schedule updates on the sheet.
7. Open the **3D Views: Front View**. Arrange the views so that you can see the 3D view and the top two schedules on the sheet, as shown in Figure 2–116.

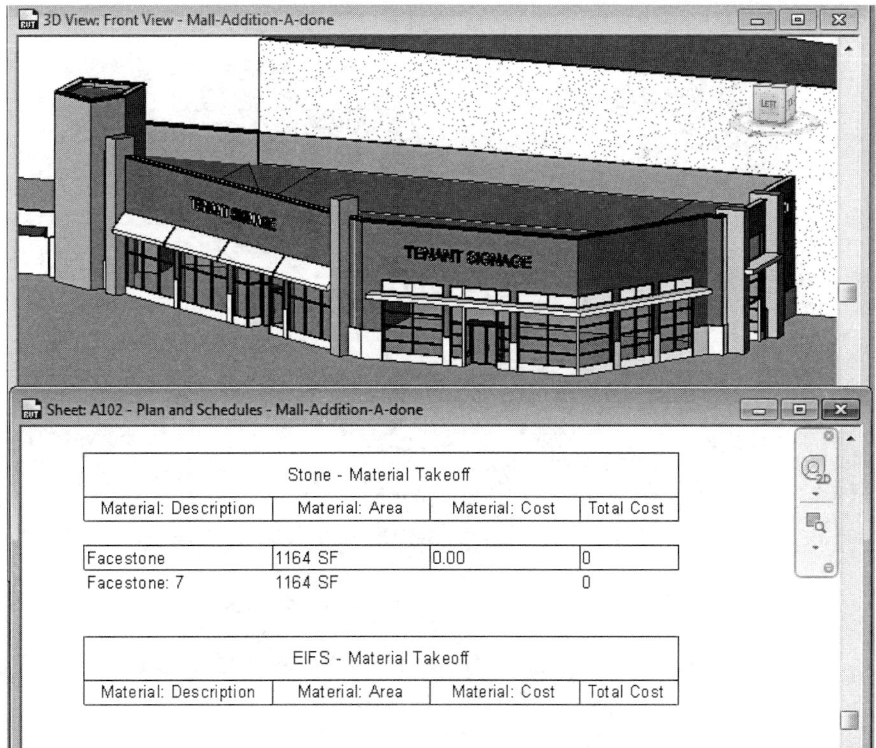

Figure 2–116

8. In the 3D view, select one of the walls on the tower to the left and change its wall type to **Basic Wall: FCE – Exterior – EIFS on Mtl.Stud**, as shown in Figure 2–117.

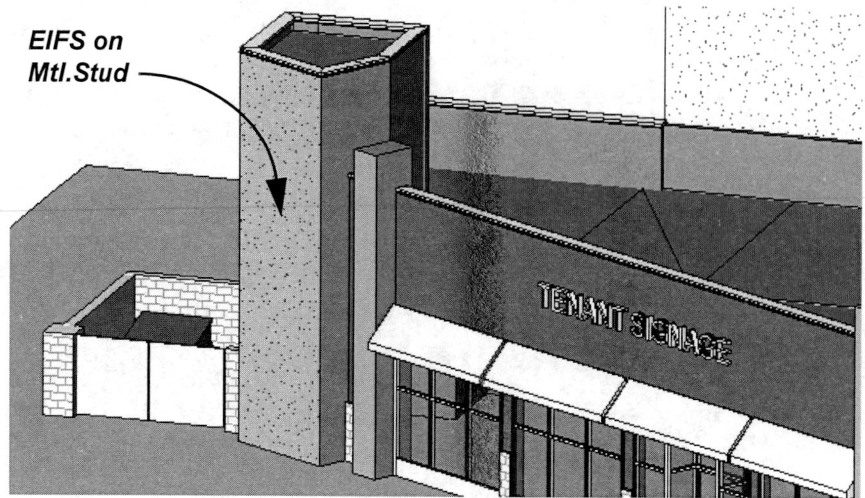

Figure 2–117

9. The schedules change to display a smaller amount of the stone used and add to the EIFS schedule that was blank before, as shown in Figure 2–118.

Stone - Material Takeoff			
Material: Description	Material: Area	Material: Cost	Total Cost
Facestone	865 SF	0.00	0
Facestone: 6	865 SF		0

EIFS - Material Takeoff			
Material: Description	Material: Area	Material: Cost	Total Cost
EIFS	298 SF	12.00	3574
EIFS: 1	298 SF		3574

Figure 2–118

10. Double-click on the Stone - Material Takeoff schedule in the sheet to open the related schedule view.

11. Change the *Material: Cost* to **10**.

12. The *Material: Cost* and *Total Cost* fields update on the sheet as well.

13. Modify these fields so that they display as currency.

Task 4 - (Optional) Create a percentage-based Calculated Value field.

1. Using (Schedule/Quantities), create a new area schedule (Rentable).
2. Add the fields *Name* and *Area*.
3. Click **Calculated Value...** and add the new *Calculated Value* field, as shown in Figure 2–119.

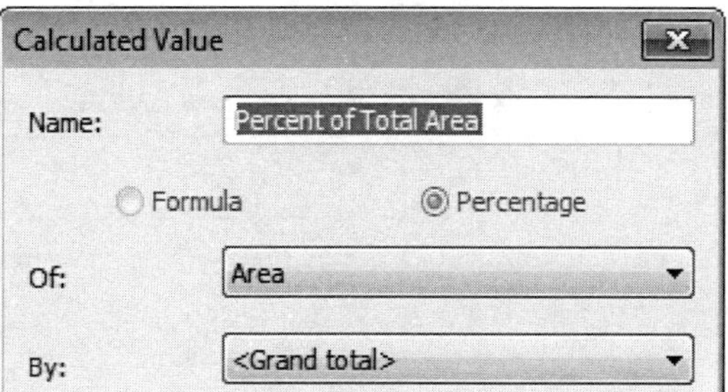

Figure 2–119

4. Click **OK** twice to create the schedule.
5. Open the **A103 – Area Plans** sheet and add the new area schedule to the sheet next to the colored legend, as shown in Figure 2–120.

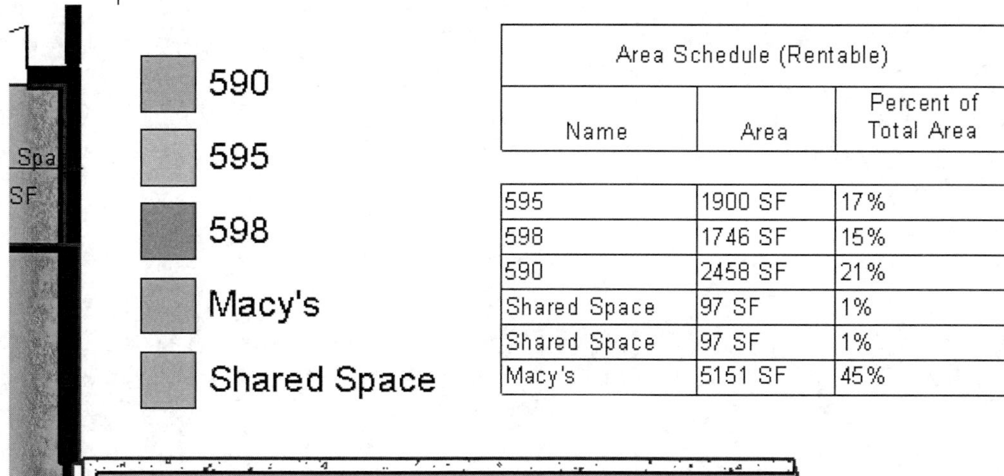

Area Schedule (Rentable)		
Name	Area	Percent of Total Area
595	1900 SF	17%
598	1746 SF	15%
590	2458 SF	21%
Shared Space	97 SF	1%
Shared Space	97 SF	1%
Macy's	5151 SF	45%

Figure 2–120

6. Save and close the project.

Practice 2h Create Complex Schedules for MEP Projects

Practice Objective

- Create a base schedule and add an embedded schedule into it.
- Test the scheduled information using conditional formatting.

Estimated time for completion: 20 minutes

In this practice you will create a base schedule including a Calculated Value and Conditional Formatting. You will then add an embedded schedule to create an Space Air Flow Check schedule, as shown in Figure 2–121.

<Space Air Flow Check>					
A	B	C	D	E	F
Space Number	Name	Condition Type	Actual Supply Airfl	Calculated Supply	Airflow Check
Diffuser ID	Diffuser Type	Diffuser Airflow			
101	Office Area 1	Heated and cooled	0 CFM	2833 CFM	-2833 CFM
102	Office Area 2	Heated and cooled	0 CFM	2546 CFM	-2546 CFM
103	Vestuble 1	Heated and cooled	0 CFM	351 CFM	-351 CFM
104	Foyer 1	Heated and cooled	0 CFM	718 CFM	-718 CFM
105	Conf. Room	Heated and cooled	800 CFM	519 CFM	281 CFM
1	Supply Diffuser	400 CFM			
2	Supply Diffuser	400 CFM			
106	Women's 1	Heated and cooled	500 CFM	203 CFM	297 CFM
13	Supply Diffuser	500 CFM			
107	Mens 1	Heated and cooled	500 CFM	201 CFM	299 CFM
14	Supply Diffuser	500 CFM			
108	Hall 1	Heated and cooled	500 CFM	864 CFM	-364 CFM
8	Supply Diffuser	500 CFM			
109	Mech. Room	Heated and cooled	100 CFM	90 CFM	10 CFM
50	Supply Diffuser	100 CFM			
110	Elec. Room	Heated and cooled	100 CFM	168 CFM	-68 CFM
49	Supply Diffuser	100 CFM			
111	Utility	Heated and cooled	0 CFM	54 CFM	-54 CFM
112	Break Room 1	Heated and cooled	500 CFM	527 CFM	-27 CFM
9	Supply Diffuser	500 CFM			
113	Vestuble 2	Heated and cooled	0 CFM	150 CFM	-150 CFM
114	Stair	Heated and cooled	0 CFM	53 CFM	-53 CFM
115	Elev.	Heated and cooled	0 CFM	25 CFM	-25 CFM
201	Office Space 3	Heated and cooled	0 CFM	7327 CFM	-7327 CFM
202	Office Space 4	Heated and cooled	0 CFM	6196 CFM	-6196 CFM
203	Foyer 2	Heated and cooled	0 CFM	867 CFM	-867 CFM
204	Womens 2	Heated and cooled	0 CFM	203 CFM	-203 CFM
205	Mens 2	Heated and cooled	140 CFM	201 CFM	-61 CFM
37	Supply Diffuser	140 CFM			
206	Hall 2	Heated and cooled	0 CFM	1003 CFM	-1003 CFM
25	Supply Diffuser	500 CFM			
207	Stair	Heated and cooled	0 CFM	238 CFM	-238 CFM
208	Break Room 2	Heated and cooled	410 CFM	497 CFM	-87 CFM
24	Supply Diffuser	205 CFM			
34	Supply Diffuser	205 CFM			
209	Mechanical Ro	Heated and cooled	0 CFM	268 CFM	-268 CFM

Figure 2–121

Task 1 - Create a Base schedule.

1. In the *...\MEP* folder, open **Office-Airflow.rvt**.

2. Create a new Schedules/Quantities schedule.

3. In the New Schedule dialog box, filter the list by *Mechanical* if needed, and select **Spaces**. Name the schedule **Space Air Flow Check** and click **OK**.

4. Add the following fields: **Number**, **Name**, **Condition Type**, **Actual Supply Airflow**, and **Calculated Supply Airflow**, as shown in Figure 2–122.

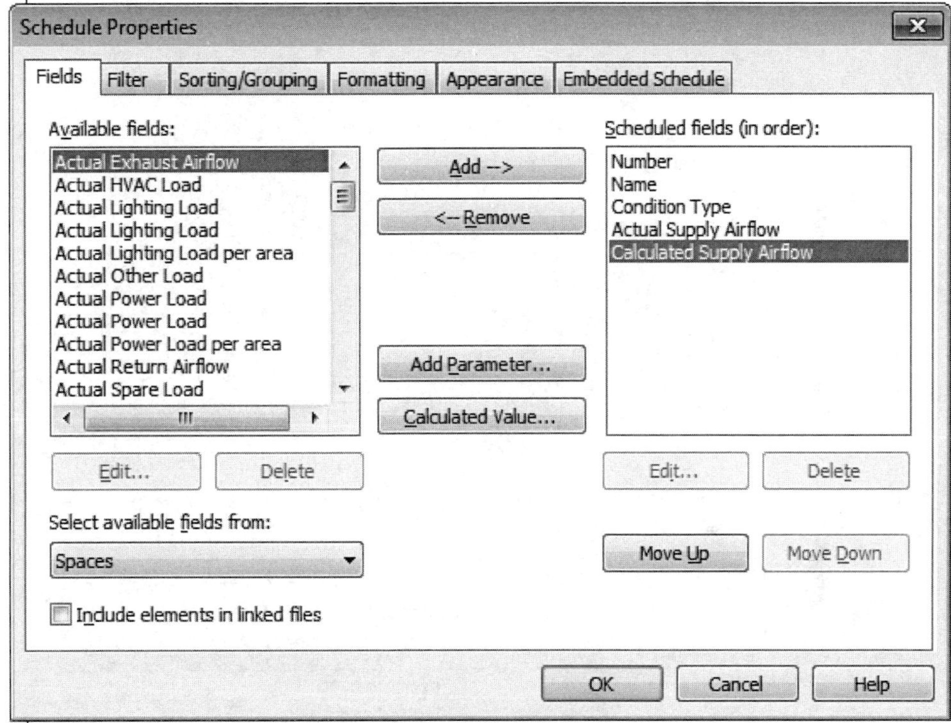

Figure 2–122

5. While still in the *Fields* tab, click **Calculated Value...**.

6. In the Calculated Value dialog box, specify the *Name* as **Airflow Check**, *Discipline* as **HVAC**, and *Type* as **Air Flow**.

By browsing for the fields, you save time by not misspelling or mis-capitalizing the field names.

7. For the Formula, click [...] and select **Actual Supply Airflow** and click **OK**. Type a minus sign (-) and then click [...] again and select **Calculated Supply Airflow**. Click **OK**.

8. In the *Sorting/Grouping* tab, sort by **Number**.

9. In the *Formatting* tab, select **Airflow Check** and click **Conditional Format...**.

10. In the Conditional Formatting dialog box, specify the different fields as shown in Figure 2–123. You can select any color you want.

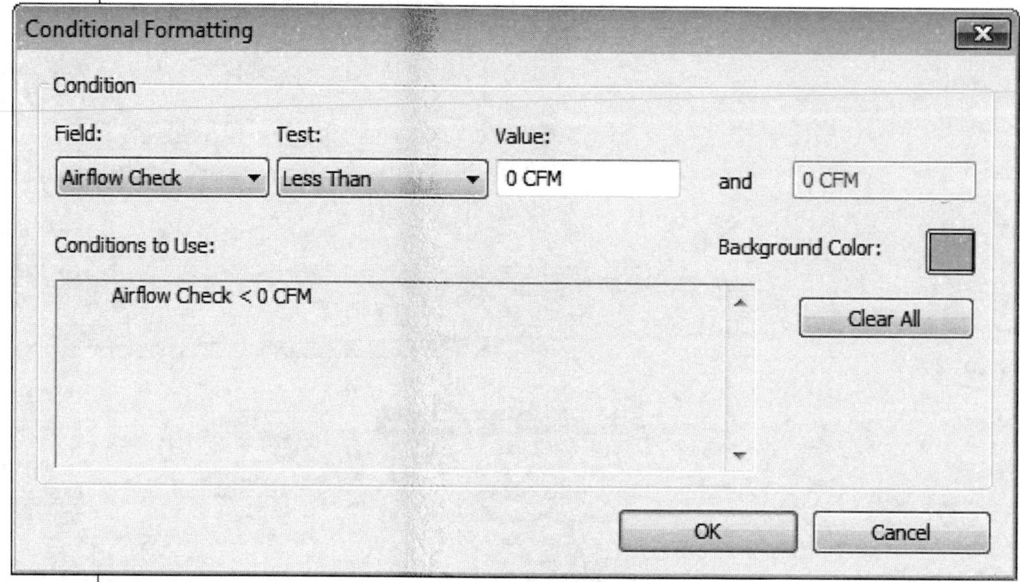

Figure 2–123

11. Click **OK** to display the schedule (shown in part) up to this point. Expand the columns as shown in Figure 2–124.

<Space Air Flow Check>

A	B	C	D	E	F
Number	Name	Condition Type	Actual Supply Airfl	Calculated Supply	Airflow Check
101	Office Area 1	Heated and cooled	0 CFM	2833 CFM	-2833 CFM
102	Office Area 2	Heated and cooled	0 CFM	2546 CFM	-2546 CFM
103	Vestuble 1	Heated and cooled	0 CFM	351 CFM	-351 CFM
104	Foyer 1	Heated and cooled	0 CFM	718 CFM	-718 CFM
105	Conf. Room	Heated and cooled	800 CFM	519 CFM	281 CFM
106	Women's 1	Heated and cooled	500 CFM	203 CFM	297 CFM
107	Mens 1	Heated and cooled	500 CFM	201 CFM	299 CFM
108	Hall 1	Heated and cooled	500 CFM	864 CFM	-364 CFM

Figure 2–124

12. Click inside the **Number** Heading and change it to **Space Number**.

13. Save the project.

Task 2 - Add an embedded schedule.

1. While still in the Space Air Flow Check schedule view, in Properties, beside *Embedded Schedule*, click **Edit...**.
2. In the Schedule Properties dialog box, in the *Embedded Schedule* tab, select **Embedded Schedule**.
3. In the Category list, select **Air Terminals**.
4. Click **Embedded Schedule Properties**.
5. In the Schedule Properties dialog box, in the *Field* tab, add the following fields:
   - *System Classification*
   - *Mark*
   - *Family and Type*
   - *Flow*
6. In the *Filter* tab, filter by: **System Classification equals Supply Air**.
7. In the *Sorting/Grouping* tab, set *Sort by as* **Mark** and *Then by* as **Family and Type**, as shown in Figure 2–125.

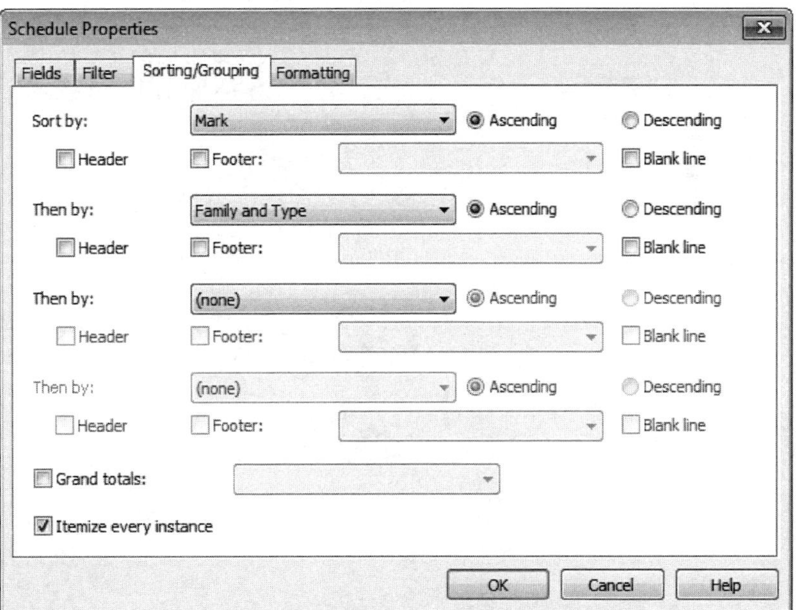

Figure 2–125

8. In the *Formatting* tab, select the *Field* **System Classification** and select **Hidden field**.
9. Select **Mark** and set *Heading* as **Diffuser ID**. Set *Heading orientation* as **Horizontal** and *Alignment* as **Right**.

10. Select **Family and Type** and set *Heading* as **Diffuser Type**. Set *Heading orientation* as **Horizontal** and *Alignment* as **Left**.

11. Select **Flow** and set *Heading* as **Diffuser Airflow**. Set *Heading orientation* as **Horizontal** and *Alignment* as **Left**.

12. Click **OK** until the schedule displays, as shown in part in Figure 2–126.

<Space Air Flow Check>					
A	B	C	D	E	F
Space Number	Name	Condition Type	Actual Supply Airfl	Calculated Supply	Airflow Check
Diffuser ID	Diffuser Type	Diffuser Airflow			
101	Office Area 1	Heated and cooled	0 CFM	2833 CFM	-2833 CFM
102	Office Area 2	Heated and cooled	0 CFM	2546 CFM	-2546 CFM
103	Vestuble 1	Heated and cooled	0 CFM	351 CFM	-351 CFM
104	Foyer 1	Heated and cooled	0 CFM	718 CFM	-718 CFM
105	Conf. Room	Heated and cooled	800 CFM	519 CFM	281 CFM
1	Supply Diffuser	400 CFM			
2	Supply Diffuser	400 CFM			
106	Women's 1	Heated and cooled	500 CFM	203 CFM	297 CFM
13	Supply Diffuser	500 CFM			

Figure 2–126

13. Save the project.

14. If you have time, work with the project to establish the correct air flow for the various spaces.

Practice 2i

Create Complex Schedules for Structural Projects

Practice Objective

- Create project parameters and modify an existing schedule using them.
- Modify and group headers in a schedule to graphically link the parameter types.

Estimated time for completion: 20 minutes

In this practice you will create project parameters for structural footing types. Then you will add the new parameters to a schedule, add an additional parameter, add headers to the schedule, and fill in the schedule as shown in Figure 2–127.

<Structural Foundation Schedule>									
A	B	C	D	E	F	G	H	I	J
	Dimensions			Reinforcement					
				Cover		Bars			
Type	Length	Width	Thickness	Top Cover	Bottom Cover	Major Bars	Minor Bars	Count	Volume
FC3.0		3' - 0"	1' - 0"	0' - 3"	0' - 1 1/2"	#4 @ 1'-0"	#4 @ 1'-0"	4	715.22 CF
FC5.0	139' - 10"	5' - 0"	1' - 0"	0' - 3"	0' - 2"	#4 @ 0'-9"	#4 @ 0'-9"	1	553.47 CF
FS6.0	6' - 0"	6' - 0"	1' - 0"	0' - 3"	0' - 1 1/2"	#4 @ 1'-0"	#4 @ 1'-0"	27	972.00 CF
FS8.0	8' - 0"	8' - 0"	1' - 0"	0' - 3"	0' - 2"	#4 @ 0'-9"	#4 @ 1'-0"	3	192.00 CF
Grand total: 35								35	2432.69 CF

Figure 2–127

Task 1 - Add Project parameters.

1. In the *...\Structural* folder, open **Century-Building.rvt**.
2. In the Project Browser, in the *Schedules/Quantities* area, open the **Structural Foundation Schedule** as shown in Figure 2–128. This schedule is limited to existing parameters and does not include any rebar information. Also, the **Foundation Thickness** parameter is recorded for continuous footing (FC) types but not for spot footings (FS) types.

<Structural Foundation Schedule>					
A	B	C	D	E	F
	Dimensions				
Type	Length	Width	Foundation Thickness	Count	Volume
FC3.0		3' - 0"	1' - 0"	4	715.22 CF
FC5.0	139' - 10"	5' - 0"	1' - 0"	1	553.47 CF
FS6.0	6' - 0"	6' - 0"		27	972.00 CF
FS8.0	8' - 0"	8' - 0"		3	192.00 CF
Grand total: 35				35	2432.69 CF

Figure 2–128

3. Open the **Structural Plans: T.O.F.** view.

4. Select one of the spot footings.

5. In Properties, click (Edit Type).

6. In the Type Properties dialog box, there are parameters for **Dimensions** and **Identity Data** (as shown in Figure 2–129), but no information about the rebar that typically needs to be included in a foundation schedule.

You can see a ***Thickness*** *parameter, but this does not link into schedules.*

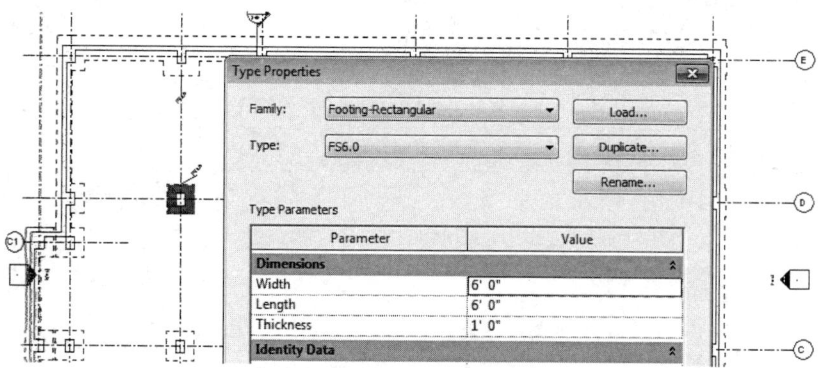

Figure 2–129

7. Click the **Cancel** button to close the dialog box and click in the view to clear the foundation selection.

8. In the *Manage* tab>Settings panel, click (Project Parameters).

9. In the Project Parameters dialog box, click **Add**.

10. In the Parameter Properties dialog box, create a *Project parameter* named **Bottom Cover**. Make it a **Type** parameter and set *Discipline* to **Common**, *Type of Parameter* to **Length**, and *Group parameter under* to **Rebar Set**. In *Categories*, set the *Filter list* to **Structure**, and select **Structural Foundations**, as shown in Figure 2–130.

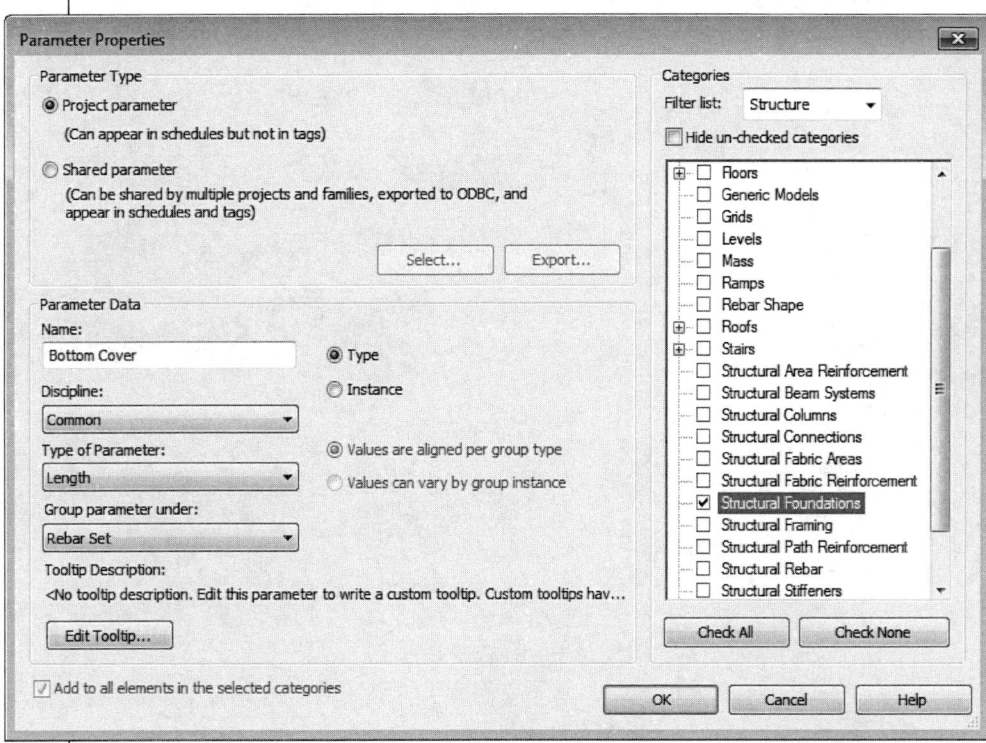

Figure 2–130

11. Click **OK**.

12. Create additional Type parameters. Set *Discipline* as **Common** and *Group parameter under* as **Rebar Set** with the following additional information:

Name	Type of Parameter	Categories
Top Cover	Length	Structural Foundations
Major Bars	Text	Structural Foundations and Walls
Minor Bars	Text	Structural Foundations and Walls

13. When you are finished, the four parameters display as shown in Figure 2–131. Click **OK**.

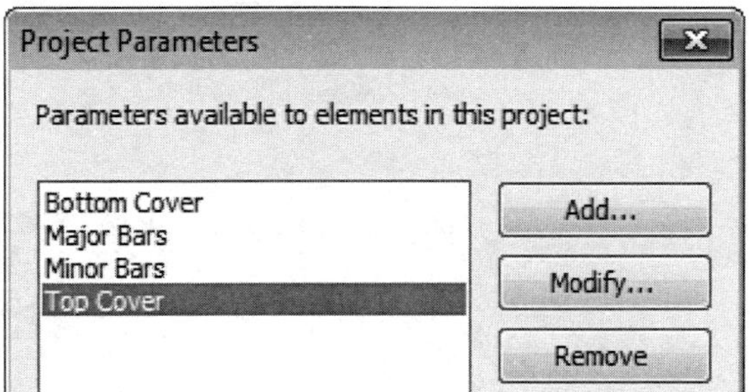

Figure 2–131

14. Select a footing again and open the Type Properties dialog box. This time the **Rebar Set** group of parameters is also displayed, as shown in Figure 2–132. Click **OK**.

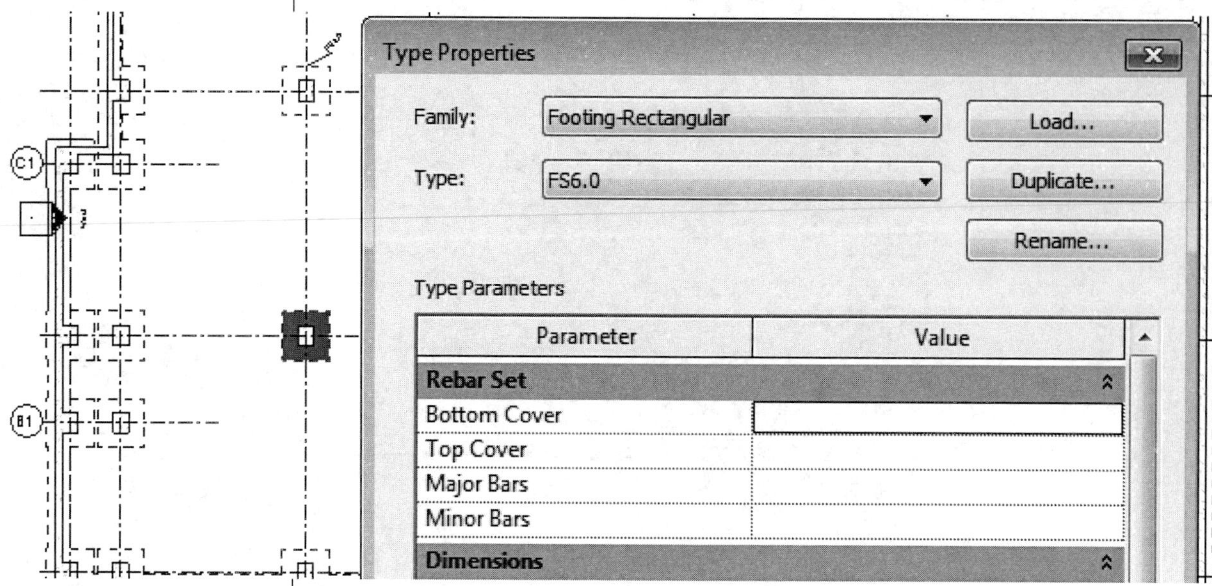

Figure 2–132

15. Save the project.

Task 2 - Modify a Foundation schedule.

1. Switch back to the Structural Foundation Schedule.
2. In Properties, in the *Other* area, next to *Fields*, click **Edit...**.
3. In the Schedule Properties dialog box, in the *Fields* tab, add the following fields: **Top Cover**, **Bottom Cover**, **Major Bars** and **Minor Bars**. Place them in the order shown in Figure 2–133.

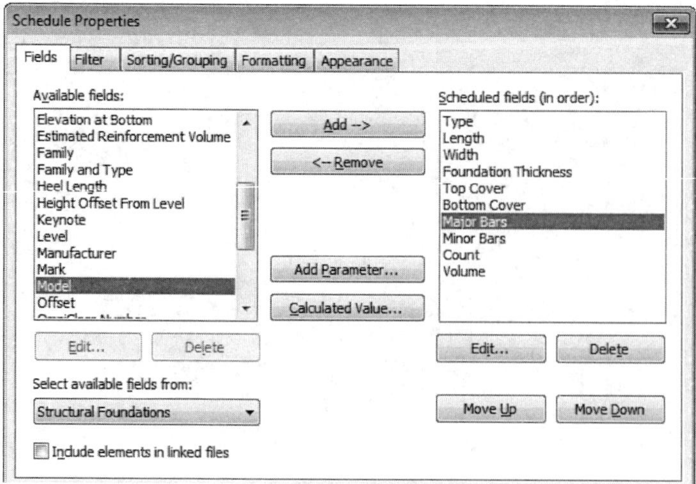

Figure 2–133

4. In the *Scheduled fields (in order)* area, select the existing **Foundation Thickness** parameter and click **<--Remove**.
5. In the *Fields* tab, click **Add Parameter...**.

This parameter is automatically applied to Structural Foundations because that is the type of schedule in which you are working.

6. In the Parameter Properties dialog box, select **Project parameter**, and enter the *Name* **FS and FC Foundation Thickness**. Set it as a **Type** parameter with *Discipline* set to **Common**, *Type of Parameter* set to **Length**, and *Group* with **Dimensions**.
7. Click **OK**.
8. Use **Move Up** to move it below the existing **Width** parameter.
9. Select it and click **Edit...**.
10. The full Parameter Properties dialog box opens. Select the Category **Walls** and **Structural Foundations**.
11. Click **OK** twice to close the dialog boxes.
12. The schedule displays as shown in Figure 2–134.

<Structural Foundation Schedule>

A	B	C	D	E	F	G	H	I	J
Type	Dimensions		FS and FC Foundation	Top Cover	Bottom Cover	Major Bars	Minor Bars	Count	Volume
	Length	Width							
FC3.0		3' - 0"						4	715.22 CF
FC5.0	139' - 10"	5' - 0"						1	553.47 CF
FS6.0	6' - 0"	6' - 0"						27	972.00 CF
FS8.0	8' - 0"	8' - 0"						3	192.00 CF
Grand total: 35								35	2432.69 CF

Figure 2–134

13. Save the project.

Task 3 - Modify and apply headers.

1. Click inside the header for FS and FC Foundation Thickness. Delete **FS and FC Foundation** so it displays as only *Thickness*.

A column cannot be added under an existing header. Therefore, delete the header and then remake it with the correct associations.

2. In the schedule view, select the *Dimensions* header and in the *Modify Schedule/Quantities* tab>Titles & Headers panel, click (Ungroup).
3. Press and drag your cursor over the columns from *Length* to *Thickness*.
4. In the Titles & Headers panel, click (Group).

5. Type a header name of **Dimensions**.

6. Repeat the process to create a header from *Top Cover* to *Minor Bars* with the name **Reinforcement**.

7. Group *Top Cover* and *Bottom Cover* and name the header **Cover**.

8. Group *Major Bars* and *Minor Bars* and name the header **Bars**. The schedule displays as shown Figure 2–135.

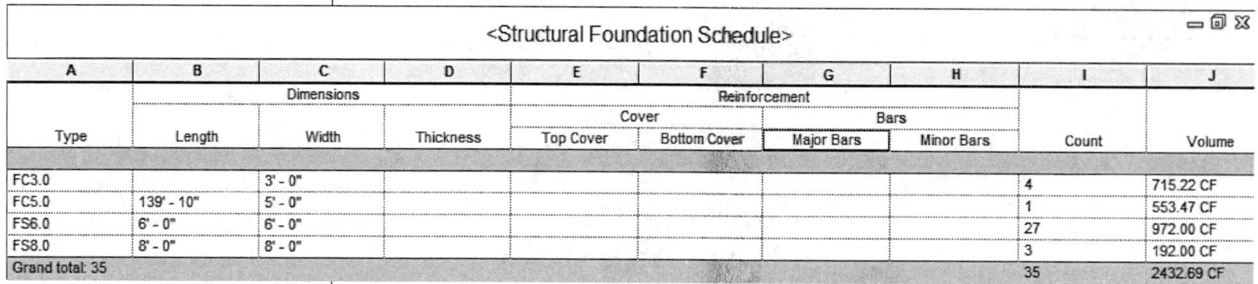

<Structural Foundation Schedule>

A	B	C	D	E	F	G	H	I	J
	Dimensions			Reinforcement					
				Cover		Bars			
Type	Length	Width	Thickness	Top Cover	Bottom Cover	Major Bars	Minor Bars	Count	Volume
FC3.0		3' - 0"						4	715.22 CF
FC5.0	139' - 10"	5' - 0"						1	553.47 CF
FS6.0	6' - 0"	6' - 0"						27	972.00 CF
FS8.0	8' - 0"	8' - 0"						3	192.00 CF
Grand total: 35								35	2432.69 CF

Figure 2–135

9. Set the *Thickness* to **1'-0"** for each of the types.

10. Open the **Structural Plans: T.O.F.** view and select one of the spot footings.

11. In Properties, click (Edit Type). The **FS and FC Foundation Thickness** parameter has been added and set to **1'-0"**. Fill out the rest of the information for the *Cover* and *Bars* (as shown in Figure 2–136) and close the Type Properties dialog box.

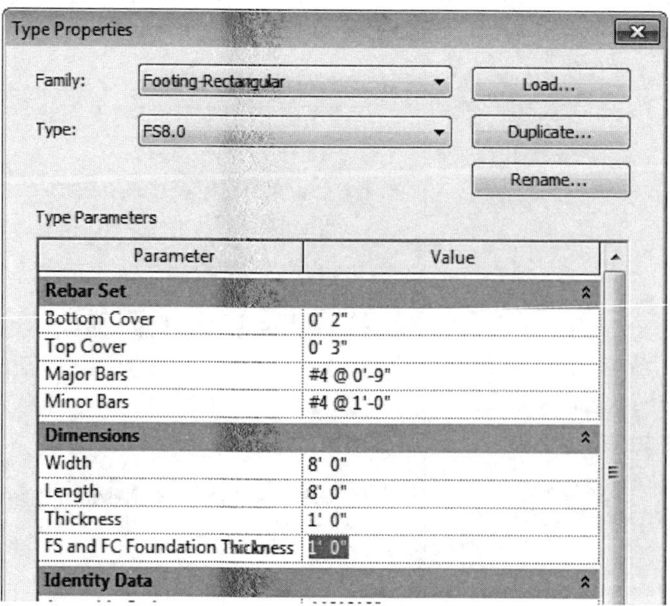

Figure 2–136

12. In the Type Properties of each foundation type or in the schedule view, fill out the rest of the information as shown in the schedule in Figure 2–137.

<Structural Foundation Schedule>

A	B	C	D	E	F	G	H	I	J
	Dimensions			Reinforcement					
				Cover		Bars			
Type	Length	Width	Thickness	Top Cover	Bottom Cover	Major Bars	Minor Bars	Count	Volume
FC3.0		3' - 0"	1' - 0"	0' - 3"	0' - 1 1/2"	#4 @ 1'-0"	#4 @ 1'-0"	4	715.22 CF
FC5.0	139' - 10"	5' - 0"	1' - 0"	0' - 3"	0' - 2"	#4 @ 0'-9"	#4 @ 0'-9"	1	553.47 CF
FS6.0	6' - 0"	6' - 0"	1' - 0"	0' - 3"	0' - 1 1/2"	#4 @ 1'-0"	#4 @ 1'-0"	27	972.00 CF
FS8.0	8' - 0"	8' - 0"	1' - 0"	0' - 3"	0' - 2"	#4 @ 0'-9"	#4 @ 1'-0"	3	192.00 CF
Grand total: 35								35	2432.69 CF

Figure 2–137

13. Save the project.

Chapter Review Questions

1. In the Schedule Properties dialog box (shown in Figure 2–138), in which tab do you define the order of the list of elements?

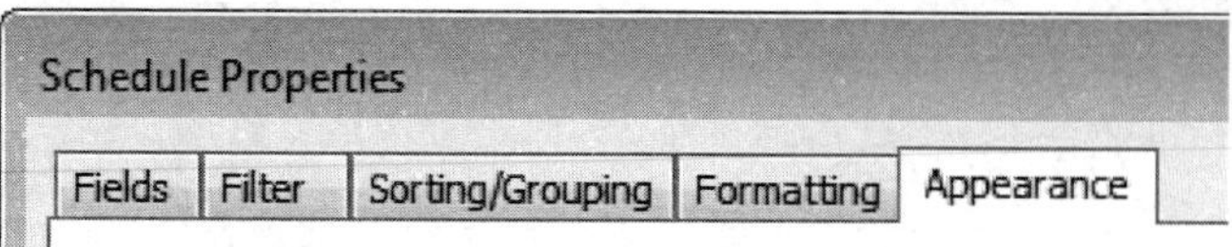

Figure 2–138

   a. *Fields*
   b. *Filter*
   c. *Sorting/Grouping*
   d. *Formatting*
   e. *Appearance*

2. Which type of schedule lists style information that is not automatically included in an element, such as the finish or frame type of a door?
   a. Material Takeoff
   b. Quantities
   c. Key
   d. Building Component

3. When creating *Calculated Value* fields (such as shown in Figure 2–139), the parameters you use must first be added to the list of fields in the schedule.

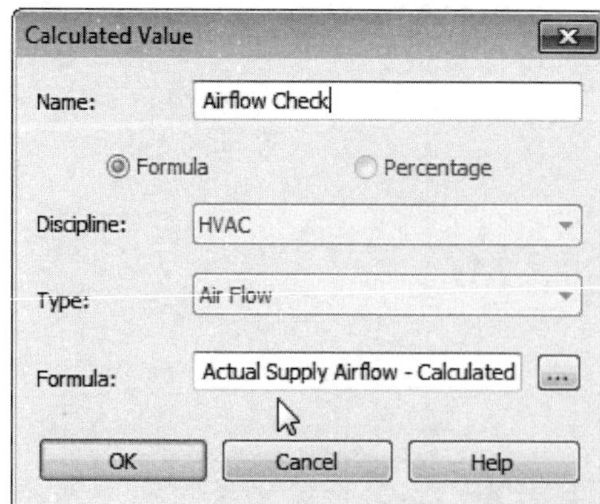

Figure 2–139

   a. True
   b. False

4. When creating a Material takeoff schedule, where do you modify information about the material, such as the cost or manufacturer, as shown in Figure 2–140? (Select all that apply.)

Stone - Material Takeoff			
Material: Description	Material: Area	Material: Cost	Total Cost
Stone	865 SF	0.00	0
Stone: 6	865 SF		0

EIFS - Material Takeoff			
Material: Description	Material: Area	Material: Cost	Total Cost
EIFS	298 SF	12.00	3574
EIFS: 1	298 SF		3574

Figure 2–140

a. In the Type Properties dialog box.

b. In the schedule, fill out the columns related to the type information.

c. In the Material Editor.

d. In the Material Browser.

5. (Autodesk Revit MEP) Which of the following should be created before you can create an embedded schedule, as shown in Figure 2–141?

A	B
SPACE NO.	NAME
DIFFUSER ID	DIFFUSER TYPE
1501	CLASSROOM
29	Supply Diffuser - Perforated - Round Neck - Ceiling Mounted:
30	Supply Diffuser - Perforated - Round Neck - Ceiling Mounted:
31	Supply Diffuser - Perforated - Round Neck - Ceiling Mounted:
32	Supply Diffuser - Perforated - Round Neck - Ceiling Mounted:
1502	CLASSROOM
45	Supply Diffuser - Perforated - Round Neck - Ceiling Mounted:
46	Supply Diffuser - Perforated - Round Neck - Ceiling Mounted:
47	Supply Diffuser - Perforated - Round Neck - Ceiling Mounted:
48	Supply Diffuser - Perforated - Round Neck - Ceiling Mounted:

Figure 2–141

a. Rooms or Spaces.

b. Elements covered in the schedule.

c. The schedule that you want to embed.

d. A key schedule.

Command Summary

Button	Command	Location
Schedule Types		
NA	**Graphical Column Schedule**	• **Ribbon:** *View* tab>Create panel> expand Schedules (*Structural Columns only*)
	Material Takeoff Schedule	• **Ribbon:** *View* tab>Create panel> expand Schedules
	Note Block	• **Ribbon:** *View* tab>Create panel> expand Schedules
	Schedule/ Quantities	• **Ribbon:** *View* tab>Create panel> expand Schedules
	Sheet List	• **Ribbon:** *View* tab>Create panel> expand Schedules
	View List	• **Ribbon:** *View* tab>Create panel> expand Schedules
Schedule Appearance		
	Align Horizontal	• **Ribbon:** *Modify Schedule/Quantities* tab>Appearance panel
	Align Vertical	• **Ribbon:** *Modify Schedule/Quantities* tab>Appearance panel
	Borders	• **Ribbon:** *Modify Schedule/Quantities* tab>Appearance panel
	Font	• **Ribbon:** *Modify Schedule/Quantities* tab>Appearance panel
	Reset	• **Ribbon:** *Modify Schedule/Quantities* tab>Appearance panel
	Shading	• **Ribbon:** *Modify Schedule/Quantities* tab>Appearance panel
	Delete (Columns)	• **Ribbon:** *Modify Schedule/Quantities* tab>Columns panel
	Hide (Columns)	• **Ribbon:** *Modify Schedule/Quantities* tab>Columns panel
	Insert (Columns)	• **Ribbon:** *Modify Schedule/Quantities* tab>Columns panel
	Resize (Columns)	• **Ribbon:** *Modify Schedule/Quantities* tab>Columns panel
	Unhide All (Columns)	• **Ribbon:** *Modify Schedule/Quantities* tab>Columns panel
	Calculated	• **Ribbon:** *Modify Schedule/Quantities* tab>Parameters panel
	Format Unit	• **Ribbon:** *Modify Schedule/Quantities* tab>Parameters panel

	Delete (Rows)	• **Ribbon:** *Modify Schedule/Quantities* tab>Rows panel
	Insert (Rows)	• **Ribbon:** *Modify Schedule/Quantities* tab>Rows panel
	Insert Data Row	• **Ribbon:** *Modify Schedule/Quantities* tab>Rows panel
	Resize (Rows)	• **Ribbon:** *Modify Schedule/Quantities* tab>Rows panel
	Clear Cell	• **Ribbon:** *Modify Schedule/Quantities* tab>Titles & Headers panel
	Group	• **Ribbon:** *Modify Schedule/Quantities* tab>Titles & Headers panel
	Insert Image	• **Ribbon:** *Modify Schedule/Quantities* tab>Titles & Headers panel
	Merge Unmerge	• **Ribbon:** *Modify Schedule/Quantities* tab>Titles & Headers panel
	Ungroup	• **Ribbon:** *Modify Schedule/Quantities* tab>Titles & Headers panel

Custom System Families

System families are embedded within an Autodesk® Revit® project or template file. They are modified by editing the type properties of the family. Walls, roofs, floors, and compound ceiling types each have an assembly of layers that can be modified along with other parameters. MEP system families include duct, pipe, conduit, and cable tray families.

Learning Objectives in this Chapter

- Create system family types for walls, roofs, floors, and ceilings.
- Modify wall types vertically to include split materials, profiles, and reveals.
- Create a vertically stacked wall type from two or more wall types.
- Create duct and pipe types using routing preferences.
- Create cable tray and conduit types using type properties.

3.1 Creating Wall, Roof, Floor, and Ceiling Types

System families are created and modified within a project or template file by duplicating an existing element type. Some of these system families (such as walls, roofs, floors, and some ceilings) are compound or layer-based. For example, to modify a compound wall, you edit the type and select the **Structure** parameter. This opens the Edit Assembly dialog box (as shown in Figure 3–1) which enables you to specify each layer of the assembly.

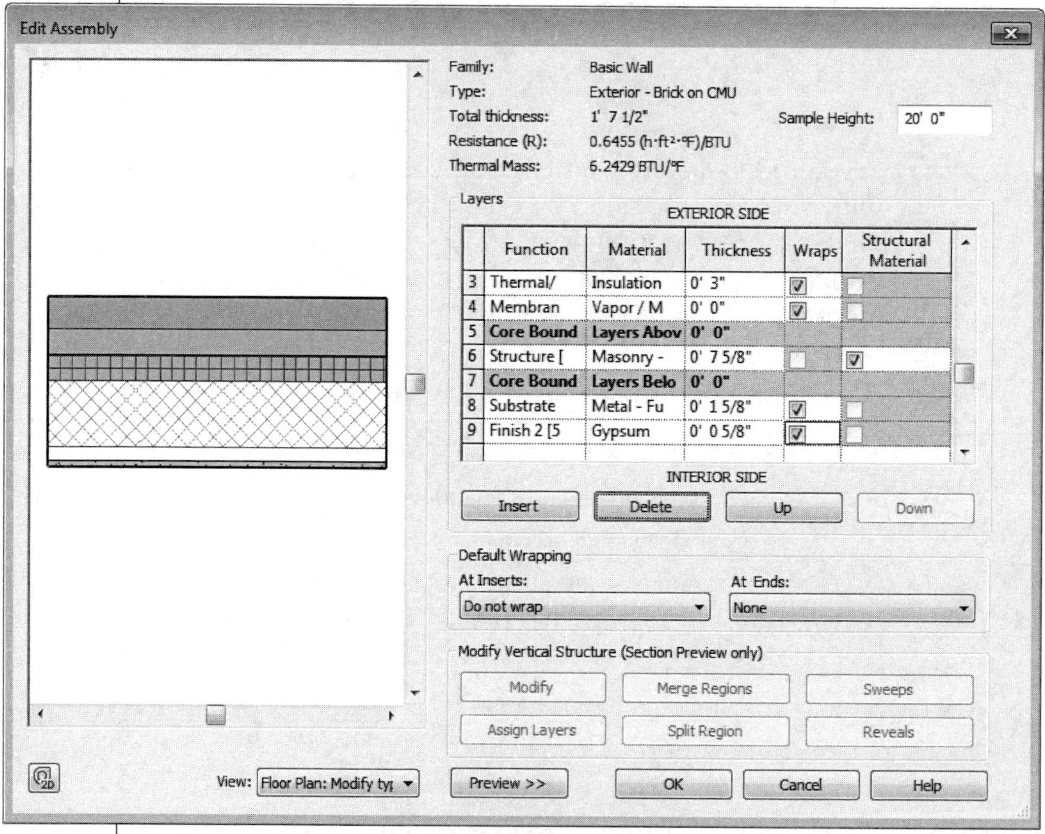

Figure 3–1

- Walls are used as the primary example, but floors, roofs, and compound ceilings follow the same pattern.
- Structural Floors often use profiles for metal decking. Creating this type of floor is covered in the profile families topic.

How To: Create a Compound Wall, Floor, Roof, or Ceiling

1. Start the wall, floor, roof or ceiling command.
2. In Properties, select a type similar to the one you want to create and click (Edit Type) .
3. In the Type Properties dialog box, click **Duplicate...**.
4. In the Name dialog box, enter a name for the new type and click **OK**.
5. Next to the **Structure** parameter, click **Edit...**.
6. In the Edit Assembly dialog box, modify the layers of the assembly as required, and then click **OK**.
7. Modify any Type Parameters in the Type Properties dialog box.
8. Click **OK** to close the dialog box.

Hint: Basic Ceilings

The basic ceiling system family does not include a structure parameter. Instead, modify the Type by specifying a *Material* for the entire thickness of the ceiling.

Editing Wall, Roof, and Floor Assemblies

In the Edit Assembly dialog box, you can define the layers that make up the compound structure, as shown in Figure 3–2.

*To better visualize the wall, click **<< Preview** to open a view of the layers in the structure. You can preview the structure in a plan or section view, and zoom or pan within the preview screen.*

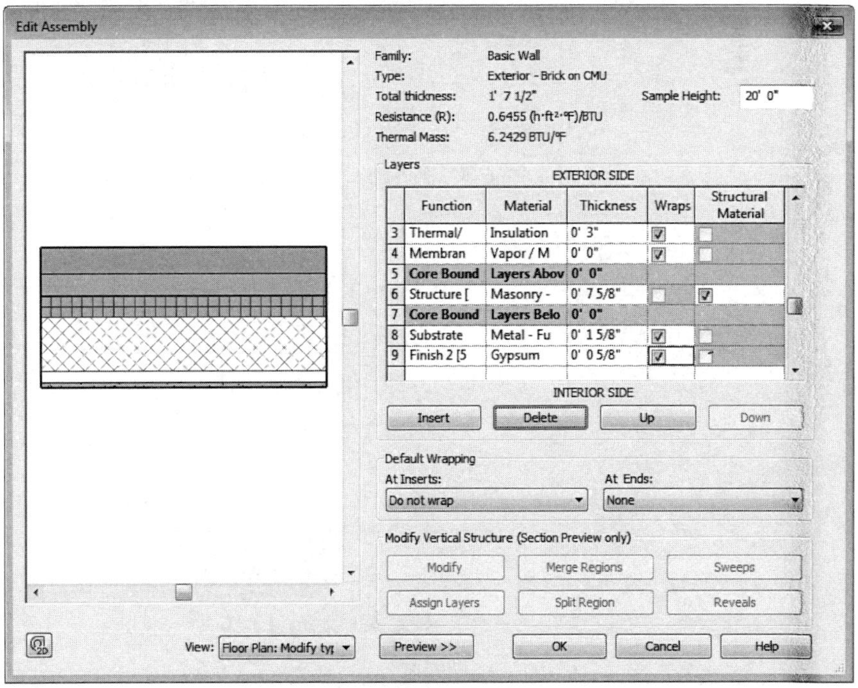

Figure 3–2

Assembly Information

The top of the dialog box lists the *Family* (such as **Basic Wall** or **Floor**), the *Type* that you gave to the new type, and the *Total thickness* (which is the sum of the layers defined in the wall), as shown in Figure 3–3. It also includes *Resistance (R)* and *Thermal Mass* which are automatically calculated from the materials assigned to the layers. You can also set a *Sample Height* for your wall design.

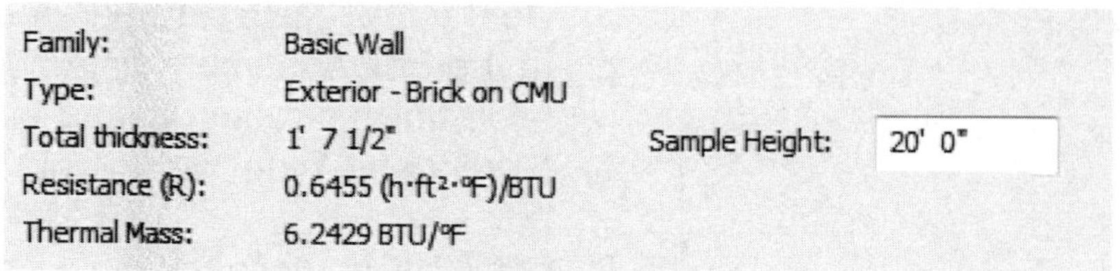

Figure 3–3

Layers

When you specify the layers for the compound element, you assign them a *Function*, *Material*, and *Thickness*, as shown in Figure 3–4.

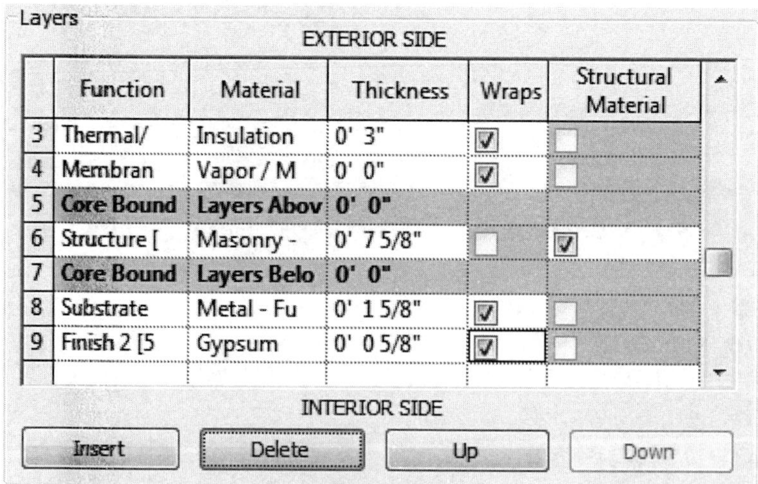

Layers
EXTERIOR SIDE
Function | Material | Thickness | Wraps | Structural Material
3 Thermal/ | Insulation | 0' 3"
4 Membran | Vapor / M | 0' 0"
5 Core Bound | Layers Abov | 0' 0"
6 Structure [| Masonry - | 0' 7 5/8"
7 Core Bound | Layers Belo | 0' 0"
8 Substrate | Metal - Fu | 0' 1 5/8"
9 Finish 2 [5 | Gypsum | 0' 0 5/8"
INTERIOR SIDE
Insert
Delete
Up
Down

Figure 3–4

- Use the buttons to insert additional layers and to rearrange them in the layer list. You can also delete layers from the list.
- The *Core Boundary* function defines the layers above and below the wrapping; a heavier line is displayed when a plan or section view is cut.

- Editing a wall assembly works from the exterior side at the top of the list to the interior side at the bottom. For floors and roofs, you work around the layers above and below the wrap of the *Core Boundary*.

Options

Function	Select from a set list of functions in the drop-down list with a priority of highest (1) to lowest (5). High priority layers connect before low priority layers.
Structure [1]	The structural support for the wall, floor, or roof.
Substrate [2]	A material that acts as a foundation for another material, such as plywood or gypsum board.
Thermal/ Air Layer [3]	An open layer for air space.
Finish 1 [4]	The exterior finish layer, such as brick for a wall.
Finish 2 [5]	The interior finish layer, such as drywall for a wall.
Membrane Layer	A vapor barrier. Typically, this is set to a zero thickness. Therefore, it does not have a priority code.
Structural Deck (1)	(Floors only) A structural support based on a Deck Profile. You can also specify the Deck Usage with a Bound Layer Above or a Standalone Deck.
Material	Select from a list of available materials. Layers clean up if they share the same material. If they do not, a line displays at the join.
Thickness	Set the thickness of the particular layer.

Wall Only Options

Sample Height	Displays the height of a wall in section when you are creating it. It does not impact the height of the wall in the project.
Default Wrapping	Set up how the heavy line style wraps around openings in walls: *at Inserts* (**Do not wrap**, **Interior**, **Exterior**, or **Both**), and *at Ends* (**None**, **Exterior**, or **Interior**). Wrapping is only visible in plan view.
Wraps	Set up individual layers to wrap—select the Wraps option at the end of each layer.

- Wall wrapping can be set in the assembly or in the Type Properties, as shown in Figure 3–5.

Wrapping at Inserts	Do not wrap
Wrapping at Ends	None

Figure 3–5

- Roofs, floors, and structural slabs have an additional parameter that relates to sloping for drains. When *Variable* is not selected, the slab is set to a constant thickness and the entire element slopes, as shown on the top in Figure 3–6. When *Variable* is selected, only the top layer slopes, as shown on the bottom in Figure 3–6.

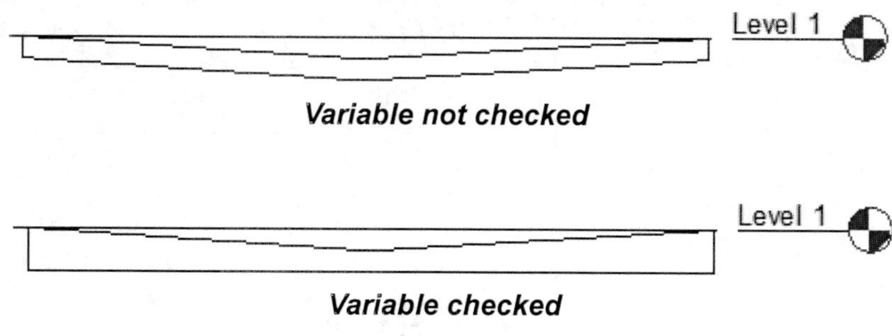

Variable not checked

Variable checked

Figure 3–6

Practice 3a Create Compound Wall Types

Practice Objectives

- Create an exterior wall with two different types of layers on the interior and the exterior.
- Create an interior wall that has the same material on either side of the structural core.

Estimated time for completion: 15 minutes

In this practice you will create interior and exterior wall types, and then draw walls using them, as shown in Figure 3–7.

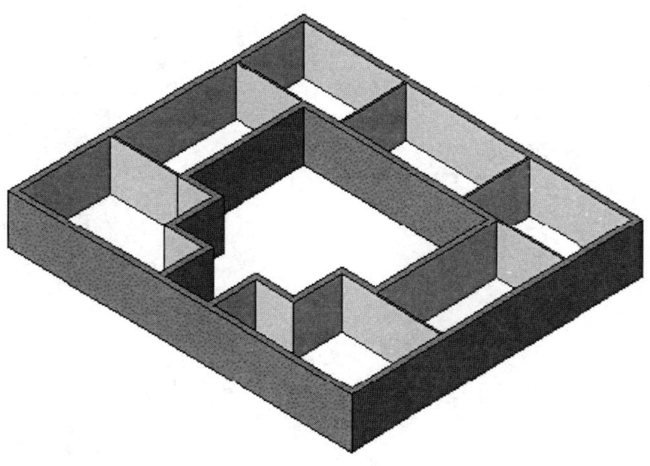

Figure 3–7

Task 1 - Create a Wall type.

1. Start a new project based on the default architectural template and save it as **SW House.rvt**.

For creating a wall type, it does not matter which command you started with.

2. In the *Architecture* tab>Build panel or *Structure* tab>Structure panel, click (Wall).
3. In Properties, click (Edit Type).
4. In the Type Properties dialog box, click **Duplicate...**.
5. In the Name dialog box, type **Exterior – Adobe** and click **OK**.
6. Next to the **Structure** parameter, click **Edit...**.
7. Next to the **Structure [1]** layer, select the *Material* column and click (Browse).

8. Expand the Library panel to display it more clearly and select **AEC Materials>Masonry>Brick, Adobe** as shown in Figure 3–8.

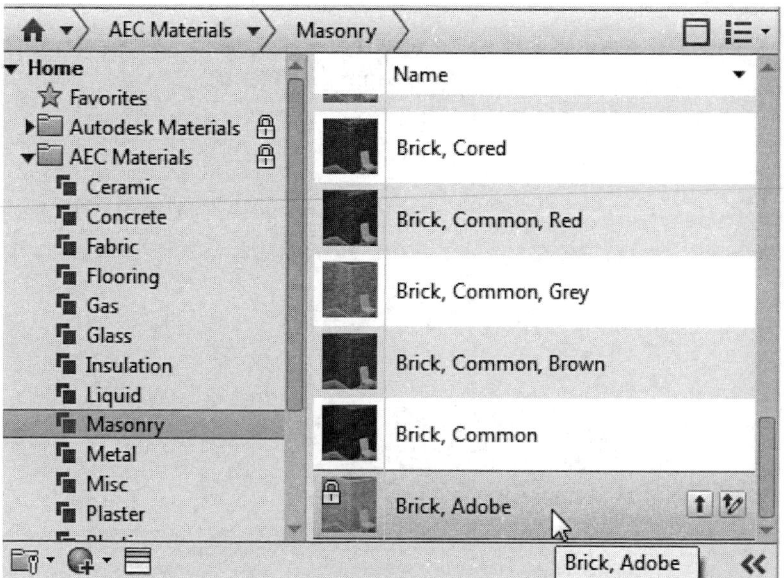

Figure 3–8

9. Double-click on **Brick, Adobe** to add it to the list of materials that are available in the project. **Brick, Adobe** is now selected in the Material Browser. Click **OK** to apply the material.

10. In the Edit Assembly dialog box, in the *Thickness* column, type **1'-0"** and verify that **Structural Material** is selected, as shown in Figure 3–9.

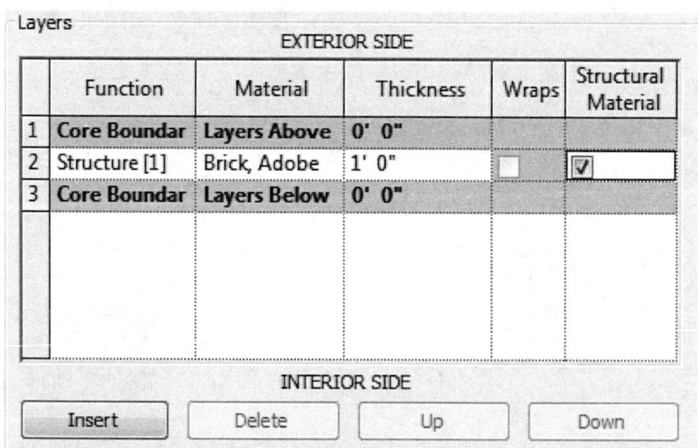

Figure 3–9

11. In the *Layers* area, click **Insert**.

12. Set this layer with *Function* set to **Substrate (2)**, *Material* to **Plywood, Sheathing**, *Thickness* to **1 5/8"**. Click **Down** until the new layer is below the Layers Below Wrap, as shown in Figure 3–10.

*In the Material Browser, type **wood** in the search field to narrow the list of materials.*

13. Add one more layer on the interior side and set *Function* to **Finish 2**, *Material* to **Cherry**, *Thickness* to **5/8"**. Select **Wraps** but not **Structural Material**, as shown in Figure 3–10.

Layers

EXTERIOR SIDE

	Function	Material	Thickness	Wraps	Structural Material
1	**Core Boundar**	**Layers Above**	**0' 0"**		
2	Structure [1]	Brick, Adobe	1' 0"	☐	☑
3	**Core Boundar**	**Layers Below**	**0' 0"**		
4	Substrate [2	Plywood, Sh	0' 1 5/8"	☑	☐
5	Finish 2 [5]	Cherry	0' 0 5/8"	☑	☐

INTERIOR SIDE

Figure 3–10

14. Click **OK** to close the Edit Assembly dialog box.

15. Repeat the process to create another new wall type named **Interior – Wood Panel**. In the Type Properties dialog box, change the *Function* to **Interior**.

16. Set up the layer structure as shown in Figure 3–11.

*The **Softwood, Lumber** material is found in the AEC Materials list if it is not in the document list.*

Layers

EXTERIOR SIDE

	Function	Material	Thickness	Wraps	Structural Material
1	Finish 2 [5]	Cherry	0' 0 5/8"	☑	☐
2	**Core Boundar**	**Layers Above**	**0' 0"**		
3	Structure [1]	Softwood, L	0' 3 5/8"	☐	☑
4	**Core Boundar**	**Layers Below**	**0' 0"**		
5	Finish 2 [5]	Cherry	0' 0 5/8"	☑	☐

INTERIOR SIDE

Figure 3–11

17. Click **OK** twice to close the dialog boxes.

Task 2 - Draw walls using the new Wall types.

*For the Interior walls, set the Location Line to **Finish Face – Interior** or **Finish Face – Exterior** to place the horizontal walls and **Wall Centerline** for the vertical walls.*

1. Set the *Detail Level* to (Medium).
2. Draw the house shown in Figure 3–12.
    - **Exterior: Exterior – Adobe** and *Height* of **10'-0"**
    - **Interior: Interior – Wood Panel** and *Height* of **10'-0"**

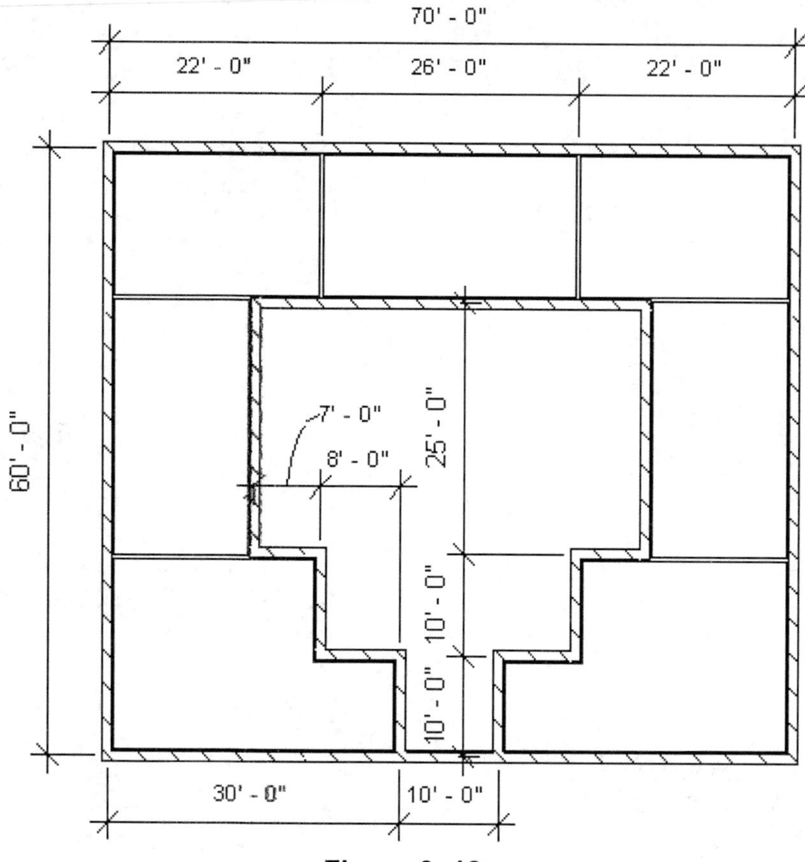

Figure 3–12

3. Save the project.

3.2 Vertically Compound Walls

Vertically Compound walls are made of regions of different materials, as well as optional permanent sweeps or reveals, as shown in Figure 3–13. Several options help you create these walls: **Modify**, **Split Region**, **Merge Regions**, **Assign Layers**, **Wall Sweeps**, and **Reveals**.

Figure 3–13

You must be have the *View:* set to **Section: Modify Type Attributes** to work with the **Modify Vertical Structure** tools, as shown in Figure 3–14.

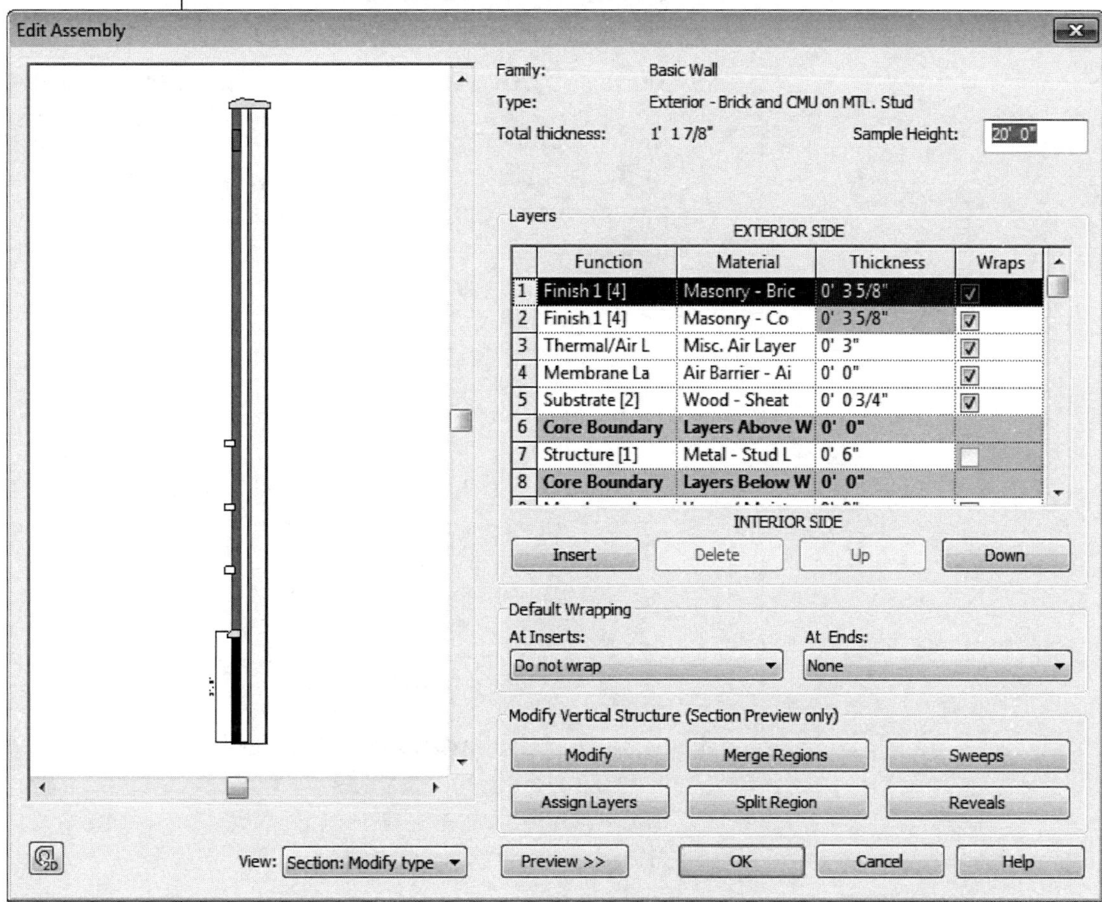

Figure 3–14

How To: Modify the Vertical Structure of a Wall Type

1. Open the wall type you want to modify and edit the structure.
2. Create any additional layers that might be needed.
3. In the Edit Assembly dialog box, open the preview and set the *View:* to **Section: Modify Type Attributes**. This activates the *Modify Vertical Structure* tools, as shown in Figure 3–15.

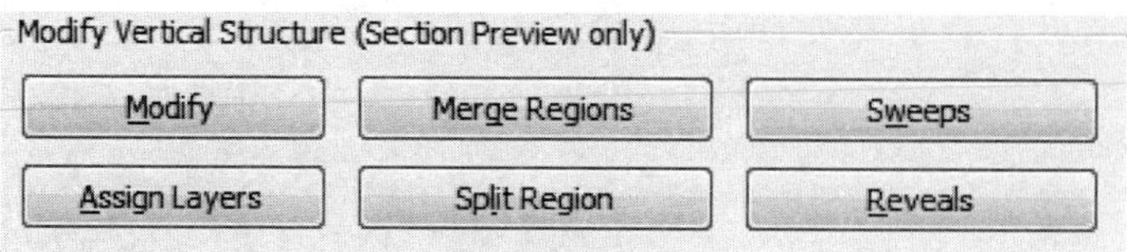

Figure 3–15

4. Zoom in as needed to see the various layers.
5. In the Layers list, select the layer you want to work with.
6. Use the various tools that are outlined below.
7. Click **OK** until all of the dialog boxes are closed to save the wall type.

- When making changes to the vertical structure of a wall type it helps to have the *Sample Height* set to the shortest expected height for the wall type so the changes are displayed more clearly.

How To: Split Regions and Assign Layers

1. In the Edit Assembly dialog box, click **Split Region**.
2. In the section preview, move your cursor up the wall to the place where you want the region to be cut. Select the wall at that point and a dimension is displayed as shown in Figure 3–16

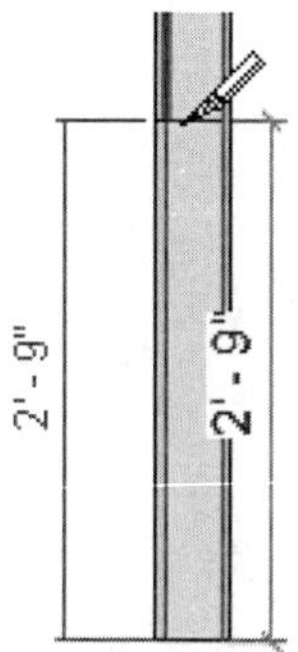

Figure 3–16

3. In the *Layers* area, highlight the layer that you want to use and click **Assign Layers**.
4. In the section view, select the wall part that you want to change and click **Modify**.

- You can create as many split regions as needed within one layer.

- To change the dimension, click **Modify** and select the border between the two regions that have been split. Then type a new distance using the temporary dimension.

- You can modify the width of the layer in the section view or in the *Thickness* column in the *Layers* area.

- You can add additional functions as needed. In the example shown in Figure 3–17, two layers have the function **Finish 1**, and each layer has a different material. The *Thickness* is set by the selected layer and cannot be modified.

Do not select more than you need. There is no ***Undo*** *option in this dialog box.*

Layers

EXTERIOR SIDE

	Function	Material	Thickness	Wraps
1	Finish 1 [4]	Masonry - Bric	0' 3 5/8"	☑
2	Finish 1 [4]	Masonry - Co	0' 3 5/8"	☑
3	Thermal/Air L	Misc. Air Layer	0' 3"	☑
4	Membrane La	Air Barrier - Ai	0' 0"	☑
5	Substrate [2]	Wood - Sheat	0' 0 3/4"	☑

Figure 3–17

Merging Regions

To link regions together, click **Merge Regions** and select the line between the layers you want to merge, as shown in Figure 3–18. Watch the arrow cursor. It tells you which way the merge takes place. The tool tip also tells you names of the layers that are being merged. You can only merge layers that are next to each other.

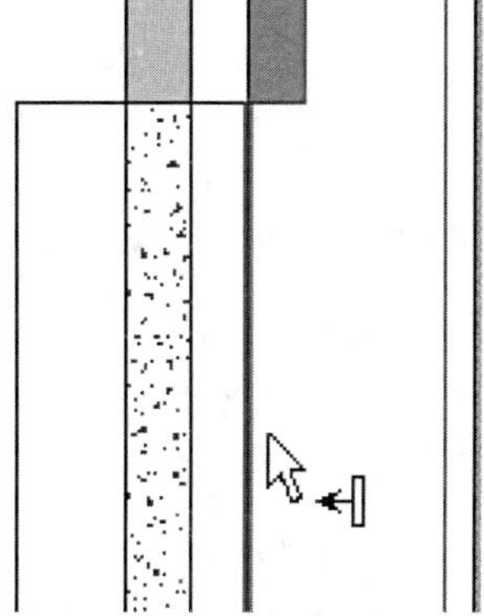

Figure 3–18

- You can merge layers vertically or horizontally. For example, to create the situation shown in Figure 3–18, you need to split the regions horizontally before merging some of the vertical lines.

Wall Sweeps and Reveals

When you split regions, the parts you create cannot have different widths. To create a protrusion or a reveal, add a wall sweep, as shown in Figure 3–19.

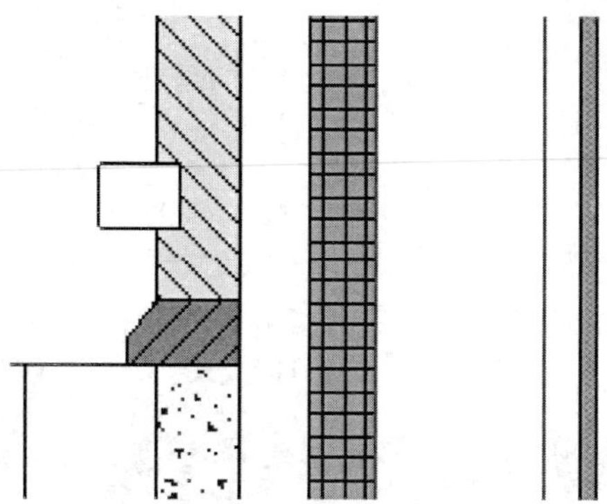

Figure 3–19

How To: Add a Sweep

1. In the Edit Assembly dialog box, click **Sweeps** to open the Wall Sweeps dialog box.
2. In the Wall Sweeps dialog box, click **Add**. A default row is added as shown in Figure 3–20.

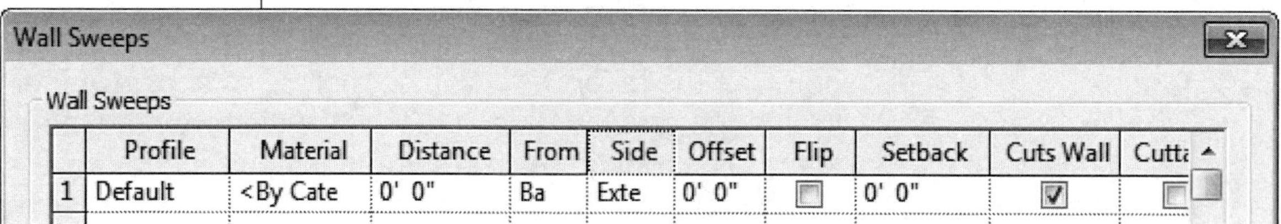

Figure 3–20

3. Select the *Profile* column and select from the list of the existing profiles in the project.

4. Continue setting up the profile by selecting a *Material* and setting the *Distance* from the top or bottom, interior or exterior *Side*, and the *Offset* from that side. If needed, place a check mark in the *Flip* column to flip the profile, as shown in Figure 3–21.

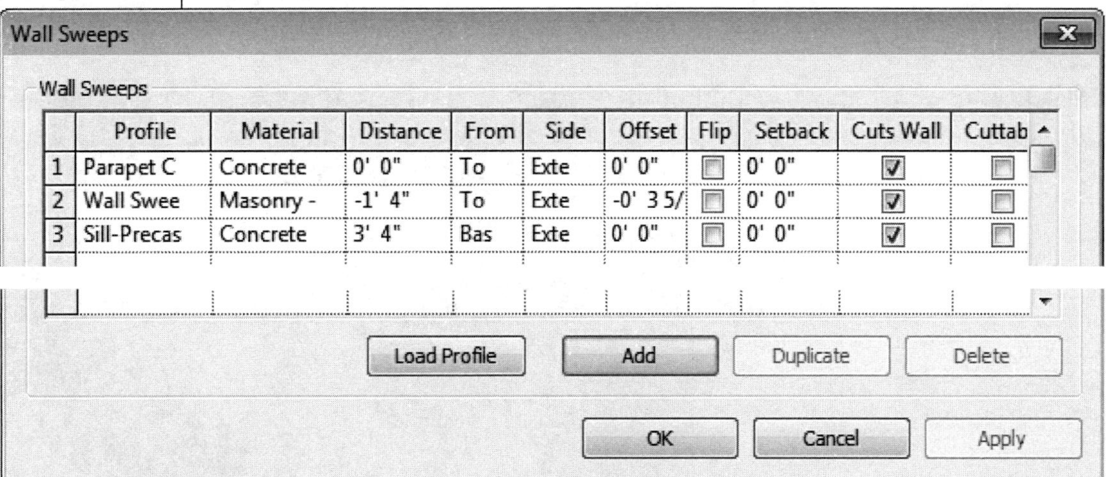

Figure 3–21

5. Click **Apply** to see the addition before you click **OK** to finish.

- Reveals work the same way, except that you do not assign a material to a reveal. The whole shape of the reveal profile is visible in the section view, but you only see the cut in the project.

- You can also click **Load Profile** to add a profile within the Wall Sweeps or Reveals dialog box.

Practice 3b Create Vertically Compound Walls

Practice Objective

- Create a vertically compound wall.

Estimated time for completion: 15 minutes

In this practice you will modify a wall type to create a wooden wainscot (**Split Region** and **Assign Layers**) below a chair rail (**Sweep**), as shown in Figure 3–22.

Figure 3–22

Task 1 - Split regions and assign layers in a wall type.

1. In the *...\Architectural* folder, open **SW-House-Vertical.rvt.**
2. Select one of the **Interior – Wood Panel** walls and in Properties, click (Edit Type).
3. Edit the *Structure* parameter.
4. In the Edit Assembly dialog box, change the *Sample Height* to **9'-0"**. This sets the correct wall height for the preview.
5. Click **<< Preview** and set the *View:* to **Section: Modify type attributes**.

*The material **Plaster** is located in the AEC Material area of the Material Browser. Add it to the document so that you can select it.*

6. In the *Layers* area, insert an additional finish on both sides of the structure. Set the material to **Plaster**. Do not specify a thickness. Once you apply it to part of the split region, it takes on the appropriate thickness.

7. In the *Modify Vertical Structure* area, click **Split Region**.

8. In the Section view, move the cursor to a point **3'-6"** above the base, as shown in Figure 3–23. Zoom in to make sure you split the finish layer and not the central structural layer.

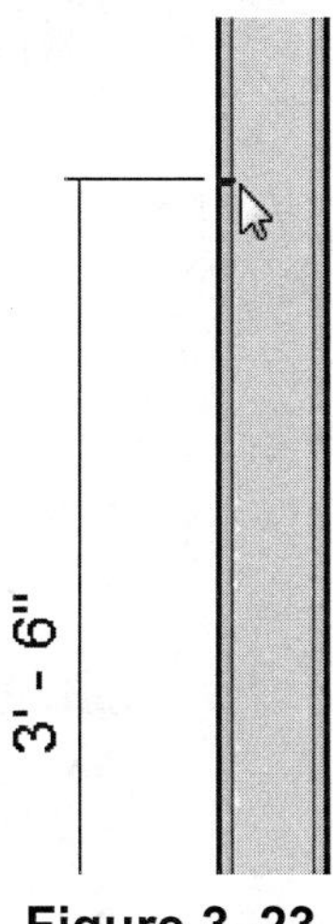

Figure 3–23

9. Repeat with the opposite finish layer.

10. Stay zoomed in, close to the place where you made the split.

11. In the *Layers* area, select the **Plaster** layer on the Exterior Side.

12. In the *Modify Vertical Structure* area, click **Assign Layers**.

13. In the preview, select the top part of the exterior wall where you want the finish applied. Click **Modify**. The color should change as shown in Figure 3–24.

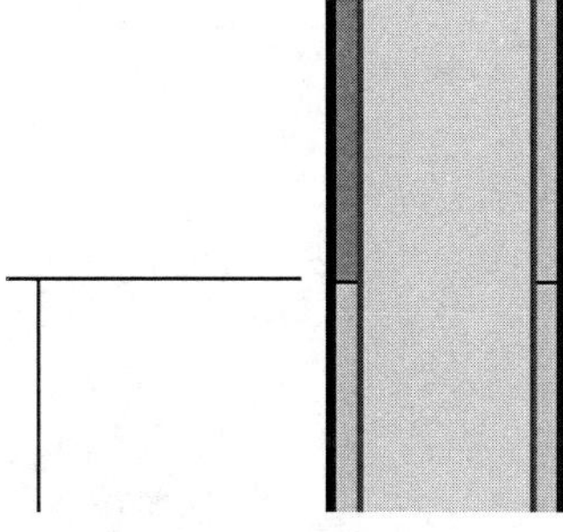

Figure 3–24

14. Repeat with the other finish and the side of the wall.

15. Click **Modify** to finish the process. The list of layers displays, as shown in Figure 3–25.

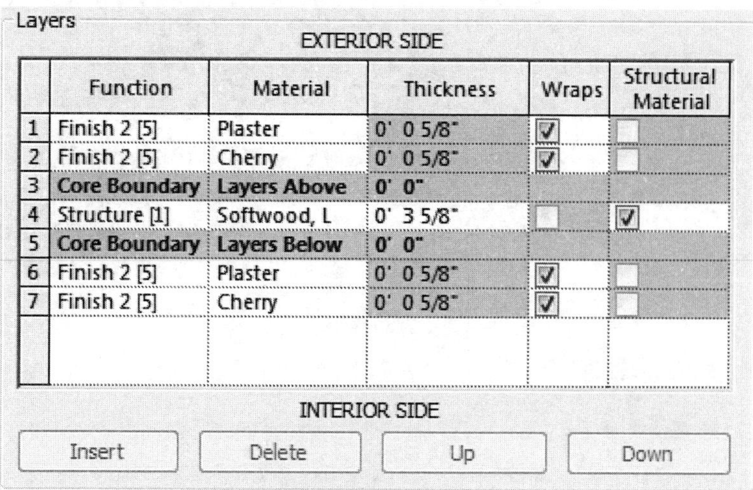

	Function	Material	Thickness	Wraps	Structural Material
1	Finish 2 [5]	Plaster	0' 0 5/8"	☑	☐
2	Finish 2 [5]	Cherry	0' 0 5/8"	☑	☐
3	Core Boundary	Layers Above	0' 0"		
4	Structure [1]	Softwood, L	0' 3 5/8"	☐	☑
5	Core Boundary	Layers Below	0' 0"		
6	Finish 2 [5]	Plaster	0' 0 5/8"	☑	☐
7	Finish 2 [5]	Cherry	0' 0 5/8"	☑	☐

Figure 3–25

16. Click **OK** twice to close the dialog boxes.

17. Look at the interior walls in a 3D or section view. Shade the view. You should see the difference above and below the paneling, as shown in Figure 3–26.

Figure 3–26

Task 2 - Add sweeps to the wall type.

1. Edit the structure of the same wall type.

2. In the Edit Assembly dialog box, *Modify Vertical Structure* area, click **Sweeps**.

3. In the Wall Sweeps dialog box, click **Load Profile**.

4. In Load Family dialog box, navigate to the *Profiles>Finish Carpentry* folder, select **Base 2.rfa** and click **Open**.

5. In the Wall Sweeps dialog box, click **Add** and set up two wall sweeps using the following information, as shown in Figure 3–27.

    - **Profile:** Base 2: 3 1/2" x 3/4" (or similar)
    - **Material:** Cherry
    - **Distance:** 3'-6"
    - **From:** Base
    - **Side: Interior** for one sweep and **Exterior** for the other

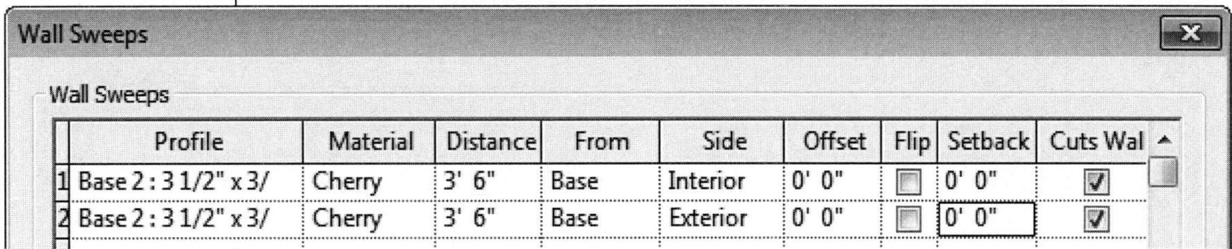

Wall Sweeps
Wall Sweeps
Profile
Material
Distance
From
Side
Offset
Flip
Setback
Cuts Wal
1 Base 2 : 3 1/2" x 3/
Cherry
3' 6"
Base
Interior
0' 0"
0' 0"
2 Base 2 : 3 1/2" x 3/
Cherry
3' 6"
Base
Exterior
0' 0"
0' 0"

Figure 3–27

6. Click **OK** to exit the Wall Sweeps dialog box.

7. In the preview of the Edit Assembly dialog box, you should see the two sweeps when you zoom in, as shown in Figure 3–28.

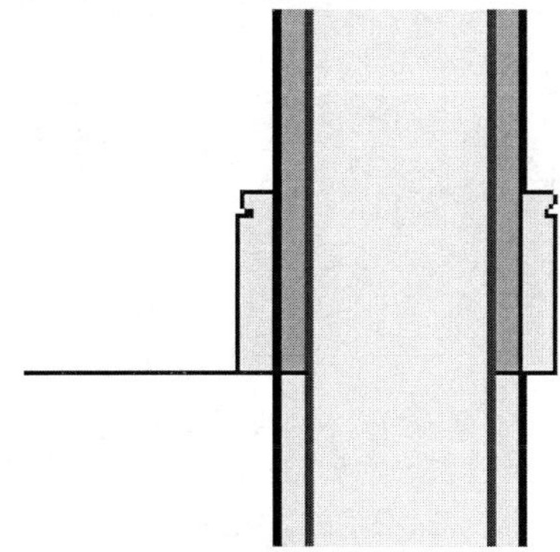

Figure 3–28

8. Click **OK** twice to close the dialog boxes. The entire set of interior walls updates with the new information.

9. Add an additional interior wall using this type so that you can see how it automatically uses the information.

10. If time permits, modify the **Exterior-Adobe** wall so that the inside of that wall matches the interior walls.

11. Save the project.

3.3 Stacked and Embedded Walls

A vertically stacked wall is a specific system family that takes two or more existing basic walls and stacks them on top of each other at specific heights, as shown in Figure 3–29. One wall must be variable in height. The basic wall types have to be in place before you create the stacked wall. These walls are created by copying and editing an existing Vertically Stacked Wall type.

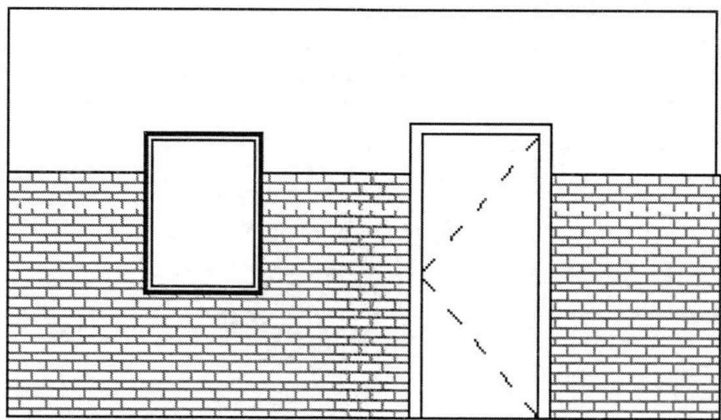

Figure 3–29

How To: Create a Vertically Stacked Wall

1. Start the **Wall** command.
2. In Properties, select an existing stacked wall type and click (Edit Type).
3. Duplicate the wall type and give it a new name.
4. In the Type Properties dialog box, next to the **Structure** parameter, click **Edit...**.
5. In the Edit Assembly dialog box, set the *Offset* for how the walls are stacked, and a *Sample Height* for the preview, as shown in Figure 3–30.

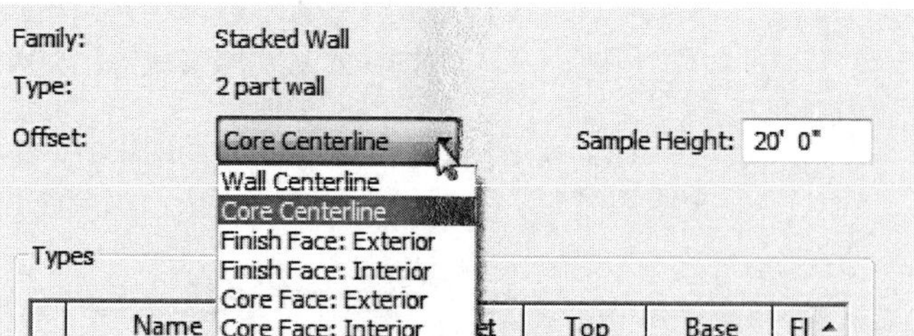

Figure 3–30

6. In the *Name* column, select the basic wall types you want to add to the stacked wall, as shown in Figure 3–31.

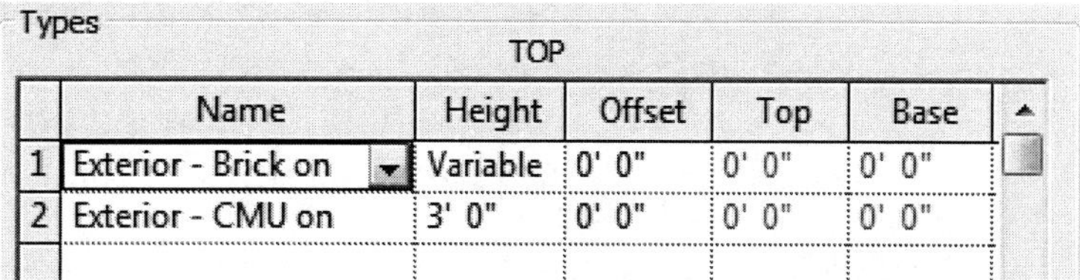

Figure 3–31

7. For each wall type, set the appropriate height and location (Up or Down) within the list. One height must be variable. Set the *Offset* of the wall as needed.
8. Click **<< Preview** to see the wall, as shown in Figure 3–32.

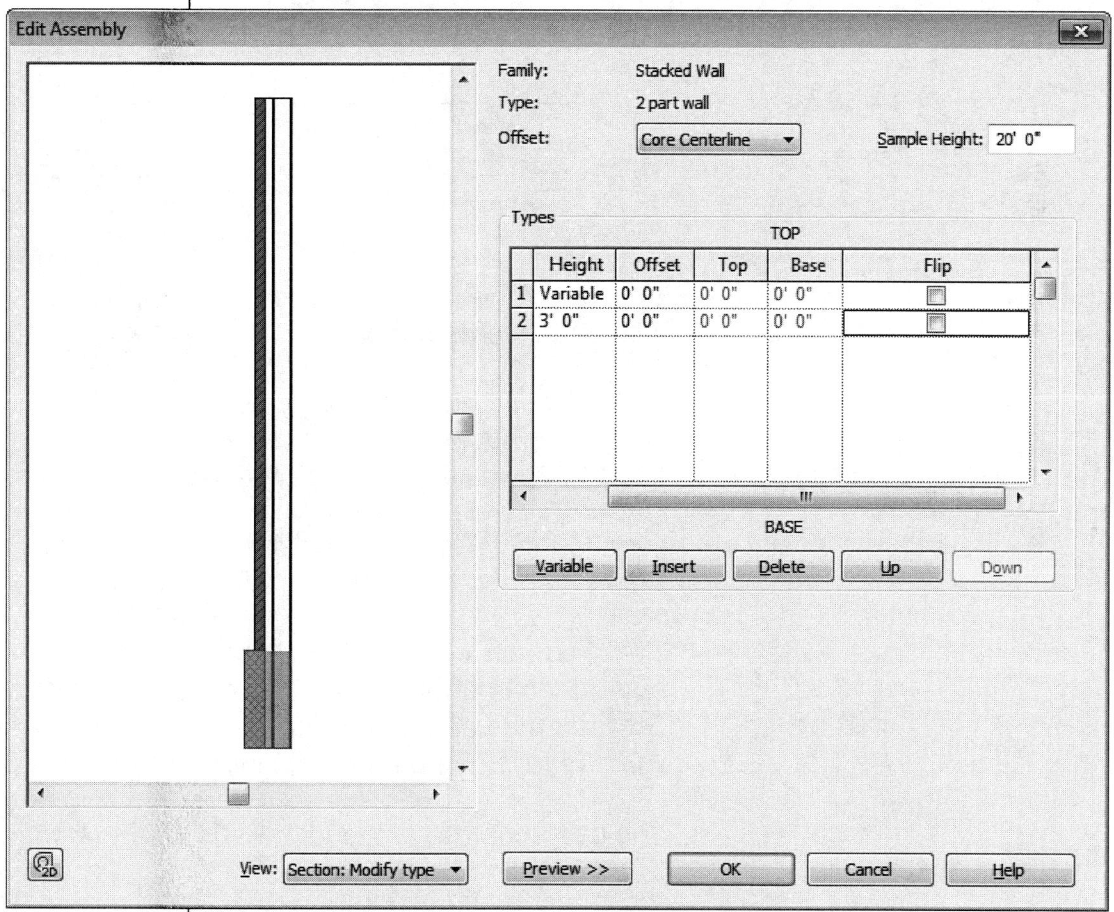

Figure 3–32

9. Click **OK** until all of the dialog boxes are closed.

Hint: Embedding a Wall Inside Another Wall

Another way of creating a compound wall is to embed one wall inside another wall, as shown in Figure 3–33. When you have drawn a host wall, draw another wall on top of, or parallel to it near the host. A warning box opens and recommends that you use **Cut Geometry** to embed the wall in the host wall.

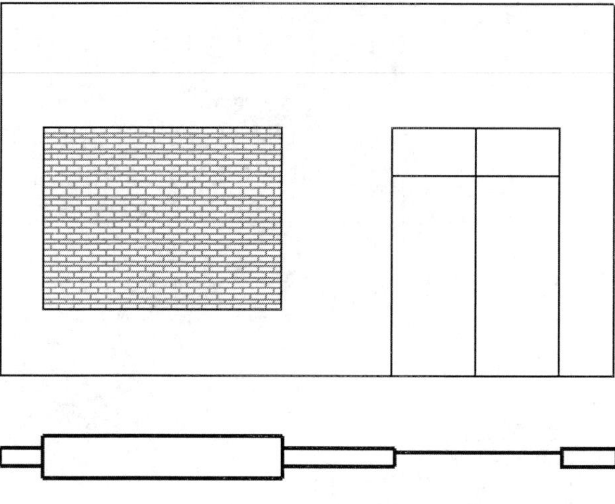

Figure 3–33

1. Add the embedded wall to the host wall.
2. Close the warning box.
3. In the *Modify* tab>Geometry panel, click (Cut).
4. Select the host wall.
5. Select the wall that cuts the host wall.

- Embedded walls work similar to windows. You can modify the embedded wall with controls.
- Some curtain wall types are created to be automatically embedded in another wall. The type **Curtain Wall: Storefront** is an example. The Type Parameter *Automatically Embed* is available for all curtain wall types.
- You can separate embedded walls. In the *Modify* tab>Geometry panel, expand (Cut) and click (Uncut Geometry).

Practice 3c Create Stacked and Embedded Walls

Estimated time for completion: 10 minutes

Practice Objectives

- Create a vertically stacked wall type.
- Embed a wall into a stacked wall.

In this practice you will create a vertically stacked wall and use it in a project. You will also embed a curtain wall and another wall type into a host wall, as shown in Figure 3–34.

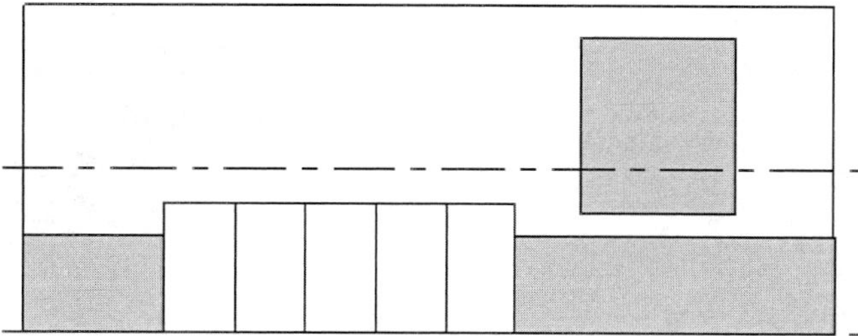

Figure 3–34

Task 1 - Create a stacked wall.

1. Start a new project based on the default architectural or structural template and save it as **Warehouse.rvt**.
2. Start the **Wall** command and select the **Stacked Wall: Exterior – Brick Over CMU w Metal Stud** type.
3. Edit the type and duplicate it to create a new wall type named **Exterior – EIFS over Brick/CMU**.
4. Edit the structure of the new wall.
5. For the top wall, select **Exterior – EIFS on Mtl. Stud**. Leave the *Height* as **Variable** and set the *Offset* to **4”**.
6. For the bottom wall, select **Exterior – Brick on CMU** and set the *Height* to **6’-0”**.
7. Click **OK** to close all of the open dialog boxes.
8. Draw a rectangular building **50’-0” x 30’-0”** using the new wall style.
9. Display the walls in 3D to verify that the Brick/EIFS are displayed on the exterior.
10. Save the project.

Task 2 - Create an embedded wall.

1. In the *Floor Plans* area, open the **Level 1** view.
2. On the south face of the building, add a wall using **Curtain Wall: Storefront** at an *Unconnected Height* of **8'-0".** Place it directly on the center line of the existing wall along only a portion of the wall. It automatically cuts the existing wall.
3. In the *Floor Plans* area, open the **Level 2** view.
4. Add another wall on the same face using **Basic Wall: Exterior – Brick on Mtl. Stud**. This time a warning box opens.
5. Close the warning box.
6. In the *Modify|Place Wall* tab>Geometry panel, click (Cut).
7. Select the host wall.
8. Select the wall that cuts the host wall.
9. In the *Elevations* area, open the **South** view.
10. Flip the orientation of the brick insert as needed.
11. Change the size of the embedded wall using the controls, but do not move it down into the lower brick wall.
12. Save the project.

3.4 Creating MEP System Families

Autodesk Revit MEP comes loaded with a variety of MEP system families for Ducts, Pipes, Cable Trays and Conduits. You can also duplicate and modify these families to create new family types. Ducts (as shown in Figure 3–35) and Pipes are set using Routing Preferences. Flex Ducts and Flex Pipes as well as Cable Tray and Conduit are set up by defining fittings in the Type Properties.

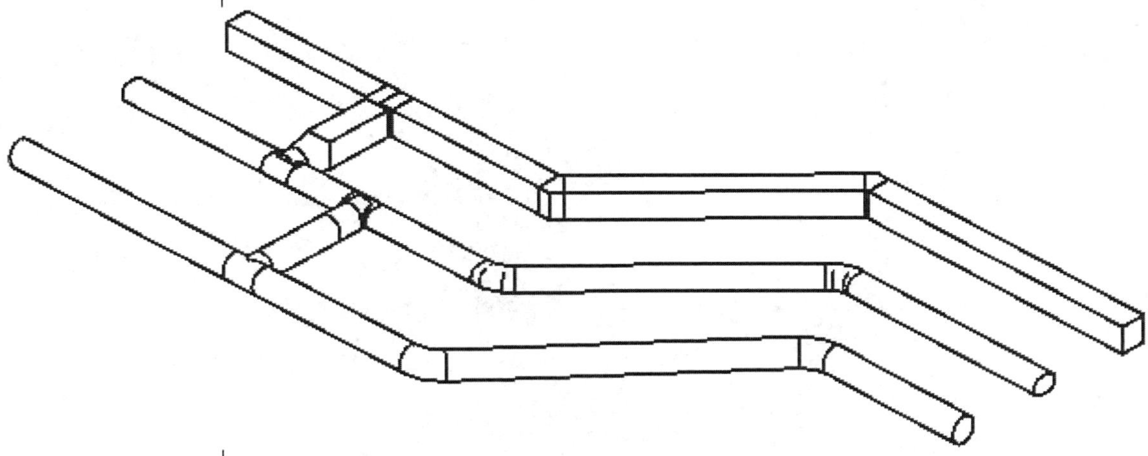

Figure 3–35

How To: Create Duct Types

*You can also right-click on a type name in the Project Browser and select **Duplicate** before opening the Type Properties.*

1. Start the **Place Duct** command, select one of the existing types, and click (Edit Type) or, in the Project Browser, expand **Families>Ducts>Oval, Rectangular** or **Round Duct** and then double-click on one of the existing types (as shown in Figure 3–36) to open the Type Properties dialog box.

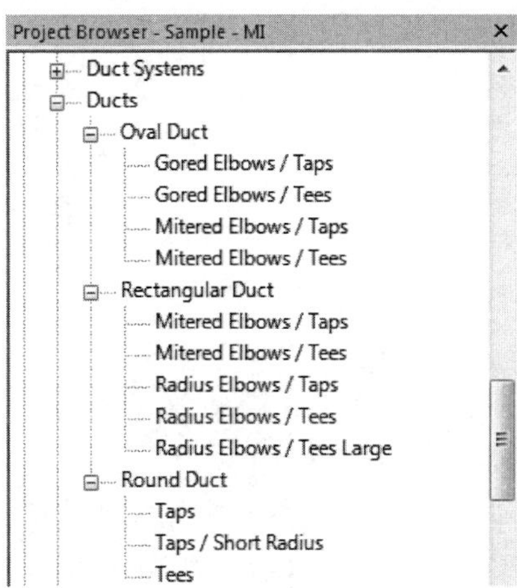

Figure 3–36

2. In the Type Properties dialog box, click **Duplicate...**, type a new name in the Name dialog box, and click **OK**.
3. Next to the **Routing Preferences** parameter, click **Edit...**.
4. The Routing Preferences dialog box, displays as shown in Figure 3–37.

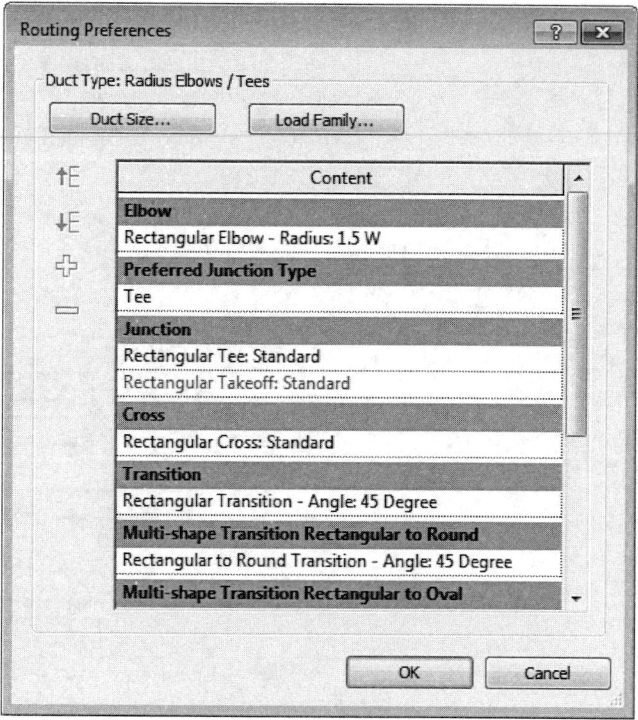

Figure 3–37

5. Click **Duct Size...** to open the Mechanical Settings dialog box to the Duct Settings so that you can add or remove the typical sizes used in the project.
6. Click **Load Family...** to add any duct fittings that you want to use in the type that are not already loaded in the project.
7. For each of the fittings, expand the drop-down list and select the size that you want use, as shown in Figure 3–38.

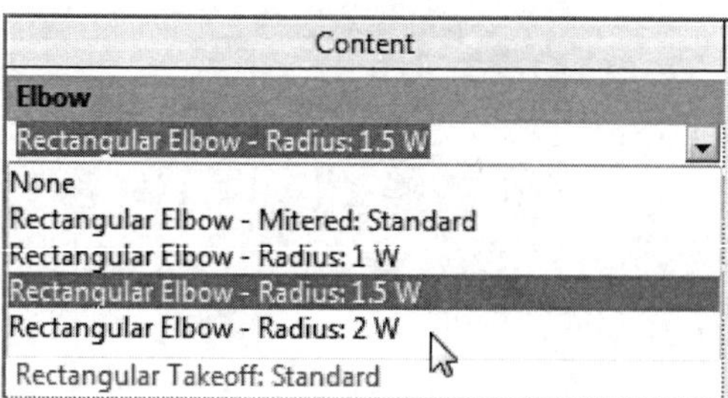

Figure 3–38

8. If you want to include more than one option for a fitting, click (Add row) and select the secondary option for the fitting. You can also use (Move row up) or (Move row down) to specify the order in which the fittings are applied by default or click (Remove Row) to delete the ones you do not need to use.
9. When you have finished assigning fittings, click **OK** to close the Routing Preference dialog box.
10. Fill in the other Parameters for the Type Properties as needed and click **OK** to finish the new duct type.

- Flex ducts do not have routing preferences, the Fitting selection takes place directly in the Type Properties, as shown in Figure 3–39.

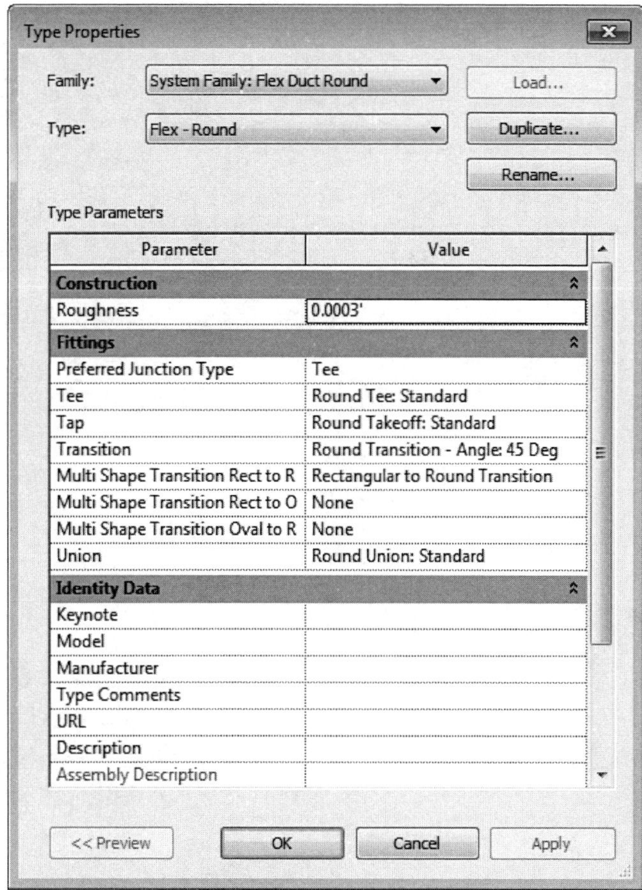

Figure 3–39

How To: Create Pipe Types

You can also right-click on a type name in the Project Browser and select ***Duplicate*** *before opening the Type Properties.*

1. Start the **Place Pipe** command, select one of the existing types, and click (Edit Type) or, in the Project Browser, expand **Families>Pipes>Pipe Types** and double-click on one of the existing types to open the Type Properties dialog box.
2. In the Type Properties dialog box, click **Duplicate...**, type a new name in the Name dialog box, and click **OK**.
3. Next to the **Routing Preferences** parameter, click **Edit...**.
4. The Routing Preferences dialog box opens, as shown in Figure 3–40.

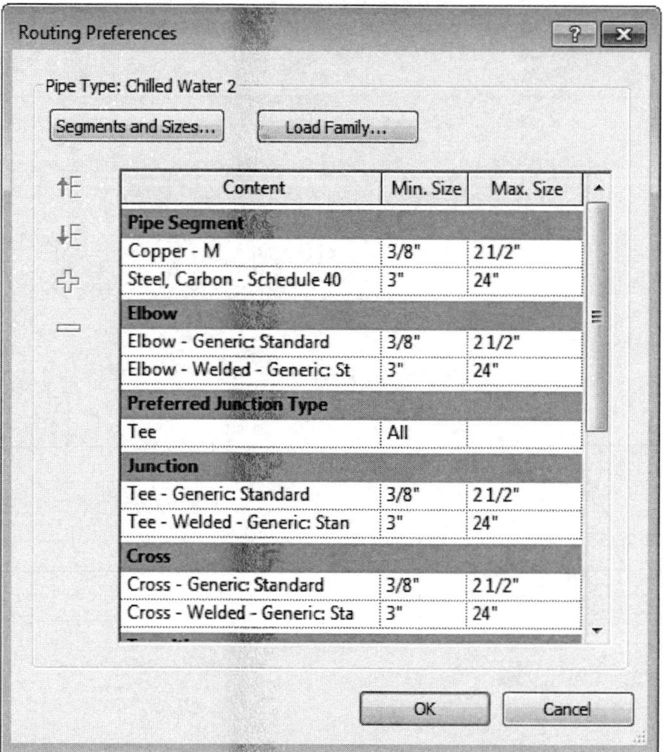

Figure 3–40

5. Click **Segments and Sizes...** to open the Mechanical Settings dialog box to the Pipe Settings. In this dialog box, specify the Segment types and sizes for the various types of pipe as shown in Figure 3–41.

Segment types can be added or removed here.

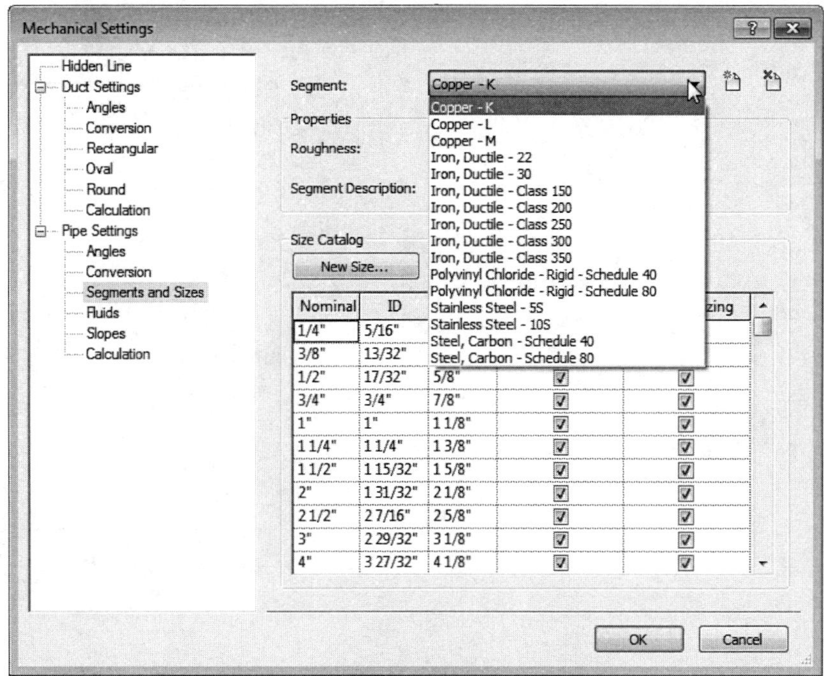

Figure 3–41

6. Click **Load Family...** to add any pipe fittings that you want to use in the type that have not already been loaded in the project.
7. For the Pipe Segment type and each of the fittings, expand the drop-down list and select the size that you want to use as shown in Figure 3–42.

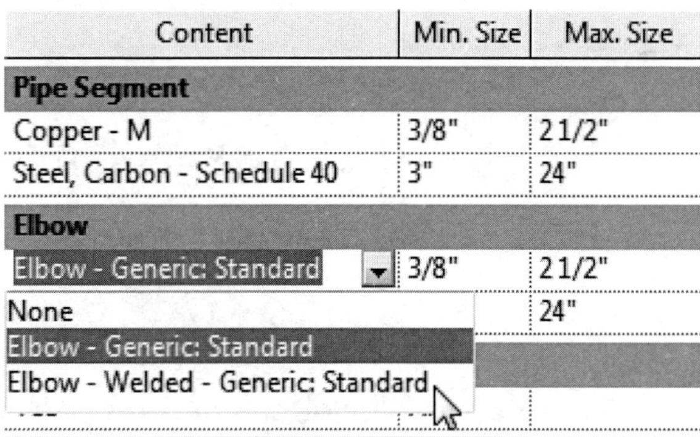

Figure 3–42

8. If you want to include more than one option for a fitting, click (Add row) and select the secondary option for the fitting. You can also use (Move row up) or (Move row down) to specify the order in which the fittings are applied by default or (Remove Row) to delete the ones you do not need to use.

9. When you have finished assigning fittings, click **OK** to close the Routing Preference dialog box.
10. Fill in the other Parameters for the Type Properties as needed and click **OK** to finish the new pipe type.

- Flex pipes do not have routing preferences, the Fitting selection takes place directly in the Type Properties as shown in Figure 3–43.

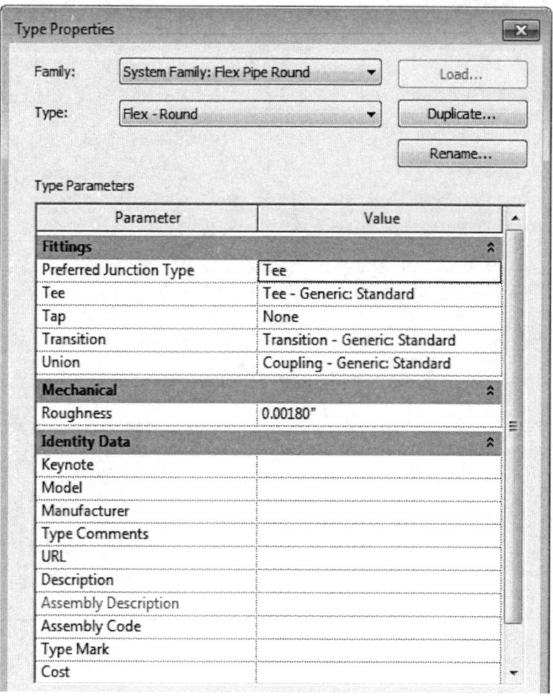

Figure 3–43

How To: Set up Cable Tray and Conduit Types

1. Start the **Cable Tray** or **Conduit** command, select one of the existing types, and click (Edit Type) or, in the Project Browser, expand **Families>Cable Trays** (or Conduits) and double-click on one of the existing types (either with or without fittings, as shown in Figure 3–44) to open the Type Properties dialog box.

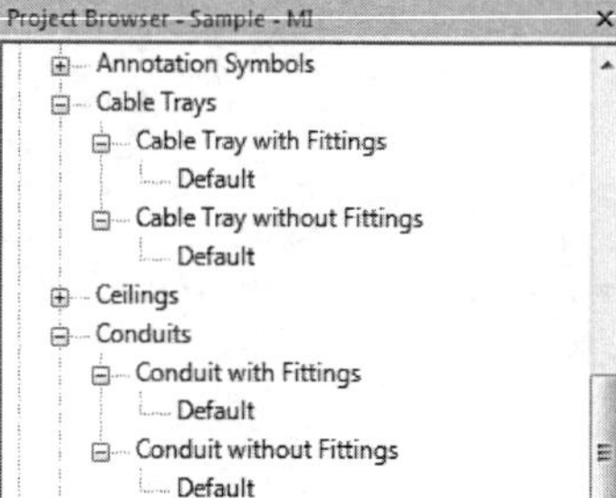

Figure 3–44

2. In the Type Properties dialog box, click **Duplicate...**, type a new name in the Name dialog box, and click **OK**.
3. Select the fittings that you want to use as shown in Figure 3–45.

The fitting families need to be in the project before you get to this point.

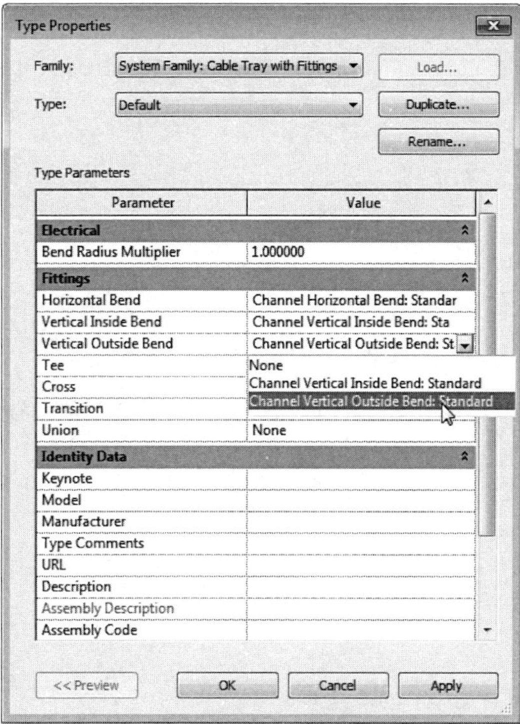

Figure 3–45

- For Conduit you first need to specify the Electrical Standard that you are using, as shown in Figure 3–46. Ensure that you also load the correct fittings that match the specific standard as well.

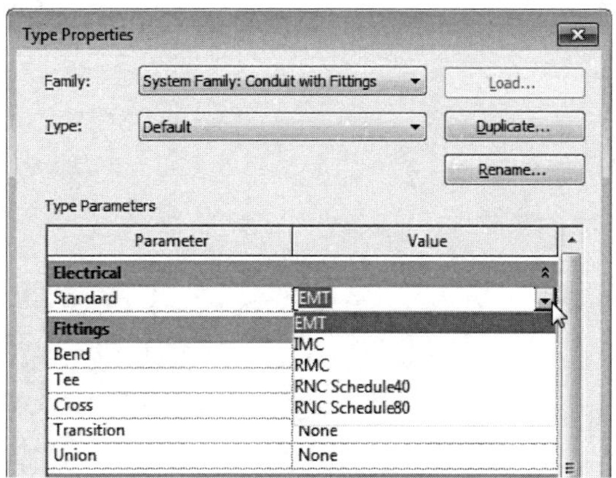

Figure 3–46

4. Fill in the values for any of the Identity Data information you might have and click **OK** to finish.

Chapter Review Questions

1. When creating a compound wall, you assign a Function for each layer. The numbers after the function names is the priority of the layers when they are joined together with other walls, as shown in Figure 3–47. Which of the following function connects first?

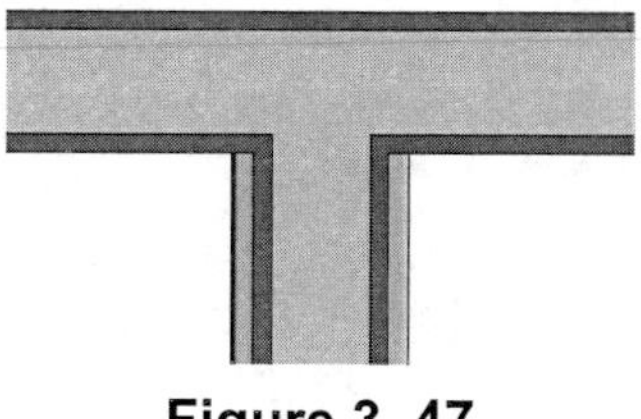

Figure 3–47

    a. Membrane Layer
    b. Substrate
    c. Structure
    d. Thermal/Air Layer
    e. Finish

2. How do you create a wall type that has more than one finish material on the vertical plane within the one type, as shown for stone and stucco in Figure 3–48?

Figure 3–48

    a. Create separate wall types for the stone and stucco. Duplicate an existing Stacked Wall type and set the wall types to stone and stucco.
    b. Draw two walls on top of each other and use (Join Geometry) to link them.
    c. Duplicate an existing Basic Wall type and use **Split Region** and **Assign Layers** to apply the stone and stucco materials to separate parts of the wall.
    d. Draw the first wall, draw the other wall on top of the first, and use (Cut Geometry) to cut the second wall out of the first wall.

3. Which command do you need to use to embed a wall inside another wall?
   a. **Wall Joins**
   b. **Cut Geometry**
   c. **Cope**
   d. **Split Face**
4. (MEP only.) When setting up a new duct or pipe type, which of the following items can be set by type?
   a. System Type
   b. Width and Height
   c. Preferred Junction Type
   d. Justification
5. (MEP only.) When creating a new conduit or cable tray type you need to assign Routing Preferences.
   a. True
   b. False

Command Summary

Button	Command	Location
	Cable Tray	• **Ribbon:** *Systems* tab>Electrical panel • **Shortcut:** CT
	Conduit	• **Ribbon:** *Systems* tab>Electrical panel • **Shortcut:** CN
	Cut Geometry	• **Ribbon:** *Modify* tab>Geometry panel
	Duct	• **Ribbon:** *Systems* tab>HVAC panel • **Shortcut:** DT
	Edit Type	• **Ribbon:** *Modify* tab>Properties panel • **Properties Palette**
	Pipe	• **Ribbon:** *Systems* tab>Plumbing & Piping panel • **Shortcut:** PI
	Uncut Geometry	• **Ribbon:** *Modify* tab>Geometry panel> expand Cut
	Wall: Architectural	• **Ribbon:** *Architecture* tab>Build panel or *Structure* tab>Structure panel>expand Wall • **Shortcut:** WA
	Wall: Structural	• **Ribbon:** *Architecture* tab>Build panel or *Structure* tab>Structure panel>expand Wall • **Shortcut:** WA

Component Family Concepts

Component families are the heart of the Autodesk® Revit® software. They include furniture; doors and windows; mechanical, electrical, and plumbing equipment; and structural framing. While some families come with the program or can be loaded from a service such as Autodesk® Seek®, it is important to know how to create custom families yourself. The basics include creating a parametric framework, adding solid geometry, and creating family types.

Learning Objectives in this Chapter

- Understand the different types of families that can be created in the software.
- Set up a parametric framework using reference planes, dimensions, and labels.
- Flex the framework to test all of the parameters.
- Create solid/void forms including extrusions, blends, revolves, sweeps, and swept blends.
- Create family types.
- Load and test families in a project.

4.1 Creating Component Families

Families are model or annotation elements that have been grouped together and set up with dimensions and other parameters. Families are parametric, meaning that they can be changed without having to recreate the elements. For example, when you change the *Height* and *Depth* of the storage pedestal shown in Figure 4–1, the other parts move with it.

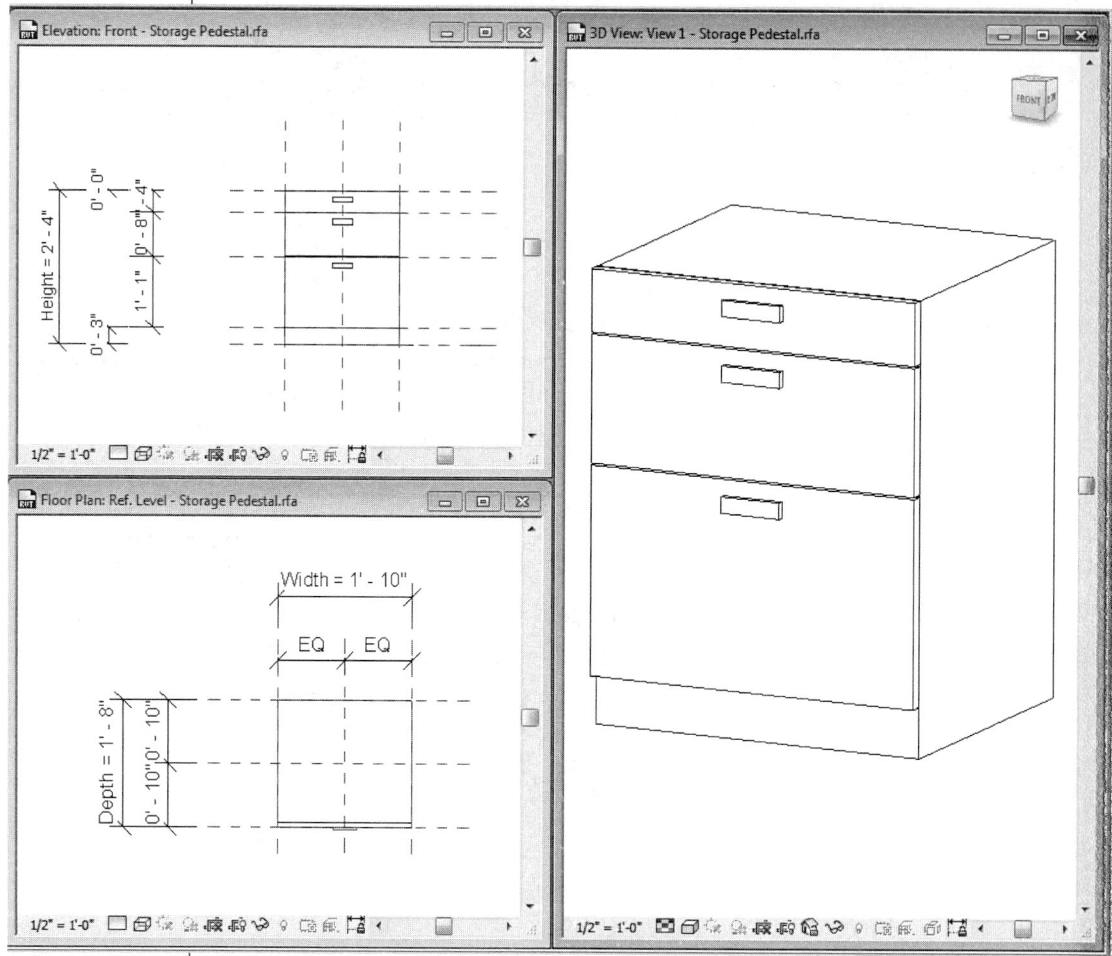

Figure 4–1

- *Component Families* (e.g., doors, air handling units, columns, etc.) are defined in the Family Editor and loaded into a project. One family can have many types.
- Many families are complex and time-consuming to create, but creating them all follows the same basic process.

How To: Create a Component Family (Overview)

1. In the Application Menu, click (New) and select **Family**.
2. In the New Family - Select Template File dialog box, select the template you want to use and click **Open**.
3. Set the parametric framework by adding reference planes/lines in plan and elevation views. This is a very important step when working with parametric families .
4. Label and dimension the reference planes/lines to control the movement of the elements and create parameters.
5. Flex the model to ensure the parameters change as expected.
6. Use the Forms tools (shown in Figure 4–2) to draw the elements.

Flexing a model: *Changing the dimensions of a model to test its parametric features.*

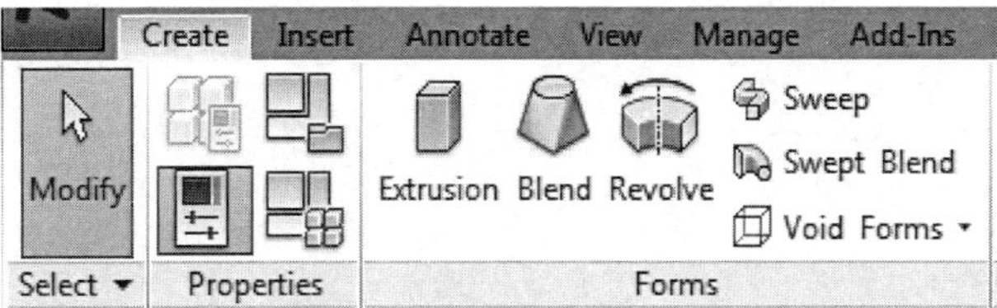

Figure 4–2

7. Lock the elements to the reference planes/lines.
8. Flex the elements again by changing the parameters to ensure that they are working.
9. Create family types with different parameters, as required.
10. Save the family.
11. Click (Load into Project) to test the family in a project..

Hint: Other Family Types

- *System Families* are preset in the templates used to create projects (e.g., walls, roofs, slabs, ducts, pipes, and conduit). You create new types by modifying existing parameters.
- *In-Place Families* are component families created directly in a project and are dependent on the model geometry (e.g., custom gutters, special trim, or built-in columns).

Preparing to Create Families

As you create families, you should plan ahead by asking a variety of questions, including:

- Is this family going to be used by one project (in-place) or by many projects (component)?
- Do you need different 2D representations in different views? What about 3D elements?
- Is this going to be a host-based or stand-alone family?
- Do you need various sizes of this family component?
- Are you going to post the family on Autodesk Seek?

Hint: Host-based vs. Stand-alone Families

Host-based families are dependent on a host. Examples include downlights (which require a ceiling) and sconces (which require a wall), as shown in Figure 4–3.

Stand-alone families do not need a host. Examples include the desk and table lamp shown in Figure 4–3.

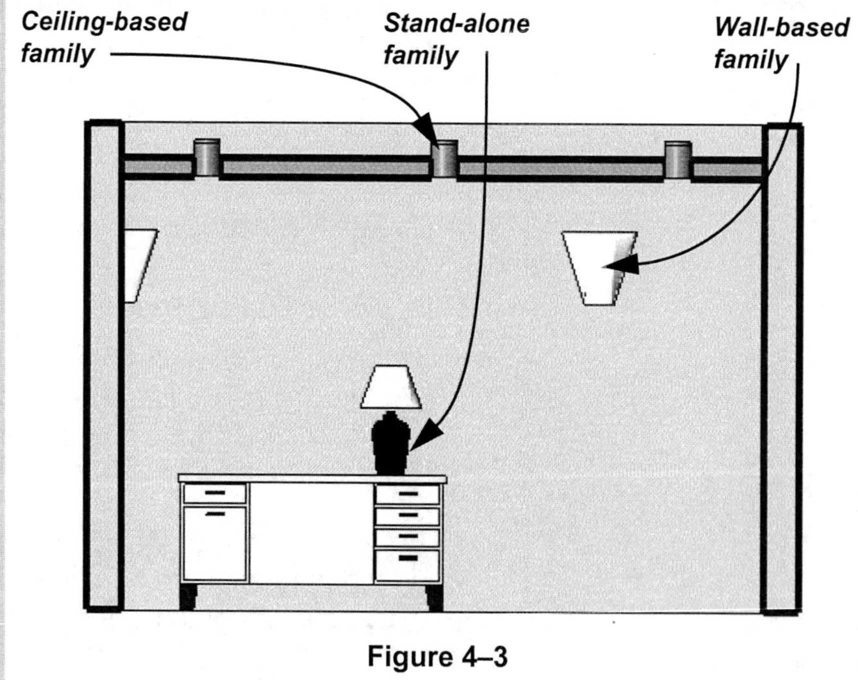

Figure 4–3

Saving Custom Family Files

When you create custom family files, you should save them to a location shared by the entire company, similar to the Autodesk Revit library. For easy access, set the custom library as the primary or secondary library available when you load a family.

How To: Specify the File Location for Custom Libraries

1. In the Application Menu, click **Options**.
2. In the Options dialog box, select the *File Locations* tab.
3. Click **Places...**.
4. In the Places dialog box, click (Add Value). A new *Library Name* is added to the list. Specify the name and path for the library, as shown in Figure 4–4.

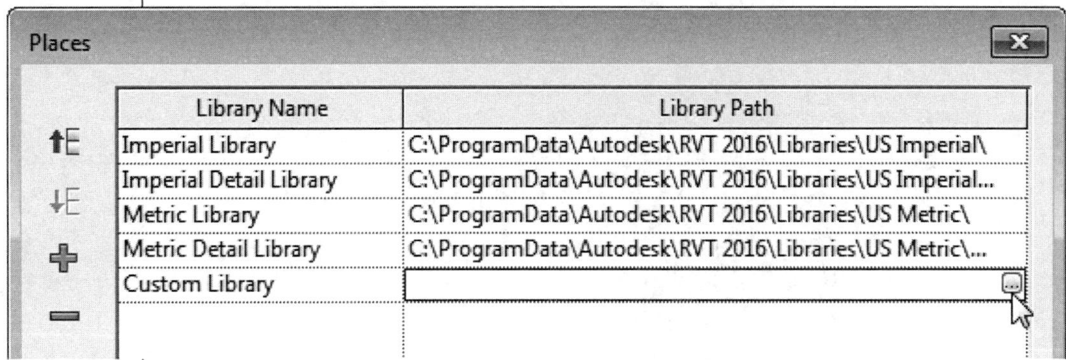

Figure 4–4

5. Click (Move Rows Up) and (Move Rows Down) to reorder the list of library names.
6. Click **OK** to close the dialog boxes.

- The library at the top of the list will be the default location when loading families.

4.2 Creating the Parametric Framework

Autodesk Revit families can be parametric (i.e., controlled by parameters). These parameters are the framework for creating family elements, and include dimensions, labels, and other options (such as materials), as shown in Figure 4–5.

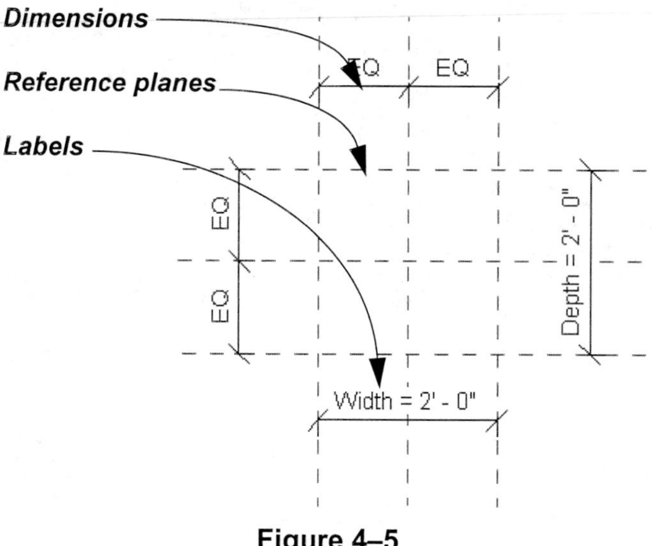

Figure 4–5

Before you draw any elements in a family, set up the parametric framework:

1. Place reference planes or reference lines at critical locations.
2. Dimension the reference planes/lines and label some of them so they can be manipulated.
3. Flex the dimensioned and labeled reference planes/lines to ensure that they function as expected.

Placing Reference Planes and Lines

Reference planes and lines shown in Figure 4–6) are the backbone of the geometry. Create as many reference planes as required before you start creating elements.

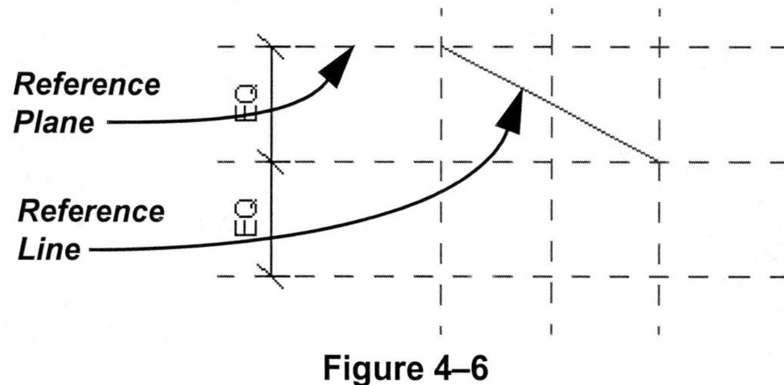

Figure 4–6

How To: Draw Reference Planes

1. In the Family Editor, in the *Create* tab>Datum panel, click (Reference Plane).
2. In the *Modify | Place Reference Plane* tab>Draw panel, click (Line) or (Pick Lines).
3. Draw or select the lines that define the location of the plane.
4. Draw or select other lines as needed.
5. Click (Modify) to finish the command.

- (Reference Lines) are similar to reference planes, but have distinct start and end points. They can be drawn using a wide variety of geometry, such as rectangles, circles, or arcs.
- The number of reference planes/lines you need depends on what you are drawing.
- Reference planes/lines display in plan and elevation views, but not in 3D views.

Reference Plane Options

- By default, the origin of a family is the center of the space where you create the family. In most templates, this is located at the intersection of the existing reference planes. You can also specify your own origin by modifying the properties of two intersecting reference planes to specify that they define the origin, as shown in Figure 4–7.

Other	
Is Reference	Front
Defines Origin	☑

Figure 4–7

- Reference planes/lines do not display in the project when you insert the family type, but might impact snap and alignment behavior.
  - In Properties, under *Other*, set the **Is Reference** parameter to the plane you want to use, as shown in Figure 4–8.
  - If you do not want a reference plane to be a snap or an alignment, set the **Is Reference** parameter to **Not a Reference**.

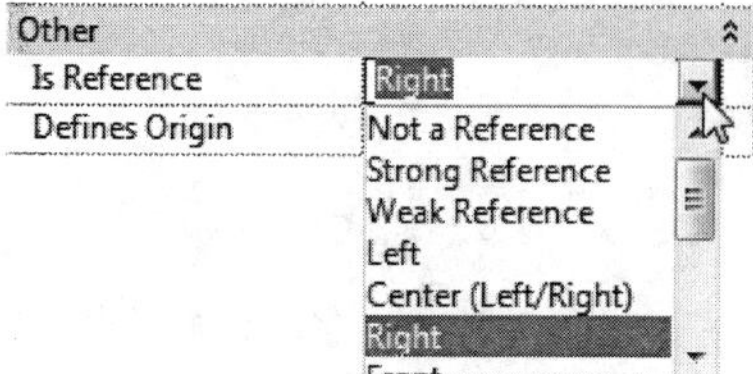

Figure 4–8

- Naming a reference plane enables you to use it to set the work plane when you are ready to add geometry to the family. In Properties, under *Identity Data*, type a *Name*. The name displays in the family file when you select the reference plane, as shown in Figure 4–9.

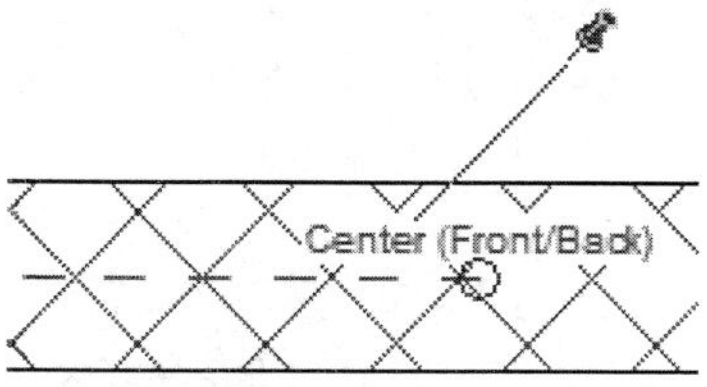

Figure 4–9

- Reference lines cannot be named.

Adding Dimensions and Labels

Add dimensions to reference planes/lines to start the constraining process. For example, you might want to run a string of dimensions and then set the distances to be equal by clicking on the associated **EQ** symbol, as shown in Figure 4–10. This moves the reference planes into place. Labeling a dimension creates a parameter that can be modified in the Family Types dialog box.

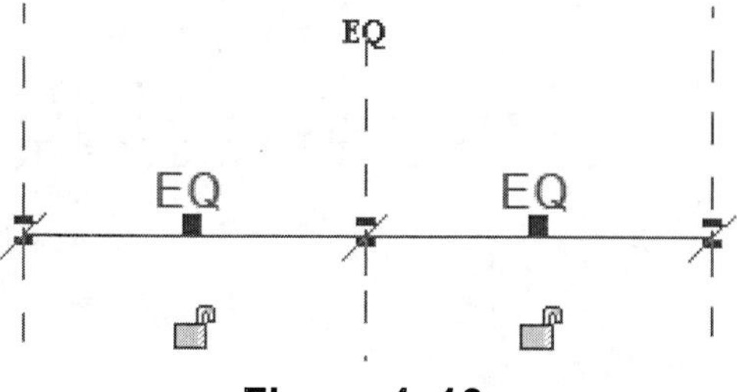

Figure 4–10

How To: Label a Dimension

1. Select the dimension you want to label.
2. In Properties or the Options Bar, select an existing label from the list. Alternatively, in the Options Bar, select **<Add parameter...>** to create a new label.
   - If you right-click on a dimension and select **Label**, the label options also display at the dimension, as shown in Figure 4–11.

When naming a parameter, it is good practice to use title case and a short name without abbreviations. Parameters are case sensitive.

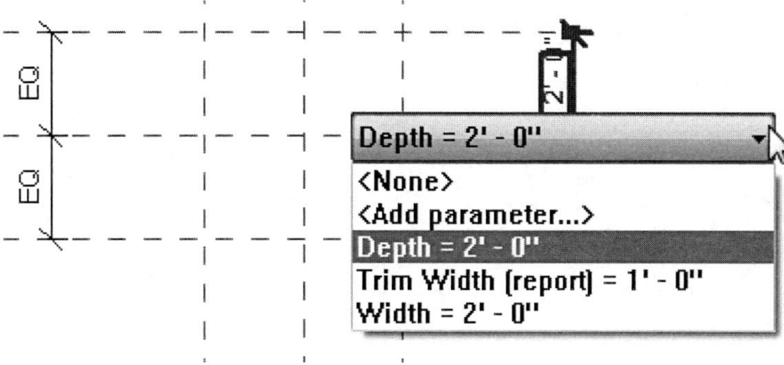

Figure 4–11

3. If you are adding a parameter, the Parameter Properties dialog box opens. In the *Parameter Type* area, select the **Family parameter** option, as shown in Figure 4–12.

Shared parameters must be created in your system to be able to use this option.

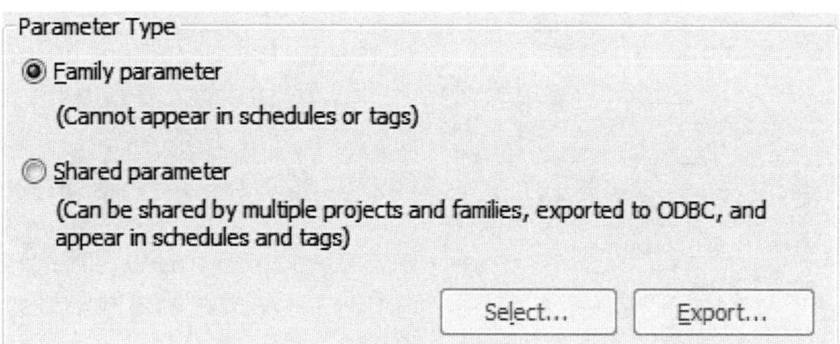

Figure 4–12

4. In the *Parameter Data* area, specify the *Name*, as shown in Figure 4–13. The *Type of Parameter* is automatically set to **Length** and the *Group parameter under* is set to **Dimensions**, because the parameter is a dimension.

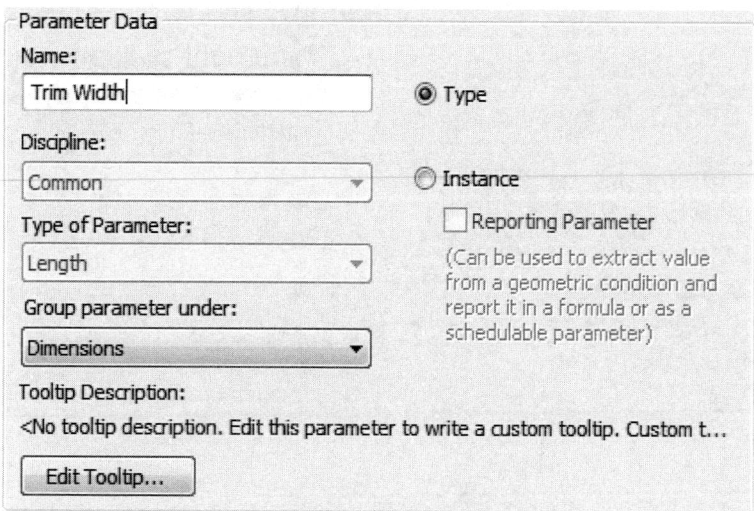

Figure 4–13

5. Select **Type** or **Instance**.
   - **Type** applies it to all instances of a family when it is inserted into a project. To modify it, you need to change the Type Properties.
   - **Instance** applies it to individual instance when inserted into a project. It can be modified directly in Properties.
6. Add a tooltip description if needed.
7. Click **OK**. The dimension is labeled and ready to be flexed.

- If you add too many labeled dimensions, the Autodesk Revit software warns that you have over-constrained the sketch. You cannot ignore this warning. The label is not created in this case.

Flexing Geometry

Once you have established dimensions and labeled them, you need to verify that the parameters work correctly with each other. To do this, flex the geometry by changing the value of a label (as shown in Figure 4–14), or using the Family Types dialog box.

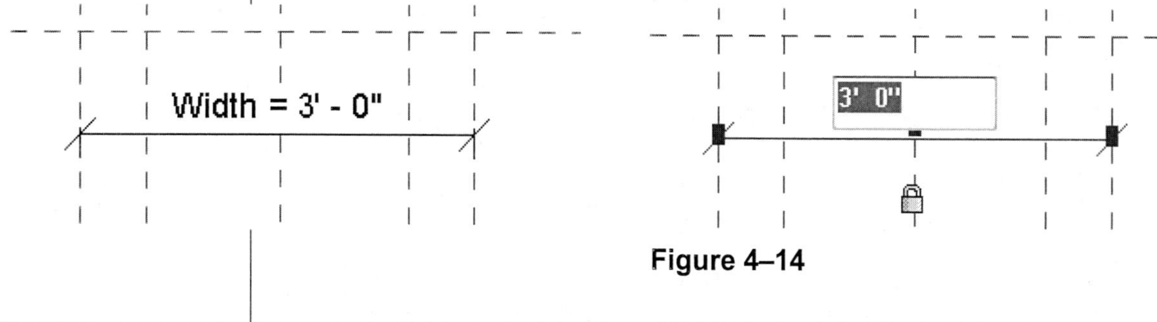

Figure 4–14

- Select the labeled dimension and then the dimension text to change the value.
- To see all of the values you can flex at one time and to add formulas and non-dimension parameters, you need to work in the Family Types dialog box.

How To: Flex the Geometry in the Family Types Dialog box

1. In the *Create* or *Modify* tab>Properties panel, click (Family Types). The Family Types dialog box displays as shown in Figure 4–15.
   - You might have to move the dialog box to display the elements you are adjusting.

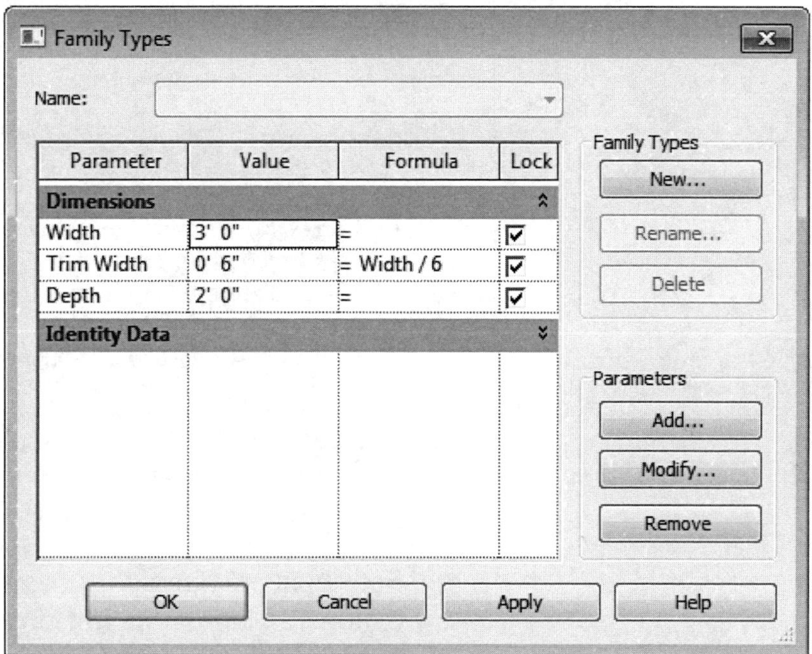

Figure 4–15

2. Change the labeled parameters and click **Apply**.
3. Do the elements move correctly? Try several modifications.
4. Set the parameters as required and click **OK** to finish the command.

Creating and Modifying Parameters

In addition to creating parameters for dimensions in a family, you can add formulas to parameters so that they change when a related parameter changes. You can also add parameters that store other information (such as the material of the element, the wattage of a light fixture, or the reinforcement cover setting in concrete).

How To: Add Formulas to Parameters

1. In the *Create* or *Modify* tab>Properties panel, click (Family Types).
2. In the Family Types dialog box, in the *Formula* column, add formulas to the parameters. In the example shown in Figure 4–16, the *Shelf Height* is defined as the **Height** divided by **4**. When you change the *Shelf Height* or *Height*, the other value updates according to the formula.

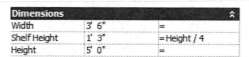

Dimensions		
Width	3' 6"	=
Shelf Height	1' 3"	=Height / 4
Height	5' 0"	=

Figure 4–16

You can use several types of formulas with families:

Arithmetic	Basic arithmetic operations in formulas include addition (+), subtraction (-), multiplication (*), division (/), exponentiation, logarithms, and square roots.
Trigonometric	Trigonometric functions include sine, cosine, tangent, arcsine, arccosine, and arctangent.
Conditional	Conditional functions include comparisons in a condition (e.g., <, >, =, etc.) and Boolean operators with a conditional statement (e.g., AND, OR, NOT, etc.). See the Autodesk Revit Help files for more information about creating conditional statements.

- You can enter numbers as integers, decimals, or fractions. Conditional statements can include numeric values, numeric parameter names, or **Yes/No** parameters.
- Formulas are case-sensitive. Therefore, if you have created a dimension label named **Height**, ensure that the formula that uses this label also has the H capitalized (e.g., **Height * 2**).

Hint: Formulas in Projects

You can use a formula to specify the length of a wall by typing it in the temporary dimension, as shown in Figure 4–17. Start with an equal sign (e.g., =12 / 3 + 4). The example shown in Figure 4–17 returns the number 8 (interpreted as 8'-0") in the dimension. Formulas also work with numerical values in Properties.

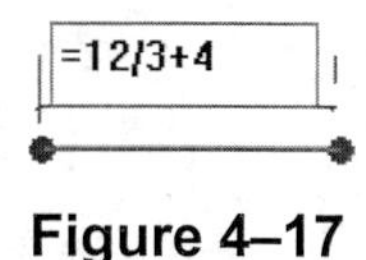

Figure 4–17

How To: Create Parameters

1. In the Family Editor, in the *Create or Modify* tab>Properties panel, click (Family Types).
2. In the Family Types dialog box, in the *Parameters* area, click **Add...**.
3. In the Parameter Properties dialog box, set up the Parameter Data, as shown in Figure 4–18.

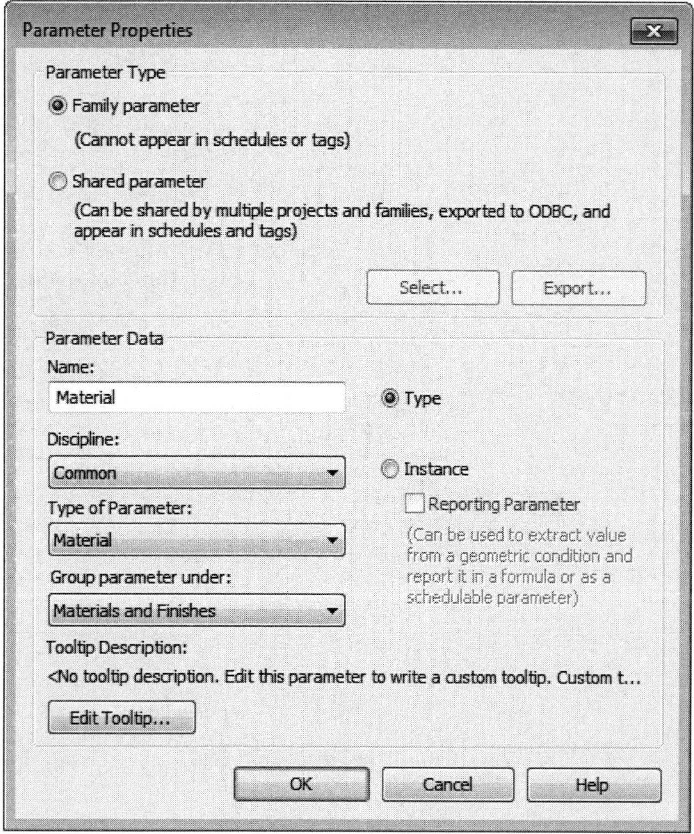

Figure 4–18

- The *Discipline* can be set to **Common**, **Structural**, **HVAC, Electrical**, **Piping**, or **Energy**. The *Type of Parameter* list changes according to the discipline.

4. Click **OK**.
5. Create or modify additional parameters. When you are finished, click **OK**.

- In the Family Types dialog box, use **Move Up** and **Move Down** to organize the parameters within each group.
- Click **Modify** to change a parameter. This displays the Parameter Properties dialog box, where you can change the name, group, or tooltip, or switch between Type and Instance. You can also change the parameter to reference a shared parameter.

- Delete parameters by clicking **Remove**.
- The *Sorting Order* can be changed by clicking **Ascending** or **Descending**. This sets the order of the parameters to alphabetical order and removes any changes you might have made.

Common Parameter Types

Text	A text parameter can host any type of information that consists of both numbers and letters. Examples could be the fabric color or part number of the element.
Integer	Any value that is always represented by a whole number. Examples could be the number of chairs that fit around a conference table or the number of shelves in a bookcase.
Number	Any numeric value: whole, decimal, or fraction. You can use formulas with **Number** parameters.
Length	When you label dimensions, you create **Length** parameters. You can use formulas with **Length** parameters.
Area	Establishes the area of an element. It is numeric and can have formulas applied to it.
Volume	Establishes the volume of an element. It is numeric and can have formulas applied to it.
Angle	Establishes the angle of an element. It is numeric and can have formulas applied to it.
Slope	Displays a slope, as set in the Field Format.
Currency	Displays the amount in dollars or other currencies, as set up in the Field Format.
Mass Density	Displays the mass per unit volume of material.
URL	Specifies a link to a web site.
Material	Provides a place to assign a material from the list of materials that are set up in the project.
Image	Creates a parameter at which you can add a raster image connected to the family.
Yes/No	Used with instance parameters where you need a **Yes** or **No** answer to a question listed in the name. The default is **Yes**.
<Family Type>	Opens a dialog box where you can select from the list of family types, such as doors, furniture, or tags.

Practice 4a

Set up a Bookcase Family

Practice Objective

- Add reference planes.
- Dimension and label reference planes.
- Add a formula to a parameter.
- Flex the framework.

Estimated time for completion: 20 minutes

In this practice you will create a family file based on a template and view the existing reference planes. You will also create reference planes, dimension them, and label the dimensions, as shown in Figure 4–19. You will add a formula to control a parameter and then test the framework by changing parameter values.

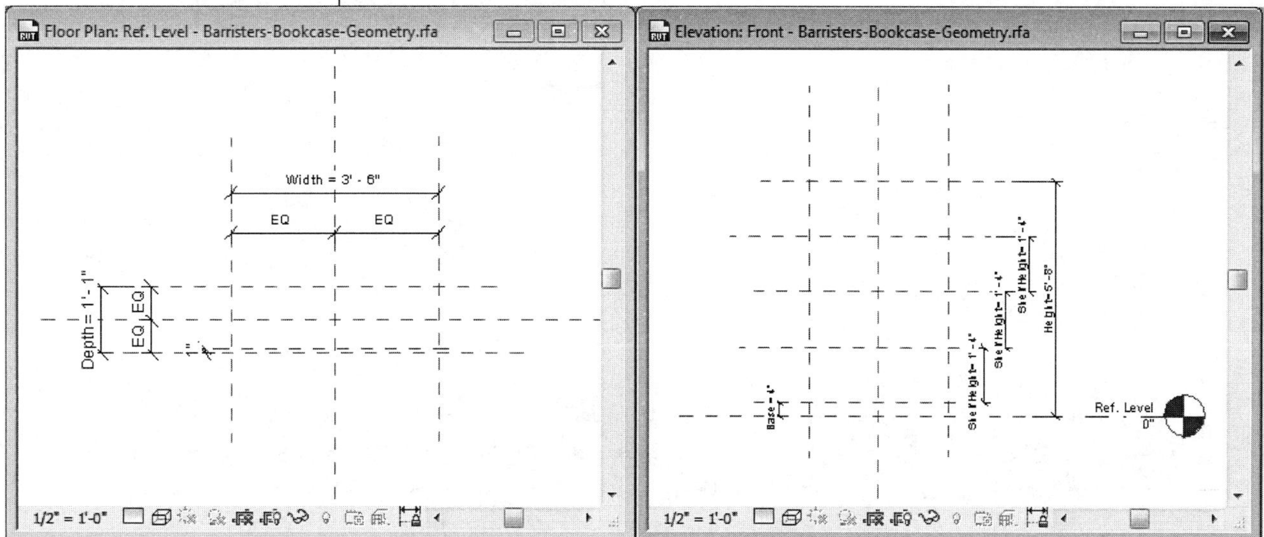

Figure 4–19

Task 1 - Open a family template.

1. Close all open projects.
2. On the Start page, under the *Families* area, select **New**.
3. In the New Family - Select Template File dialog box, select the template **Furniture.rft** from the Revit Library and click **Open**.
4. In the *View* tab>Windows panel, click (Tile) (or type **WT**).
5. The four open views display so you can see each of them. Type **ZA** (the shortcut for **Zoom All to Fit**) to fit the view in all of the windows.

6. Several existing reference planes are included in the file, as well as **Ref.Level**.

7. Save the family in the ...*Architectural* folder as **Barrister Bookcase.rfa**.

Task 2 - Create plan view reference planes.

1. Maximize the **Ref. Level** view.

2. Create four new reference planes, one on each side of the existing planes, as shown in Figure 4–20. Do not worry about the exact location yet.

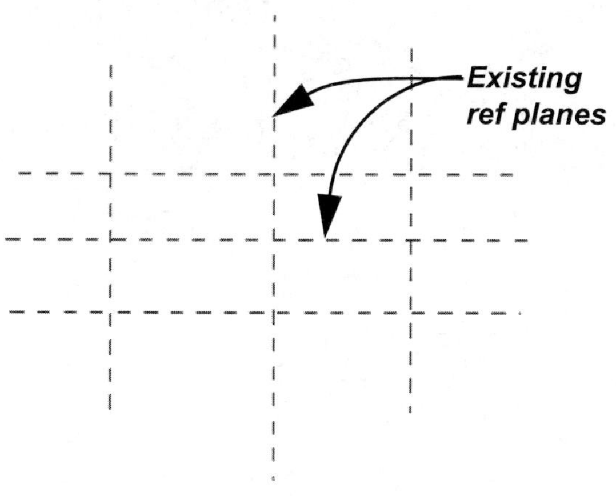

Figure 4–20

3. Add dimensions between each reference plane and across the top and side, as shown in Figure 4–21. Set the interior dimensions to **Equal**.

Your dimensions might vary depending on where you originally placed the reference planes.

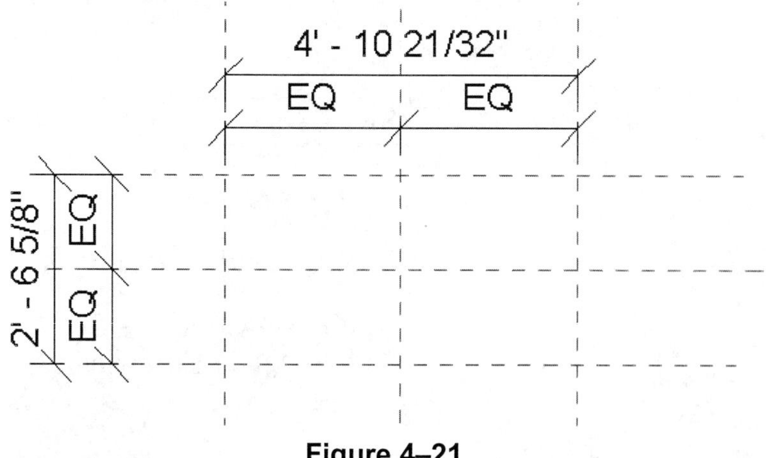

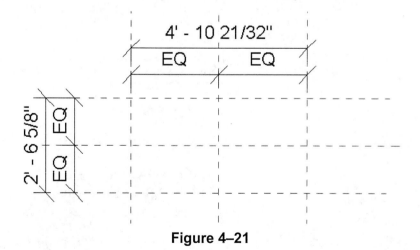

Figure 4–21

4. Select the overall horizontal dimension. In the Options Bar, expand the **Label** list and select **<Add parameter...>**.

5. In the Parameter Properties dialog box, in the *Parameter Data* area, ensure that **Type** is selected and then type the name **Width**. Click **OK**.

6. Repeat the process for the overall vertical dimension and name the parameter **Depth**.

7. Select the *Width* dimension and click on the text. Change it to **3'-0".** Repeat the process and set the *Depth* to **1-'4"**, as shown in Figure 4–22.

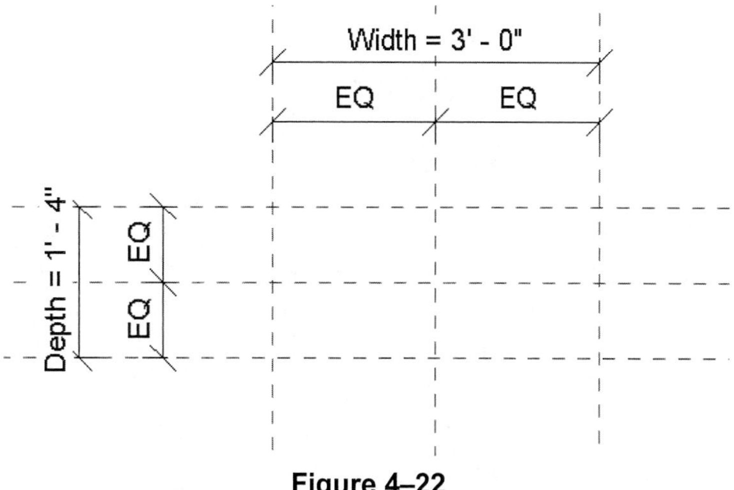

Figure 4–22

8. Create one more horizontal reference plane **1"** inside the front reference plane. Dimension and lock it in place, as shown in Figure 4–23.

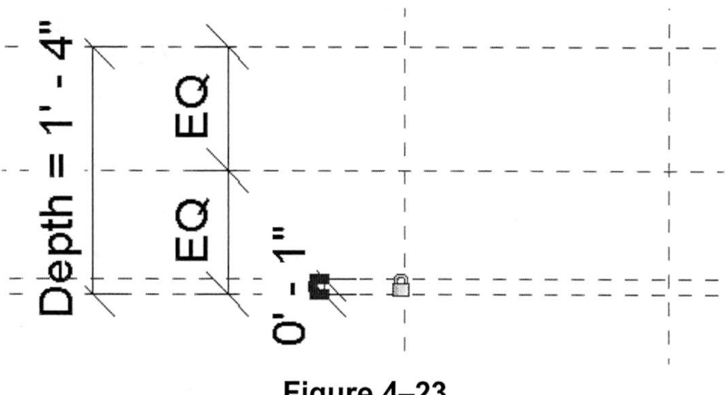

Figure 4–23

9. Select the new reference plane. In Properties, set the *Name* to **Door Face**.

10. Flex the framework by changing the *Width* dimension to **3'-6"** and the *Depth* dimension to **1'-1"**.

11. Verify that the other reference planes have remained in place.

12. Save the family file.

Task 3 - Create front view reference planes.

1. Open the **Elevations: Front** view.

2. Create the reference planes and dimensions and label them, as shown in Figure 4–24.

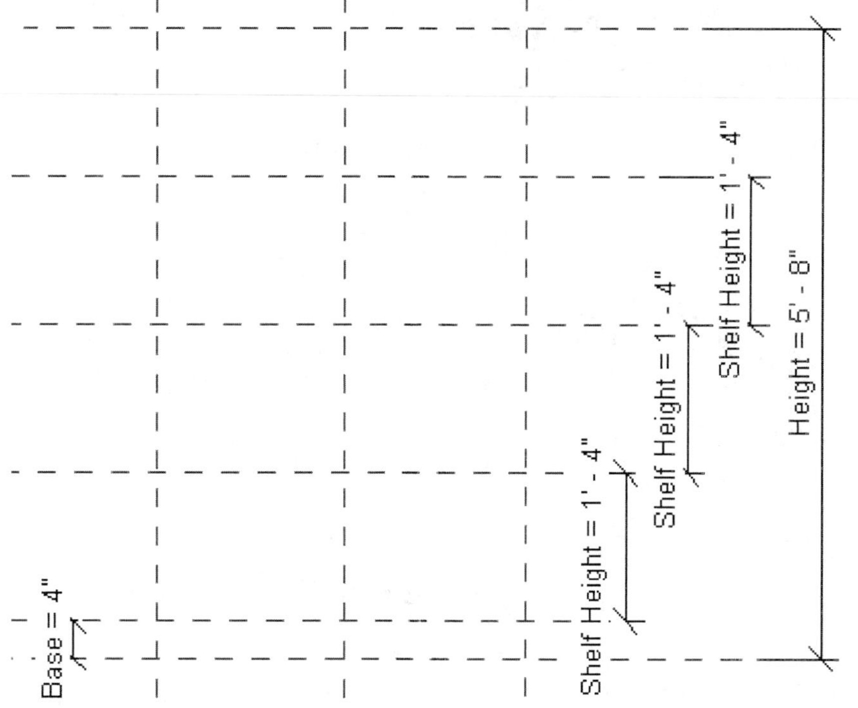

Figure 4–24

- Each of the shelf height dimensions must be separate. Once you have created one shelf height parameter, you can apply the same label to the other dimensions.

3. In the *Modify* (or *Create*) tab>Properties panel, click (Family Types).

4. In the Family Types dialog box, set up a formula for the *Shelf Height*, as shown in Figure 4–25. Lock the **Height** parameter and the **Shelf Height** parameter also locks.

The parameters automatically display in alphabetical order. You can reorganize them as needed.

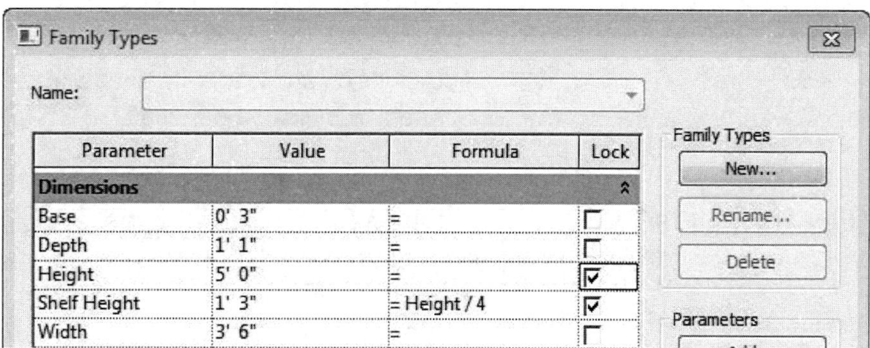

Figure 4–25

5. Move the dialog box to one side and try several different values. Click **Apply** to apply the changes to the family. Verify that all of the reference planes move together. Click **OK**.

6. Save the family.

Practice 4b Set up a Heat Pump Family

Practice Objectives

- Add reference planes.
- Dimension and label reference planes.
- Add a formula to a parameter.
- Flex the framework.

Estimated time for completion: 20 minutes

In this practice you will create a family file based on a template and view the existing reference planes. You will also create reference planes, dimension them, label the dimensions and use a formula in parameters, as shown in Figure 4–26.

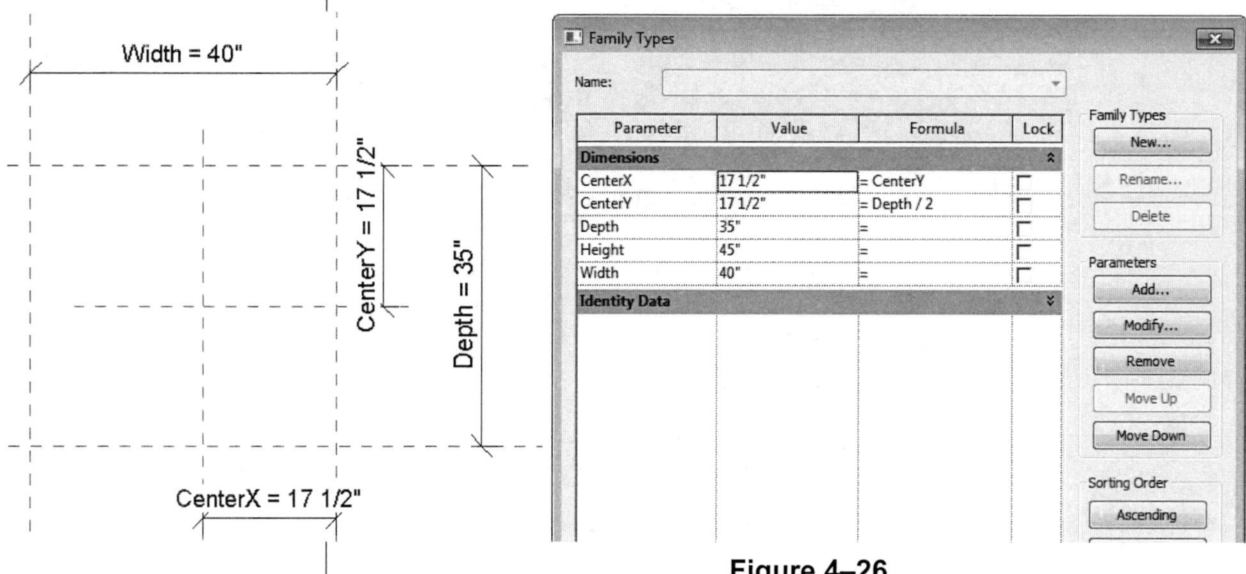

Figure 4–26

Task 1 - Open a family template.

1. Close all open projects.
2. On the Start page, under the *Families* area, click **New**.
3. In the New Family - Select Template File dialog box, select the template **Mechanical Equipment.rft** from the Revit Library and click **Open**.
4. In the *View* tab>Windows panel, click (Tile) or type **WT.**
5. The four open views display. Type **ZA** (the shortcut for **Zoom All to Fit**) to fit the view in all of the windows.
6. Several existing reference planes are included in the file, as well as **Ref.Level**.

Changing the units so that they display as inches makes it easier to work with the dimensions.

7. Type **UN** to open the Project Units dialog box. Click the button under *Format* for *Length* and change the *Units* to **Fractional inches**.

8. Click **OK** until all of the dialog boxes are closed. Save the family in the ...*MEP* folder as **Heat Pump.rfa**.

Task 2 - Create plan view reference planes.

1. Maximize the **Ref. Level** view.

2. Create two new reference planes, above and to the right of the existing planes, as shown in Figure 4–27. Do not worry about the exact location yet.

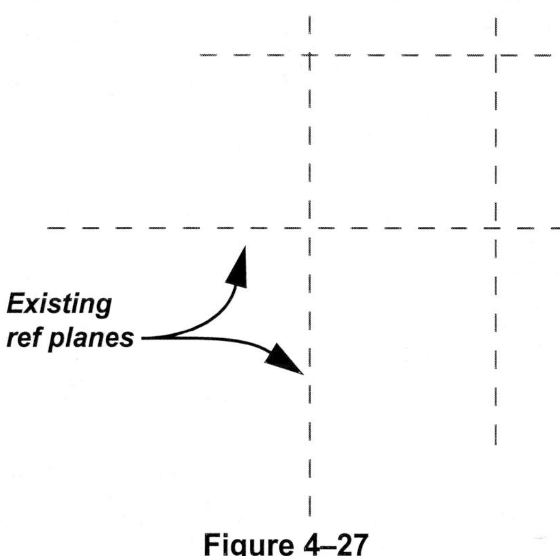

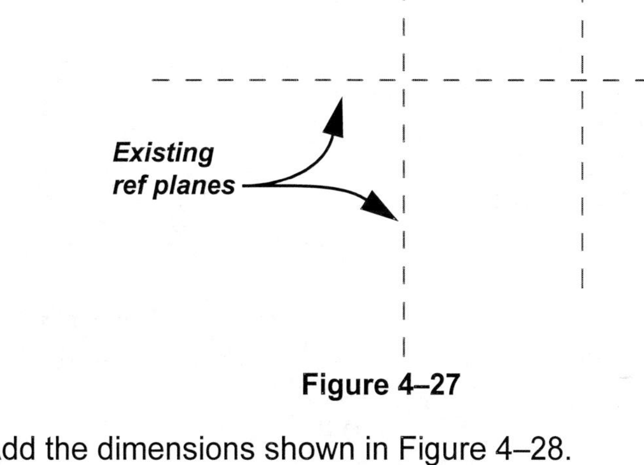

Figure 4–27

3. Add the dimensions shown in Figure 4–28.

Your dimensions might vary depending on where you originally placed the reference planes.

The scale in the graphics have been changed to display the dimensions more clearly.

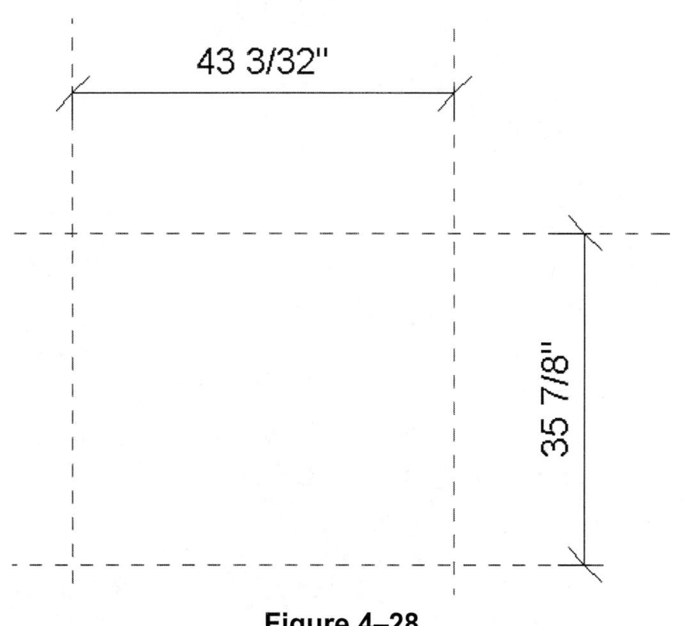

Figure 4–28

4. Select the horizontal dimension. In the Options Bar, expand the **Label** list and select **<Add parameter...>**.

5. In the Parameter Properties dialog box, in the *Parameter Data* area, ensure that **Type** is selected and then type the name **Width**. Click **OK**.

6. Repeat the process for the overall vertical dimension and name the parameter **Depth**.

7. Add two more reference planes inside the other reference planes. Dimension and label them, as shown in Figure 4–29.

These new dimensions are not at the center of the unit but a specific distance from the right side and the top reference planes. The exact distance is not required at this time.

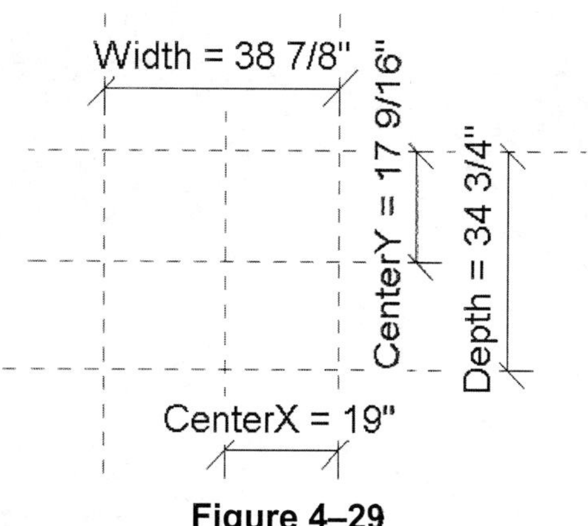

Figure 4–29

8. Open the Elevations**: Front** view.

9. Create the reference planes and dimensions. Label the *Height* dimension, as shown in Figure 4–30, and lock the **1"** and **EQ** dimensions into place. You might need to zoom in to do this.

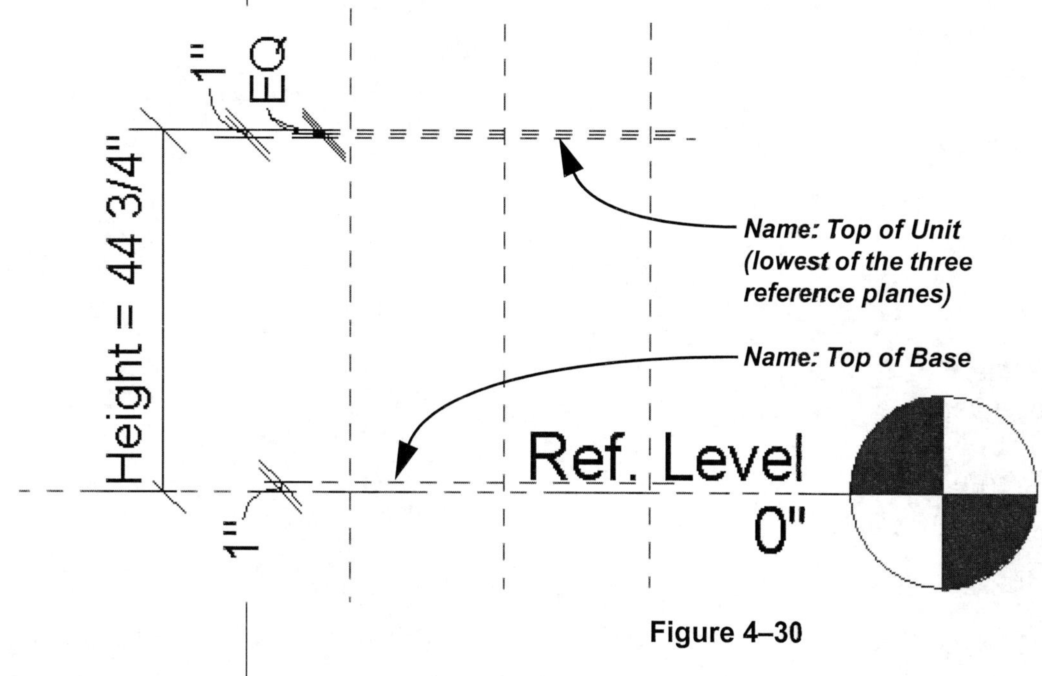

Figure 4–30

10. In Properties, name the **Top of Unit** and **Top of Base** reference planes, as shown in Figure 4–30.

11. Save the family.

Task 3 - Flex the parameters.

1. Remain in elevation view.
2. Select the **Height** parameter and change it to a different height. Verify that the top dimensions moves with the height and the bottom one stays in place.
3. Try a different height to fully check it.
4. Return to the **Floor Plans: Ref. Level** view.
5. In the *Modify* (or *Create*) tab>Properties panel, click (Family Types).
6. In the Family Types dialog box, set up a formula for *CenterY* and *CenterX*, as shown in Figure 4–31. Lock one of the center parameters and the related parameters also lock.

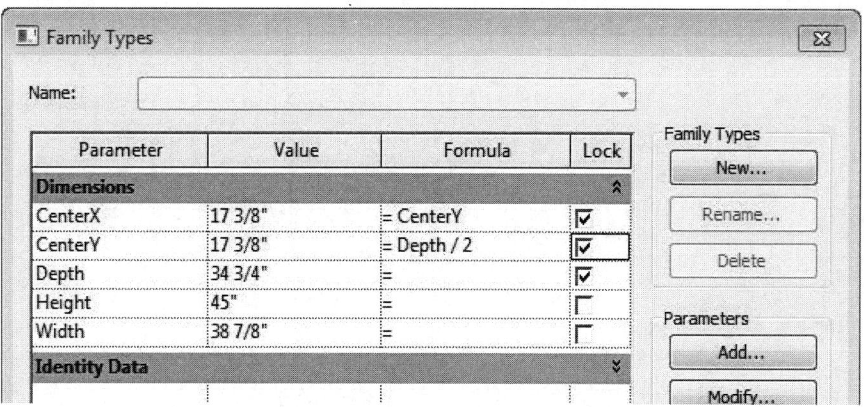

Figure 4–31

7. Move the dialog box to one side and try several different values. Click **Apply** to apply the changes to the family. Verify that all of the reference planes move together. Click **OK**.
8. Save the family.

Practice 4c Set up a Structural Column Family

Practice Objectives

- Add reference planes and reference lines.
- Add dimensions and labels.
- Add a formula to a parameter.

Estimated time for completion: 10 minutes

In this practice you will create a family file based on a template and view the existing reference planes. You will also create reference lines and reference planes, dimension them, label the dimensions, as shown in Figure 4–32, and use a formula in parameters.

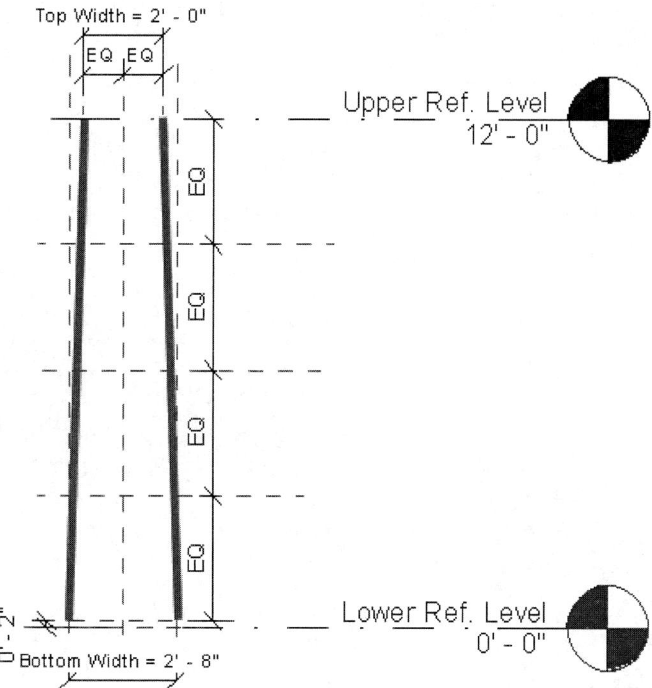

Figure 4–32

Task 1 - Open a family template.

1. Close all open projects.
2. On the Start page, in the *Families* area, select **New**.
3. In the New Family - Select Template File dialog box, select the template **Structural Column.rft** from the Revit Library and click **Open**.
4. Open a 3D view and the **Elevations: Front** views.

5. In the *View* tab>Windows panel, click (Tile) or type **WT.** The three open views display so you can see each of them. Type **ZA** (the shortcut for **Zoom All to Fit**) to fit the view in all of the windows.

6. Several existing reference planes and some dimensions are included in the file, as well as **Lower Ref.Level** and **Upper Ref. Level** which set the height of the column once it is used in a project.

7. Save the family in the ...*Structural* folder as **Trellis Column.rfa**.

Task 2 - Create reference planes.

1. Maximize the Elevations**: Front** view.

2. Change the *Upper Ref. Level* to **12'-0"**.

3. Extend the reference planes past the upper level marker.

4. Draw four new horizontal reference planes between the Lower and Upper Ref. Levels, as shown in Figure 4–33. Do not worry about the exact location yet but the lowest one is close to the Lower Ref. Level and the others are more or less equally spaced in the remaining area.

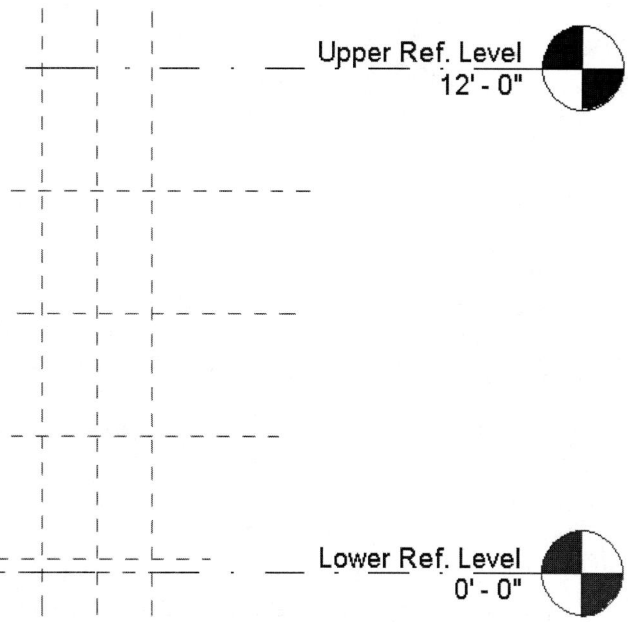

Figure 4–33

5. Dimension the lowest reference plane to the Lower Ref. Level. Set the dimension to **2"** and lock the dimension as shown in Figure 4–34.

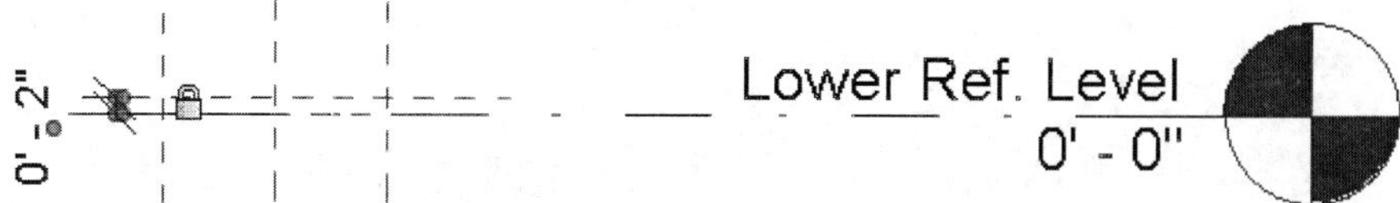

Figure 4–34

6. Add a string of dimensions between the evenly spaced reference planes and the lowest reference plane, as shown in Figure 4–35. Set the dimension string to **Equal**.

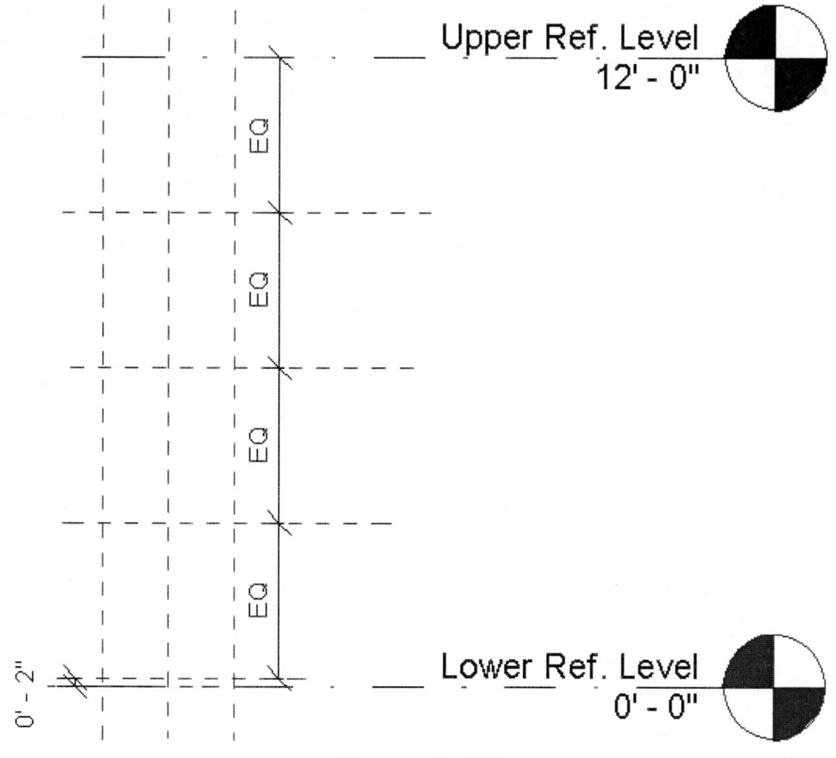

Figure 4–35

7. Add two short vertical reference planes crossing the Upper Ref. Level.

8. Dimension and set them as equal, as shown in Figure 4–36.

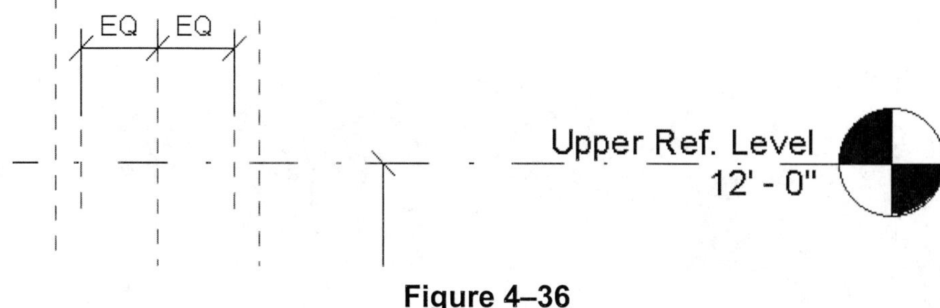

Figure 4–36

9. Add an overall dimension to the new short reference planes.

10. Select the overall dimension.

11. In the Options Bar, expand the *Label* list and select **<Add parameter...>**.

12. In the Parameter Properties dialog box, name it **Top Width** and click **OK**. The new parameter displays with the dimension, as shown in Figure 4–37.

Your dimensions might vary depending on where you originally placed the reference planes.

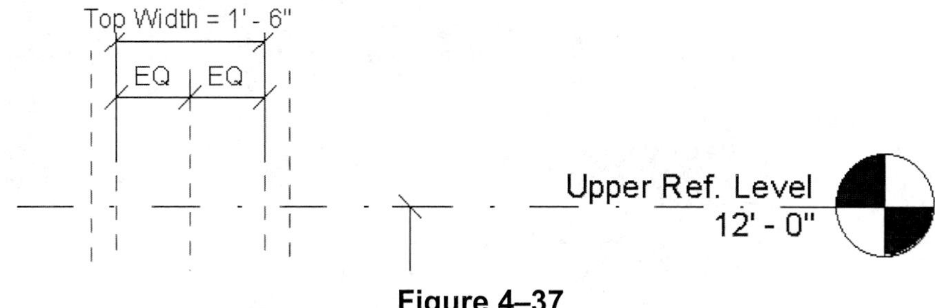

Figure 4–37

13. Add another dimension to the overall vertical reference planes below the Lower Ref. Level and label it **Bottom Width**, as shown in Figure 4–38.

14. In the *Create* tab>Datum panel, click (Reference Line).

15. Draw reference lines, as shown in Figure 4–38. Ensure that you snap to the intersections of the reference planes. The bottom intersection is with the lower reference plane, not the level.

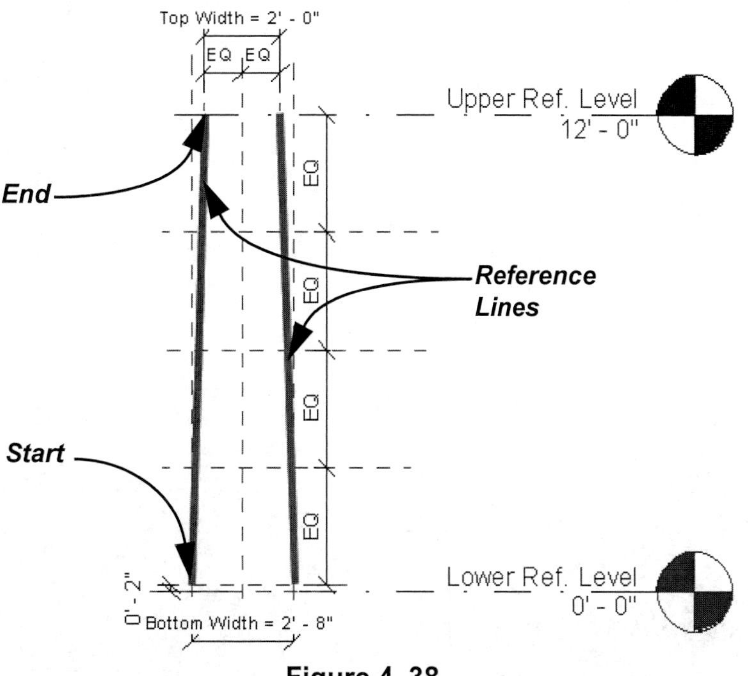

Figure 4–38

16. In the *Modify* (or *Create*) tab>Properties panel, click (Family Types).

17. Set up a formula for the *Top Width*, as shown in Figure 4–39.

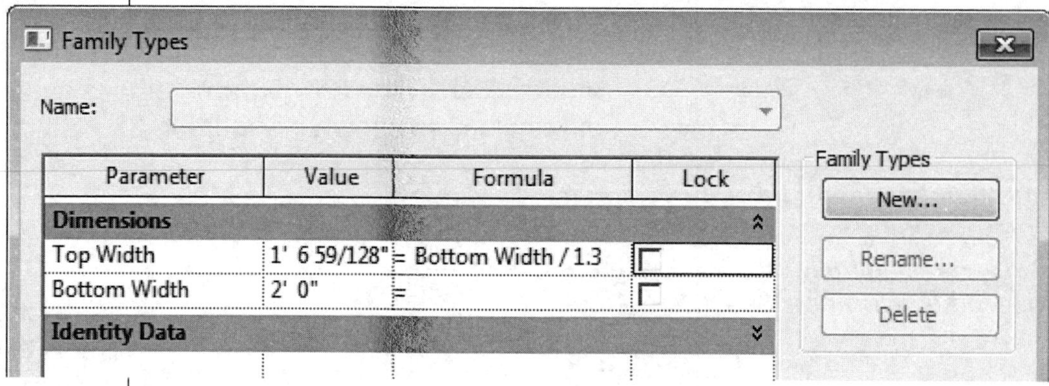

Figure 4–39

18. Move the dialog box to one side and try several different values for the bottom width. Click **Apply** to apply the changes to the family. Verify that all of the reference planes and lines move together. Click **OK**.

19. Save the family.

4.3 Creating Family Elements

When you have built the parametric framework, the next step in creating families is to add the actual geometry to the family file, as shown in Figure 4–40. You can also add controls, openings, and components, such as the sink shown in the countertop in Figure 4–40. These elements can be aligned and locked to reference planes and other geometry.

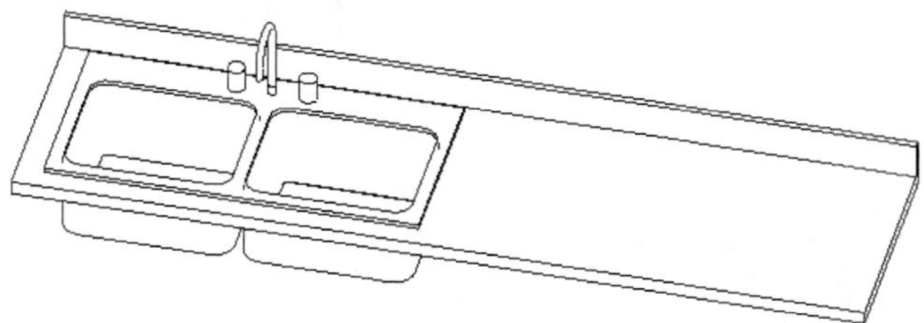

Figure 4–40

The actual components in a family can include both 2D and 3D elements. 3D elements are either solids or voids, model text and model lines. 2D elements include symbolic lines and text. All of these tools are located in the Family Editor.

Creating 3D Elements

There are five methods of creating solid forms, as shown in Figure 4–41. Each method uses a sketched 2D profile as the basis of the 3D shape.

Figure 4–41

Void forms use the same creation methods.

- **Extrusion** pushes the profile out in one direction.
- **Blend** links two profiles together.
- **Revolve** rotates the profile around an axis.
- **Sweep** extends a profile along a path.
- **Swept Blend** connects two different profiles along a path.

Extrusions

Extrusions are the simplest elements to create. All you need to do is draw a closed profile using Autodesk Revit sketch tools and assign a depth for the extrusion, as shown in Figure 4–42.

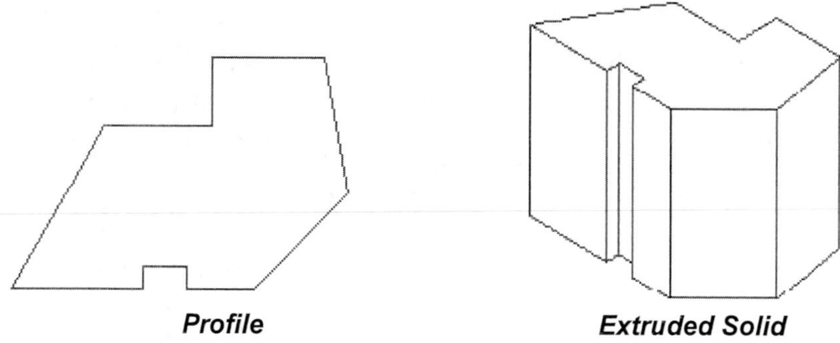

Profile *Extruded Solid*

Figure 4–42

How To: Create Solid Extrusions

1. In the *Create* tab>Forms panel, click (Extrusion).
2. In the Options Bar, set the *Depth* for the extrusion, as shown in Figure 4–43.

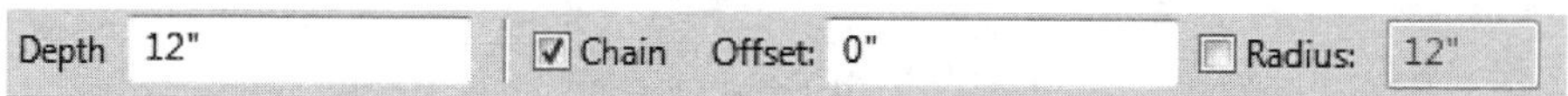

Figure 4–43

   - A positive value for the depth extrudes up or toward you depending on the work plane. A negative value extrudes down or away from you.

3. In the *Modify | Create Extrusion* tab>Draw panel, use the sketch tools to create the profile for the extrusion. You can add dimensions and reference planes as needed to create the profile.
4. In the Mode panel, click (Finish Edit Mode) to create the extrusion.
5. In the *Modify | Extrusion* tab, make changes to the extrusion as needed.
6. In Properties, you can adjust the *Extrusion Start* and *Extrusion End*, the *Visible* and *Visibility/Graphics Overrides*, and *Material*. You can also change it from a **Solid** to a **Void** or vice versa, and place it in the *Subcategory* of **Hidden Lines** or **Overhead Lines**.

*By default, **Start** is at the work plane and **End** is at the depth.*

- In the *Create* tab>Work Plane panel, click (Set) to select the plane on which you want to draw a profile (e.g., on top of another extrusion).

Blends

If the entire blended element is not displayed in plan view, select the View name in the Project Browser, and then in Properties, modify the View Range Cut Plane.

Blends are defined by two profiles: one for the base (bottom) and one for the top. The two profiles are connected by the solid element. If the base and top do not have the same number of corners, adjust the vertex connection, as shown in Figure 4–44.

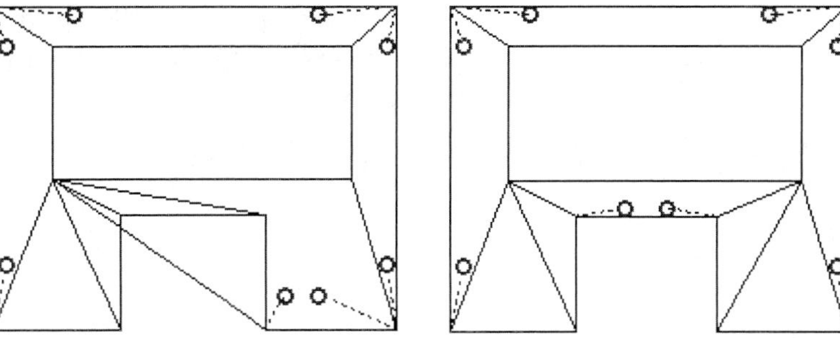

Figure 4–44

How To: Create Solid Blends

1. In the *Create* tab>Forms panel, click (Blend).
2. In the *Modify | Create Blend Base Boundary* tab>Draw panel, use the sketch tools to draw the base profile.
3. In the Mode panel, click (Edit Top).
4. In the Options Bar, set the *Depth* for the top.
5. Draw the top profile using the sketch tools.
6. If the intersections are complex, click (Edit Vertices) and make sure all of the connections work correctly.
7. Click (Finish Edit Mode) to create the blend.
8. The *Modify | Blend* tab displays, where you can edit the base or the top. You also have access to the Properties of the blend.

- When you click (Edit Vertices), open blue dots are displayed on the dotted lines. These lines are suggested connections. Select the dot to toggle the connection on or off, and repeat with other corners to obtain the required connections. Use the tools in the *Edit Vertices* tab>Vertex Connect panel, as shown in Figure 4–45, to modify the vertices.

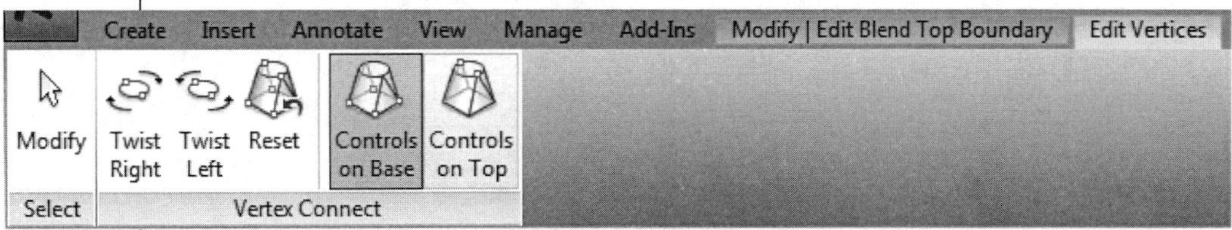

Figure 4–45

Revolves

A revolved solid form requires a profile and an axis around which it is revolved, as shown in Figure 4–46. The profile often needs to be sketched in a work plane perpendicular to the *ground*. Before sketching the profile, create a reference plane that you can use for the work plane. For example, if you are creating a dome, create two intersecting reference planes at the center of the dome and name them. Then, when you start the revolution, you can select one reference plane for your work plane and the other for your axis.

You can name a reference plane in Properties.

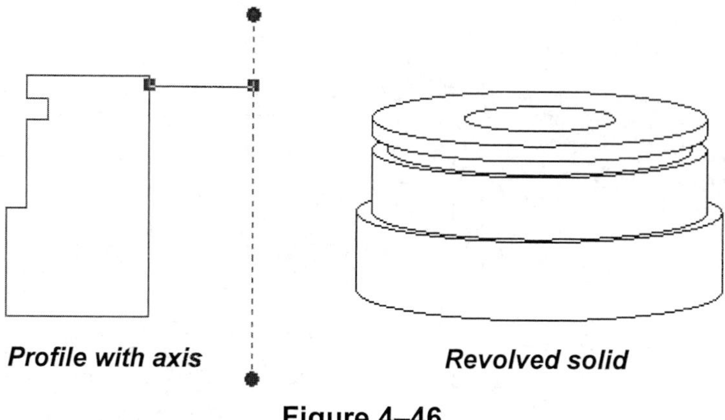

Figure 4–46

How To: Create Solid Revolves

1. In the *Create* tab>Forms panel, click (Revolve).
2. In the *Modify | Create Revolve* tab>Work Plane panel, click (Set). In the Work Plane dialog box, select the plane on which you want to draw the profile. A prompt warns you if you need to change views to draw in that work plane.
3. In the Draw panel, click (Boundary Line) and draw the profile.
4. In the Draw panel, click (Axis Line) and draw or select the axis about which the profile will rotate.
5. Click (Finish Edit Mode) to create the solid.
6. The *Modify | Revolve* tab displays, where you can edit the revolve. You can also change the Properties (such as the *End Angle* and *Start Angle*).

The axis can be an edge of the profile or some distance away if there is an opening.

Sweeps

A sweep is similar to an extrusion. However, instead of only extruding in one direction, it follows a path that can have multiple segments, as shown in Figure 4–47. You define the path first and then create the profile on the path.

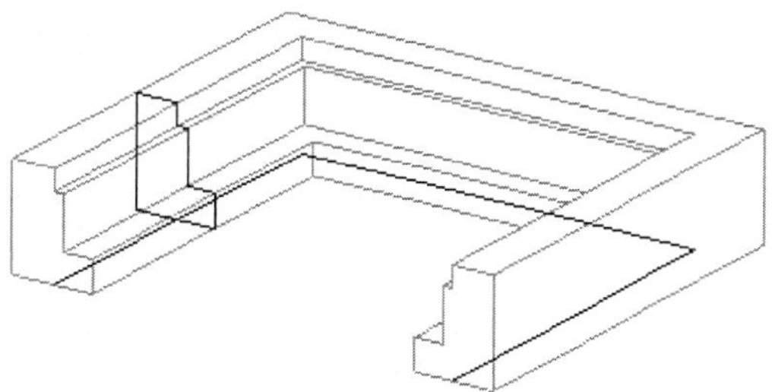

Figure 4–47

How To: Create Solid Sweeps

1. In the *Create* tab>Forms panel, click (Sweep).
2. In the *Modify* | *Sweep* tab>Sweep panel, click (Sketch Path) or (Pick Path). Draw or select the path you want the sweep to follow.
3. Click (Finish Edit Mode).
4. In the Sweep panel, select a profile from the list or click (Select Profile), and then click (Edit Profile) to sketch a profile.
    - The Go To View dialog box opens if you need to change views to draw the profile.Select the view in which you want to draw the profile and click **Open View**.
5. Draw the profile for the sweep on the profile plane, as shown in Figure 4–48.

Click (Load Profile) to load additional profiles into the family file for use with the sweep.

The profile plane is automatically drawn on the first line of the path. It is the red dot with the green line through it, as shown in a 3D view in Figure 4–48. The red dot is the place where the profile and path intersect.

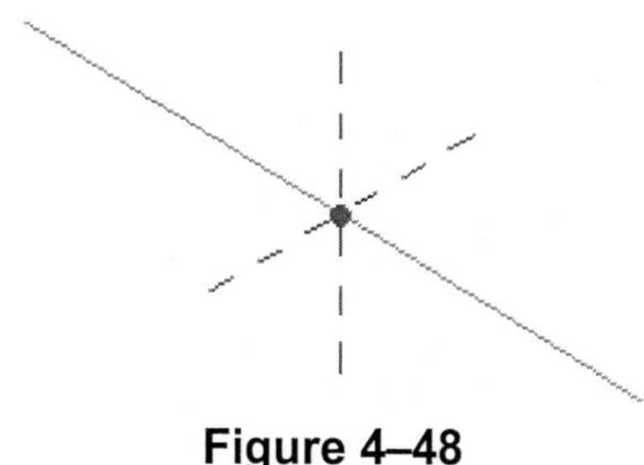

Figure 4–48

6. Click (Finish Edit Mode) twice to complete the process.
7. The *Modify* | *Sweep* tab displays, where you can edit the sweep. In Properties, you can change several options of the profile along with the other standard options.

Swept Blends

Swept blends consist of a path and two profiles, as shown in Figure 4–49. The profiles can have different shapes. The path must be one segment, which can be a line, an arc, or a spline.

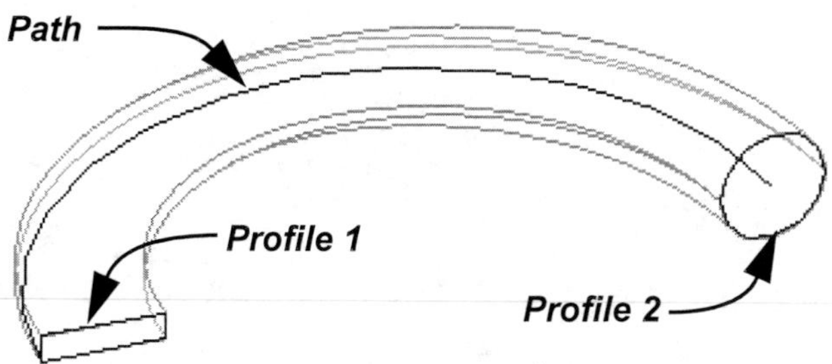

Figure 4–49

How To: Create Solid Swept Blends

1. In the *Create* tab>Forms panel, click (Swept Blend).
2. In the *Modify | Swept Blend* tab>Swept Blend panel, click (Sketch Path) or (Pick Path). Draw or select the path you want the swept blend to follow. You can only have one segment in the path.
3. Click (Finish Edit Mode).
4. In the Swept Blend panel, click (Select Profile 1).
5. Select a profile from the list or create a new profile using (Edit Profile).
6. Click (Select Profile 2) and repeat the process. It does not need to be the same size or shape as the first profile.
7. In the Swept Blend panel, click (Edit Vertices) and align the vertices. This controls the twist of the swept blend.
8. Click (Finish Edit Mode).
9. In the *Modify | Swept Blend* tab, make any changes as needed.
10. In Properties, set any options as needed.

Aligning and Locking

As you draw elements in families, it is important to lock and align them to reference planes so that they can be moved parametrically. The padlock symbol displays when you snap to reference planes while drawing. Select the padlock to lock the element to the reference plane, as shown in Figure 4–50.

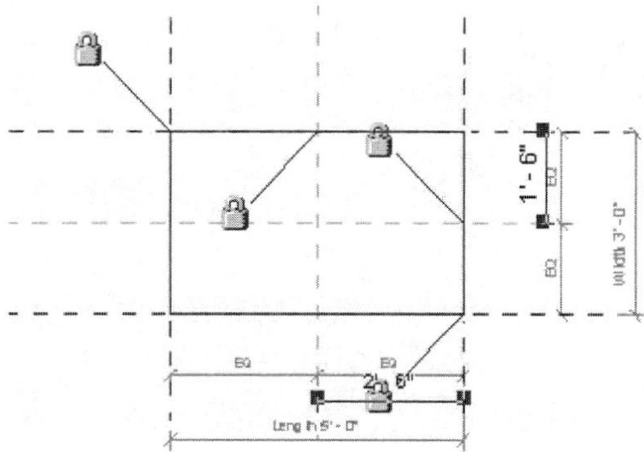

Figure 4–50

- When using the (Pick Line) option, in the Options Bar you can select **Lock** to automatically lock the sketch line to the selected reference plane.
- You can also use the **Align** command to align and lock elements to reference planes.

Hint: Locking Sketches to Reference Planes

It is important to lock sketches to reference planes (rather than other elements in a family) so that they will flex correctly. To ensure that you have the right intersection, press <Tab> until the appropriate elements are highlighted, and then snap in the required location, as shown in Figure 4–51.

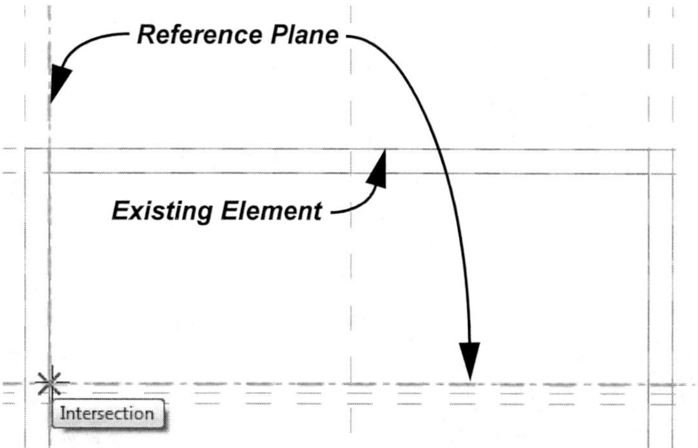

Figure 4–51

- You can temporarily hide other elements before sketching to ensure that you only work with the reference planes.

Practice 4d Create Family Geometry for the Bookcase

Practice Objectives

- Create solid extrusion elements.
- Align and lock the elements to reference planes.
- Join the geometry of elements.

Estimated time for completion: 20 minutes

In this practice you will create extruded solids for the base, frame, shelves, and back of a bookcase, and lock them in place with dimensions. You will also join the solid elements together to form one unit, as shown in Figure 4–52.

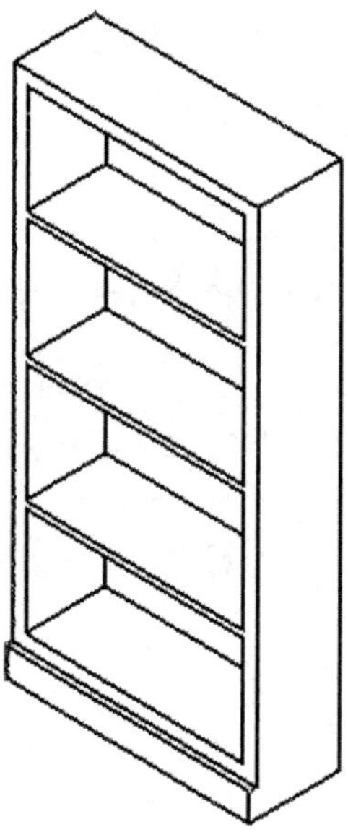

Figure 4–52

Task 1 - Create a base for the bookcase.

1. In the ...*Architectural* folder, open **Barrister-Bookcase-Geometry.rfa**.
2. Open the **Floor Plans: Ref. Level** view.
3. In the *Create* tab>Forms panel, click (Extrusion). In the Options Bar, set the *Depth* to **4"**.

4. In the *Modify | Create Extrusion* tab>Draw panel, use the sketch tools to draw a rectangle across the outside intersections of the reference planes. Lock all four lines to the reference planes, as shown in Figure 4–53.

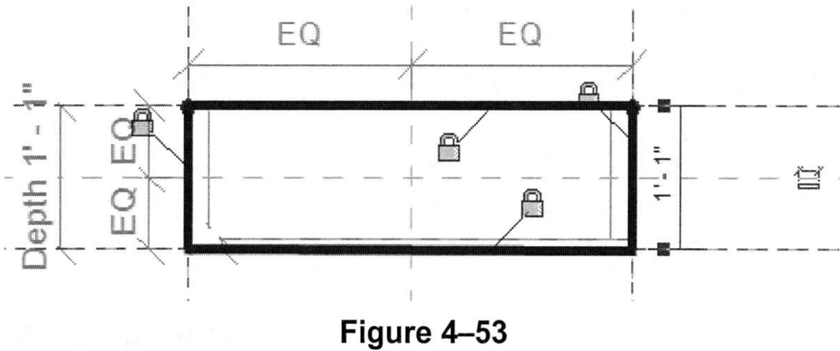

Figure 4–53

5. Click (Finish Edit Mode).

6. Open the **Elevations: Front** view.

7. In the *Modify* tab>Modify panel, click (Align). Align the top of the extrusion to the *Base* reference plane and lock it.

8. Save the family.

Task 2 - Create the frame and shelves for the bookcase.

1. In the *Create* tab>Work Plane panel, click (Set).

2. In the Work Plane dialog box, set the *Name* to **Reference Plane : Door Face** and click **OK**.

3. Start the **Extrusion** command.

Drawing the reference planes inside the extrusion helps to keep the primary views from getting cluttered.

4. Draw four reference planes inside the existing reference planes. Dimension and label them as shown in Figure 4–54.

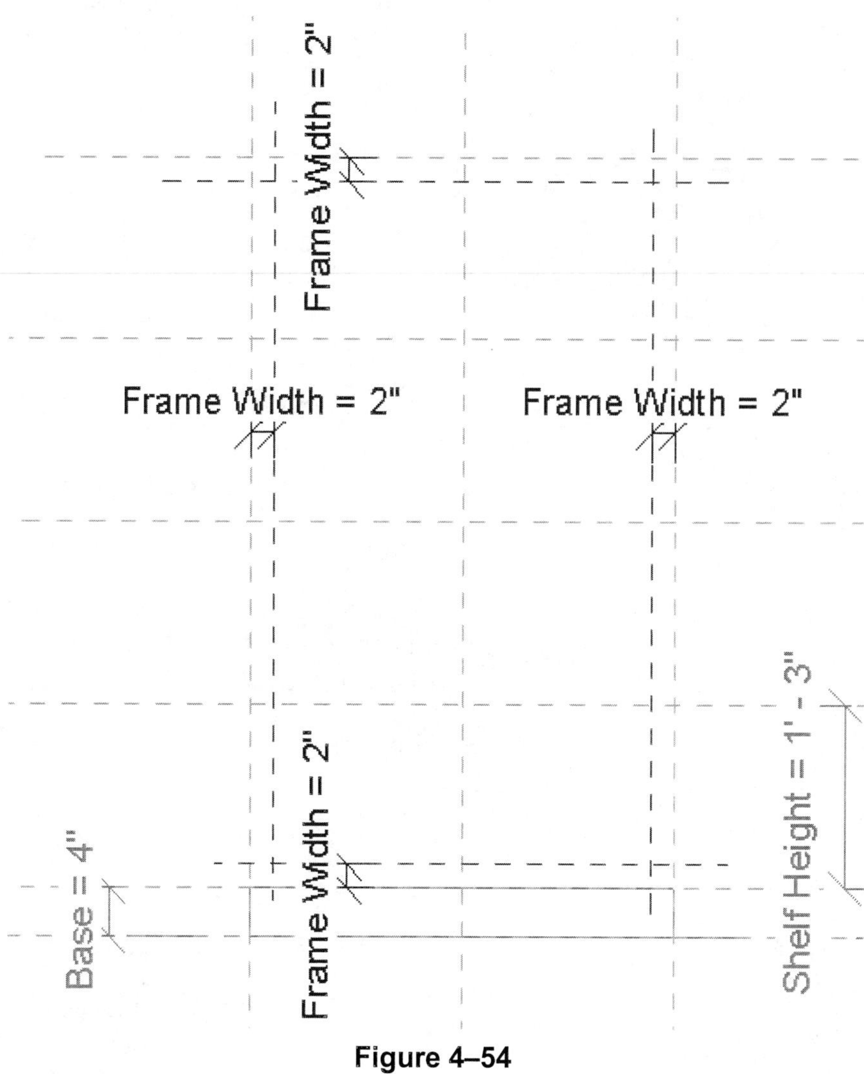

Figure 4–54

5. In the Family Types dialog box, flex the *Height* and *Width* of the overall bookcase to verify that the new **Frame Width** parameters move as expected.

6. Return to the *Modify | Create Extrusion* tab and select the **Rectangle** option.

7. In the Options Bar, set the *Depth* to **1'-0"**.

*Hint: type **SI** to force the snap to the intersection of the reference planes.*

8. Draw and lock two rectangular sketches to the reference planes, as shown in Figure 4–55.

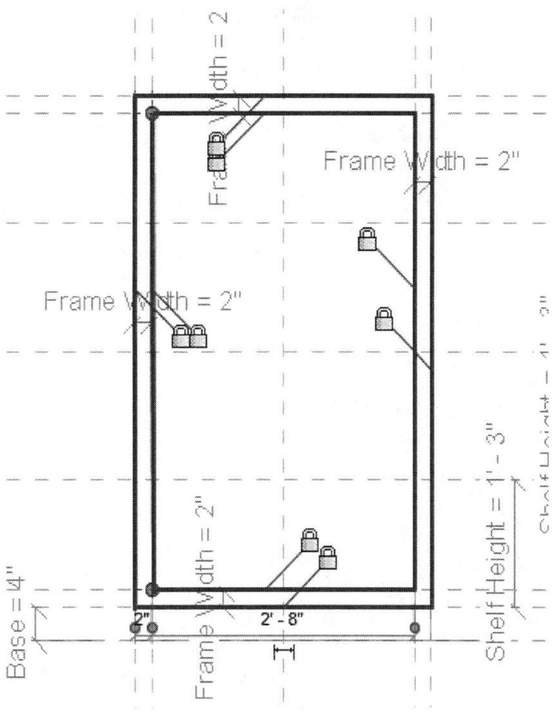

Figure 4–55

9. Click (Finish Edit Mode).

10. Start the **Extrusion** command again.

11. Draw two reference planes on either side of the shelf height reference plane. Dimension and label them as shown in Figure 4–56. Repeat with the other two shelf height locations.

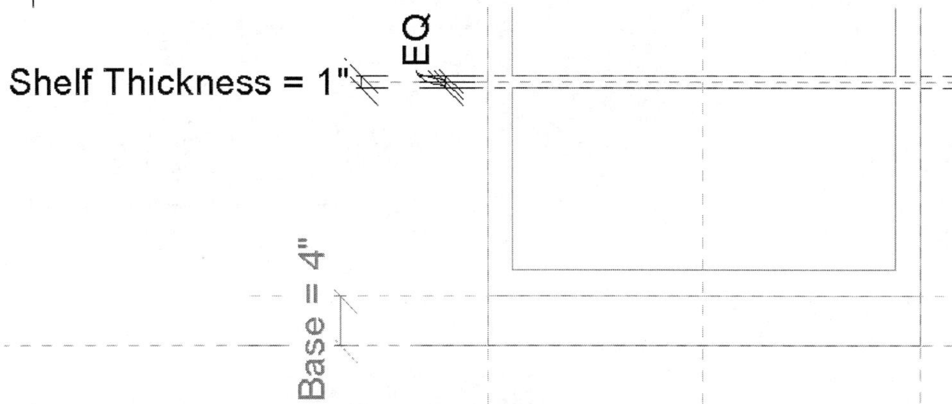

Figure 4–56

- To make the labels more readable, change the scale to **1"=1'-0"**. You can also create an additional dimension type with the *Read Convention* set to **Horizontal** and use it for the **Shelf** parameters, as shown in Figure 4–56.

12. In the *Modify | Create Extrusion* tab>Draw panel, select the **Rectangle** option.

13. In the Options Bar, ensure that the *Depth* is set to **1'-0"**.

14. Draw the sketch around each shelf, as shown in Figure 4–57. Lock the sketches to the reference planes.

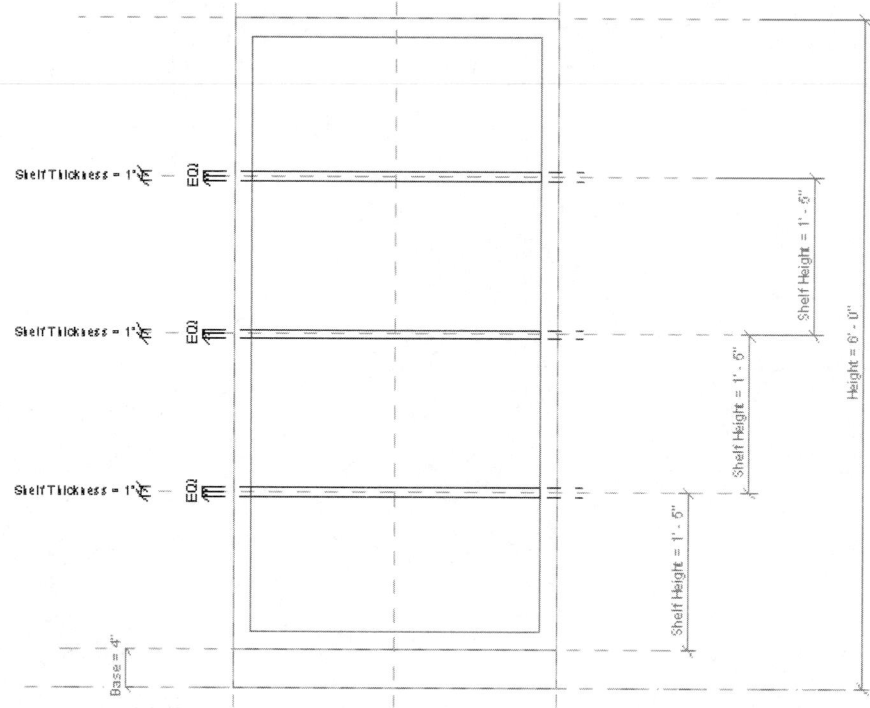

Figure 4–57

15. Click (Finish Edit Mode).

16. Open the Family Types dialog box and use **Move Up** and **Move Down** to move the *Height*, *Width*, and *Depth* to the top of the parameters list.

17. Flex the *Height* and *Width* of the overall bookcase to verify the new **Shelf Thickness** parameters move as expected. Finish the flexing with the *Height* set to **5'-0"** and the *Width* to **3'-0"**.

18. Save the family.

Task 3 - Create the back of the bookcase.

1. Open the **Elevations: Right** view.

2. Align and lock the back of the frame and shelf extrusions to the back reference plane.

3. Open the Floor Plans: **Ref. Level** view.
4. Start the **Extrusion** command again.
5. Add a reference plane **1/4"** from the back of the bookcase. Dimension and lock it in place, as shown in Figure 4–58.

Create a label if you want to change the size of the back.

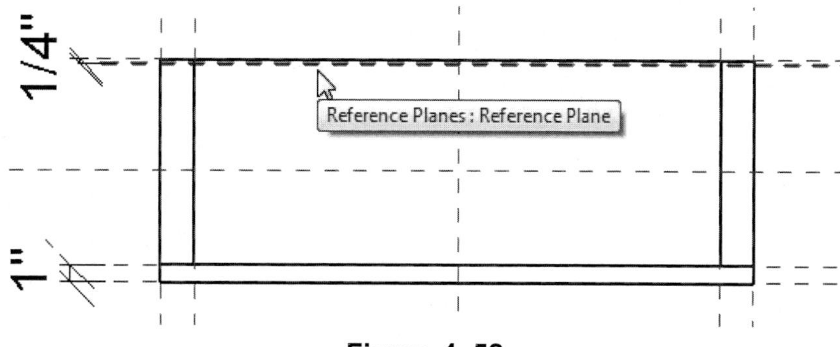

Figure 4–58

6. Return to the *Modify | Create Extrusion* tab and select the **Rectangle** option.
7. In the Options Bar set the *Depth* to **2'-6"**. (This is less than the bookcase height. You will adjust it later.)
8. Draw a rectangular sketch across the back and lock it to the reference planes.
9. Click ✔ (Finish Edit Mode).
10. Open the **Elevations: Back** view. Switch the Visual Style to (Hidden Line), if needed.
11. Align and lock the top of the back panel with the top reference plane. Select the top reference plane first, and then select the top of the panel to align with it, as shown in Figure 4–59.

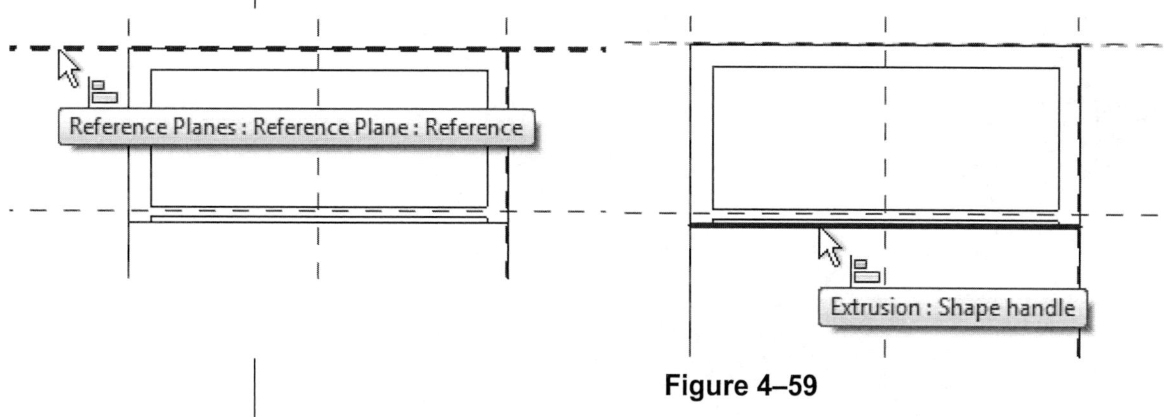

Figure 4–59

12. Switch to a 3D view. The base, frame, shelves, and back are all separate elements, as shown in Figure 4–60 .

13. In the *Modify* tab>Geometry panel, click (Join) and join the four elements. The bookcase should display as shown in Figure 4–61.

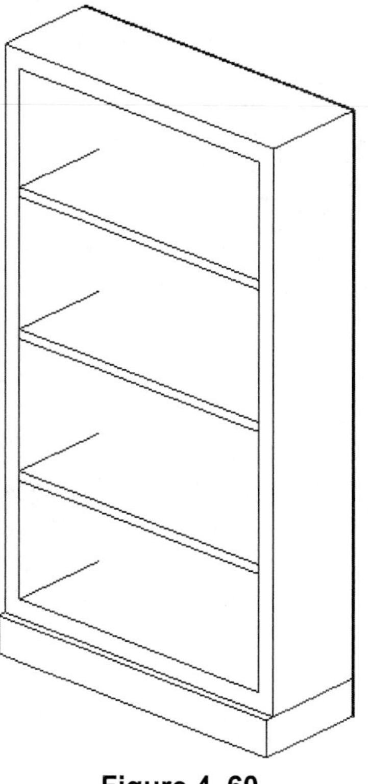

Figure 4–60

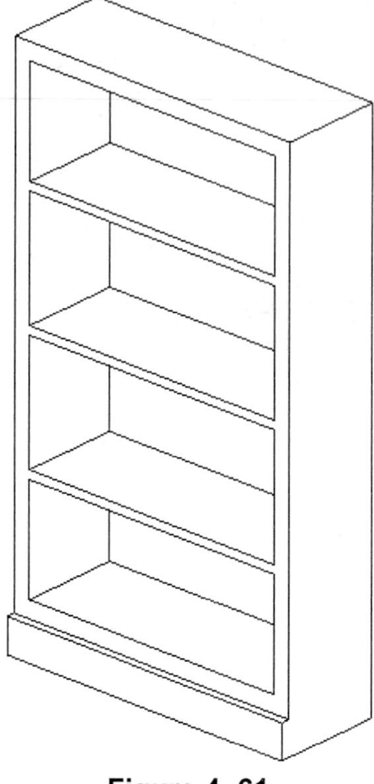

Figure 4–61

14. Save the family.

Practice 4e

Create Family Geometry for the Heat Pump

Estimated time for completion: 20 minutes.

Practice Objective

- Create extrusions and blends.

In this practice you will create extruded and blended solids for the base, cabinet, top, and fan unit of a heat pump, as shown in Figure 4–62.

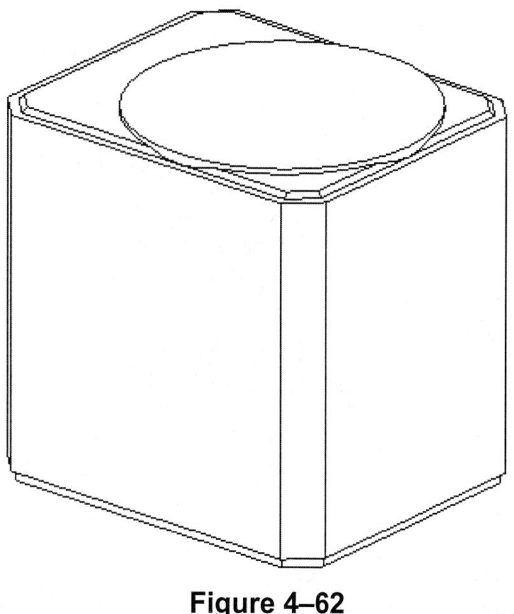

Figure 4–62

Task 1 - Create the main cabinet of a heat pump.

1. In the ...*MEP* folder, open **Heat-Pump-Geometry.rfa.**
2. Ensure that the **Floor Plans: Ref. Level** view is open.
3. In the *Create* tab>Datum panel, click (Reference Line).
4. In the Options Bar, turn off **Chain** if needed.

5. Draw a reference line at each of the four corners, as shown in Figure 4–63. Ensure that you lock any open padlocks that display as you draw each line.

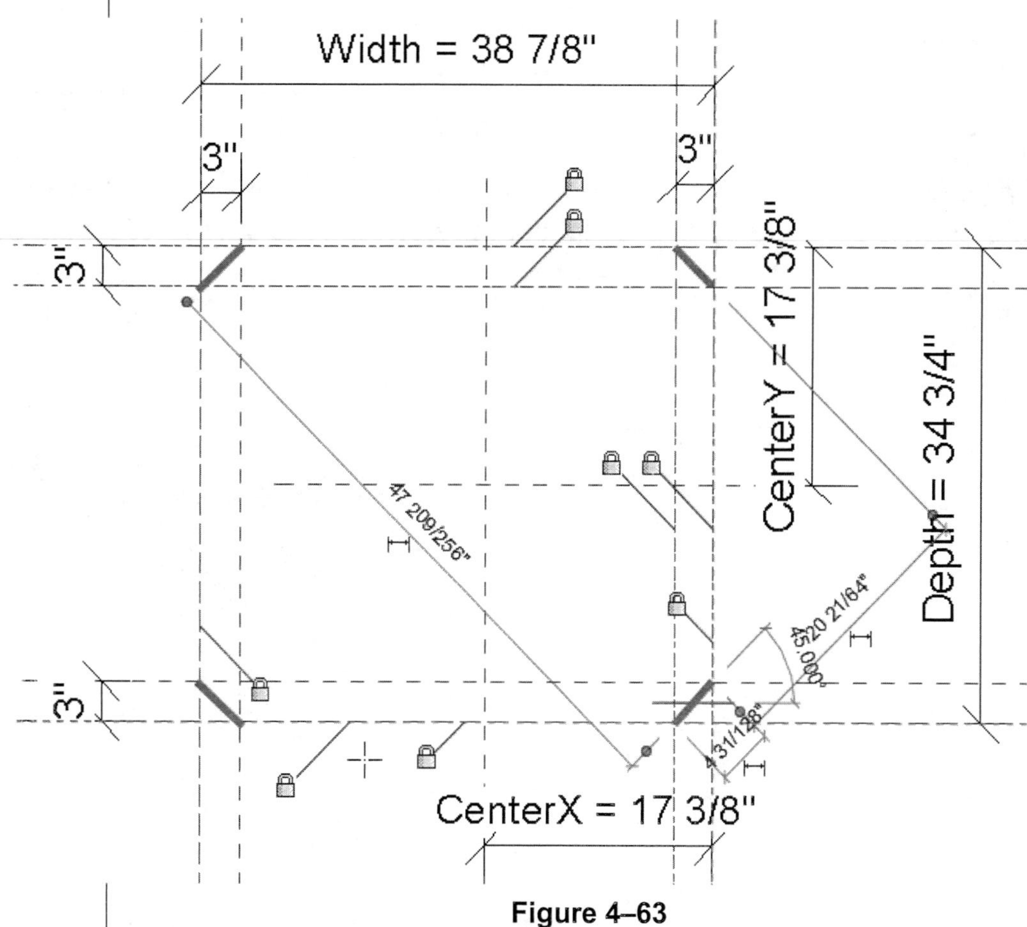

Figure 4–63

6. Flex the framework by changing the *Width* and *Depth*.

7. In the *Create* tab>Work Plane panel, click (Set).

8. In the Work Plane dialog box, set *Name* as **Reference Plane : Top of Base** and click **OK**.

9. In the *Create* tab>Forms panel, click (Extrusion). In the Options Bar, set the *Depth* to **30"**.

10. Sketch and lock the shape to the reference lines shown in Figure 4–64. (Hint: use **Pick Lines** with the **Lock** option set and then trim the excess, as required.)

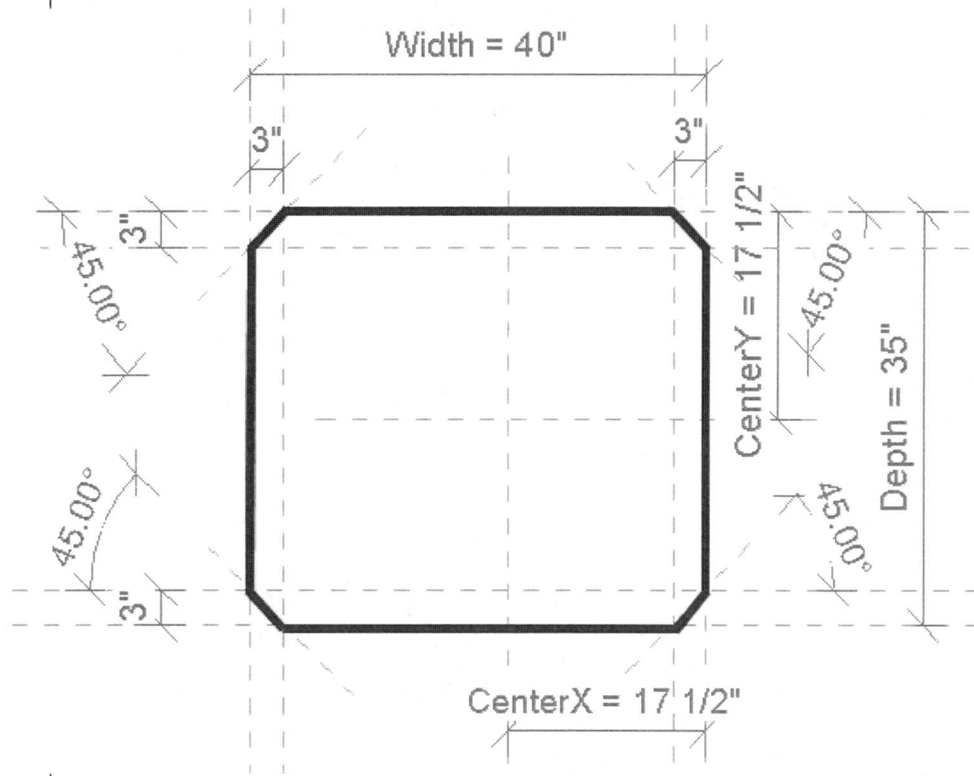

Figure 4–64

11. Click (Finish Edit Mode).

12. Open the **Elevations: Front** view. Align and lock the top of the unit extrusion to the **Top of Unit** reference plane, as shown in Figure 4–65.

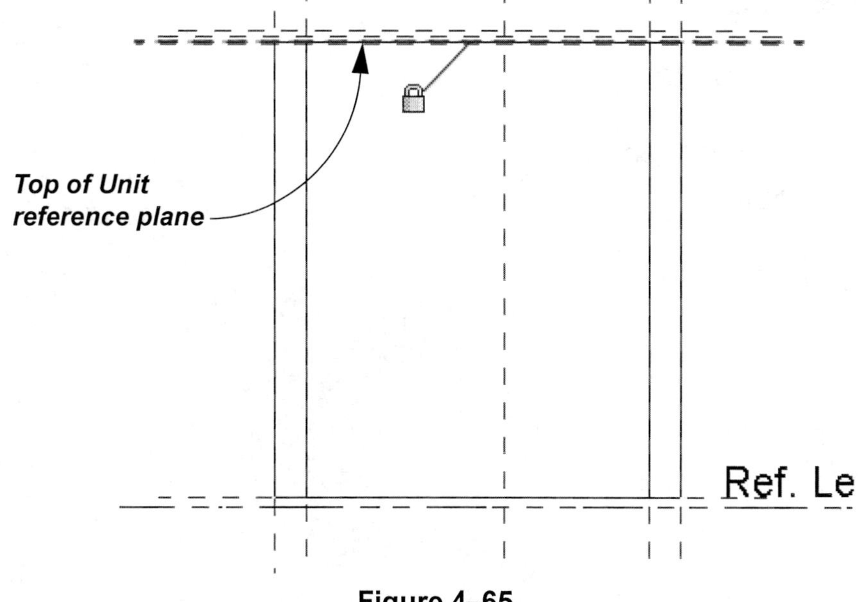

Figure 4–65

13. Save the family.

Task 2 - Create the top and base of the heat pump.

1. Return to the **Floor Plans: Ref. Level** view.
2. Temporarily hide the existing extrusion.
3. Set the Work Plane to the **Top of Unit** reference plane.
4. In the *Create* tab>Forms panel, click (Blend).
5. In the *Modify | Create Blend Base Boundary* tab>Draw panel, click (Pick Lines).
6. In the Options Bar, set the *Depth* to **1/2"**, the *Offset* to **1"**, and select **Lock**.
7. Click on the outer-most reference planes and reference lines. Use **Trim/Extend to Corner** to clean up the sketch for the base, as shown in Figure 4–66.
8. Click (Edit Top). Select **Pick Lines**, and then set the *Depth* to **1/2"** and the *Offset* to **1-1/2"**.
9. Use and trim the same reference planes and reference lines to create a sketch for the top, as shown in Figure 4–66.

Using reference planes to establish the sketch of the blend, rather than other elements, ensures the accuracy of the elements. It is also required if you are posting your families to Autodesk Seek.

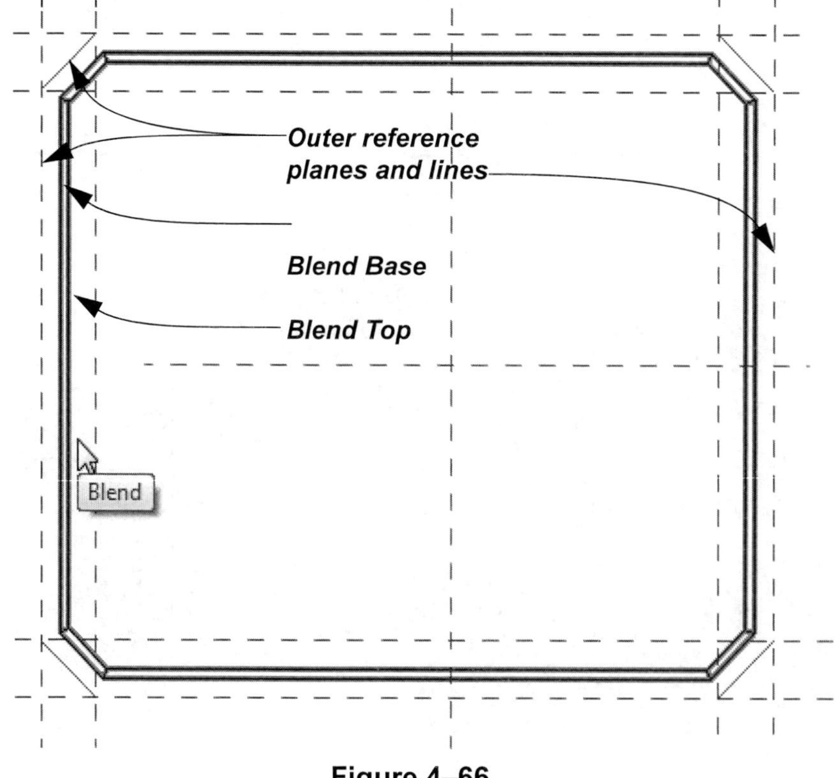

Figure 4–66

10. Click (Finish Edit Mode).

11. Flex the framework to ensure that the blend is working.

12. Select the blend and temporarily hide it.

13. Set the Work Plane to **Level: Ref. Level**.

14. In the *Create* tab>Forms panel, click (Extrusion). In the Options Bar, set the *Depth* to **1"**.

15. In the *Modify | Create Extrusion* tab>Draw panel, use the **Pick Lines** tool to draw a boundary at **1/2"** inside the unit reference planes, as shown in Figure 4–67.

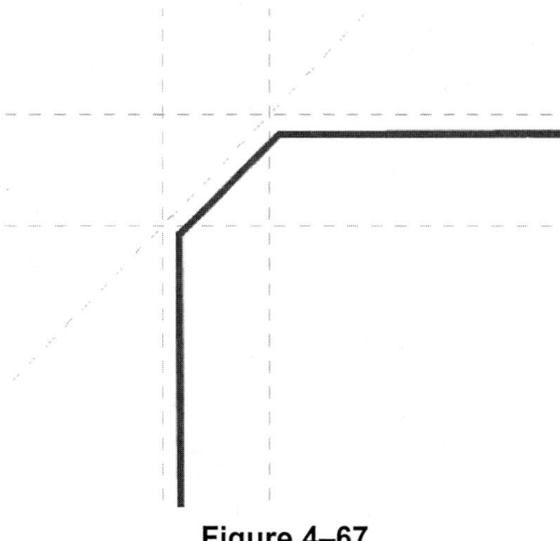

Figure 4–67

16. Click (Finish Edit Mode).

17. Reset Temporary Hide/Isolate.

18. Open the **Elevations: Front** view.

19. Temporarily hide the main cabinet extrusion.

20. In the *Modify* tab>Modify panel, click (Align) and align the top of the base extrusion to the *Top of Base* reference plane and lock it, as shown in Figure 4–68.

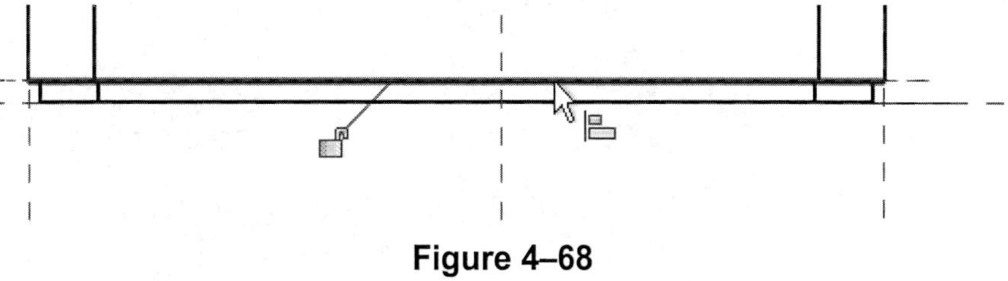

Figure 4–68

21. Reset Temporary Hide/Isolate.

22. Open a 3D view.

23. Open the Family Types dialog box and use **Move Up** to reorder the parameters so that *Width*, *Height,* and *Depth* are at the top.

24. Flex the dimensions to check that everything is working as expected. Save the family.

Task 3 - Add the fan.

1. Return to the **Floor Plans: Ref. Level** view.

2. Set the Work Plane to the **Top of Unit** reference plane.

3. Start the **Extrusion** command and in the Options Bar, set the *Depth* to **1"**.

4. Draw a circle from the intersection of the CenterX and CenterY reference planes with a radius close to the outside, as shown in Figure 4–69. The exact dimension does not matter for this step.

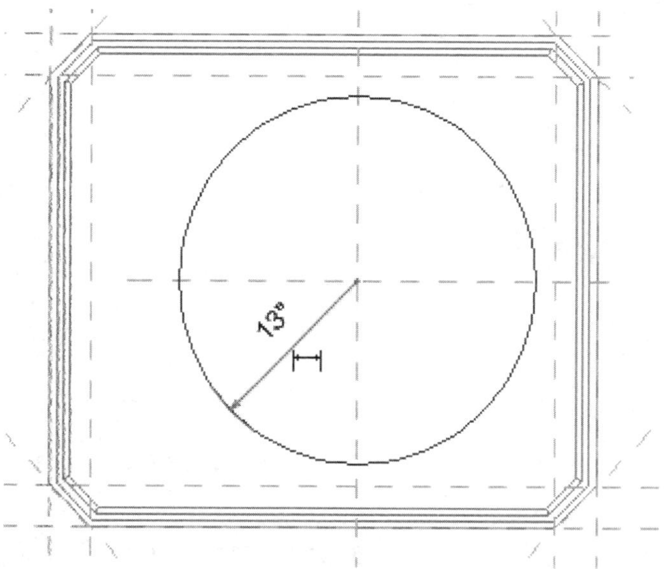

Figure 4–69

5. Dimension the radius of the circle and create a new Label based on the dimension named **Radius of Fan**.

6. Open the Family Types dialog box and move the **Radius of Fan** parameter below all of the others.

7. Next to **Radius of Fan**, in the *Formula* area, type **Depth / 2 - 1"**.

8. Click **Apply**.

9. Flex the element by changing the *Depth*. When it works as expected, click **OK**.

10. Click (Finish Edit Mode).

11. Open the **Elevations: Front** view. Align and lock the top of the fan to the top reference level, as shown in Figure 4–70.

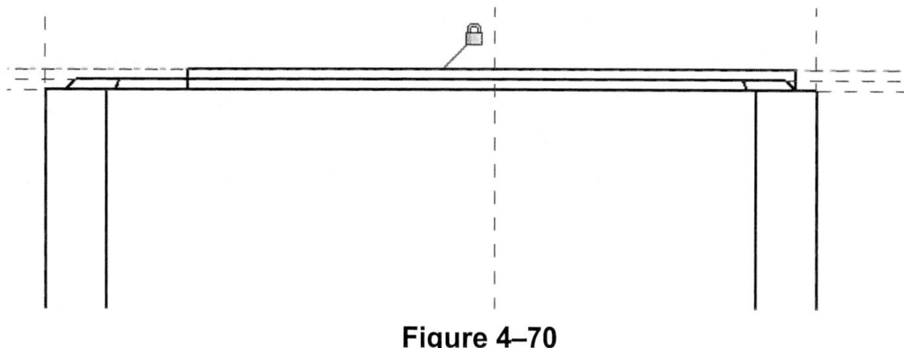

Figure 4–70

12. In the *Modify* tab>Geometry panel, click (Join).

13. Join the unit and the beveled blend together. Then join the fan to the unit and the base to the unit.

14. View the unit in 3D, as shown in Figure 4–71.

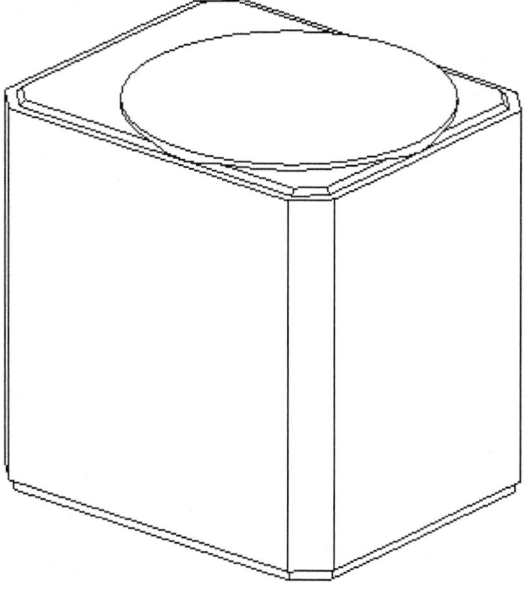

Figure 4–71

15. Save the family.

Practice 4f Create Family Geometry for the Structural Column

Estimated time for completion: 10 minutes.

Practice Objectives

- Create extrusions.
- Add dimensions and labels.

In this practice you will create extruded solids for the base and open framework of the column. You will lock them in place with the **Align** tool. You will also test the geometry by modifying the height and widths of the columns. The final version is shown in Figure 4–72.

Figure 4–72

Task 1 - Create a base for the column.

1. In the *...\Structural* folder, open **Trellis-Column-Geometry.rfa**.
2. Open the **Floor Plans: Lower Ref. Level** view.
3. In the *Create* tab>Forms panel, click (Extrusion). In the Options Bar, set the *Depth* as **2"**.

4. Draw four reference planes to the outside of the existing reference planes. Dimension and lock them at 2", as shown in Figure 4–73.

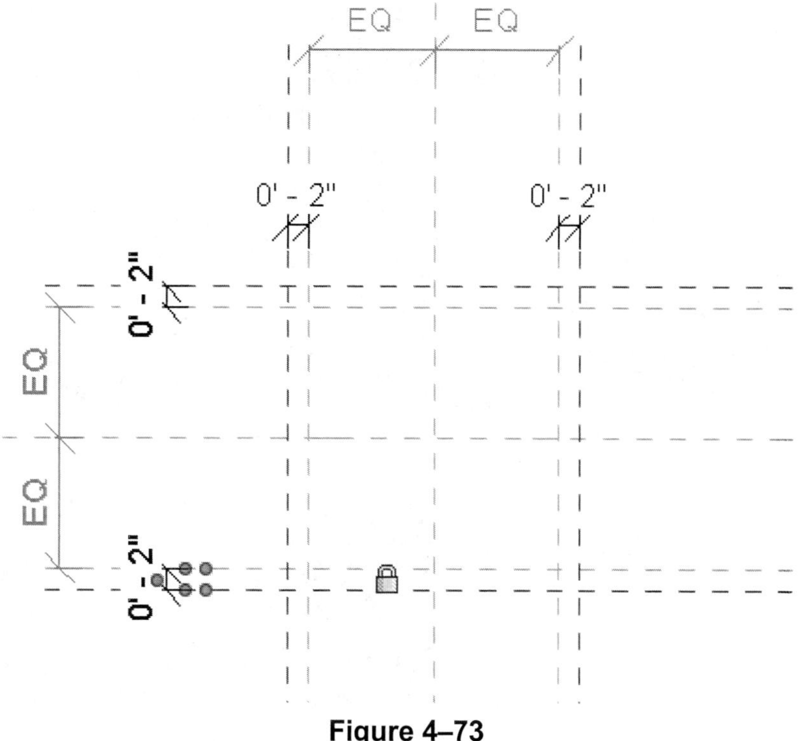

Figure 4–73

5. In the *Modify | Create Extrusion* tab>Draw panel, use the sketch tools to draw a rectangle across the outside intersections of the reference planes. Lock all four lines to the reference planes, as shown in Figure 4–74.

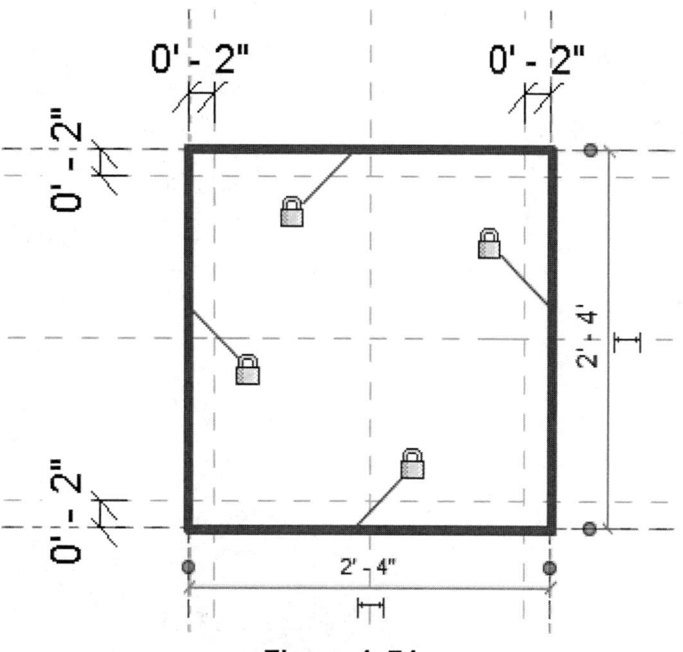

Figure 4–74

6. Click (Finish Edit Mode).
7. Open the **Elevations: Front** view.
8. In the *Modify* tab>Modify panel, click (Align). Align the top of the extrusion to the lower reference plane and lock it, as shown in Figure 4–75.

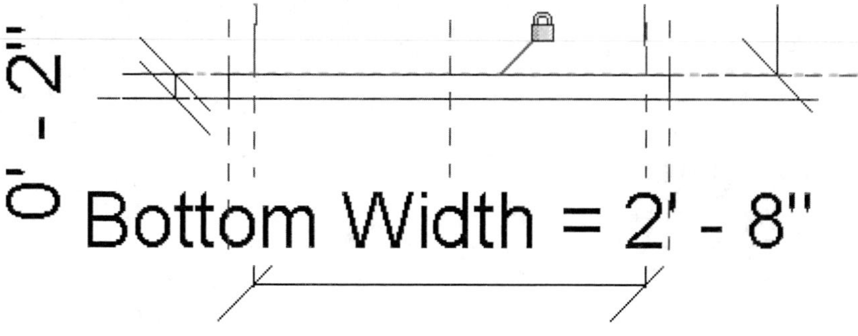

Figure 4–75

9. Save the family.

Task 2 - Create the open work frame for the column.

1. Continue working in the **Elevations: Front** view.
2. In the *Create* tab>Work Plane panel, click (Set).
3. In the Work Plane dialog box, set *Name* as **Reference Plane : Front**, as shown in Figure 4–76, and click **OK**.

All of these reference planes were added and named in the family template file.

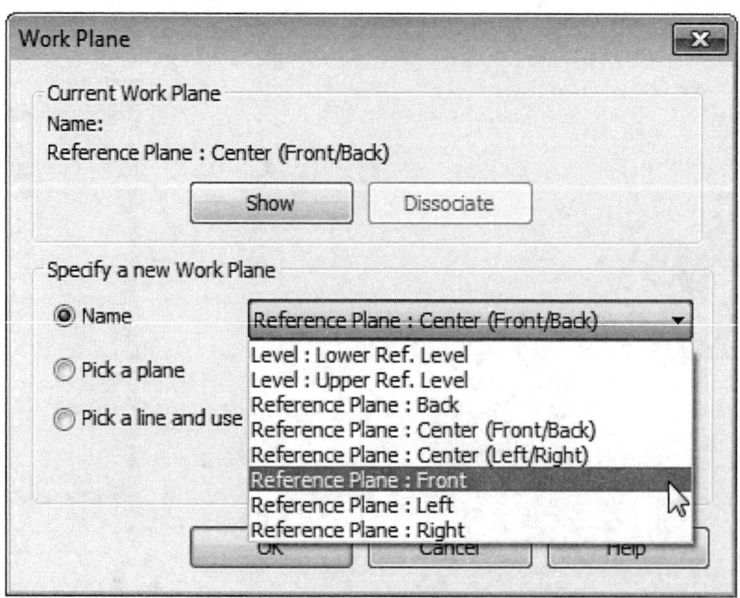

Figure 4–76

4. In the *Create* tab>Datum panel, click (Reference Line).
5. Draw four angled reference lines, as shown in Figure 4–77.
   - Make sure you are snapping to the intersections of the horizontal reference planes and the angled reference lines that establish the taper.
   - Use the shortcut **SI** to snap to the intersections and not the end points.

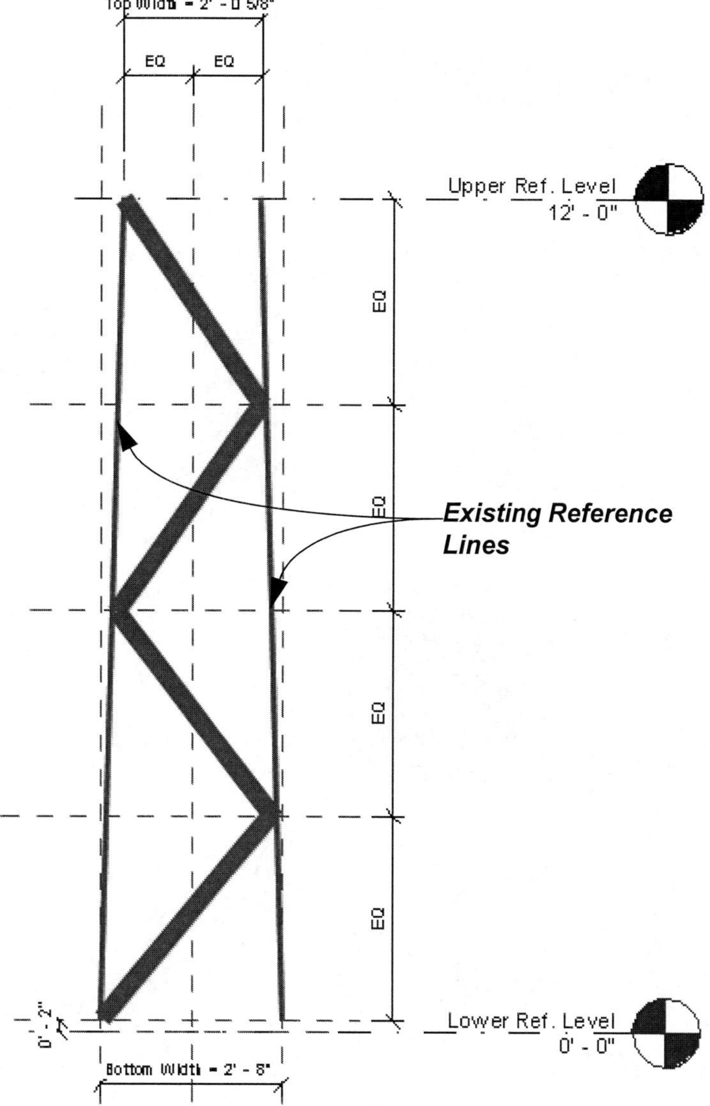

Figure 4–77

6. Change the height of the Upper Ref. Level and the Bottom Width and verify that all of the reference planes and lines are working as expected. End with the **Upper Ref. Level** set to **12'-0"** and the **Bottom Width** set to **2'-8**.

7. Start the **Extrusion** command.
8. Draw the sketch shown in Figure 4–78. On the Options Bar, use (Pick Lines) with Lock selected.
   - Use offsets, (Split Element), and (Trim Element) to create the lines, as required.
   - Make sure the sketch is completely closed.
   - Zoom in to make adjustments.
   - All dimensions are **2"**.

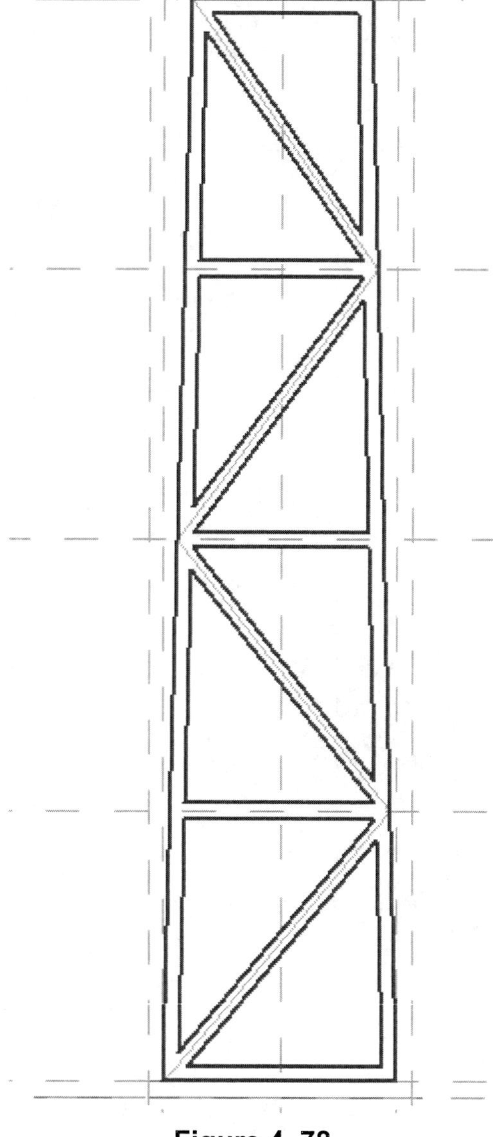

Figure 4–78

9. Click (Finish Edit Mode).
10. Open the **Floor Plans: Lower Ref. Level** view.

11. Align and lock the extrusion to the Front and Back Reference planes as shown in Figure 4–79.

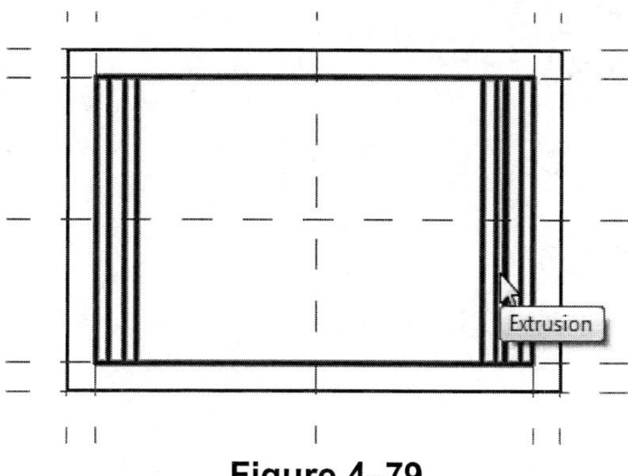

Figure 4–79

12. Open the **3D Views: View 1** view.

13. In the View Control Bar, set the Visual Style to (Shaded).

14. Rotate the view so you can see the open framework.

15. Save the family.

Task 3 - Dimension the framework for flexing.

1. Open the **Elevations: Front** view.

2. Open the Family Types dialog box and move it so you can see the column.

3. In the Family Types dialog box, change the **Top Width** *Formula* to **Bottom Width /1.5** and click **Apply**.

4. You can see how the connections of the open framework deform, as shown in the detail in Figure 4–80.

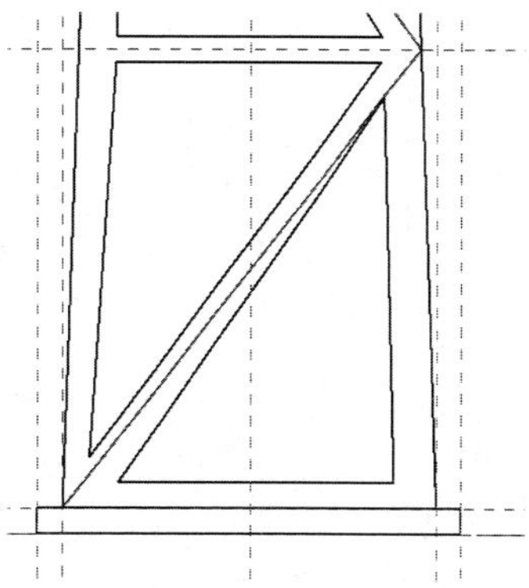

Figure 4–80

5. Close the dialog box and undo the change.

It is important to work with the exact sizes when you create the extrusion, otherwise the angled lines are not exactly parallel and you cannot dimension them.

6. Verify that the **Upper Ref. Level** is set as **12'-0"**.
7. Select the extrusion. In the *Modify | Extrusion* tab>Mode panel, click (Edit Extrusion).
8. Dimension each part of the framework, as shown in part in Figure 4–81.

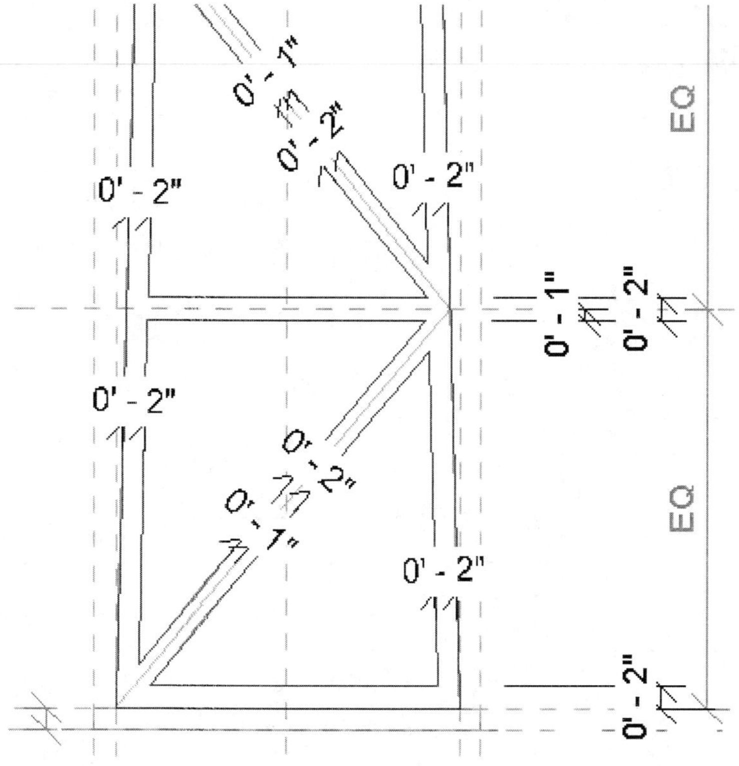

Figure 4–81

- The dimensions around the outside of the framework are dimensioned to a reference line, but the ones that are offset on both sides of the reference line are not. Therefore, you have to add a 1" dimension to the reference line for the constraints to be satisfied.

9. Open the Family Types dialog box and change the formula again. Does everything work as expected? If not, click the **Cancel** button to close the dialog box and fix the dimensions.
10. Click (Finish Edit Mode).
11. View the family in 3D.
12. Save the family.

4.4 Creating Family Types

An important aspect of Family Types is that it helps you *flex* or test the parametric dimensions you set up, as well as helps you create formulas for parameters. A further use for Family Types is the ability to create preset sizes for insertion into a project. When you select the related command, the types display in the Type Selector, as shown in Figure 4–82 for a desk component.

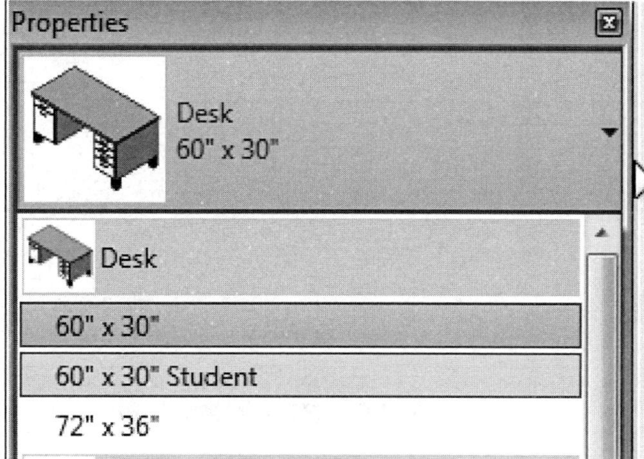

Figure 4–82

How To: Create Family Types

1. In the *Create* or *Modify* tab>Properties panel, click (Family Types).
2. In the Family Types dialog box, in the *Family Types* area, click **New...**.
3. In the Name dialog box, type a name for the Family Type (as shown in Figure 4–83) and then click **OK**.

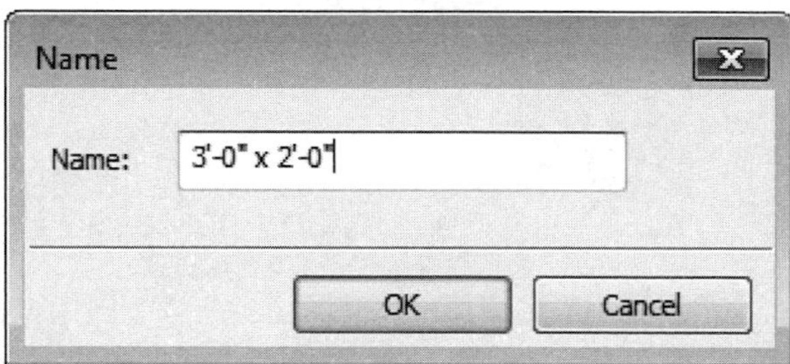

Figure 4–83

- Select a name that is useful for the entire set of types, typically a size.
- The name of the family always precedes the name of the type. Therefore, you do not need to re-enter that information.

4. In the Family Types dialog box, set the *Value(s)* for that size (as shown in Figure 4–84) and then click **Apply**.

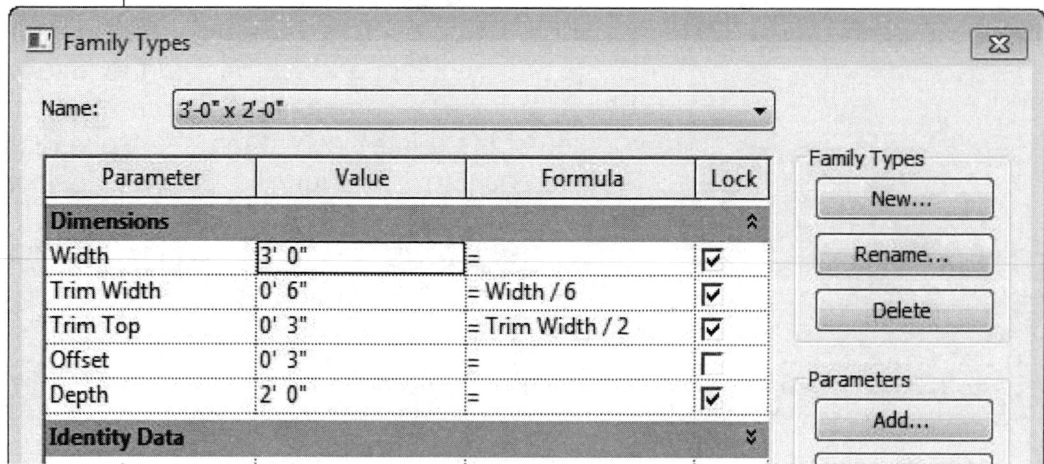

Figure 4–84

5. Repeat the process. You can create as many different types as you need.
6. Verify that all of the types are functioning correctly.
7. Click **OK** to finish the command.
8. Save the family.
9. Load the family into a project and test the new types you have created.

- In the Family Types area click **Rename...** and **Delete** as required to modify the list.
- To edit a type, select the type name, change the value(s), and click **Apply**.

Hint: Exporting and Importing Family Types

Family Types can be exported to a text file and then imported into a related family. This can save you time when creating different versions of the same basic category.

- In the Application Menu, expand (Export), scroll down and click (Family Types). In the Export As dialog box, select a location and name. Click **Save**.
- In the *Insert* tab>Import panel, click (Import Family Types). In the Import Family Types dialog box, select the required text file, and click **Open**.

Working with Families in Projects

To continue testing your family it helps to see how it works in a project. You can load a family into one or more projects while in the Family Editor. You must have a project open before doing this.

How To: Load a Family into a Project

1. In the Family Editor panel, click (Load into Project). The panel is displayed in all of the Family Editor tabs.
2. If only one project is open, the family is automatically loaded into it. If you have more than one project open, the Load into Projects dialog box opens. Select the project(s) to load into, as shown in Figure 4–85.

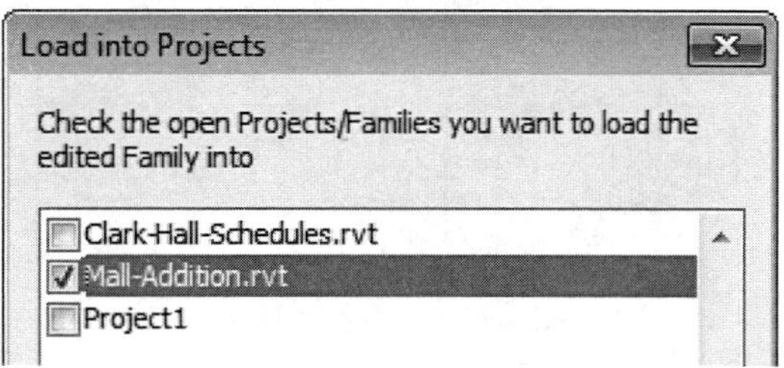

Figure 4–85

3. The project opens with the family loaded. Often, the related command is started with the new family selected in the Type Selector.

- If you have finished working on the family you can click (Load into Project and Close).

Practice 4g Create Family Types for the Bookcase

Practice Objectives

- Create family types of different sizes.
- Test the family in a project.

Estimated time for completion: 10 minutes

In this practice you will create Family Types for different sizes of the bookcase. Then you will test the bookcase family in a project by placing each of the different types, as shown in Figure 4–86.

Figure 4–86

Task 1 - Create Family Types.

1. In the *...\Architectural* folder, open **Barristers-Bookcase-Types.rvt**.
2. In the *Create* or *Modify* tab>Properties panel, click (Family Types).
3. In the Family Types dialog box, in the *Family Types* area, click **New...**.
4. In the Name dialog box, type **3'-0" x 5'-0"** and click **OK**.
5. In the *Dimensions* area, verify or set the following values for the parameters as shown in Figure 4–87:
   - **Width:** 3'-0"
   - **Height:** 5'-0"
   - **Depth:** 1'-1"
   - **Base:** 4"

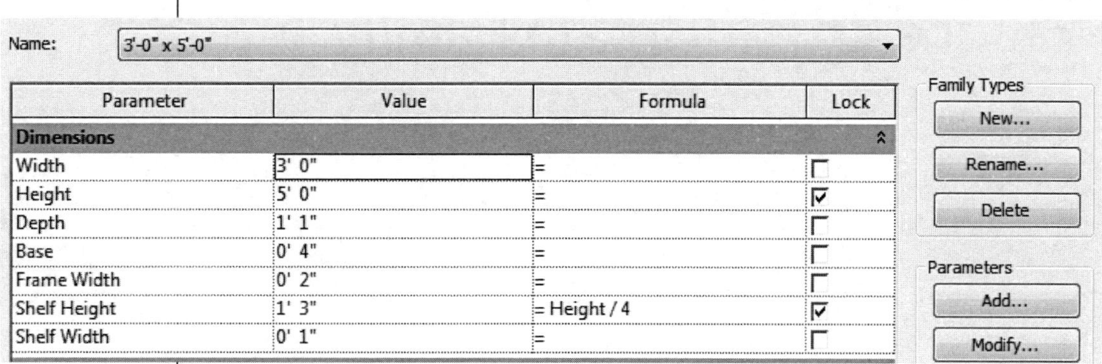

Figure 4–87

6. Click **Apply**.
7. Repeat these steps for the following sizes:
   - **3'-0" x 6'-0"**
   - **3'-6" x 5'-0"**
   - **3'-6" x 6'-0"**
8. In the Family Types dialog box, test each of the new named types.
9. Click **OK**.
10. Save the family.

Task 2 - Use the family in a project.

1. In the *...\Architectural* folder, open **Barristers-Bookcase-Project.rvt**.
2. In the *View* tab>Windows panel, expand (Switch Windows). Select a Barrister-Bookcase view.
3. In the Family Editor panel, click (Load into Project).
   - If there is only one project open, the family file is automatically loaded into that project.
   - If more than one project is open, the Load into Projects dialog box opens. Select the project you want to load and click **OK**.
4. The **Component** command is automatically started with the Barristers Bookcase family selected.
5. Place one of each type in the project.
6. Create a 3D view to see the differences.
7. Save the project.

Practice 4h Create Family Types for the Heat Pump

Practice Objectives

- Create family types of different sizes.
- Test the family in a project..

Estimated time for completion: 10 minutes.

In this practice you will create and apply Family Types of different sizes to a heat pump and test the family in a project, as shown in Figure 4–88.

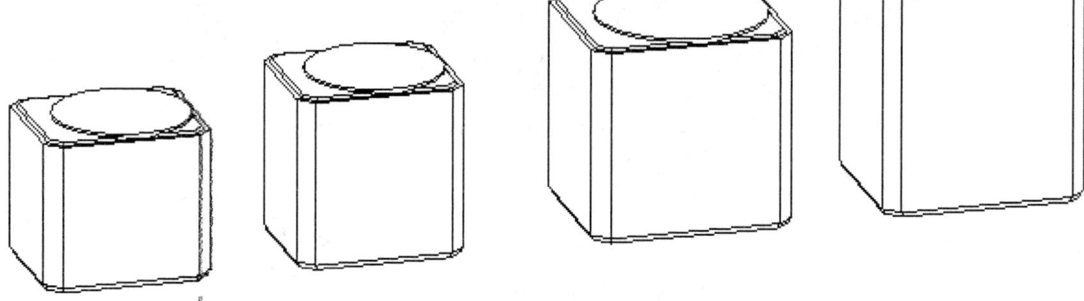

Figure 4–88

Task 1 - Create Family Types.

1. In the ...*MEP* folder, open **Heat-Pump-Types.rfa**
2. In the *Create* tab>Properties panel, click (Family Types).
3. In the Family Types dialog box, in the *Family Types* area, click **New...**.
4. In the Name dialog box, type **HP060C** as the name.
5. In the *Dimensions* area, verify or set the following values for the parameters, as shown in Figure 4–89:
   - **Width:** 38 7/8"
   - **Height:** 44 3/4"
   - **Depth:** 34 3/4"

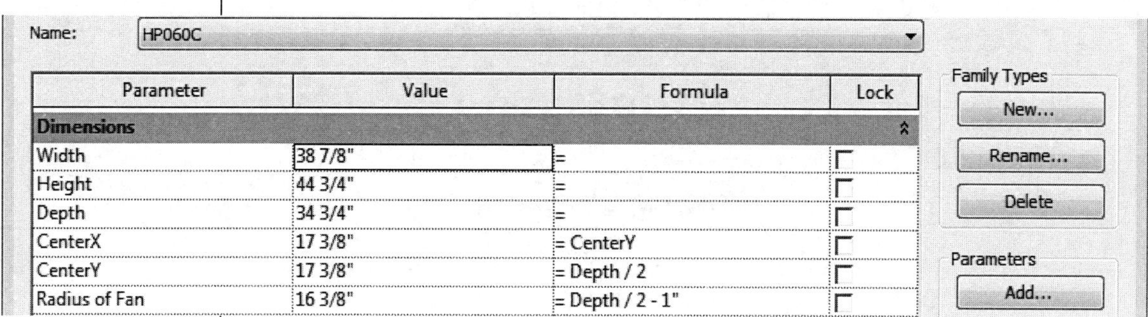

Figure 4–89

6. Click **Apply**. The family resizes to match the new information.

7. Create the other models as follows:

Model	Width	Height	Depth
HP048C	38 7/8"	37"	34 3/4"
HP030C	32 7/8"	32 3/4"	28 3/4"
HP024C	32 7/8"	28 5/8"	28 3/4"

8. In the Family Types dialog box, test the family types by selecting the *Name* and clicking **Apply** after each selection.

9. Click **OK**.

10. Save the family.

Task 2 - Use the family in a project.

1. In the ...*MEP* folder, open **Heat-Pump-Project.rvt**.

2. Switch back to one of the Heat Pump family views.

3. In the Family Editor panel, click (Load into Project).
   - If there is only one project open, the family file is automatically loaded into that project.
   - If more than one project is open, the Load into Projects dialog box opens. Select the project you want to load and click **OK**.

4. The **Component** command is automatically started with the new family selected.

5. Place one of each type in the project.

6. Create a 3D view to see the differences

7. Save the project and the family.

Practice 4i Create Family Types for the Structural Column

Practice Objectives

- Create family types.
- Use the family in a project.

Estimated time for completion: 10 minutes

In this practice you will create and apply Family Types for the column. You will add the column family to a project and test the columns by modifying levels, as shown in Figure 4–90.

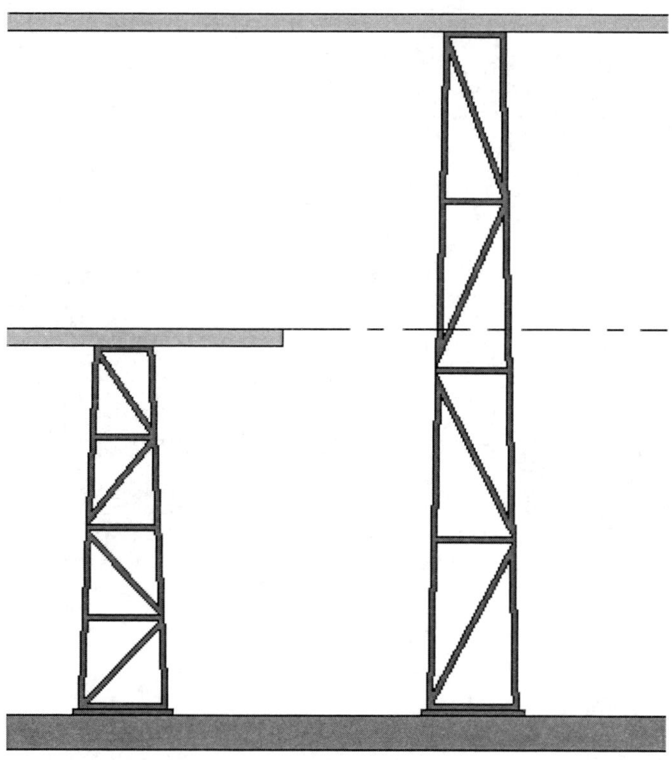

Figure 4–90

Task 1 - Create Family Types.

1. In the ...*Structural* folder, open **Trellis-Column-Types.rfa**.
2. In the *Create* or *Modify* tab>Properties panel, click (Family Types).
3. In the Family Types dialog box, in the *Family Types* area, click **New...**.

4. In the Name dialog box, set the name as **Over 12'-0"** and click **OK**.
5. In the *Dimensions* area, set the formula for the Top *Width* as **Bottom Width / 1.3**.
6. Click **Apply**.
7. Create another new Family Type with the name **Under 12'-0"**.
8. In the *Dimensions* area, set the formula for the *Top Width* as **Bottom Width / 1.5**.
9. Click **OK**.
10. Save the family.

Task 2 - Use the family in a project.

1. In the ...*Structural* folder, open **Trellis-Column-Project.rvt**
2. Open the **Structural Plans: Level 2** view.
3. In the *View* tab>Windows panel, expand (Switch Windows). Select a Trellis column view.
4. In the Family Editor panel, click (Load into Project).
   - If there is only one project open, the family file is automatically loaded into that project.
   - If more than one project is open, the Load into Projects dialog box opens. Select the project you want to load and click **OK**.
5. The **Structural Column** command is automatically started with the Trellis Column family selected. Verify that (Vertical Column) is selected.
6. Place one of each type in the project using the default information.
7. Open the **Elevations: South** view.

Change the Top Level first, so that the column does not go in the wrong direction.

8. Select the **Under 12'-0"** column. In Properties, set the *Top Level* to **Level 2,** the *Top Offset* to (negative) **-6", a**nd the *Base Level* to **Level 1** with no *Base Offset.*

9. Select the **Over 12'-0"** column. In Properties, set the *Top Level* to **Level 3**, the *Top Offset* to (negative) **-6"**, and the *Base Level* to **Level 1** with no *Base Offset*, as shown in Figure 4–91.

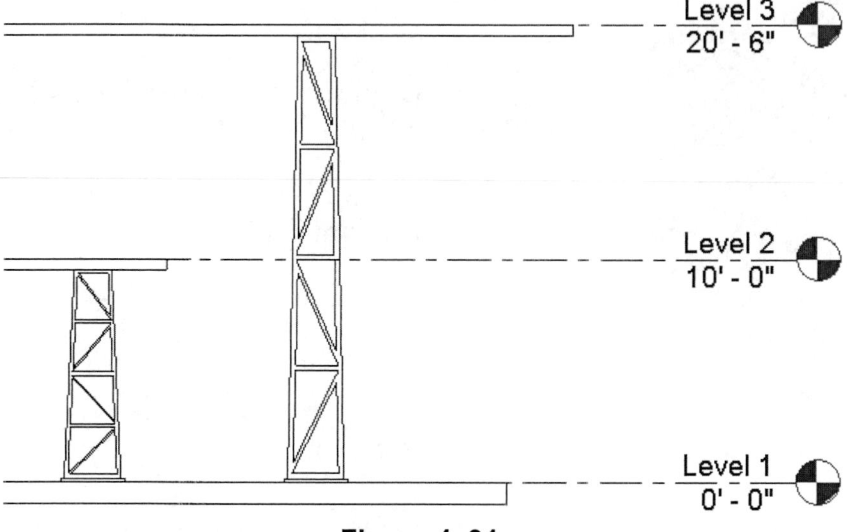

Figure 4–91

10. Change the height of the levels to see how the look of the column changes.

11. Save the project.

Chapter Review Questions

1. In which of the following views do the Reference planes, as shown in Figure 4–92, display? (Select all that apply.)

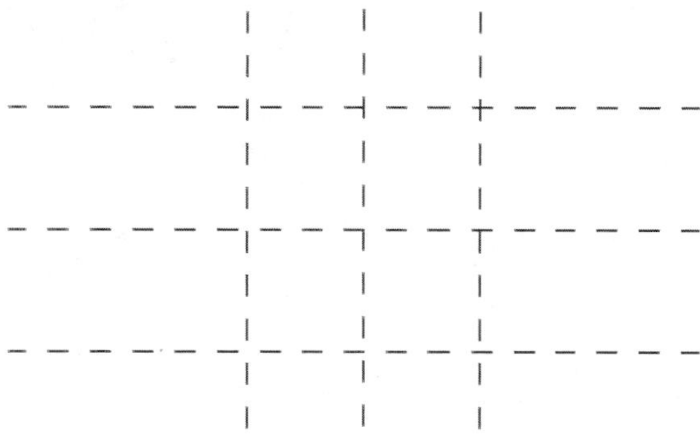

Figure 4–92

   a. Plan
   b. Elevation
   c. 3D
   d. Section

2. Which of the following is created when you label a dimension?
   a. Family
   b. Type
   c. Parameter
   d. Value

3. Which of the following commands, shown in Figure 4–93, require profiles instead of sketches? (Select all that apply.)

Figure 4–93

   a. **Extrusion**
   b. **Blend**
   c. **Revolve**
   d. **Sweep**
   e. **Swept Blend**

4. Which of the following commands creates the element shown in Figure 4–94?

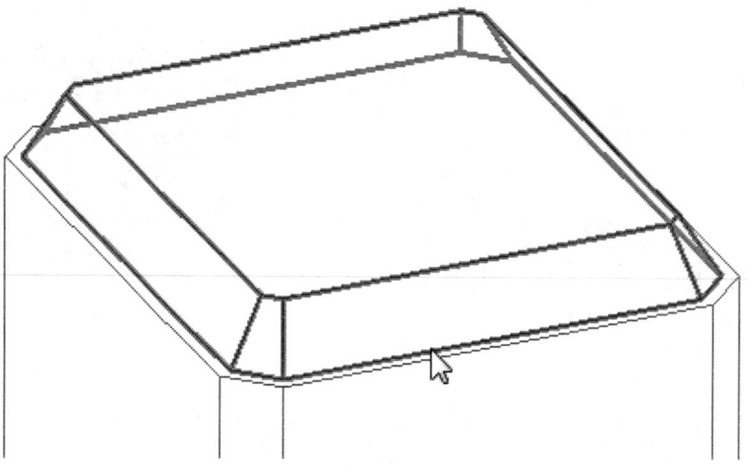

Figure 4–94

   a. **Extrusion**
   b. **Blend**
   c. **Revolve**
   d. **Sweep**
   e. **Swept Blend**

5. In which dialog box do you specify sizes for component types?
   a. Family Editor
   b. Family Types
   c. Family Categories
   d. Family Properties

Command Summary

Button	Command	Location
Forms		
	Blend	• **Ribbon**: *Create* tab>Forms panel
	Extrusion	• **Ribbon:** *Create* tab>Forms panel
	Revolve	• **Ribbon:** *Create* tab>Forms panel
	Sweep	• **Ribbon:** *Create* tab>Forms panel
	Swept Blend	• **Ribbon:** *Create* tab>Forms panel
	Void Forms	• **Ribbon:** *Create* tab>Forms panel
Other Tools		
	Align	• **Ribbon:** *Modify* tab>Modify panel
	Family Types	• **Ribbon:** *Create* or *Modify* tab> Properties panel
	Load Family	• **Ribbon:** *Insert* tab> Load from Library panel
	Load into Project	• **Ribbon**: *all tabs*> Family Editor panel
Reference and Work Plane		
	Reference Line	• **Ribbon:** *Create tab*>Datum panel
	Reference Plane	• **Ribbon:** *Create* tab>Datum panel
	Set	• **Ribbon:** *Create* tab>Work Plane panel

Advanced Family Techniques

Once you have created basic families with parametric framework, geometry, and family types, you can enhance the families with additional parameters, elements and settings. Possible enhancements include controls for flipping the family, or connectors for MEP system elements. You can also nest families into other families, and then coordinate the parameters between the two. Visibility Display settings enable you to specify the elements you want to display at various Detail Levels.

Learning Objectives in this Chapter

- Add controls, room calculation points, connectors, and openings.
- Insert components and associate the parameters.
- Control the display of elements in views at various detail levels using visibility settings.
- Create masking regions to block elements that would otherwise display through a family.
- Add Model and Symbolic lines.

5.1 Additional Tools for Families

Once you have created the basic framework of your family, added solid elements, established family types and tested it in a project, you can start to add items that will make the family even more powerful. Some family-specific functions include the following:

- Controls that flip and mirror the family.
- A room calculation point that makes a family room-aware.
- Connectors (Autodesk Revit MEP only) that contain information such as pipe or duct size and electrical currents.
- Openings in a host, such as a wall or ceiling.
- Nested families made of multiple families as shown in Figure 5–1.

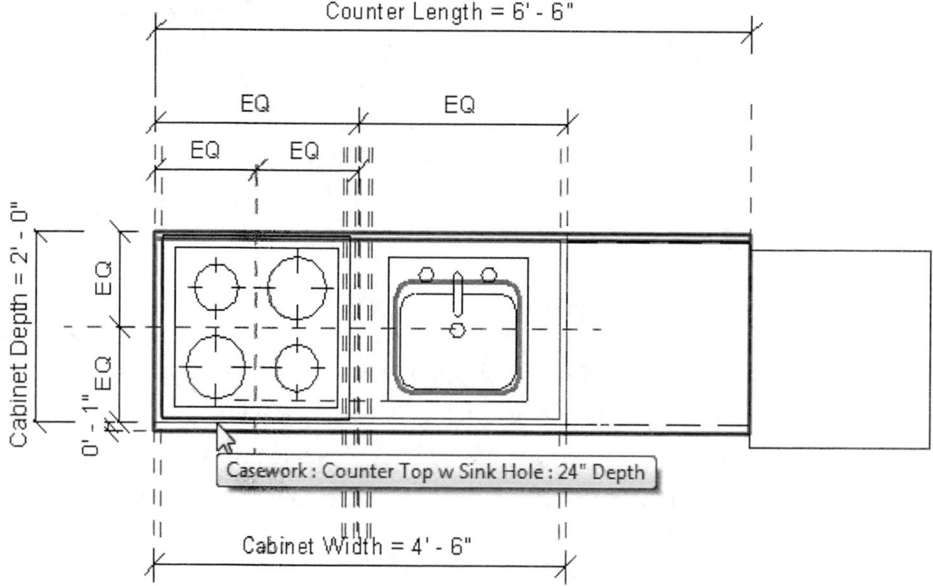

Figure 5–1

Adding Controls

You can place control arrows to flip or mirror geometry in the family, as shown in Figure 5–2. The controls also display on doors and other elements. Some family templates already include controls and you can add others.

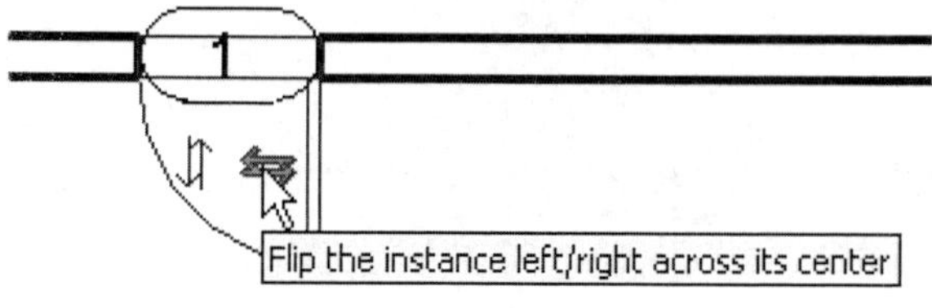

Figure 5–2

How To: Add Controls

1. In the *Create* tab>Control panel, click (Control).
2. In the *Modify | Place Control* tab>Control Type panel, click one of the following controls:
    - (Single Vertical)
    - (Single Horizontal)
    - (Double Vertical)
    - (Double Horizontal)
3. Click on the screen to place the control in the view. Move the controls once you have placed them, as required.

Setting Room Calculation Points

Some families (e.g., lighting fixtures, specialty equipment, casework, furniture, etc.) can be placed so that the default center of the element is not in the room where it needs to be counted, as shown in Figure 5–3. You can adjust this by modifying the Room Calculation Point, making the family a room-aware family.

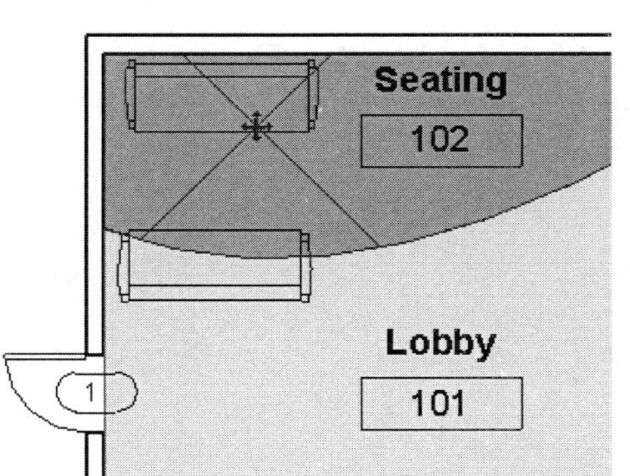

<Furniture Schedule>

A	B	C	D
Family and Type	Room: Number	Room: Name	Count
Sofa-Pensi	101	Lobby	1
Sofa-Pensi	102	Seating	1
Grand total: 2			

Before room calculation point

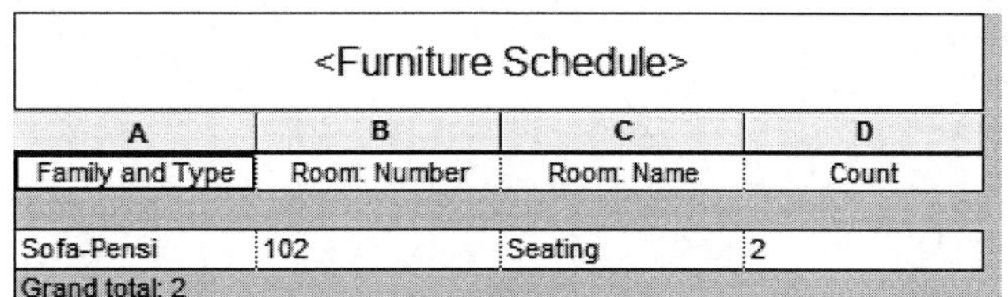

<Furniture Schedule>

A	B	C	D
Family and Type	Room: Number	Room: Name	Count
Sofa-Pensi	102	Seating	2
Grand total: 2			

After room calculation point

Figure 5–3

- Room Calculation Points must be set in the family.

How To: Set a Room Calculation Point

1. Open the Family file.
2. In Properties, select Room Calculation Point, as shown on the left in Figure 5–4.
3. Move the room calculation point to the required location, as shown on the right in Figure 5–4.

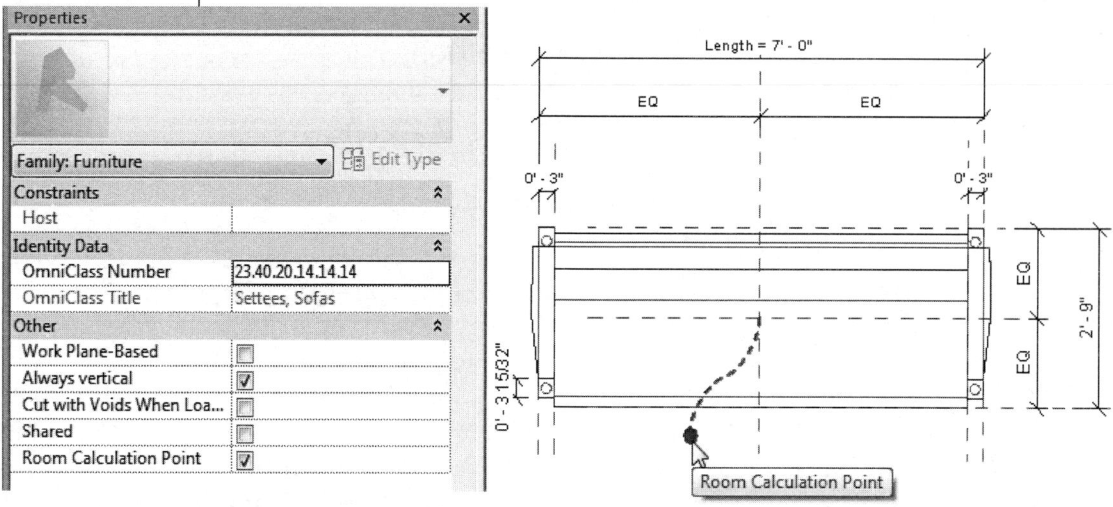

Figure 5–4

4. Save and load (or reload) the family file into the project.

Adding Connectors

In MEP families there is an additional option for creating connectors for the various systems. For example, you can add a duct connector to an air terminal where the duct element attaches to the air terminal, as shown in Figure 5–5.

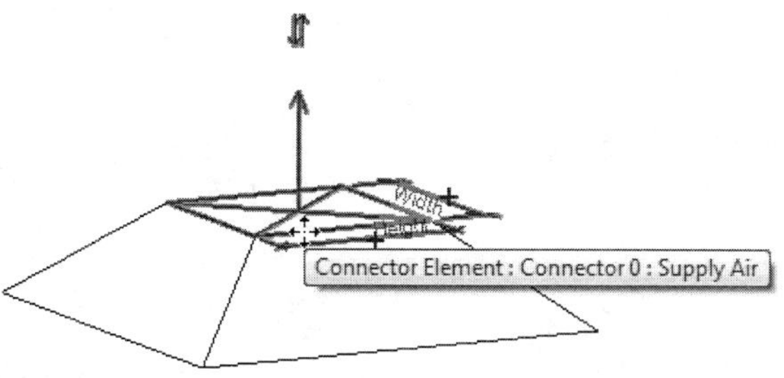

Figure 5–5

These tools are found on the *Create* tab>Connectors panel, as shown in Figure 5–6.

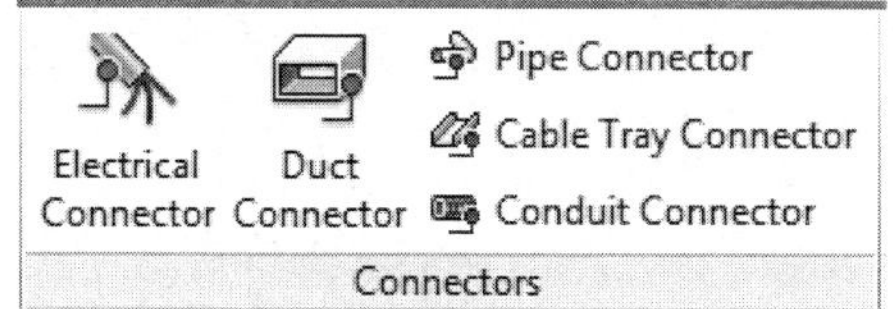

Figure 5–6

How To: Add a Connector

1. In the *Create* tab>Connectors panel, select the type of connector you want to use.
2. In the Options Bar, specify the type of connector (if applicable), as shown for an electrical connector in Figure 5–7.

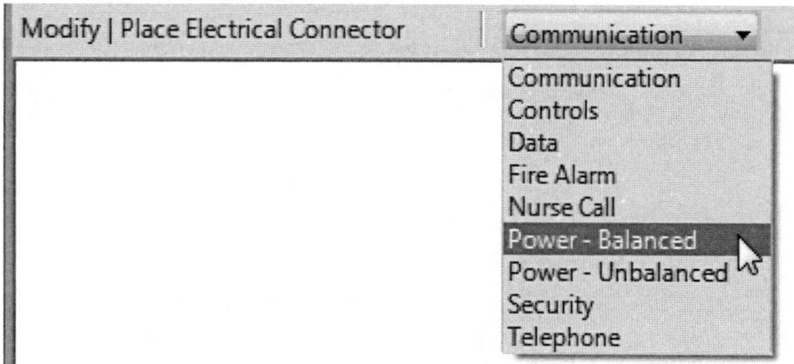

Figure 5–7

3. In the *Modify | Place Connector* tab>Placement panel, click (Face) or (Work Plane).
4. Select a face or workplane that matches the location and direction of the connector.
5. Select the connector and make any required adjustments in Properties.

Adding Openings

The **Opening** command enables you to sketch an opening in a host element. You need to have started the family using one of the host-based templates.

How To: Add an Opening to a Host Element

1. In the *Create* tab>Model panel, click (Opening).
2. In the *Modify | Create Opening Boundary* tab>Draw panel, use the sketch tools to create the opening. Use reference planes and dimensions as required.
3. Click (Finish Edit Mode).
4. In the *Modify | Opening cut* tab, make changes as needed.
5. In the Options Bar, select when you want the cut to display transparently (in **3D**, **Elevation**, or both), as shown in Figure 5–8.

Figure 5–8

Adding Components

Adding a component to a family produces a nested family. In most cases, you want to do this with elements that are used multiple times in other families, such as the hardware shown in Figure 5–9.

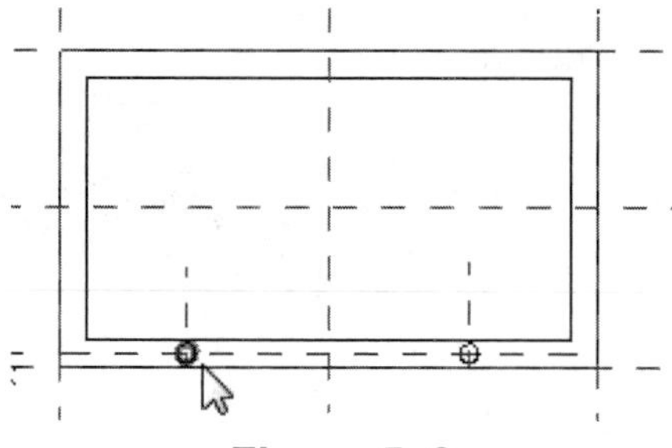

Figure 5–9

- Nesting families increase the file size and can impact your computer's performance.

There are two types of components:

(Component): In the *Create* tab>Model panel. places model components that display in all views.

(Detail Component): In the *Annotate* tab>Detail panel, places detail components that display only in the current view.

Associating Family Parameters

When you are working with nested families, you might need to associate parameters in the host family with the inserted family.

How To: Associate Family Parameters

1. Select the element(s) whose parameters need to be associated.
2. In Properties, in the row where you want to connect the parameters, click (Associate Family Parameter), as shown in Figure 5–10.

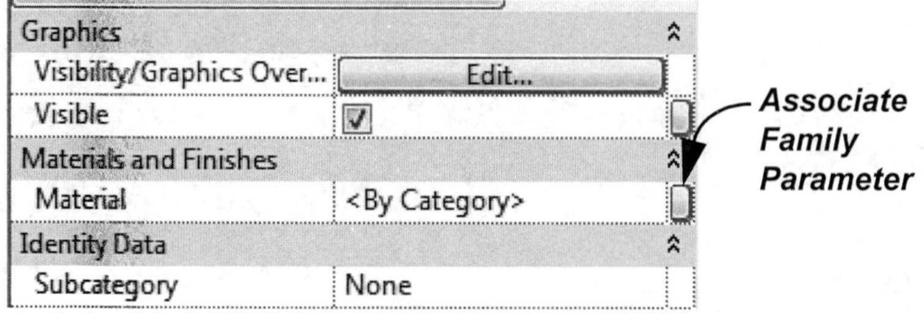

Figure 5–10

3. In the Associate Family Parameter dialog box, select the related parameter (as shown in Figure 5–11) and click **OK**.

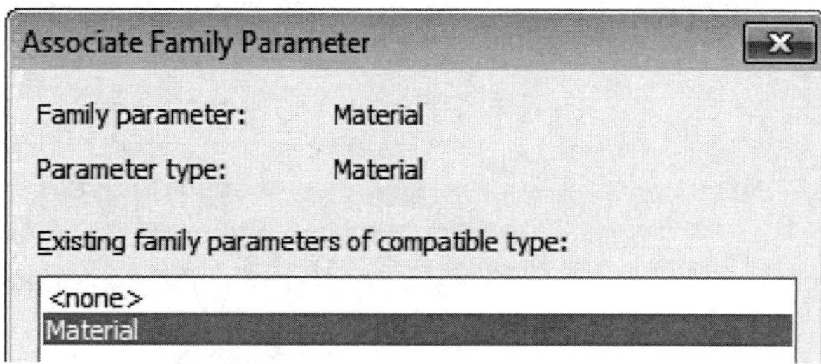

Figure 5–11

- The parameter is now grayed out and marked with the **=** sign, as shown in Figure 5–12.

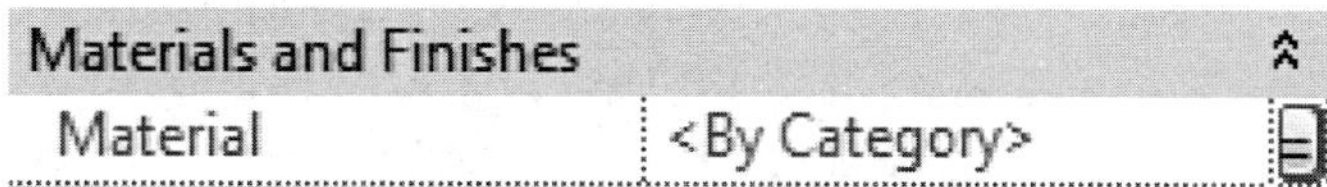

Figure 5–12

Practice 5a Add a Component to the Bookcase

Practice Objectives

- Add a component to a family.
- Associate parameters between the component and the family.
- Copy and align the component.

Estimated time for completion:15 minutes

In this practice you will insert a glass door component into the bookcase family. You will fit the component to the correct location on one shelf by aligning it and associating door width and height parameters with the related parameters in the family. You will then copy and align that door to other shelves, as shown in Figure 5–13. Optionally, you will add a Material parameter and test the options in a project.

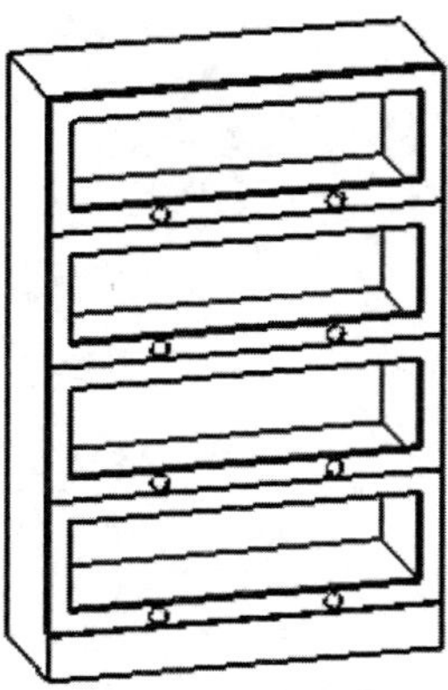

Figure 5–13

Task 1 - Add door components.

1. In the ...*Architectural* folder, open **Barrister-Bookcase-Component.rfa**.
2. Open the **Floor Plans: Ref. Level** view.
3. In the *Insert* tab>Load from Library panel, click (Load Family).
4. In the Load Family dialog box, in the ...*Architectural* folder, select **Shelf-Door-Glass.rfa** and click **Open**.
5. In the *Create* tab>Model panel, click (Component).

6. Place a copy of the door in the view aligned to the **Center L/R** reference plane, as shown in Figure 5–14. The width might not match the bookcase width.

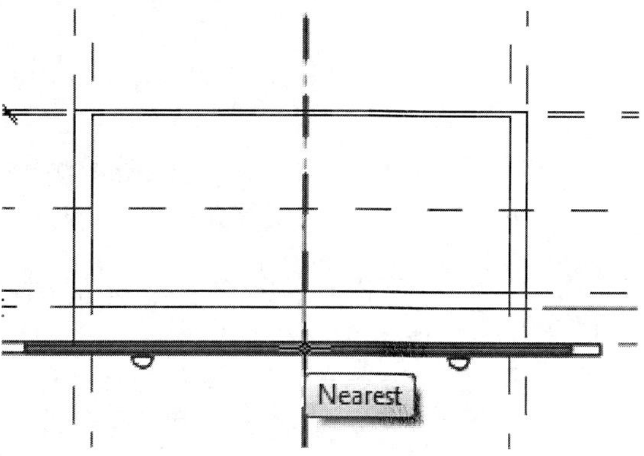

Figure 5–14

7. Use (Align) to lock the center of the bookcase with the center reference plane, as shown in Figure 5–15.

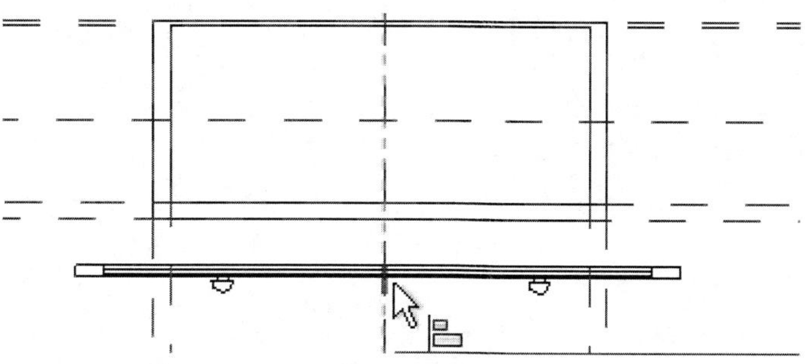

Figure 5–15

Select the Door Face reference plane first when you are aligning.

8. Use (Align) to line up the back of the door with the **Door Face** reference plane and lock it, as shown in Figure 5–16.

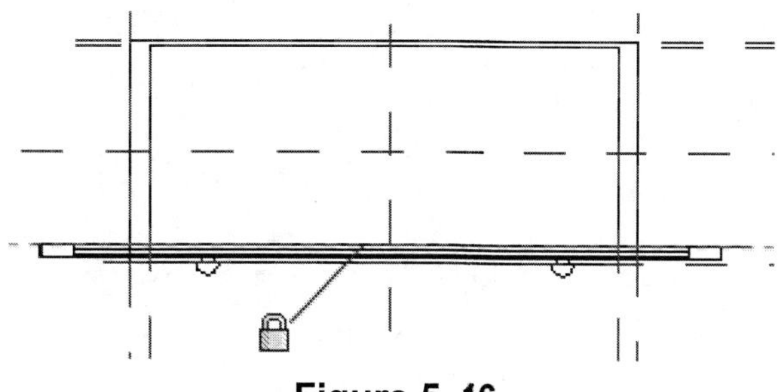

Figure 5–16

9. Save the family.

Task 2 - Associate parameters.

1. Click (Modify) and select the door.
2. In Properties, scroll down to the *Other* heading.
3. Beside **Door Width**, at the end of the row. click (Associate Family Parameter).
4. In the Associate Family Parameter dialog box, select **Width** and then click **OK**.
5. Repeat the process to associate **Door Height** with **Shelf Height**. The parameters in Properties gray out and an equal sign displays in the button, as shown in Figure 5–17. The door should now fit within the shelf, as shown in Figure 5–18.

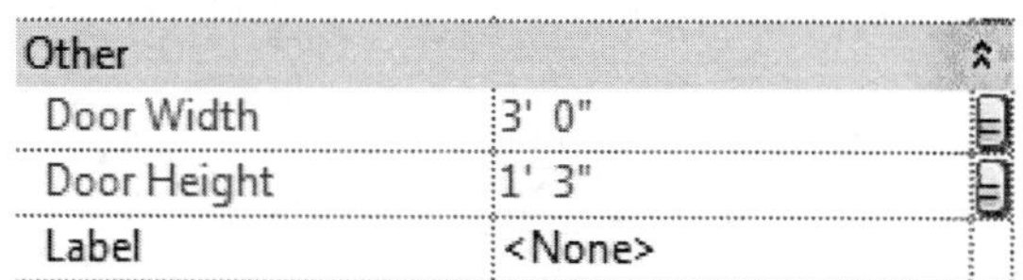

Figure 5–17

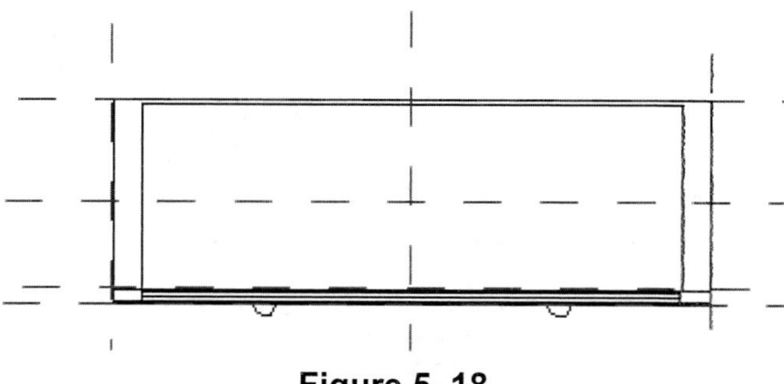

Figure 5–18

6. Change the *Width* dimension to flex the model to ensure that it works correctly.
7. Save the family.

Task 3 - Align and copy the doors

1. Open the **Elevations: Front** view and set the Visual Style to (Hidden Line). The door needs to be aligned vertically.

2. Align the top, as shown in Figure 5–19, and bottom of the door with the shelf height reference planes for the bottom shelf. Lock them in place.

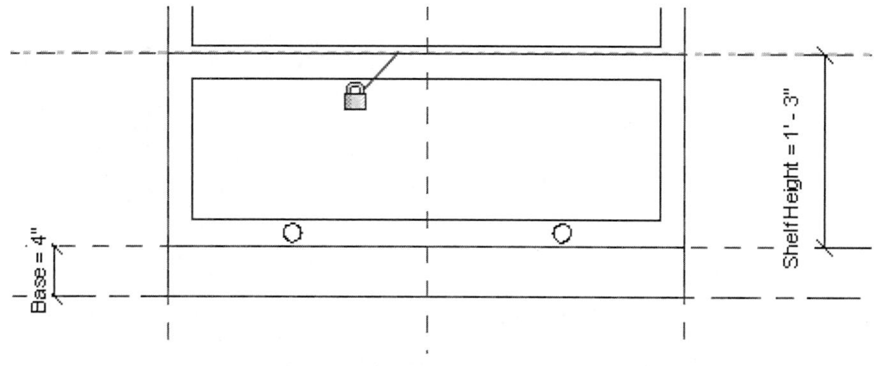

Figure 5–19

3. Copy the door to the other shelves.

4. Align and lock the top and bottom of each copied door with the associated reference plane, as shown in Figure 5–20.

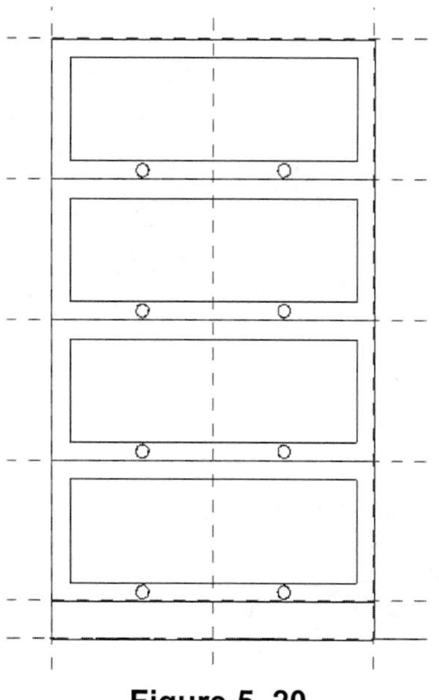

Figure 5–20

5. Change the *Height* dimension to flex the model and check the door alignments.

6. Open the **Elevations: Left** view and align the back of the top three door faces with the **Door Face** reference plane.

7. Switch to a 3D view, as shown in Figure 5–21.

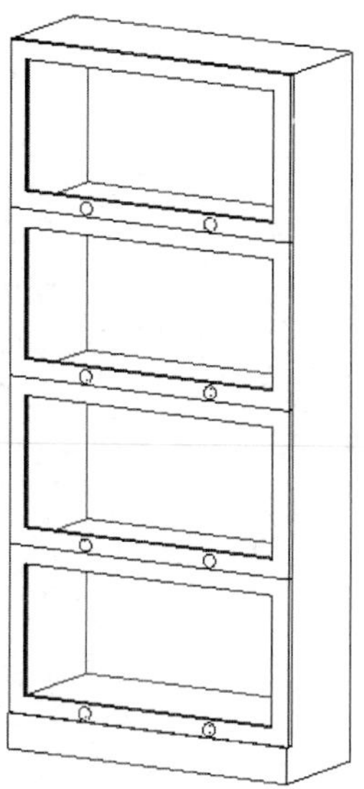

Figure 5–21

8. Use Family Types to flex the model again to make sure everything is working together.
9. Save the family.

Task 4 - Add and test a Material parameter. (Optional)

1. In the *Create* tab>Properties panel, click (Family Types).
2. In the Family Types dialog box, in the *Parameters* area, click **Add...**.
3. Create a family parameter with the *Parameter Data* set as shown in Figure 5–22. Click **OK**.

*Setting the type as an **Instance** makes modifying this parameter in the project file easier.*

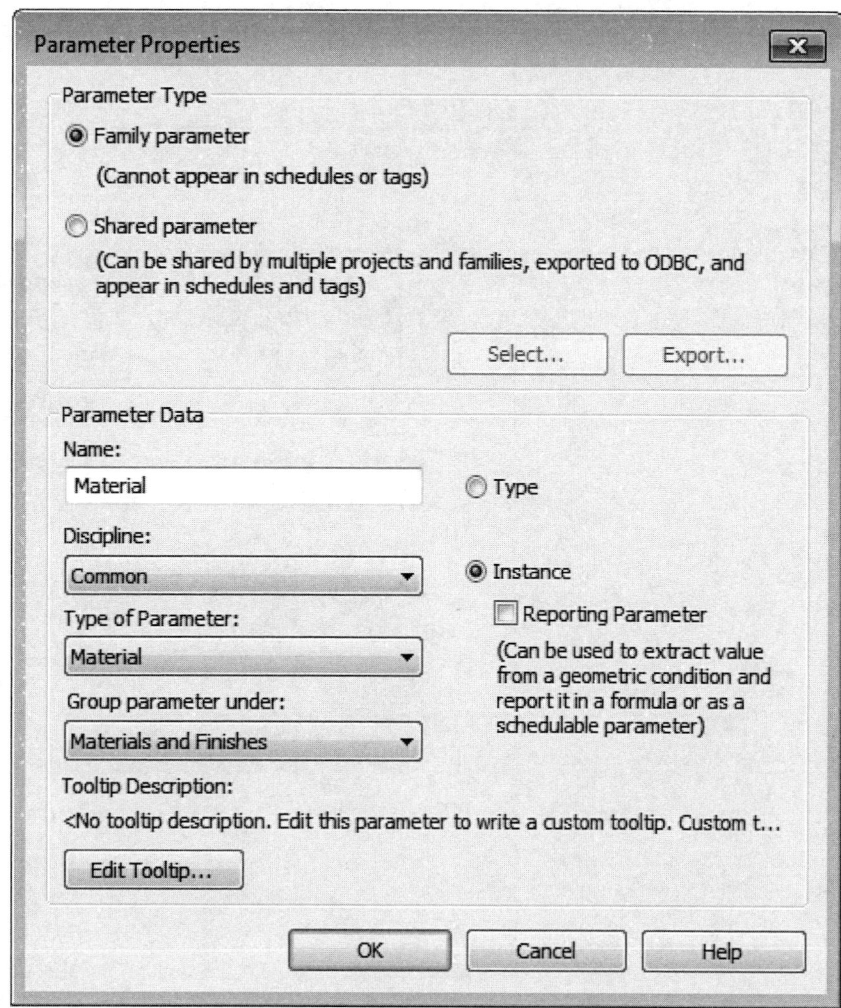

Figure 5–22

4. In the Family Types dialog box, leave the **Material** parameter set to **<By category>**, as shown in Figure 5–23, and click **OK** to close the dialog box.

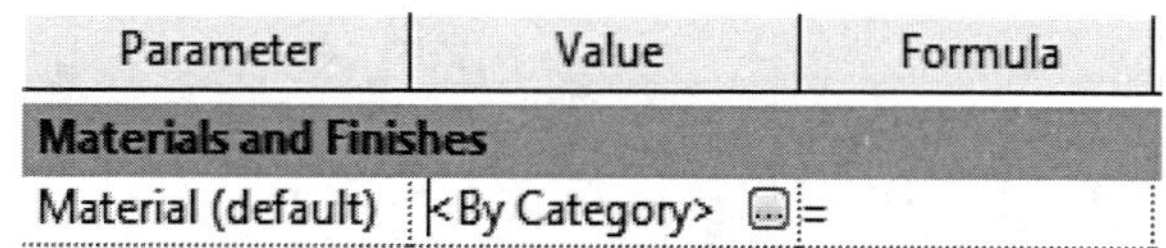

Figure 5–23

5. In the 3D view, select all of the bookcase components, but not the doors.

6. In Properties, a new instance parameter labeled **Material** displays under *Materials and Finishes*.

7. In the *Material* row, click (Associate Family Parameter), as shown in Figure 5–24.

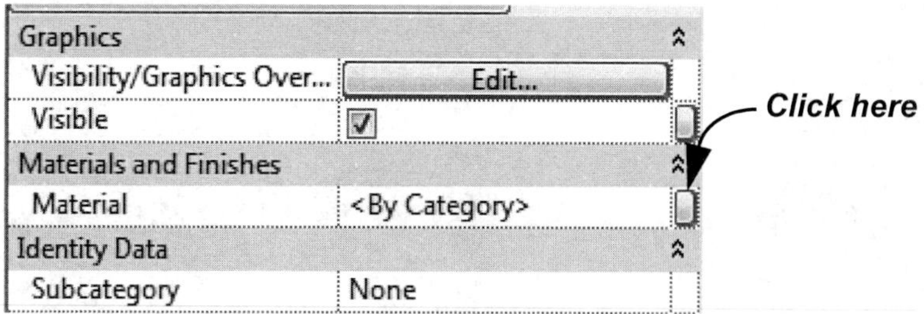

Figure 5–24

8. In the Associate Family Parameter dialog box, select the **Material** parameter, as shown in Figure 5–25, and click **OK**.

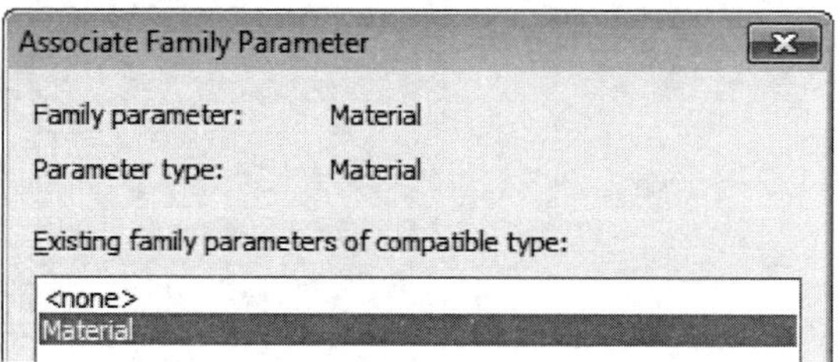

Figure 5–25

- The Material is now marked with the **=** sign, as shown in Figure 5–26.

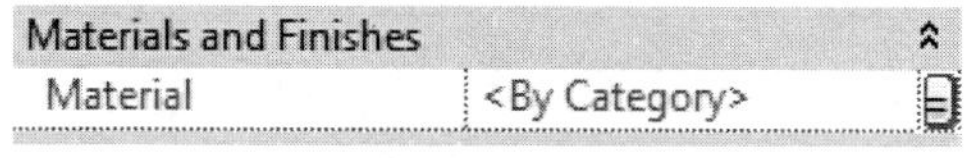

Figure 5–26

9. Select one of the bookcase doors.
10. In Properties, click (Edit Type).
11. In the Type Properties dialog box, associate the **Door Material** parameter with the **Material** parameter.
12. Click **OK**.
13. Open the **Barristers-Bookcase-Project.rvt** file, if it is not already open.
14. Switch to the bookcase family and load it into the project.
15. Place one of each type in the project.
16. Create a 3D view to see the differences.
17. In the View Status Bar, set the *Visual Style* to (Consistent Colors).

18. Select a bookcase. In Properties, in the *Materials and Finishes* area, next to the **Material** parameter, click **...** (Browse).

19. In the Material Browser, select a material and click **OK** to apply it to the bookcase. (Several materials starting with the word **Bookcase** are available for you to use.)

20. Repeat with the other bookcases using different materials, such as those shown in Figure 5–27.

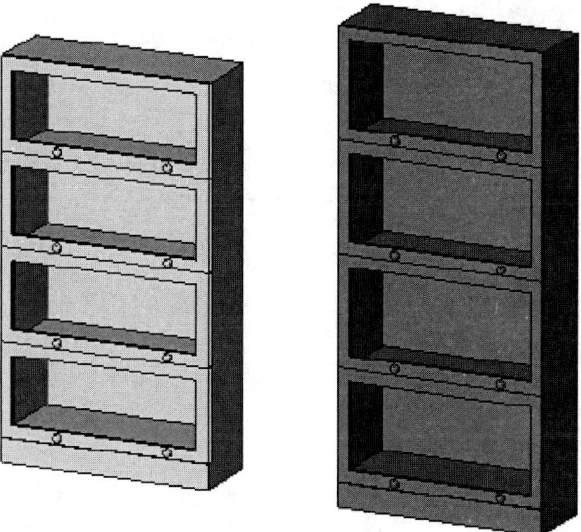

Figure 5–27

21. Save the project and family.

Practice 5b

Add a Component and Connectors to the Heat Pump

Practice Objectives

- Add a void element.
- Add a component to a family.
- Add electrical and piping connectors
- Associate parameters.

Estimated time for completion: 20 minutes.

In this practice you will add a void element to cut out part of the main unit of the heat pump. Then you will insert a panel component into that void. You will create additional extrusions and add connectors to these and other extrusions in the panel component. Finally, you will create and associate parameters to set pipe sizes. The completed family is shown in Figure 5–28.

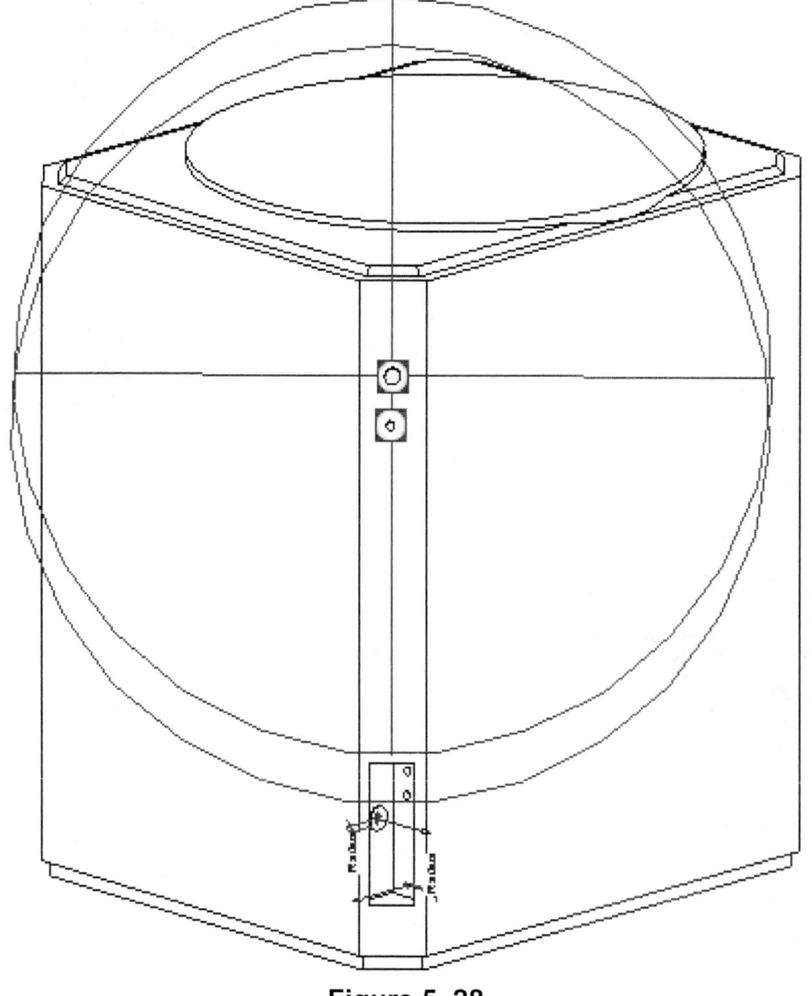

Figure 5–28

Task 1 - Add a void.

1. In the ...*MEP* folder, open **Heat-Pump-Connectors.rfa**.
2. Open the **3D**, **Ref. Level**, and **Front** views. Tile the views so that all three are visible on the screen.
3. In the **Front** elevation view, add two references planes to the lower left corner of the unit and dimension them as shown in Figure 5–29.
4. Select the lower reference plane and, in Properties, name it **Bottom of Void,** as shown in Figure 5–29.

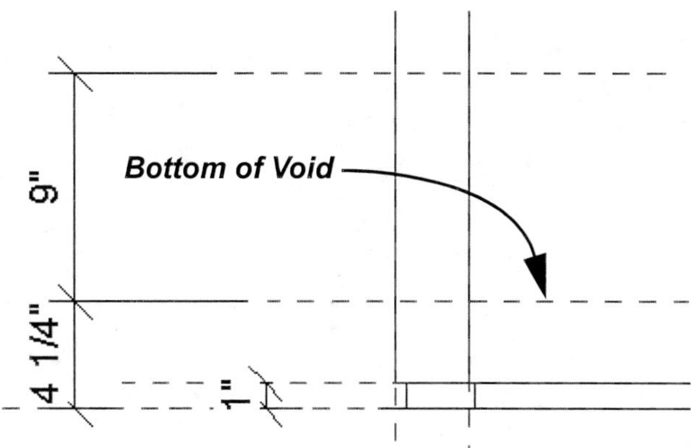

Figure 5–29

5. Open the **Ref. Level** view and zoom in on the lower left corner of the equipment.
6. Draw and dimension the reference planes shown in Figure 5–30.

Existing 3" dimensions shown for clarity.

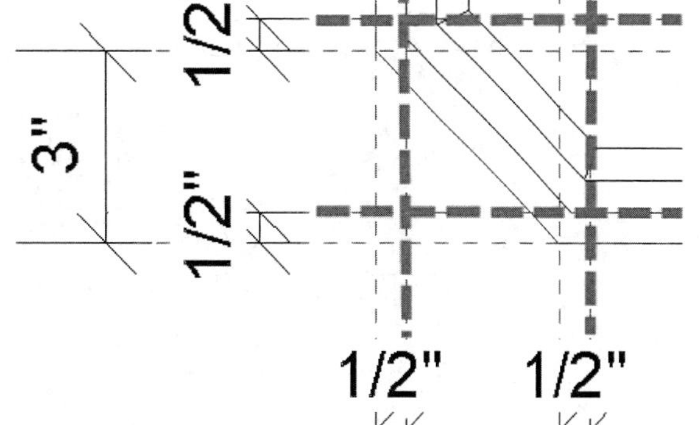

Figure 5–30

7. in the *Create* tab>Forms panel, expand (Void Forms) and click (Void Extrusion).
8. In the *Modify | Create Void Extrusion* tab>Work Plan panel, click (Set).
9. In the Work Plane dialog box, select the *Name* **Reference Plane: Bottom of Void** and click **OK**.
10. Draw and lock the sketch, as shown in Figure 5–31.

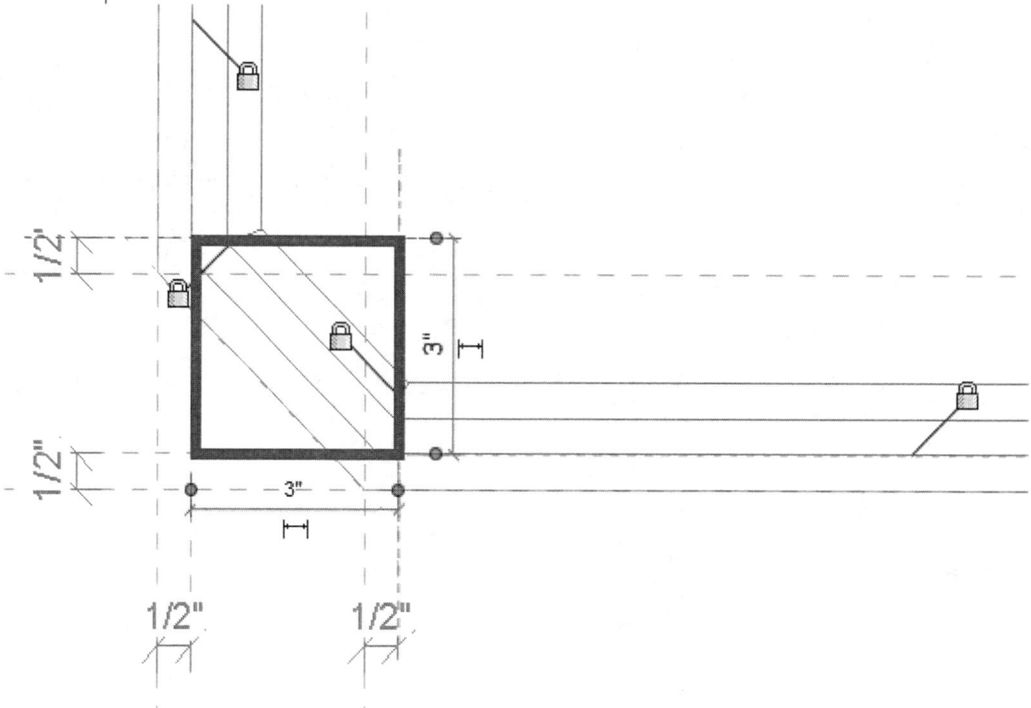

Figure 5–31

11. In Properties, set the *Extrusion End* to **9"**.
12. Click (Finish Edit Mode) and click in empty space to release the selection.
13. Rotate the 3D view so you can see the new void, as shown in Figure 5–32.

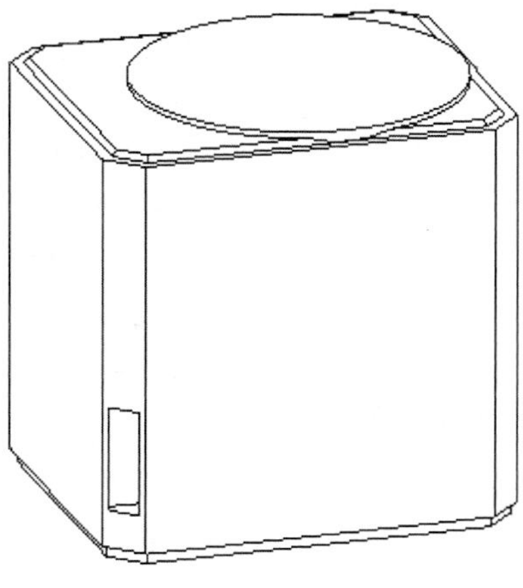

Figure 5–32

14. Save the project.

Task 2 - Insert a component.

1. Open the **Floor Plan: Ref. Level** view.
2. In the *Create* tab>Model panel, click (Component).
3. When prompted to load a family, click **Yes**.
4. Navigate to the ...*MEP* folder and open **Connector Panel.rfa**.
5. Insert the component as shown in Figure 5–33.

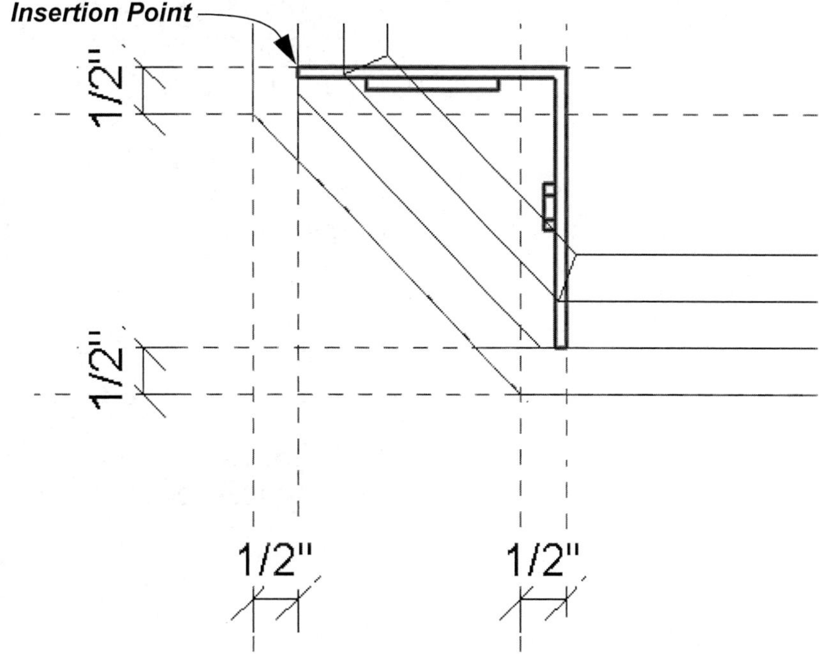

Figure 5–33

6. Align and lock the back and side of the component to the reference planes.

7. Open the **Elevations: Front** view. Align and lock the top of the component to the top of the void reference plane, as shown in Figure 5–34.

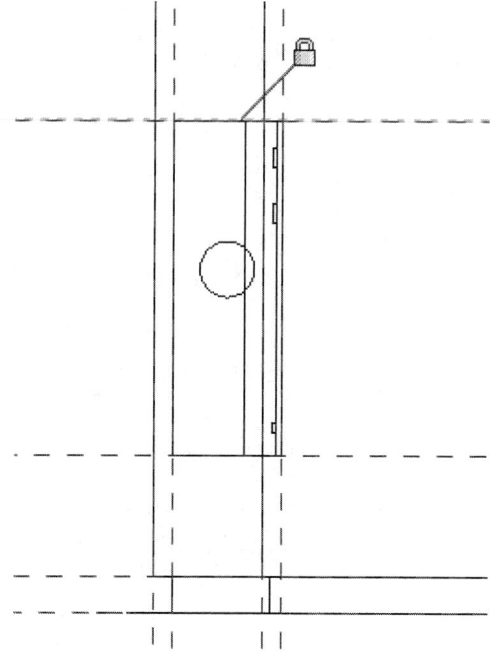

Figure 5–34

8. Save the family.

Task 3 - Add extrusions for piping connections.

1. Open the **Floor Plan: Ref. Level** view.

2. Draw a section in front of the angled reference plane at 45 degrees and open it. (This section is not cutting anything, but acts similar to an elevation.)

3. Start the **Extrusion** command.

4. In the Work Plane dialog box, select **Pick a plane**, and then click **OK**. Select the edge of the plane shown in Figure 5–35.

5. Add the reference planes, dimensions, and the extrusion shown in Figure 5–35. Set the extrusion *Depth* to **1/4"**.

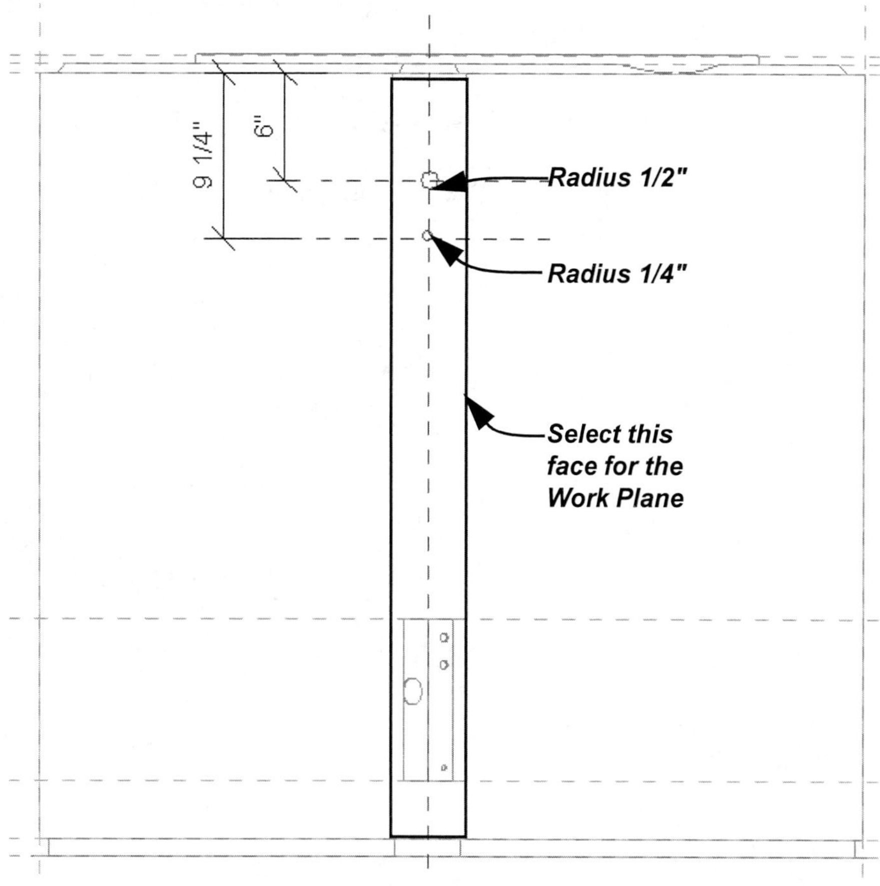

Figure 5–35

6. Join the extrusion to the rest of the unit.
7. Save the family.

Task 4 - Add connectors.

1. Open the 3D view and modify the view to display the corner of the unit where the connectors need to be applied.
2. In the *Create* tab>Connectors panel, click (Electrical Connector).
3. In the Options Bar, select **Power-Balanced**.

4. Select the face of the top circular extrusion, as shown in Figure 5–36.

5. Click (Modify) and select the new connector.

6. In Properties, in the *Identity Data* area, in the *Connector Description*, type **Electric Power Supply**.

7. Continue adding the other connectors as shown in Figure 5–36. For the Pipe Connectors, before selecting the location, in Properties, set the *Round Connector Dimension* to **Use Radius**.

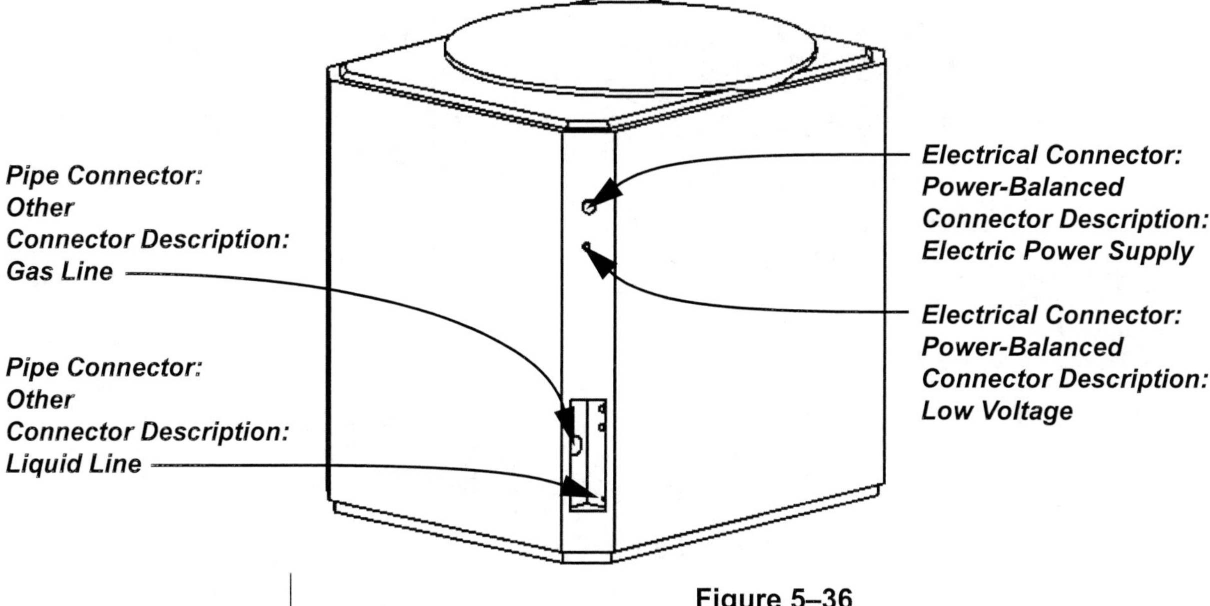

Figure 5–36

8. Save the project.

Task 5 - Create and assign new parameters for piping connectors.

1. In the Properties panel, click (Family Types).

2. Add the following Type parameters:

Name	Discipline	Type	Group
Gas Line Service Valve	Piping	Pipe Size	Mechanical
Liquid Line Service Valve	Piping	Pipe Size	Mechanical
Gas Line Radius	Piping	Pipe Size	Dimensions
Liquid Line Radius	Piping	Pipe Size	Dimensions

- The **Radius** parameters are going to be linked to **Radius** parameters in the connectors. The Autodesk Revit MEP software reads this and sets the correct diameter for the connector in the family when it is inserted into a project.

3. Assign the following values:
    - *Gas Line Service Valve:* **7/8"**
    - *Liquid Line Service Valve*: **3/8"**
4. Type in the following formulas:
    - *Gas Line Radius*: **Gas Line Service Valve / 2**.
    - *Liquid Line Radius:* **Liquid Line Service Valve / 2**.
5. Move the two radius parameters to the end of the Dimensions section. The final set of parameters is shown in Figure 5–37.

Parameter	Value	Formula
Mechanical		
Gas Line Service Valve	7/8"	=
Liquid Line Service Valve	3/8"	=
Dimensions		
Width	38 7/8"	=
Height	44 3/4"	=
Depth	34 3/4"	=
CenterX	17 3/8"	= CenterY
CenterY	17 3/8"	= Depth / 2
Radius of Fan	16 3/8"	= Depth / 2 - 1"
Gas Line Radius	7/16"	= Gas Line Service Valve / 2
Liquid Line Radius	3/16"	= Liquid Line Service Valve / 2

Figure 5–37

6. Click **OK**.
7. Select the Gas Line connector.
8. In Properties, in the *Dimensions* area, in the **Radius** parameter, click ❑ (Associate Family Parameter).

9. In the Associate Family Parameter dialog box, select **Gas Line Radius** (as shown in Figure 5–38) and click **OK**.

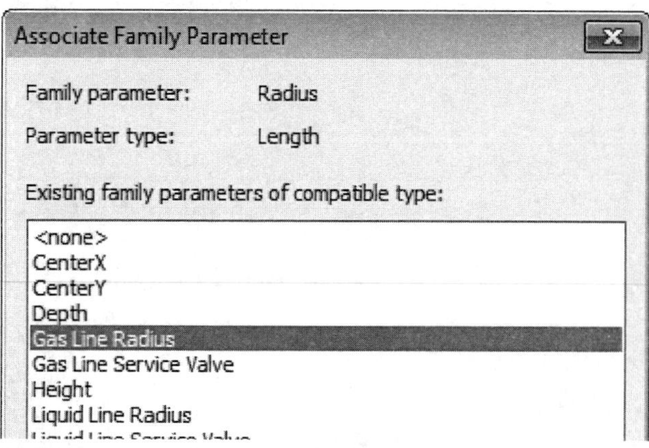

Figure 5–38

10. Repeat the process for the Liquid Line connector and associate the **Liquid Line Radius** to it.

11. When it is finished the connectors should be sized accordingly in the family, as shown in Figure 5–39.

Select the connector and click the Flip control if the Z-axis is pointing into the heat pump.

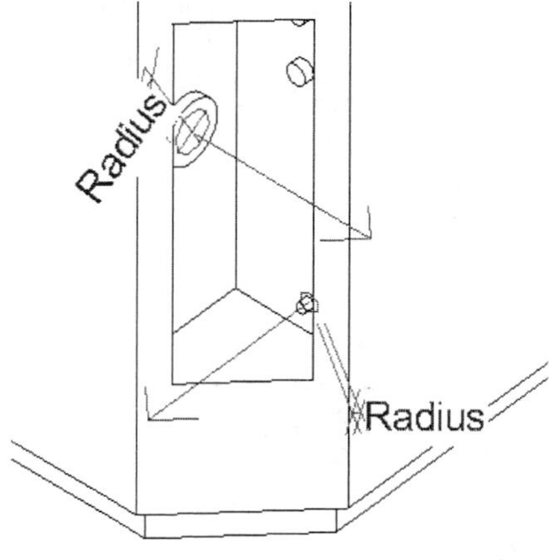

Figure 5–39

12. Save the family.

Practice 5c Modify the Structural Column

Practice Objectives

- Add and test material parameters.
- Associate parameters.
- Add parameters to an existing extrusion.

Estimated time for completion: 15 minutes

In this practice you will add a material parameter and test it in a project, as shown in Figure 5–40. You will also modify the trellis extrusion and create parameters so that the side and horizontal members can be a different size than the diagonal members.

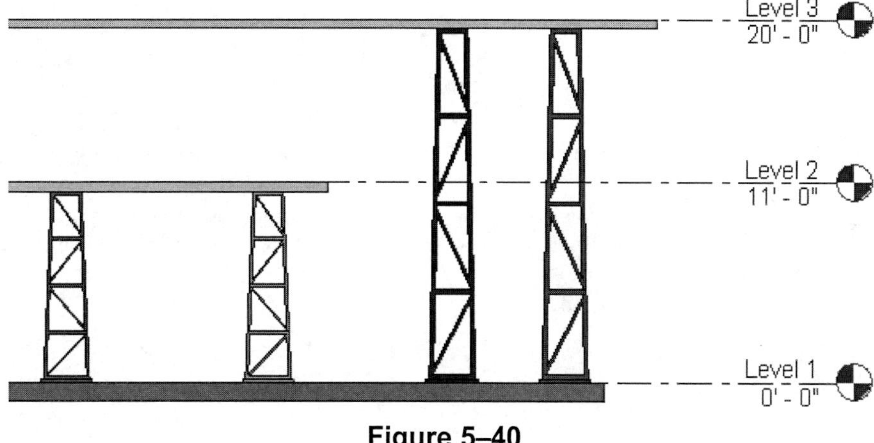

Figure 5–40

Task 1 - Add a material parameter.

1. In the ...*Structural* folder, open **Trellis-Column-Advanced.rfa**
2. In the *Create* tab>Properties panel, click (Family Types).
3. In the Family Types dialog box, in the *Parameters* area, click **Add...**.

*Setting the type as an **Instance** makes modifying this parameter in the project file easier.*

4. Create a **Family parameter** and set the *Parameter Data,* as shown in Figure 5–41. Click **OK**.

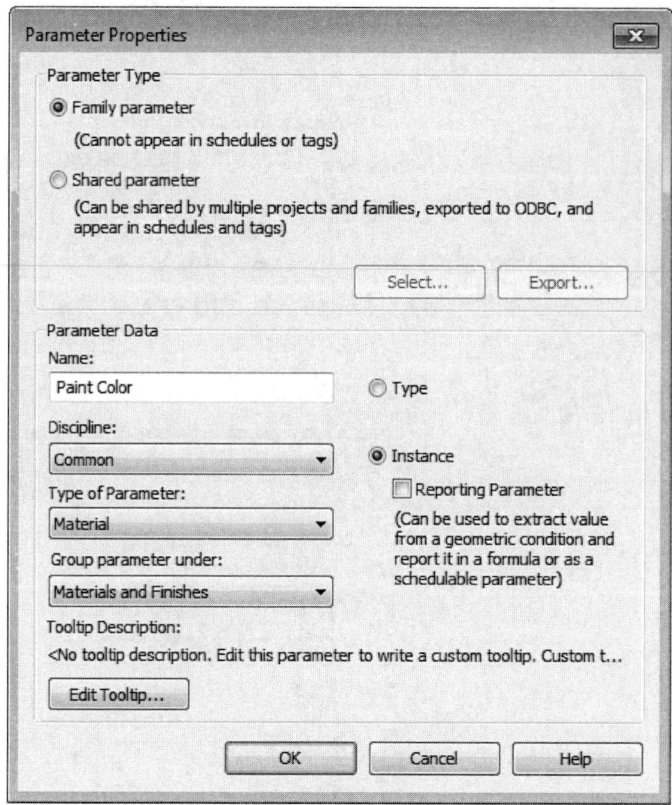

Figure 5–41

5. In the Family Types dialog box, ensure that the **Paint Color** parameter is set to **<By category>**, as shown in Figure 5–42. Click **OK** to close the dialog box.

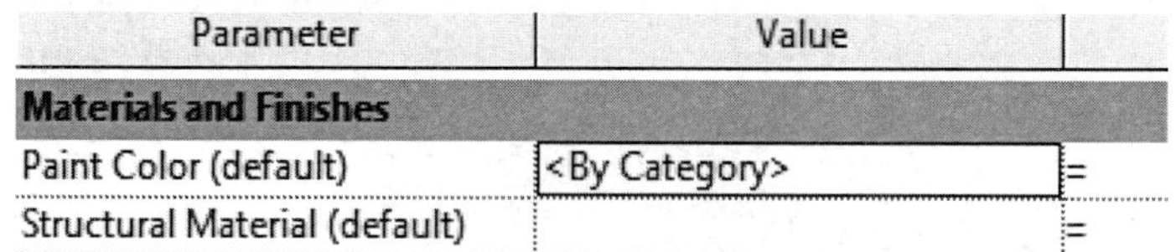

Parameter	Value	
Materials and Finishes		
Paint Color (default)	<By Category>	=
Structural Material (default)		=

Figure 5–42

6. In the 3D view, select the column extrusion.

7. In Properties, in the *Materials and Finishes* area, a parameter labeled **Material** displays.

8. In the *Material* row, click (Associate Family Parameter), as shown in Figure 5–43.

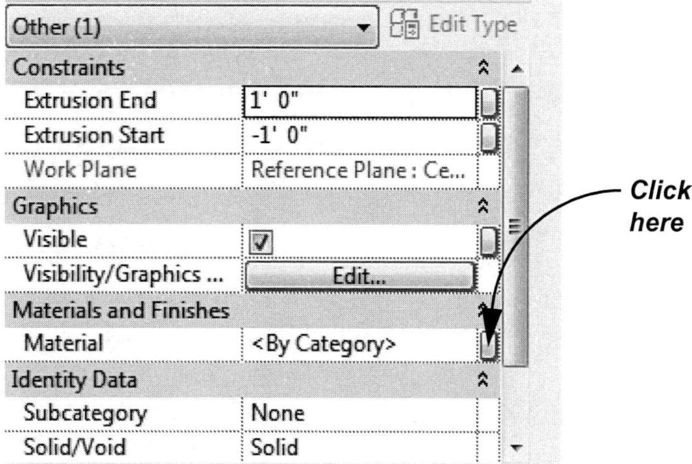

Figure 5–43

9. In the Associate Family Parameter dialog box, select **Paint Color** (as shown in Figure 5–44), and then click **OK**.

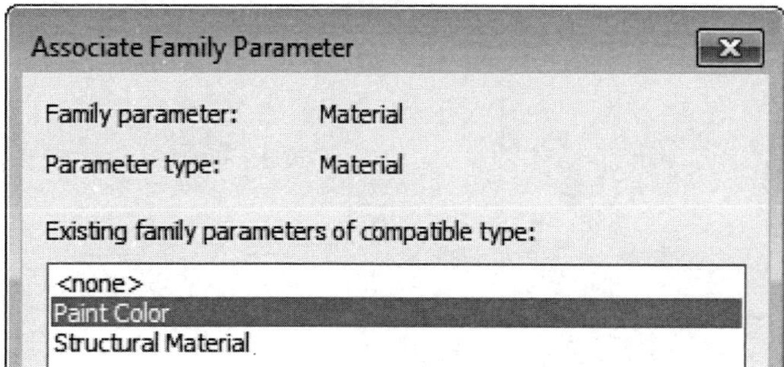

Figure 5–44

- The Material is now marked with the **=** sign, as shown in Figure 5–45.

Figure 5–45

10. Select the base extrusion and repeat this process.

11. Save the family.

Task 2 - Modify the material of the columns.

1. Open the **Trellis-Column-Project.rvt** file, if it is not already open.
2. Switch to the column family and load it into the project.
3. Add several columns at different heights. Set the base and top of each column in properties.
4. Open the **Elevations: South** view.
5. In the View Status Bar, set the *Visual Style* to (Consistent Colors).
6. Select one of the columns.
7. In Properties, under the *Materials and Finishes* heading, beside *Paint Color* in the <By Category> box, click (Browse) to open the Material Browser.
8. In the Material Browser, select a material and click **OK** to apply it to the column.
    - Several materials of different colors (starting with the words **Metal - Paint Finish**) are available for you to use.
9. Repeat with the other columns using different materials.
10. Save the project.

Task 3 - Create parameters for changing the size of the structural members.

1. Return to the column family.
2. Open the **Elevations: Front** view and edit the extrusion.
3. In the Family Types dialog box, add the following Length parameters (Type):
    - **Side Bar**
    - **Angled Bar**
    - **Horizontal Bar**
    - **Angled Bar Halfwidth**
    - **Horizontal Bar Halfwidth**

4. Reorder the list and add the formulas shown in Figure 5–46.

Parameter	Value	Formula	Lock
Materials and Finishes			
Paint Color (default)	<By Category>	=	
Structural Material (default)		=	
Dimensions			
Bottom Width	2' 8"	=	☐
Depth	2' 0"	=	☐
Side Bar	0' 2"	=	☑
Angled Bar	0' 2"	=	☑
Horizontal Bar	0' 2"	= Side Bar	☑
Angled Bar Halfwidth	0' 1"	= Angled Bar / 2	☑
Horizontal Bar Halfwidth	0' 1"	= Horizontal Bar / 2	☑
Top Width	2' 0 79/128"	= Bottom Width / 1.3	☐

Figure 5–46

5. Apply the parameters to the appropriate dimensions, as shown in part in Figure 5–47.

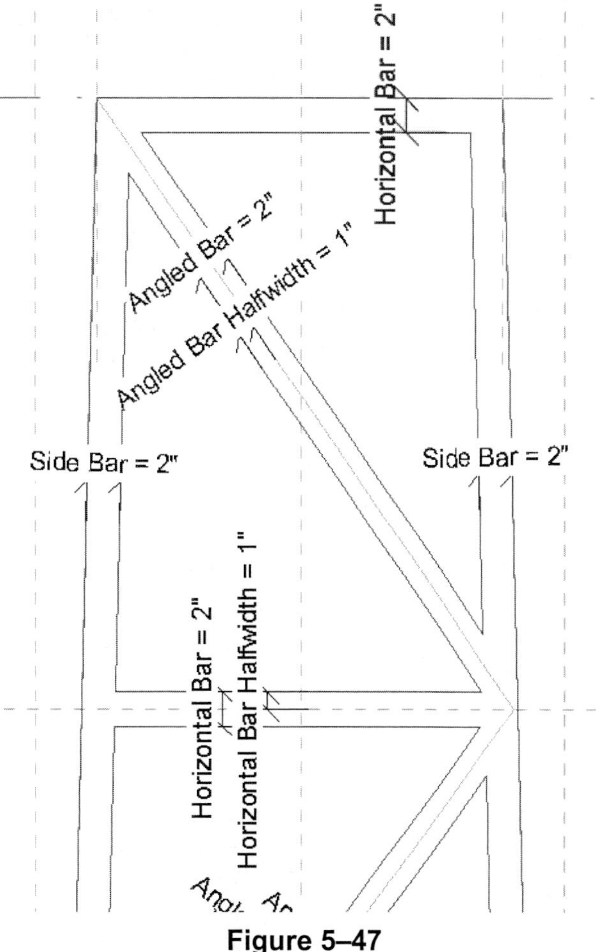

Figure 5–47

- Change the scale of the view to better display the parameters.

6. Change some of the dimensions either directly in the sketch or through the Family Types dialog box to flex the design.
7. Finish the sketch.
8. Open the 3D view and the Family Types dialog box and flex the model.
9. Save the family.

5.2 Visibility Display Settings

You can make families more versatile by controlling what is displayed in various views and at different detail levels. For example, you can set the Coarse view to display only 2D symbolic lines while the Fine detail view displays the full 3D model, as shown in Figure 5–48. To do this, change the visibility of the elements. You can also add Masking Regions to elements in a family to obscure other elements.

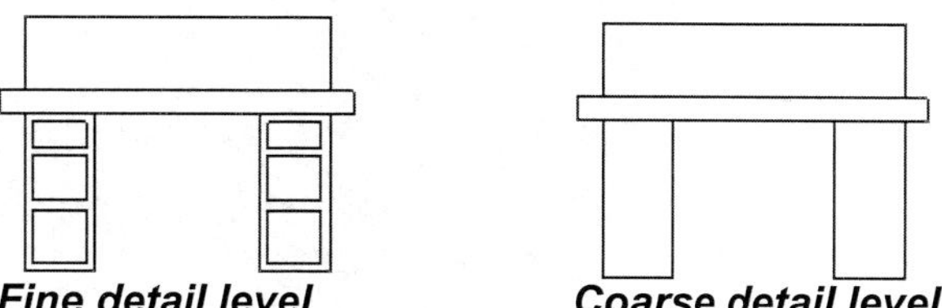

Figure 5–48

How To: Control the Display of Elements

1. Select the elements that you want to modify.
2. In the *Modify* contextual tab>Mode panel, click (Visibility Settings).
3. In the Family Element Visibility Settings dialog box, specify the views and detail level in which you want the elements to display, as shown in Figure 5–49.

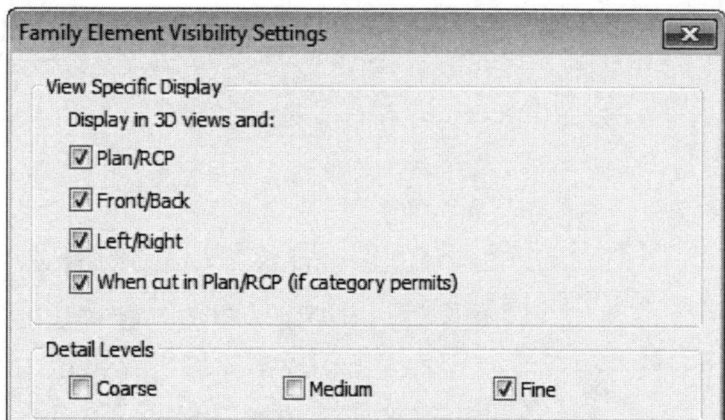

Figure 5–49

4. Click **OK**.

- All model elements (not detail elements) typically display in 3D views. You can select how they display in plan and elevation views.

- The *Detail Levels* control whether or not the element displays at various scales. If you clear the checkmark from a level, the lines defining the element turn gray in the family file. They do not display at that detail level when inserted into the project.

- The plan is cut at the cut plane specified by the *View Range*. Some elements, such as furniture and mechanical equipment, cannot be cut in plan view, while structural columns can.

Adding Lines

Lines are often used to display the model elements in plan and elevation views at the Coarse (and sometimes Medium) detail level. There are two types of lines:

- (Model Line) in the *Create* tab>Model panel, shown in all views.
- (Symbolic Line) in the *Annotate* tab>Detail panel, are visible in views parallel to the view in which they were drawn.

Lines can also be used in other situations. For example, you could use model lines to draw door hardware that is not three-dimensional, or use symbolic lines to draw the door swing (as shown in Figure 5–50) so that it does not display in 3D views.

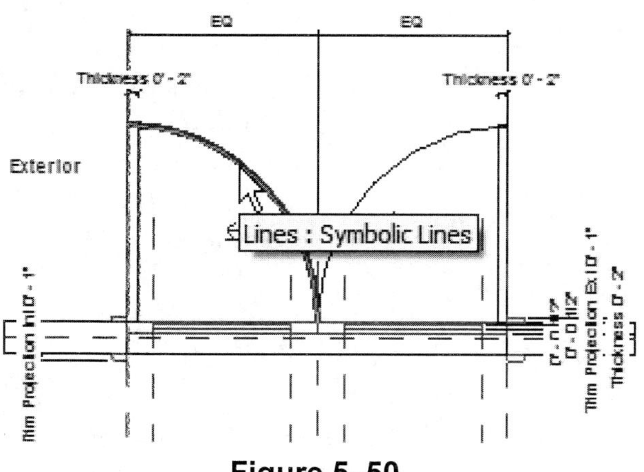

Figure 5–50

- Use the standard sketch tools in the Draw panel to draw model and symbolic lines. You can also specify a subcategory with an associated line type (as shown in Figure 5–51) for a door family.

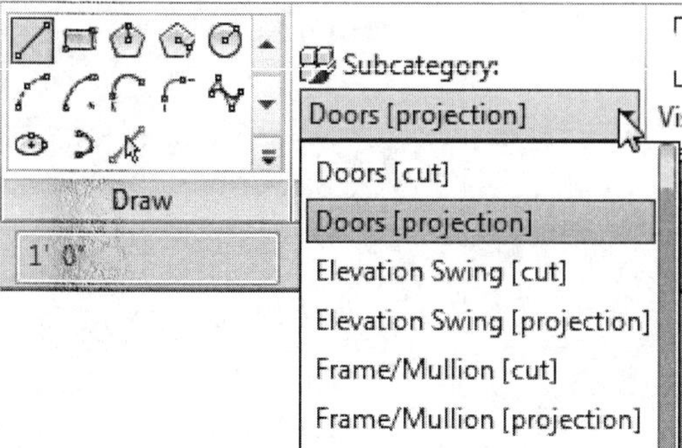

Figure 5–51

- You can create entire families of model and symbolic lines if they are only going to be used in a 2D view. For example, create a 2D family for a hospital bed, as shown in Figure 5–52, and use it in a large hospital layout where most of the instances do not need to be full 3D.

2D families can still be robust in the number and types of parameters they hold for use in schedules and quantity takeoffs.

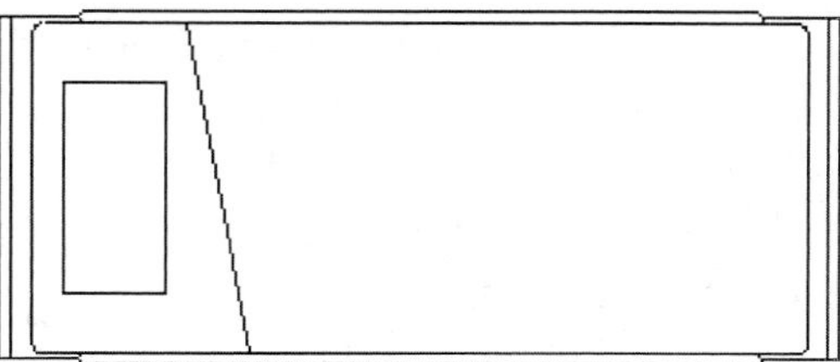

Figure 5–52

Creating Masking Regions

A masking region obscures other elements in a project. They can be placed in a family or project. For example, in the Desk family shown in Figure 5–53, a chair placed under the desk is masked but the lamp on top of the desk is not.

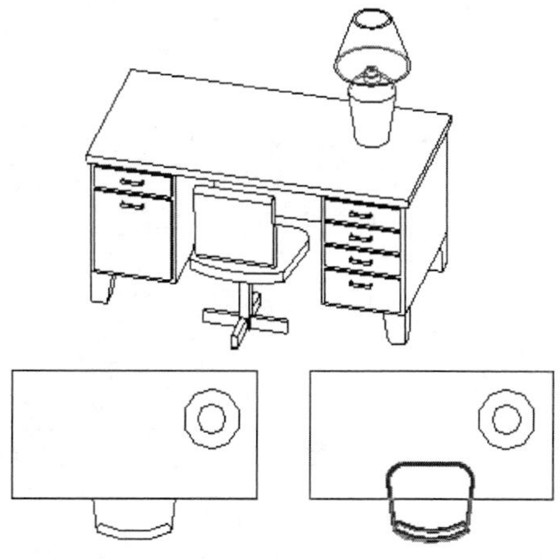

Figure 5–53

How To: Create Masking Regions

1. Open a plan or elevation view.
2. In the *Annotate tab*>Detail panel, click (Masking Region).
3. In the *Modify | Create Masking Region Boundary* tab> Subcategory panel, select the subcategory you want to use for the boundary of the region.
4. In the Draw panel, select the tools to sketch the region. You can also use most of the Modify options as you draw.
5. Click (Finish Edit Mode) and click in empty space to finish the command.

Practice 5d Modify the Visibility of Elements in the Bookcase

Estimated time for completion: 10 minutes

Practice Objectives

- Create symbolic lines.
- Modify visibility settings.

In this practice you will create a simple extrusion and sketch symbolic lines to display at the Coarse Detail Level. You will then modify the visibility settings of the main 3D elements so that they display at the Medium and Fine Detail Levels, as shown for an elevation in Figure 5–54.

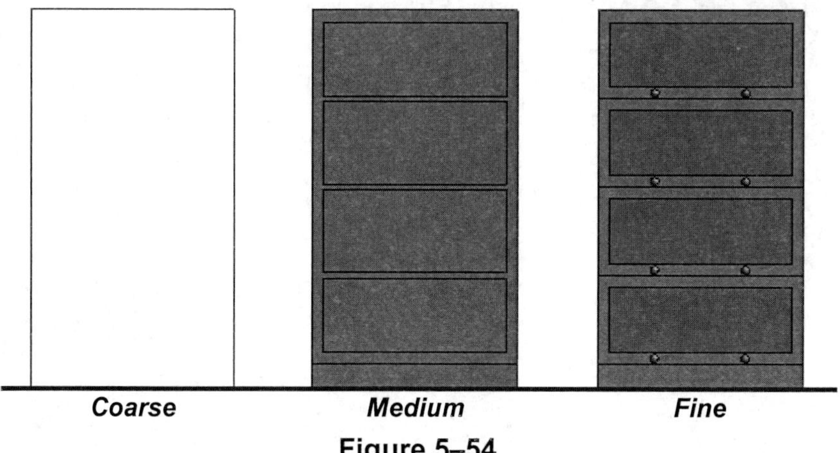

Figure 5–54

Task 1 - Create an extrusion for coarse detail in 3D views.

1. In the ...*\Architectural* folder, open **Barrister-Bookcase-Visibility.rfa**.
2. Open the **Floor Plans: Ref. Level** view.
3. Temporarily hide the bookcase elements:
    - To select the extrusion elements, press <Tab> on the edge of the elements until you see the **Joined Solid Geometry** tooltip, as shown in Figure 5–55.
    - Select the doors separately.
    - Do not use **Hide Category**.

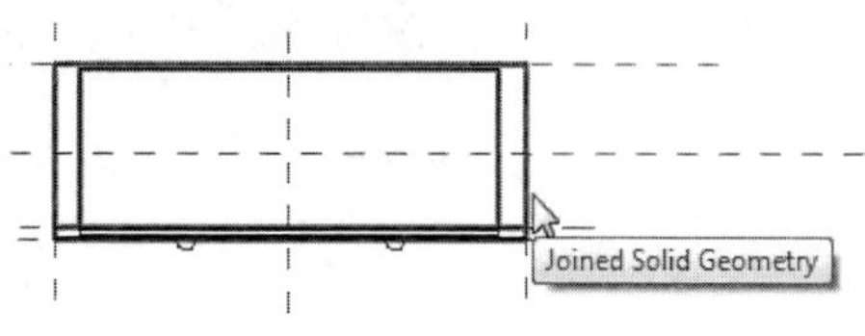

Figure 5–55

4. Create a new extrusion with the footprint shown in Figure 5–56. Set the *Depth* to **2'-6"**

Lock all sketch lines.

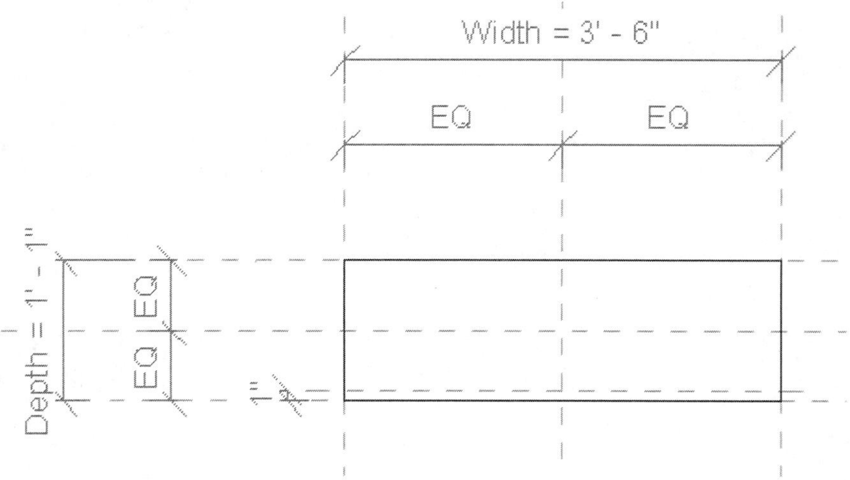

Figure 5–56

5. Switch to the **Elevations: Front** view.

6. Align and lock the top of the new extrusion to the top reference plane, as shown in Figure 5–57.

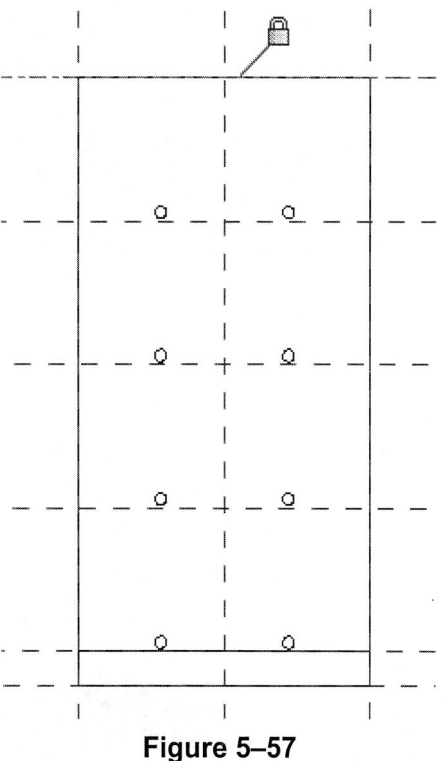

Figure 5–57

7. Tab to select the extrusion.

8. In the *Modify* | *Extrusion* tab>Mode panel, click (Visibility Settings).

You do not want this element displaying in the 2D views at any detail level.

9. In the Family Element Visibility Settings dialog box, clear all of the options except **Coarse**, as shown in Figure 5–58.

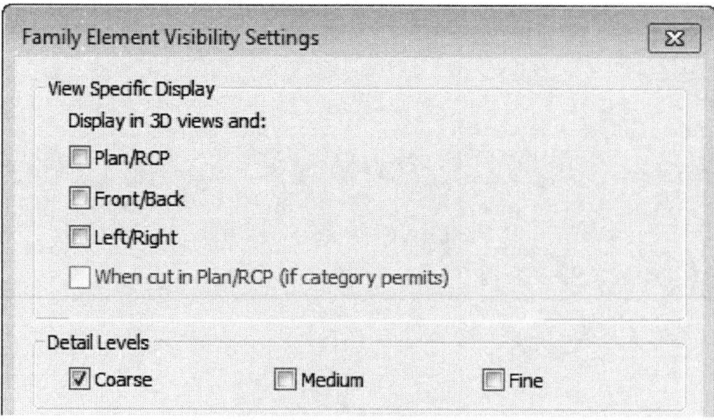

Figure 5–58

10. Save the family.

Task 2 - Add symbolic lines to elevation and plan views.

1. Continue working in the **Elevations: Front** view.
2. Temporarily hide all of the 3D geometry.
3. In the *Annotate* tab>Detail panel, click (Symbolic Line).
4. Draw a rectangle around the outside of the bookcase and lock it to the reference planes, as shown in Figure 5–59.

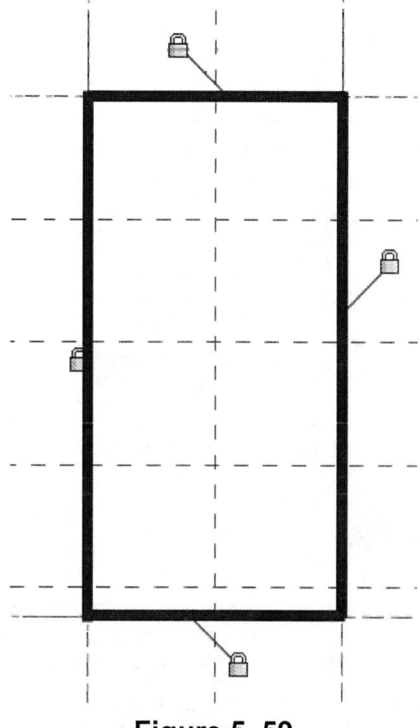

Figure 5–59

5. Select all of the lines.

6. In the *Modify* | *Lines* tab>Visibility panel, click (Visibility Settings).

7. In the Family Element Visibility Settings dialog box, in the *Detail Levels* area, select **Coarse** and clear **Medium** and **Fine**, as shown in Figure 5–60. Click **OK**.

Symbolic lines display in views parallel to the view where they were drawn. Therefore, the lines in the Front view will automatically display in the Back view.

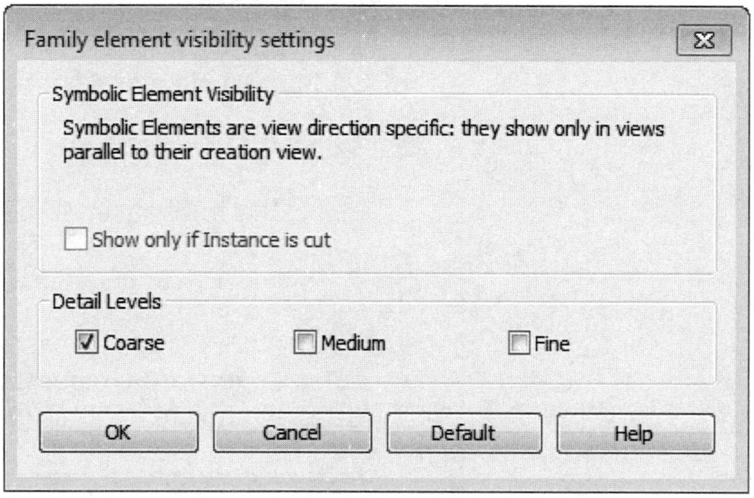

Family element visibility settings

Symbolic Element Visibility

Symbolic Elements are view direction specific: they show only in views parallel to their creation view.

☐ Show only if Instance is cut

Detail Levels

☑ Coarse ☐ Medium ☐ Fine

OK | Cancel | Default | Help

Figure 5–60

8. Reset Temporary Hide/Isolate.

9. Open the **Floor Plans: Ref. Level** view.

10. Select all of the bookcase elements and the new simple extrusion and temporarily hide them.

11. In the *Annotate* tab>Detail panel, click (Symbolic Line).

12. Draw a rectangle around the outside of the bookcase and lock it to the reference planes, as shown in Figure 5–61.

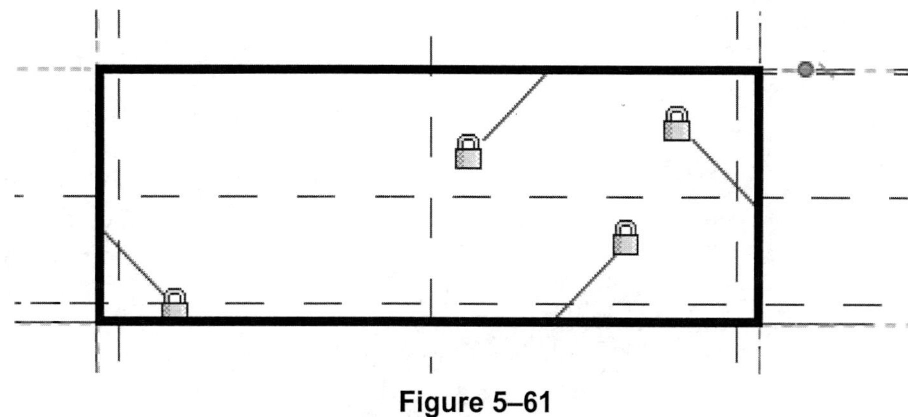

Figure 5–61

13. Select all of the lines and set the *Visibility Settings* to **Coarse**, as done in the previous elevation view.

14. Reset Temporary Hide/Isolate.

15. Switch to the **Elevations: Left** view and repeat the process.

16. Save the family.

Task 3 - Modify the visibility settings of the 3D elements.

1. Open a 3D view and set the *Visual Style* to (Hidden.Line).

2. Select the simple extrusion and temporarily hide it.

3. Select the bookcase elements (but not the doors) and click (Visibility Settings).

4. In the Family Element Visibility Settings dialog box, in the *Detail Levels* area, clear **Coarse** and select **Medium** and **Fine**, as shown in Figure 5–62, and click **OK**.

This has to be done in two steps because the doors are a different element (a nested family) from the rest of the elements.

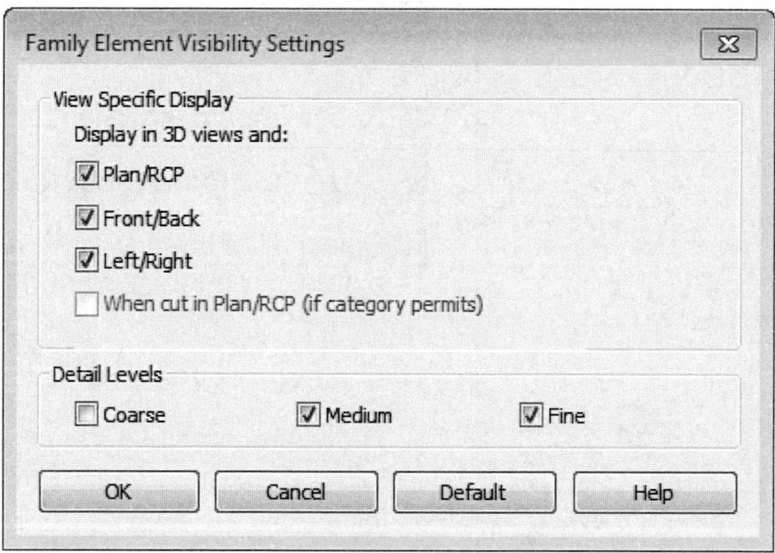

Figure 5–62

5. Select the door elements

6. In the Family Element Visibility Settings dialog box, in the *Detail Levels* area, clear **Coarse** and **Medium** and select **Fine**. Click OK. The doors will only display at the Fine detail level.

7. Save the family.

Task 4 - Load and test the family in a project.

1. Open the **Barristers-Bookcase-Project.rvt** file.
2. Open the **Floor Plans: Level 1** view.
3. Switch to the bookcase family file and load it into the project.
4. Add at least one copy of the new bookcase.
5. In the View Status Bar, set the *Visual Style* to **Shaded**.
6. Change the detail levels to see the differences, as shown in Figure 5–63.

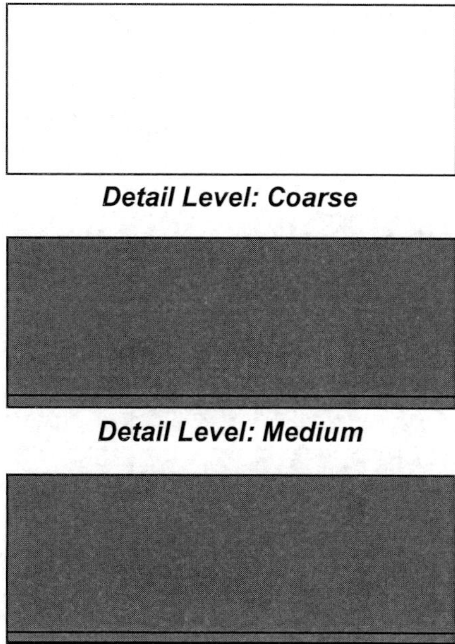

Figure 5–63

7. Test in other views to ensure that the detail levels display the way you want them to.
8. Save and close the project and the family.

Practice 5e Modify the Visibility of Elements in the Heat Pump

Practice Objectives

- Create symbolic lines.
- Modify visibility settings.

Estimated time for completion: 10 minutes

In this practice you will create elements that display at a Coarse detail level, including a simple extrusion for 3D views and symbolic lines for plan and elevation views. You will modify the visibility settings of the lines and 3D elements so that they display differently at various levels of detail, as shown in Figure 5–64.

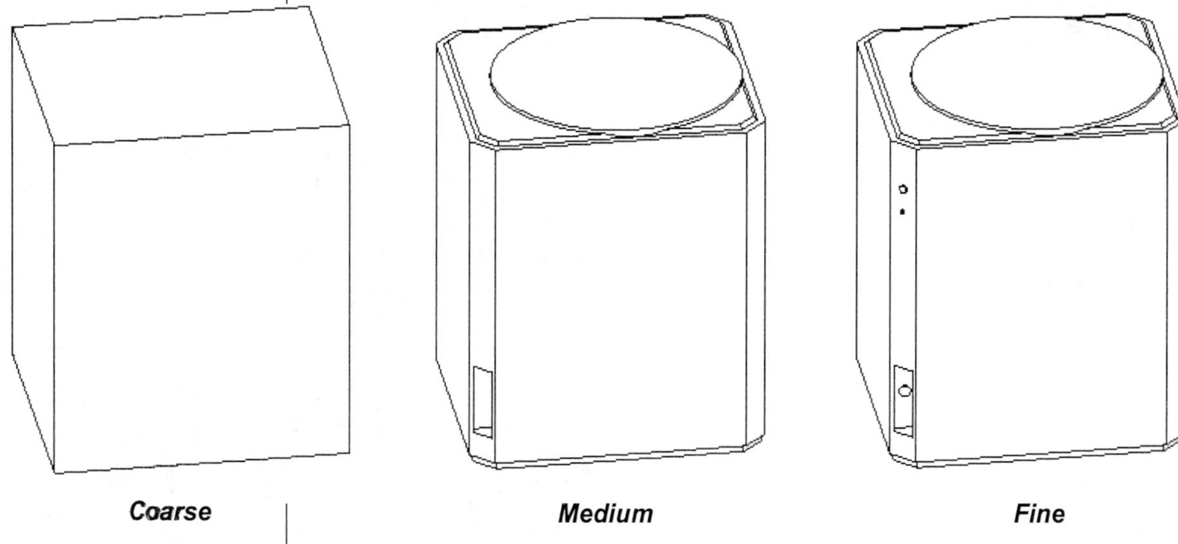

Figure 5–64

Task 1 - Sketch elements for the Coarse detail view.

1. In the ...*MEP* folder, open **Heat-Pump-Visibility.rfa**
2. Open the **Floor Plans: Ref. Level** view.
3. Set the Workplane to the **Ref. Level.**
4. Temporarily Hide the heat pump elements and connectors.
    - To select the extrusion elements, press <Tab> on the edge of the elements until you see the **Joined Solid Geometry** tooltip, as shown in Figure 5–65.
    - Select the connectors and the panel component separately.

Do not use Hide Category.

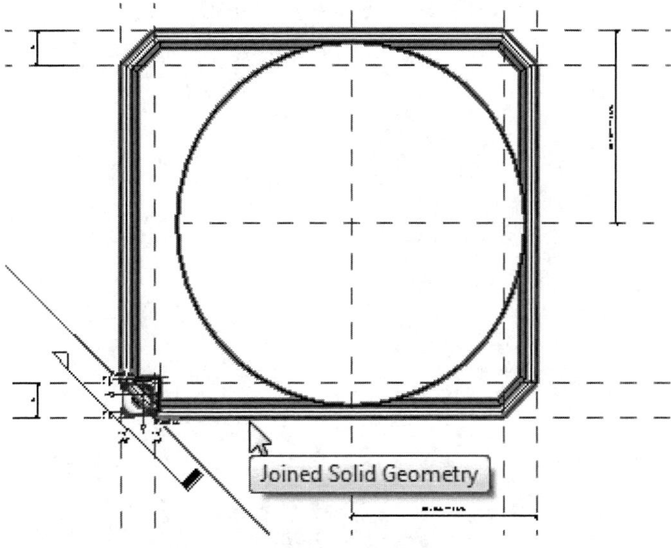

Figure 5–65

5. Create a new extrusion with the footprint shown in Figure 5–66. Set the *Depth* to **40"**

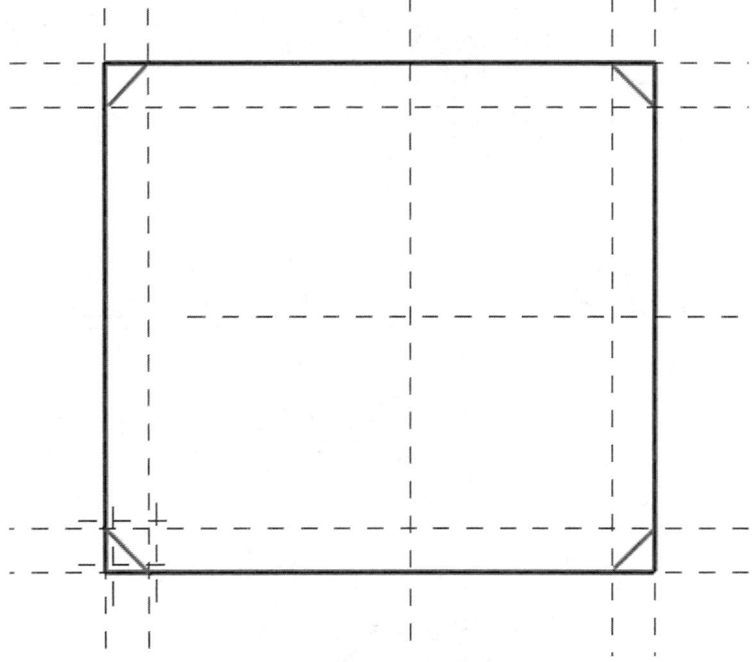

Figure 5–66

6. Switch to the **Elevations: Front** view.

7. Temporarily hide the elements that clutter the view and prevent the new extrusion from displaying.

8. Set the *Visual Style* to **Shaded** to display the extrusion.

9. Align and lock the top and bottom of the new extrusion to the reference planes, as shown in Figure 5–67.

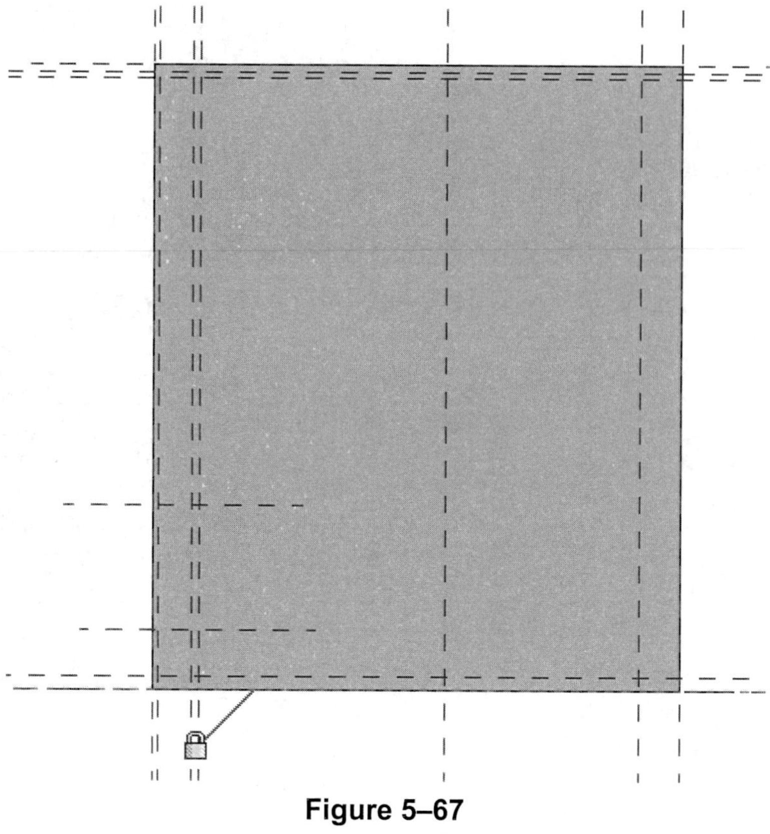

Figure 5–67

10. Select the extrusion.

11. In the *Modify | Extrusion* tab>Mode panel, click (Visibility Settings).

12. In the Family Element Visibility Settings dialog box, clear all of the options except **Coarse**, as shown in Figure 5–68.

You do not want this element displaying in the 2D views at any detail level.

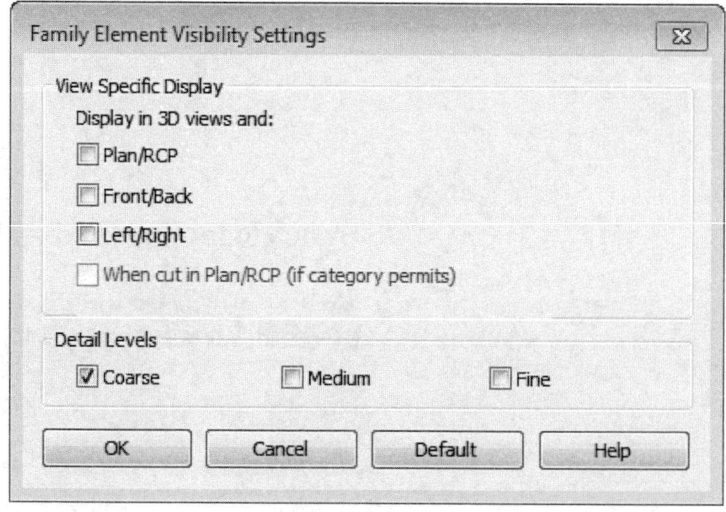

Figure 5–68

13. Temporarily Hide the new extrusion.

14. In the *Annotate* tab>Detail panel, click (Symbolic Line).

15. Draw a rectangle using the outermost reference planes of the heat pump and lock it to the reference planes,

16. Select all of the lines.

17. In the *Modify | Lines* tab>Visibility panel, click (Visibility Settings).

18. In the Family Element Visibility Settings dialog box, in the *Detail Levels* area, select **Coarse** and clear **Medium** and **Fine**, as shown in Figure 5–69. Click **OK**.

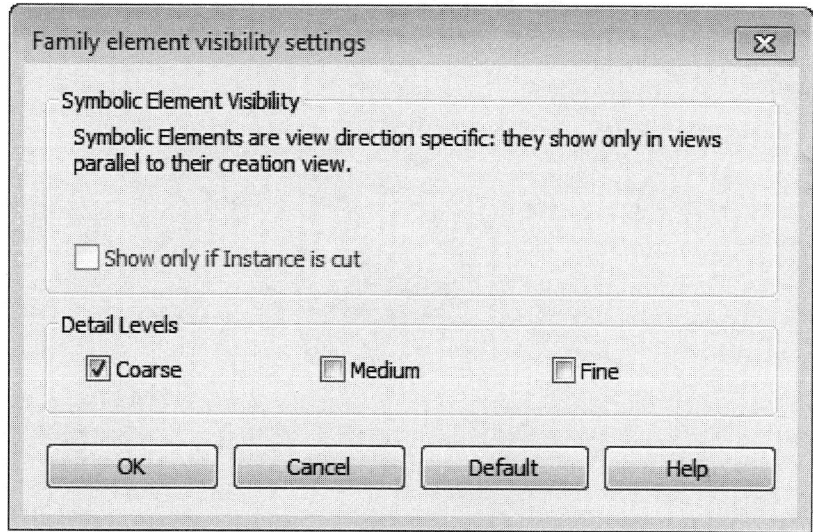

Figure 5–69

- Symbolic lines display in views parallel to the view where they were drawn. Therefore the lines in the **Front** view will automatically display in the **Back** view.

19. Reset Temporary Hide/Isolate.

20. Return to the **Ref. Level** view and draw and lock symbolic lines to the outer reference planes using the same footprint as the extrusion. Set the Visibility Settings as shown in Figure 5–69.

21. Switch to the **Elevations: Left** view and repeat the process.

22. Save the family.

Task 2 - Modify the visibility settings of the 3D elements.

1. Open a 3D view and set the *Visual Style* to (Hidden.Line).
2. Temporarily hide the simplified extrusion.
3. Select the heat pump elements (but not the panel component) and click (Visibility Settings).
4. In the Family Element Visibility Settings dialog box, in the *Detail Levels* area, clear **Coarse** and select **Medium** and **Fine**, as shown in Figure 5–70. Click **OK**.

This has to be done in separate steps because the component is a different element (i.e., a nested family) from the rest of the elements.

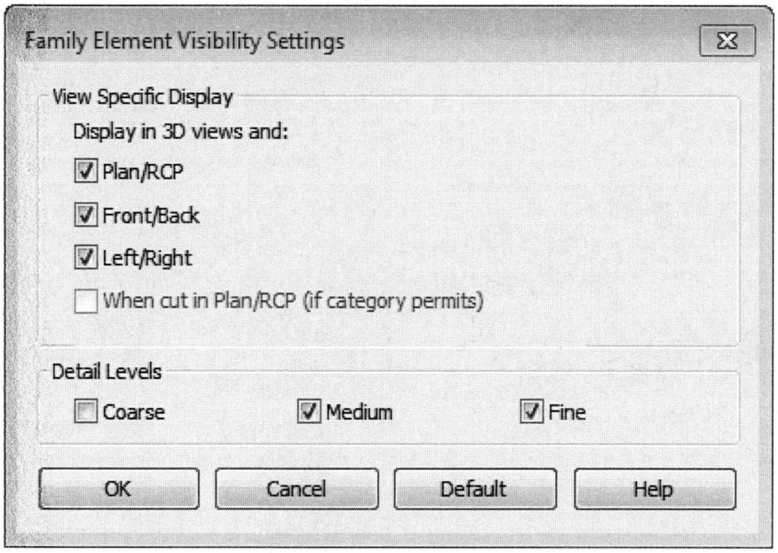

Family Element Visibility Settings

View Specific Display

Display in 3D views and:

- [x] Plan/RCP
- [x] Front/Back
- [x] Left/Right
- [ ] When cut in Plan/RCP (if category permits)

Detail Levels

- [ ] Coarse
- [x] Medium
- [x] Fine

OK | Cancel | Default | Help

Figure 5–70

5. Select the panel component, as shown in Figure 5–71.
6. In the Family Element Visibility Settings dialog box, in the *Detail Levels* area, clear **Coarse** and **Medium**, select **Fine**, and then click **OK**.
7. Select the connector extrusions on the main unit (as shown in Figure 5–71) and set the Visibility to **Fine** only. The connector extrusions will only display at the **Fine** detail level.

The electrical and piping connectors were hidden in this view for clarity.

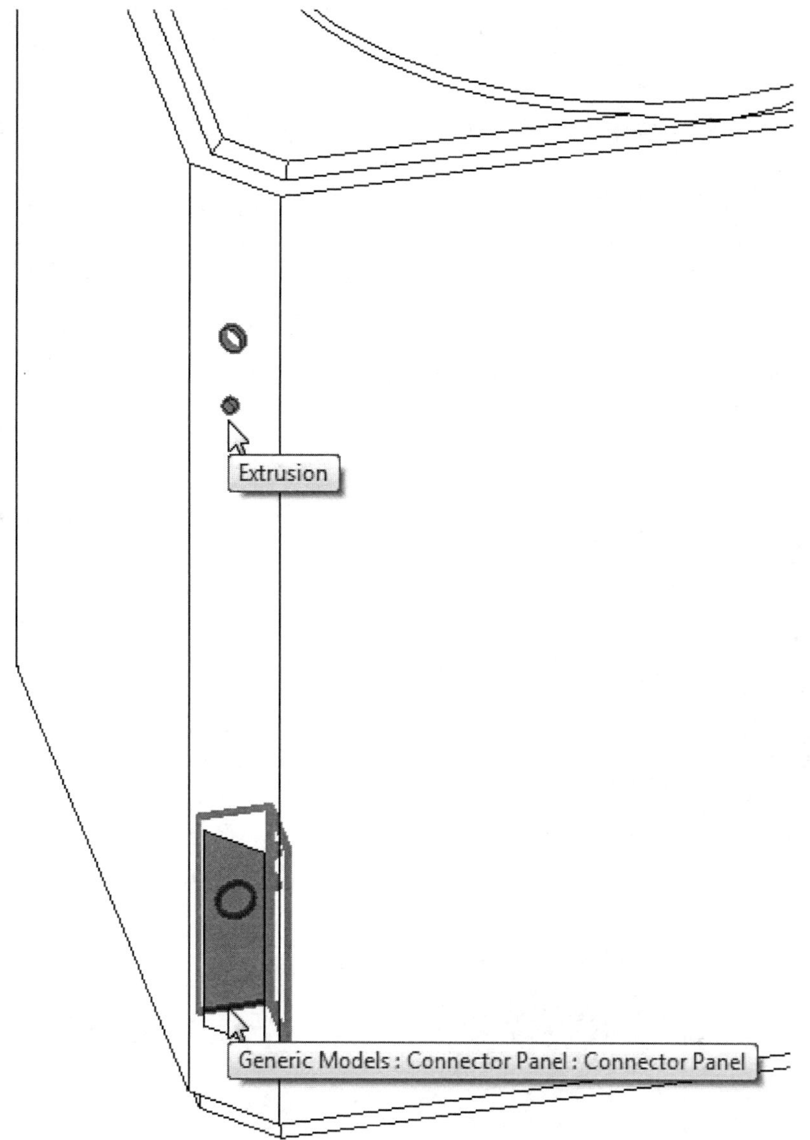

Figure 5–71

8. Save the family.

Task 3 - Load and test the family in a project.

1. Open the **Heat-Pump-Project.rvt** file, if not already open.
2. Open the Mechanical>HVAC>**Floor Plans: 1-Mech** view.
3. Switch to the heat pump family file and load it into the project.
4. Add at least one copy of the new heat pump.
5. In the View Status Bar, set the *Visual Style* to **Shaded**.

6. Change the detail levels to display the differences, as shown in Figure 5–72.

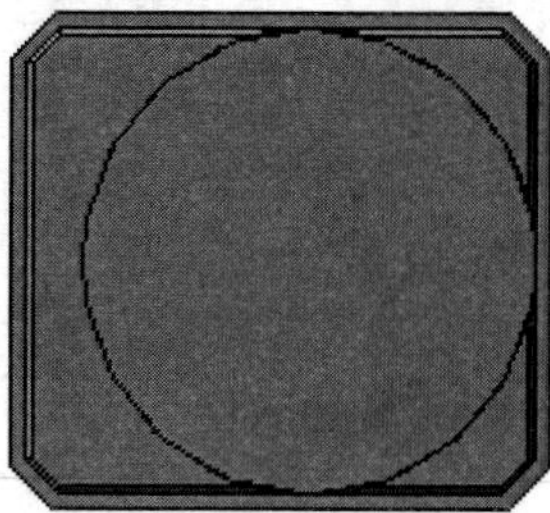

Detail Level: Coarse ***Detail Level: Medium and Fine***

Figure 5–72

7. Open a 3D view, rotate it so that the connector locations display and test the detail levels.

8. Save and close the project and the family.

Practice 5f Modify the Visibility of Elements in the Structural Column

Practice Objectives

- Create a simple extrusion.
- Add symbolic lines.
- Modify visibility settings.

Estimated time for completion: 10 minutes

In this practice you will create elements that display at a Coarse detail level, including a simple extrusion for 3D views and symbolic lines for plan and elevation views. You will modify the visibility settings of the lines and 3D elements so that they display differently at various levels of detail, as shown in Figure 5–73.

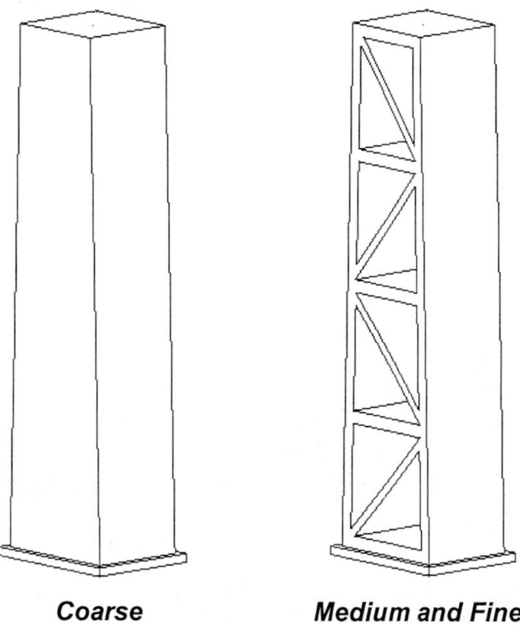

Figure 5–73

Task 1 - Sketch elements for the Coarse detail view.

1. In the ...*Structural* folder, open **Trellis-Column-Visibility.rfa**.
2. Open the **Elevations: Front** view.
3. Temporarily hide the solid column elements, but not the base.
    - Do not use Hide Category.

4. Create a new extrusion with the footprint shown in Figure 5–74. Set the *Depth* to **2'-0"**.

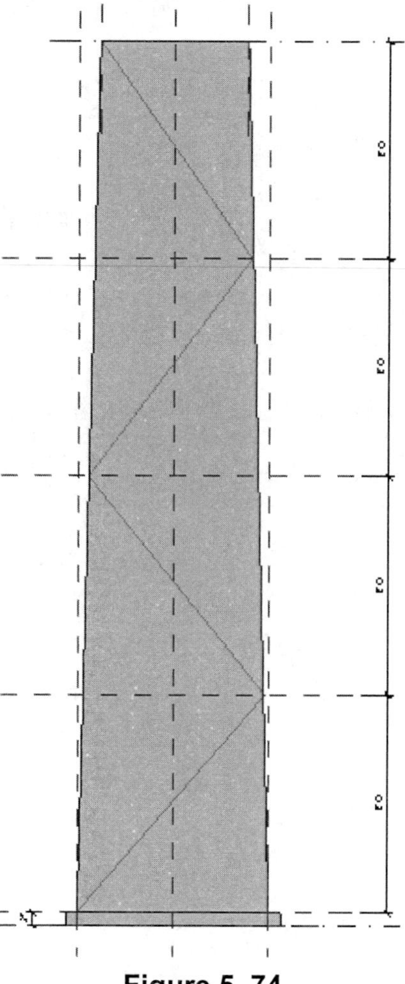

Figure 5–74

5. Select the extrusion.
6. In the *Modify | Extrusion* tab>Mode panel, click (Visibility Settings).
7. In the Family Element Visibility Settings dialog box, clear all of the options except **Coarse** and **When cut in Plan/RCP**, as shown in Figure 5–75.

You do not want this element displaying in the 2D views at any detail level.

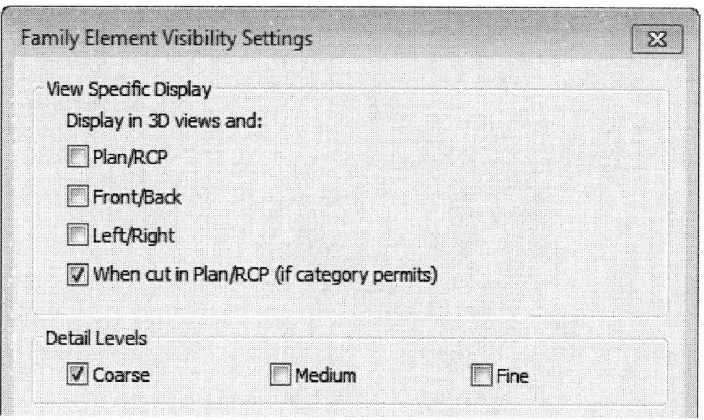

Figure 5–75

8. Temporarily Hide the new extrusion.

9. In the *Annotate* tab>Detail panel, click (Symbolic Line).

10. Create the same outline as the new extrusion. Ensure that you lock the extrusion to the reference planes and lines.

11. Select all of the lines.

12. In the *Modify | Lines* tab>Visibility panel, click (Visibility Settings).

13. In the Family Element Visibility Settings dialog box, in the *Detail Levels* area, select **Coarse**, and then clear **Medium** and **Fine**, as shown in Figure 5–76. Click **OK**.

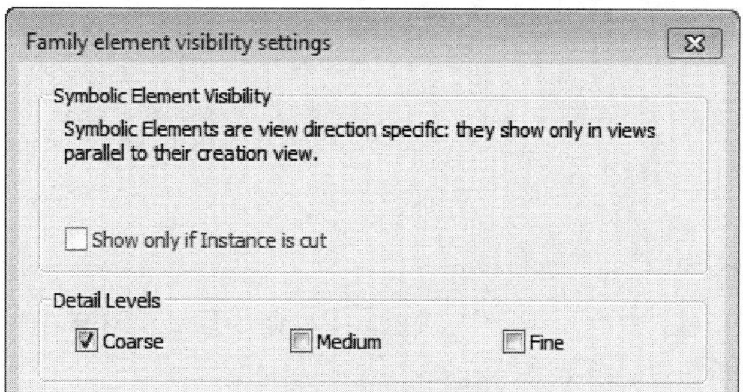

Figure 5–76

- Symbolic lines display in views parallel to the view where they were drawn. Therefore the lines in the **Front** view will automatically display in the **Back** view.

14. Open the **Floor Plans: Lower Ref. Level** view.

15. Temporarily hide all of the extrusions except the new one.

16. Align and lock the extrusion to the back reference plane, as shown in Figure 5–77.

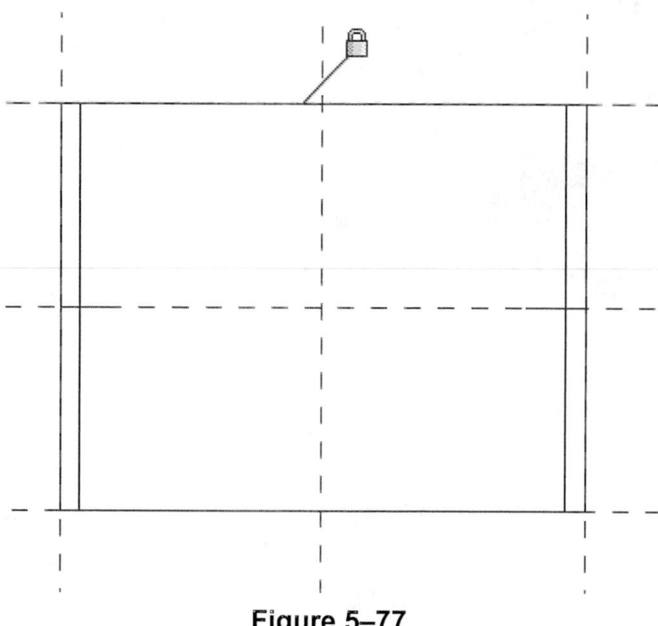

Figure 5–77

17. Temporarily hide the extrusion and reference lines (but not the reference planes). Sketch and lock Symbolic Lines around the same footprint.

18. Select the symbolic lines and set the visibility settings as shown previously in Figure 5–76.

19. Open the **Elevations: Left** view and add symbolic lines around the edge of the column (but not the base) and set the Visibility Settings to match the other lines.

20. Reset Temporary Hide/Isolate in all the views.

21. Save the family.

Task 2 - Modify the visibility settings of the 3D elements.

1. Open a 3D view.
2. Temporarily hide the simplified extrusion.
3. Select the main column elements, excluding the base, and click (Visibility Settings).
4. In the Family Element Visibility Settings dialog box, in the *Detail Levels* area, clear **Coarse** and select **Medium** and **Fine**, as shown in Figure 5–78. Click **OK**.

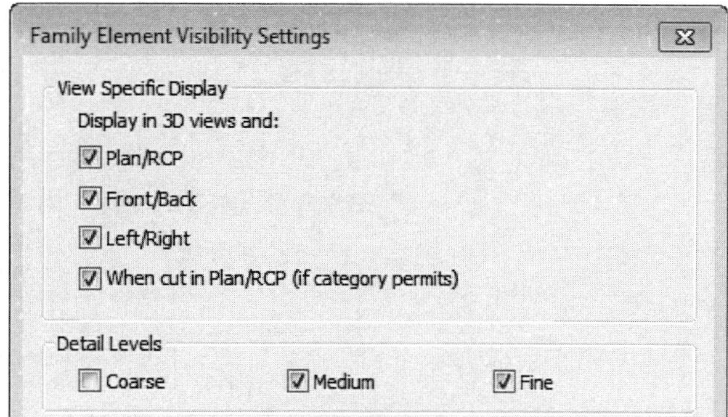

Figure 5–78

5. Save the family.

Task 3 - Load and test the family in a project.

1. Open **Trellis-Column-Project.rvt**, if not already open.
2. Open the **Structural Plans: Level 2** view.
3. Switch to the trellis column family file and load it into the project.
4. Add at least one copy of the new column.
5. Switch to the **Elevations: South** view.
6. Change the detail levels to display the differences, as shown in Figure 5–79.

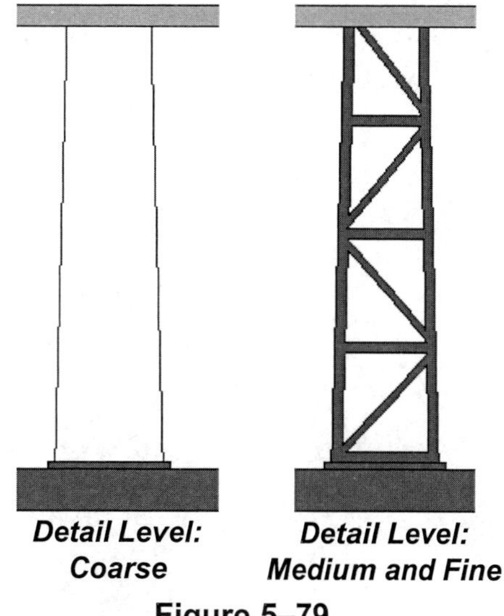

Figure 5–79

7. Open a 3D view and test the detail levels.
8. Save and close the project and the family.

Chapter Review Questions

1. If your family can be switched from one side to another (such as a door swing) which type of icon would you add?
    a. Mirror
    b. Connector
    c. Control
    d. Calculation
2. (MEP only) When adding an Electrical Connector, where do you specify the System Type (such as Power-Balanced or Security)?
    a. In the Options Bar
    b. In the right-click menu
    c. In the expanded Connectors panel
    d. In the Type Properties
3. In which of the following situations would you use a nested family? (Select all that apply.)

    When the family is used multiple times...
    a. in one family
    b. in other families
    c. in one project
    d. in several projects.
4. The main reason to associate a family parameter is when:
    a. The parameter will be shared in several projects.
    b. The parameter will be used in a schedule.
    c. You want a user-created parameter to control another parameter.
    d. You want the parameter to reference a parameter in the project.

5. For the elements in a family to display differently at Coarse and Fine detail levels (as shown in Figure 5–80), you must draw everything in 2D.

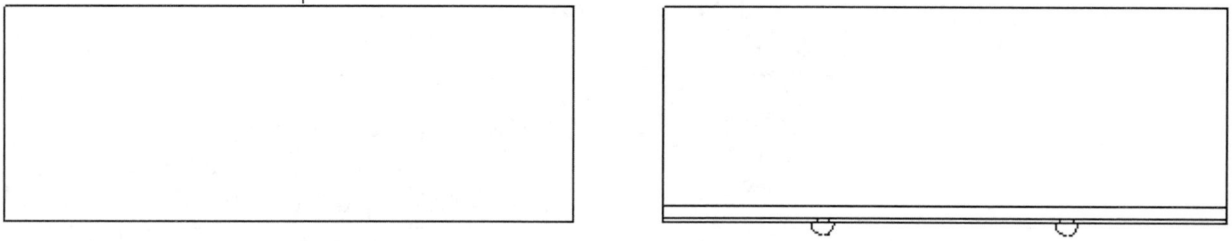

Detail Level: Coarse

Detail Level: Fine

Figure 5–80

   a. True
   b. False

Command Summary

Button	Command	Location
Connectors		
	Cable Tray Connector	• **Ribbon**: *Create* tab>Connectors panel> expand Connector
	Conduit Connector	• **Ribbon:** *Create* tab>Connectors panel> expand Connector
	Duct Connector	• **Ribbon**: *Create* tab>Connectors panel> expand Connector
	Electrical Connector	• **Ribbon:** *Create* tab>Connectors panel> expand Connector
	Pipe Connector	• **Ribbon:** *Create* tab>Connectors panel> expand Connector
Other Tools		
	Component	• **Ribbon:** *Create* tab>Model panel
	Control	• **Ribbon:** *Create* tab>Control panel
	Detail Component	• **Ribbon**: *Annotate tab*>Detail panel
	Masking Region	• **Ribbon:** *Annotate tab*>Detail panel
	Model Line	• **Ribbon:** *Create* tab>Model panel
	Opening	• **Ribbon:** *Create* tab>Model panel
	Symbolic Line	• **Ribbon:** *Annotate* tab>Detail panel
	Visibility Settings	• **Ribbon:** Various contextual tabs> Mode panel

Additional Family Types

Beyond the basics of family creation, there are several special cases. These include in-place families that are only available in the current project, 2D profile families used with various elements (such as gutters and sweeps), and annotation families for tags and symbols. Shared parameters are frequently required when you create tags with information you want to include in schedules.

Learning Objectives in this Chapter

- Create in-place families directly in a project.
- Create 2D profiles that are used to create 3D wall sweeps, reveals, railings, nosings, and other elements.
- Create 2D annotation symbol families using drawing tools, filled regions, text, and labels.
- Create shared parameters that can be used in both tags and schedules in multiple projects.

6.1 Creating In-Place Families

In-place families use standard family tools to create elements that are specific to the project in which you create them. These might include built-in furniture, sloped walls, as shown in Figure 6–1, custom door or window openings, and roofs.

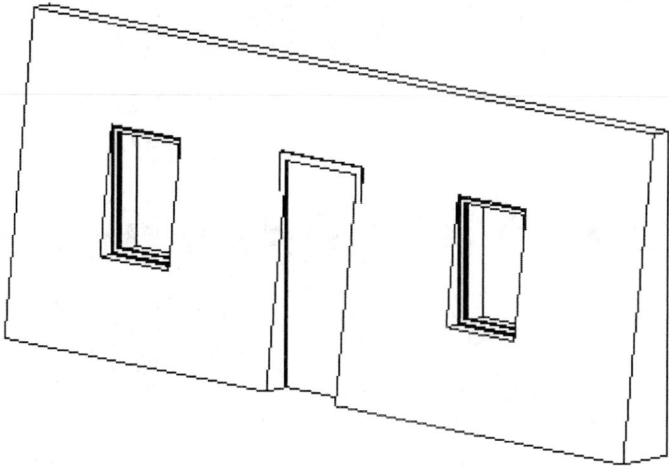

Figure 6–1

In-place family elements become the actual element you specify, such as a wall, door, roof, column, ceiling fixture, or floor.

- If the element is a host element, such as a wall, you can add elements to it, such as doors or windows.
- Elements, such as in-place doors and windows, can be copied and moved within their host.
- In-place family elements can be aligned and locked to project elements so they move together.
- Make sure reference planes are drawn in the family. If you tie an in-place family to an external reference plane and if the reference plane is deleted, you lose the family.
- In-place families are rarely used in the Autodesk® Revit® MEP software.

How To: Create an In-Place Family

1. In a project, in the *Architecture* tab>Build panel, or *Structure* tab>Model panel, or *Systems* tab>Model panel, expand (Component) and click (Model In-Place).
2. The Family Category and Parameters dialog box opens, as shown in Figure 6–2. You can filter the list to display only the disciplines needed.

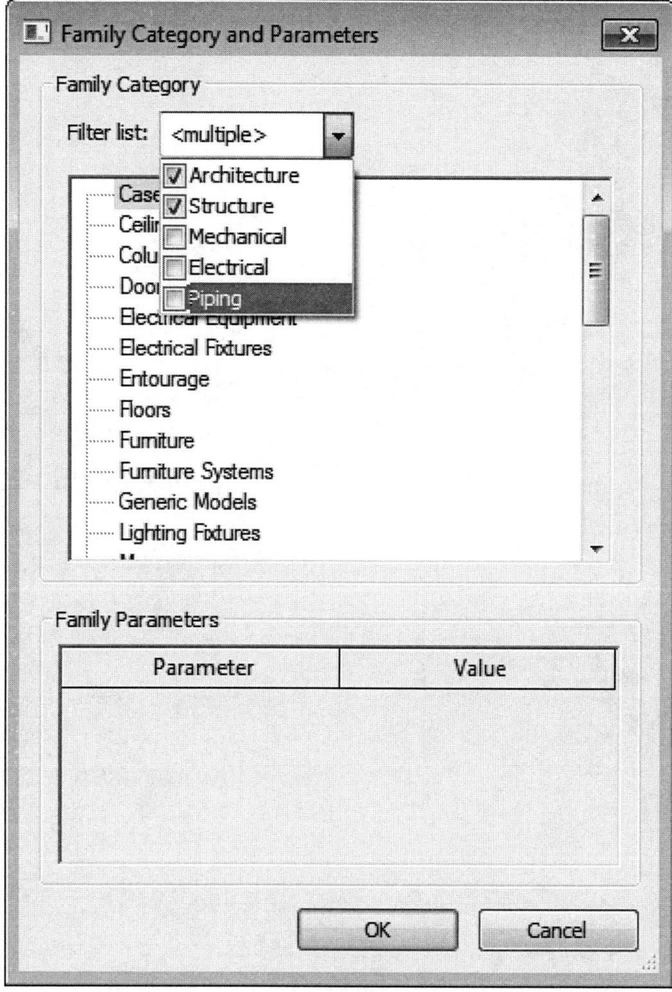

Figure 6–2

3. Select a *Family Category* and click **OK**. In the Name dialog box, enter a name for the new family and click **OK**.
4. The Family Editor tabs display in the Ribbon. Use the tools as needed to create the elements you want to have in the family.
5. In the In-Place Editor panel, click (Finish Model) to create the in-place element.

6.2 Creating Profiles

Profiles are 2D elements that are used in a variety of commands to create 3D elements. For example, you need a profile to create wall sweeps (as shown in the Base profile in Figure 6–3) and reveals, gutters, fascias, and floor slab edges. They are also used in families for railings, stair nosings, and solid sweeps.

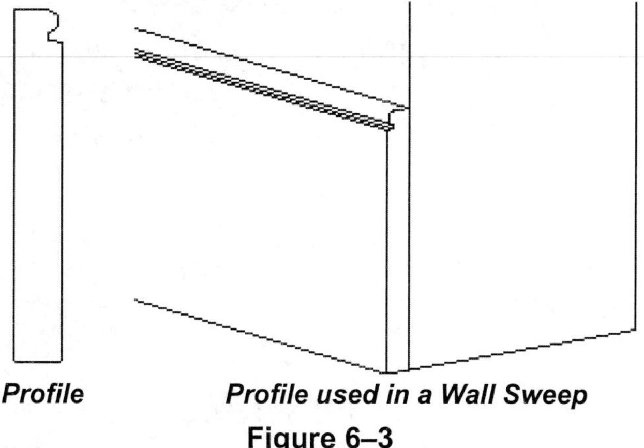

Profile *Profile used in a Wall Sweep*

Figure 6–3

Several profile templates are supplied with the Autodesk® Revit® software: **Profile**, **Profile-Hosted**, **Profile-Mullion** (Autodesk® Revit® Architecture only), **Profile-Rail**, **Profile-Reveal**, and **Profile-Stair Nosing**. These are created with existing reference planes to help you place the elements correctly on the host element.

- All profiles must be closed shapes.
- Profile families can include detail components.

Practice 6a

In-Place Families and Profiles: Door Opening

Learning Objectives

- Create an in-place door family and opening.
- Add trim for the in-place door family using a sweep based on a 2D profile.

Estimated time for completion: 15 minutes

In this practice you will create a door in-place family and add an odd-shaped opening. Using a Sweep, you will add a profile for door trim, as shown in Figure 6–4.

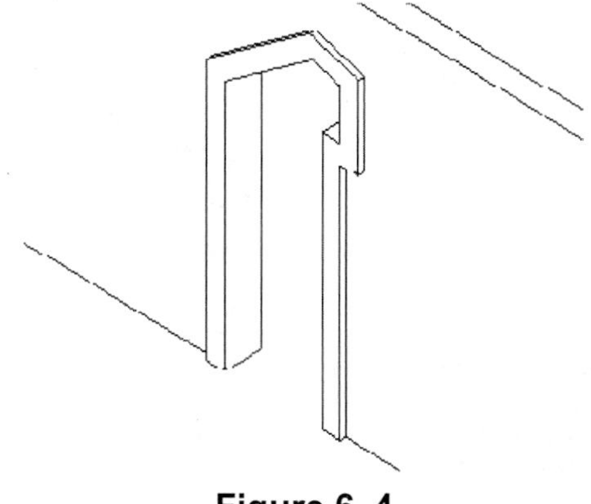

Figure 6–4

Task 1 - Create an in-place door family.

1. Start a new project based on the default architectural template.
2. Open the **Floor Plans>Level 1** view and draw a **10'-0"** high wall in the project.
3. In the *Architecture* tab>Build panel, expand (Component) and click (Model In-Place).
4. In the Family Category and Parameters dialog box, Set the *Filter list* to **Architecture** and then set the *Family Category* to **Doors** and click **OK**.
5. Name the in-place family **Odd Shaped Opening**.
6. The Family Editor tools display in the Ribbon. In the *Create* tab>Datum panel, click (Ref Plane) and draw a reference plane along the center of the wall.

The reference plane is used to specify the work plane. It does not impact the shape of the wall.

7. Select the reference plane and, in Properties, set the *Name* to **Center of Wall**. The plane displays as shown in Figure 6–5. Click in empty space to release the selection of the reference plane.

Figure 6–5

8. Switch to the **Elevations>South** view and zoom in on the wall.
9. In the *Create* tab>Model panel, click (Opening).
10. If you have more than one wall in the project you are prompted to select a host for the opening. Click **OK** and select the wall.
11. Draw an opening similar to that shown in Figure 6–6.

The exact design does not matter; ensure that it is at least the width and height of a standard door opening. The sketch must be closed.

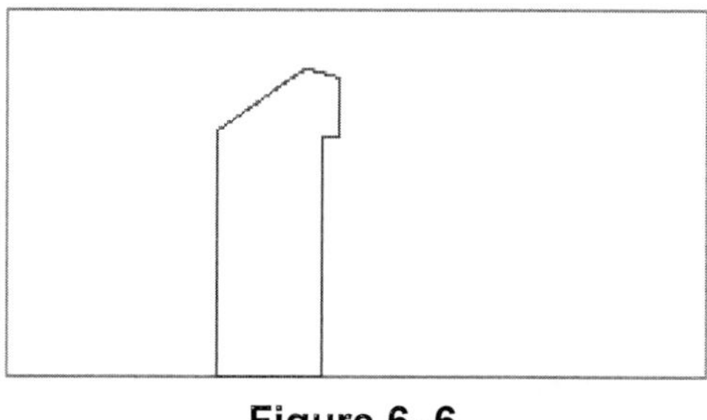

Figure 6–6

12. Click (Finish Edit Mode) but do not finish the model.

Task 2 - Add a sweep for the trim and sketch a profile.

1. In the *Create* tab>Work Plane panel, click (Set).
2. In the Work Plane dialog box, set the *Name* to **Reference Plane : Center of Wall** and click **OK**.
3. In the *Create* tab>Forms panel, click (Sweep).
4. In the *Modify | Sweep* tab>Sweep panel, click (Sketch Path).

5. Draw a path around the edge of the opening. Do not close this sketch.

   - Click (Pick Lines) and, in the Options Bar, select **Lock**. This locks the sketch to the existing geometry so the sketch updates if the opening is modified, .

6. Click (Finish Edit Mode).

7. In the *Modify | Sweep* tab>Sweep panel, click (Edit Profile).

8. In the Go To View dialog box, select **Floor Plan: Level 1** and click *Open View.*

9. Zoom in on the red dot. Draw the profile shown in Figure 6–7. The profile must be closed. You do not need to add the dimensions.

Drawing the profile within the in-place model keeps the size of the trim directly connected to the selected wall and opening.

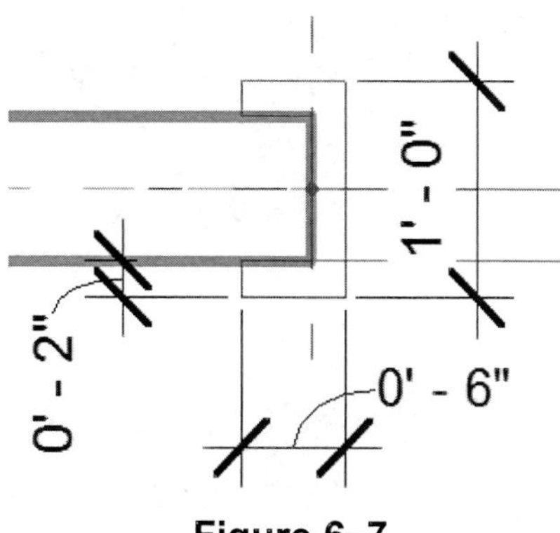

Figure 6–7

10. Click (Finish Edit Mode) to finish the profile.

11. Click (Finish Edit Mode) again to finish the sweep.

12. Click (Finish Model).

13. View the new cased opening in several views.

14. Move the opening. You can see that the opening works just like a door.

15. Save the project as **Sample Opening.rvt**.

Practice 6b In-Place Families and Profiles: Concrete Corbeling

Practice Objectives

- Create a 2D profile and use it in an in-place family to create a concrete corbel.
- Join the corbeling to the wall.

Estimated time for completion: 15 minutes

In this practice you will draw a profile of a corbel and then use it within an in-place family to create a sweep, as shown in Figure 6–8.

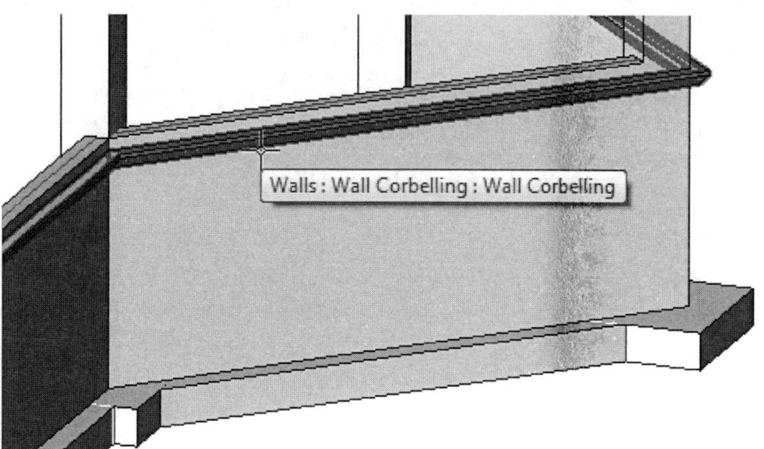

Figure 6–8

Corbel: A stepped bracket, designed to support a weight. It projects from a wall and is created from a building material, such as concrete, stone, or wood.

Task 1 - Create the profile.

1. Create a new family based on the **Profile-Hosted.rft** template found in the Autodesk Revit Library.
2. Save the family to your practice folder as **Corbel.rfa**.

The dimensions are for information only.

3. Draw the four reference planes as shown in Figure 6–9.

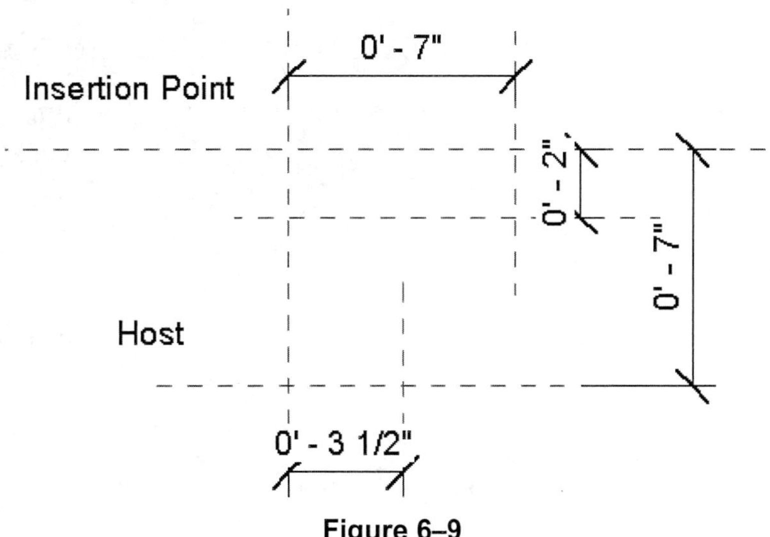

Figure 6–9

4. In the *Create* tab>Detail panel, click (Line). Draw the profile shown in Figure 6–10.

The Family Editor does not display as many Ribbon tabs or panels because fewer commands are needed to create a 2D profile.

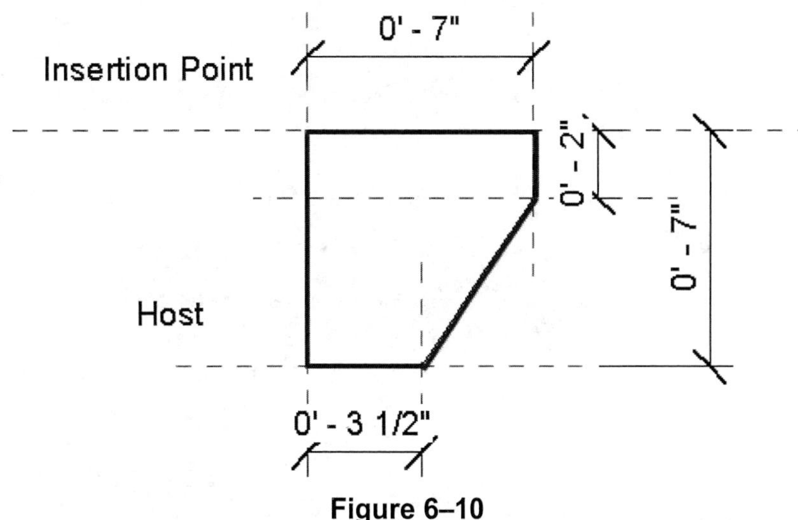

Figure 6–10

5. Save the family.

Task 2 - Add a new wall sweep using the profile.

1. In the *...\Structural* folder, open **Concrete-Structure-S.rvt**.
2. Open the Structural Plans>**Level 1** view.
3. In the *Structure* tab>Model panel, expand (Component) and click (Model In-Place).

4. In the Family Category and Parameters dialog box, set the *Filter list* to **Structure** only, select **Walls**, and click **OK**.
5. In the Name dialog box, type **Wall Corbelling** and click **OK**.
6. In the In-Place Editor, in the *Create* tab>Forms panel, click (Sweep).
7. In the *Modify* | *Sweep* tab>Sweep panel, click (Sketch Path).
8. Use the Draw tools to sketch the lines shown in Figure 6–11.

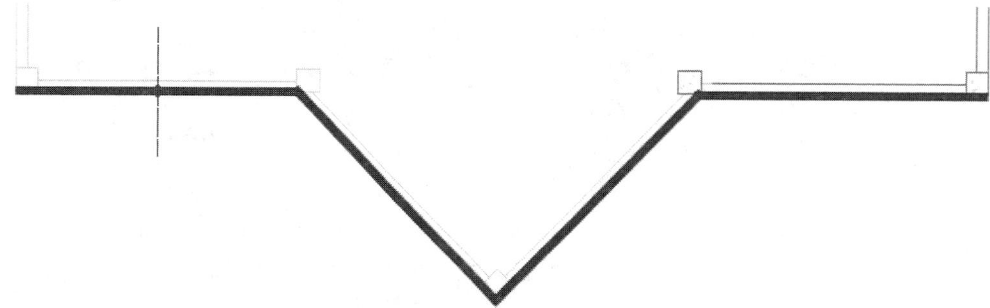

Figure 6–11

9. Click (Finish Edit Mode).
10. In the *Modify* | *Sweep* tab>Sweep panel, click (Load Profile).
11. In the Load Family dialog box, navigate to the folder where you saved the **Crobel.rfa** profile created in the previous task. Select it and click **Open**.
12. In the Sweep panel, click (Select Profile).
13. In the *Profile* list, select **Corbel**.
14. In the Quick Access Toolbar, click (Default 3D View).
15. Zoom in on the profile location, it is facing into the wall as shown in Figure 6–12.

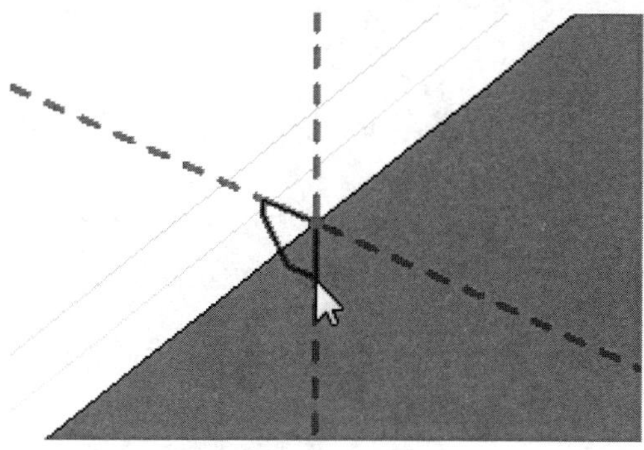

Figure 6–12

16. In the Options Bar, click **Flip**.

17. In the Options Bar or using the temporary dimension, lower the corbel (negative) **-6"** in the Y-direction, as shown in Figure 6–13.

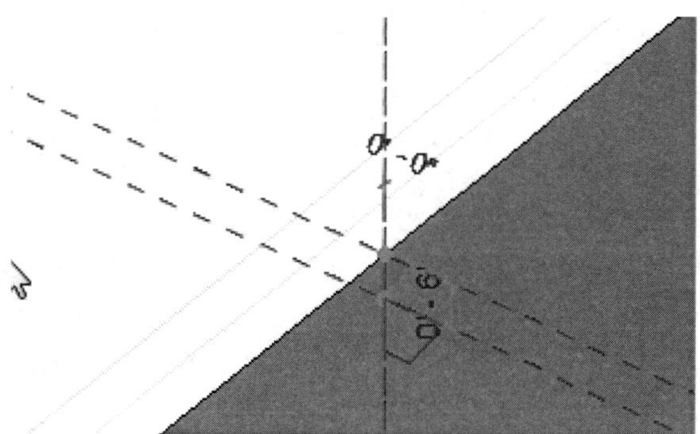

Figure 6–13

18. Click (Finish Edit Mode).

19. With the new Corbel sweep still selected, in Properties, in the *Materials and Finishes* area, next to Material, click

    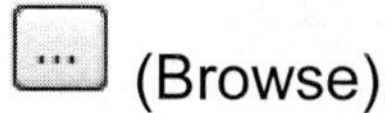 (Browse).

20. In the Materials Browser, select **Concrete - Cast-in-Place Concrete** and click **OK**.

21. In the *Modify* tab>In-Place Editor panel, click (Finish Model).

Task 3 - Join the corbel to the wall.

1. Open the Structural Plans>**Level 1** view.
2. Draw a Wall Section across the wall and corbel, as shown in Figure 6–14.

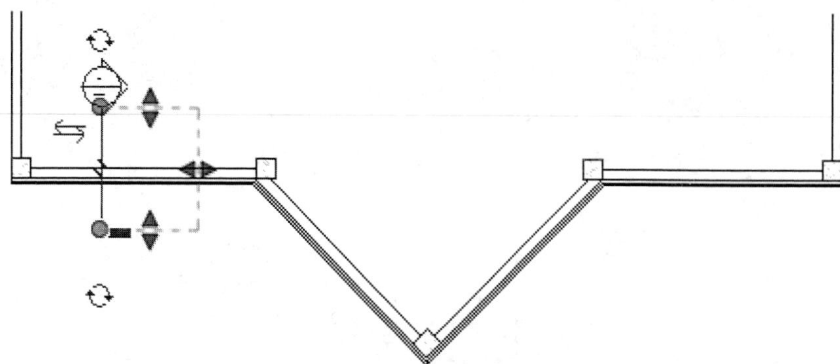

Figure 6–14

3. Open the section. You can see that the wall and corbel are still considered separate entities, as shown in Figure 6–15.
4. In the *Modify* tab>Geometry panel, click (Join). Select the wall and then the corbel. It becomes part of the existing wall, as shown in Figure 6–16.

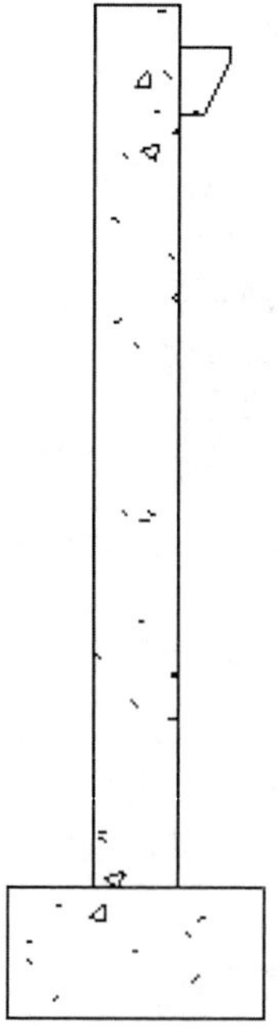

Figure 6–15

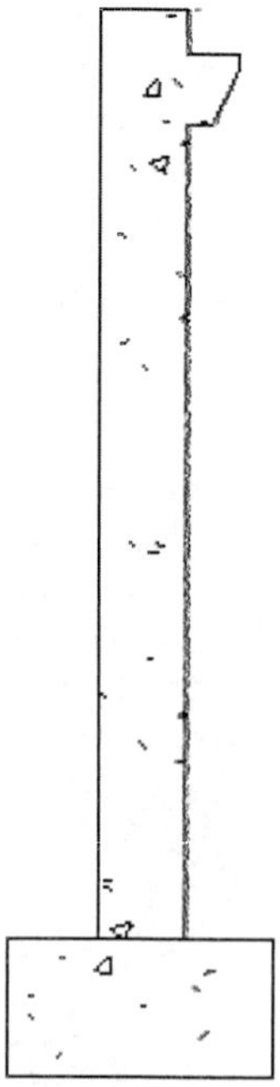

Figure 6–16

5. Save the project.

Practice 6c Use a Profile to Create a Structural Floor Type

Practice Objectives

- Create a profile to use in defining a composite metal deck.
- Load the profile into a project and create a new structural floor type using the profile.
- Test the new floor and change the span direction.

Estimated time for completion: 20 minutes

In this practice you will create a profile for a composite metal deck and save it to a library. You will then load the profile into a project, use it to create a new Metal Deck Floor type, as shown in Figure 6–17, and test the new floor type.

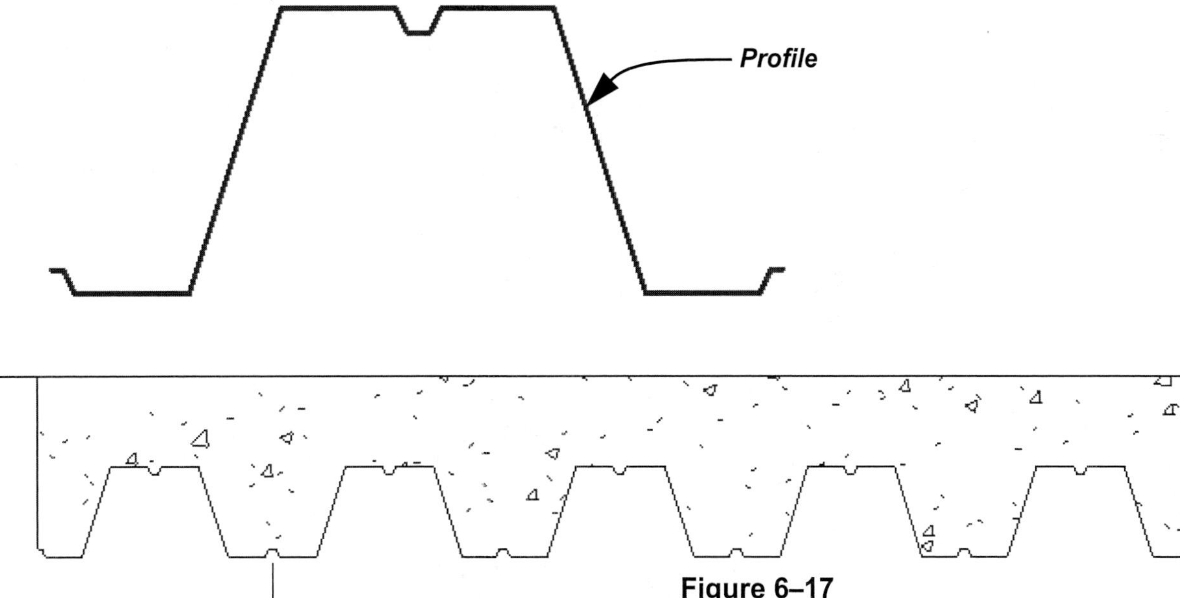

Figure 6–17

Task 1 - Create a composite Metal Deck profile.

1. Create a new family based on the **Profile.rft** template found in the Autodesk Revit Library.
2. Save the family to your practice folder as **Composite-Metal-Deck.rfa**.
3. Type **UN** to open the Project Units dialog box.
4. Next to *Length Units*, select the **Format** button.
5. In the Format dialog box, change the units to **Fractional inches**. (This makes the dimensions in the family read in inches only so that it is easier to create the profile.)

6. Click **OK** twice to close the dialog boxes.
7. Set the *Scale* to **3" = 1'-0"**.
8. In the *Create* tab>Datum panel, click (Reference Plane).
9. Add reference planes as shown in Figure 6–18. They help you draw the elements (the dimensions are shown for reference only).

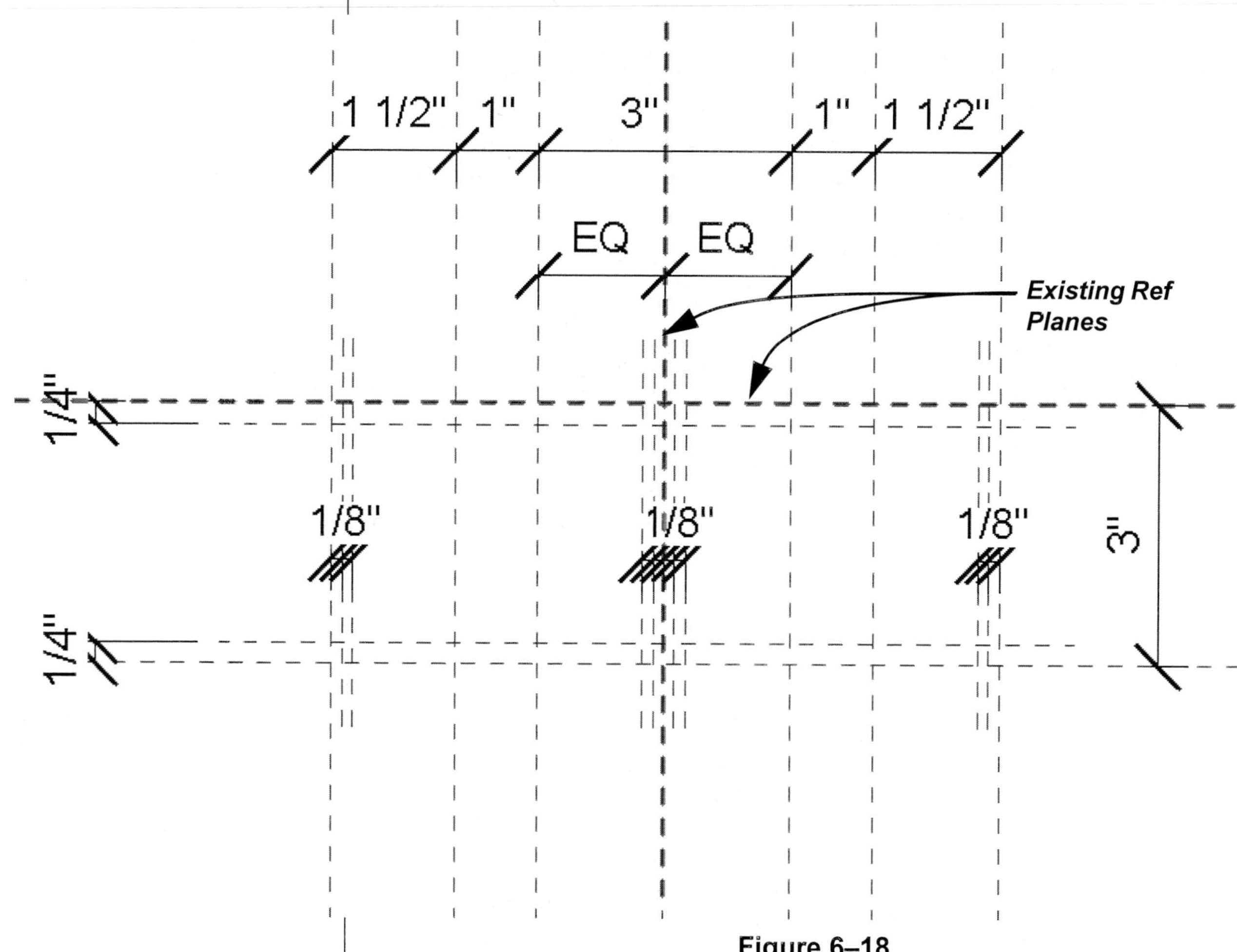

Figure 6–18

10. In the *Create* tab>Detail panel, click (Line). In the Options Bar, verify that the **Chain** option is selected.

11. Sketch the profile as shown in Figure 6–19.

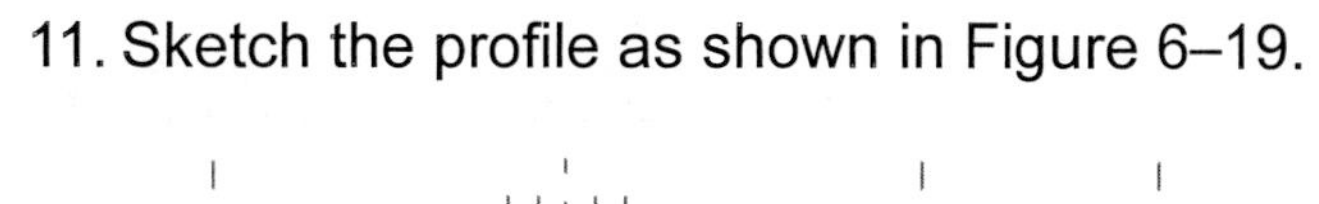

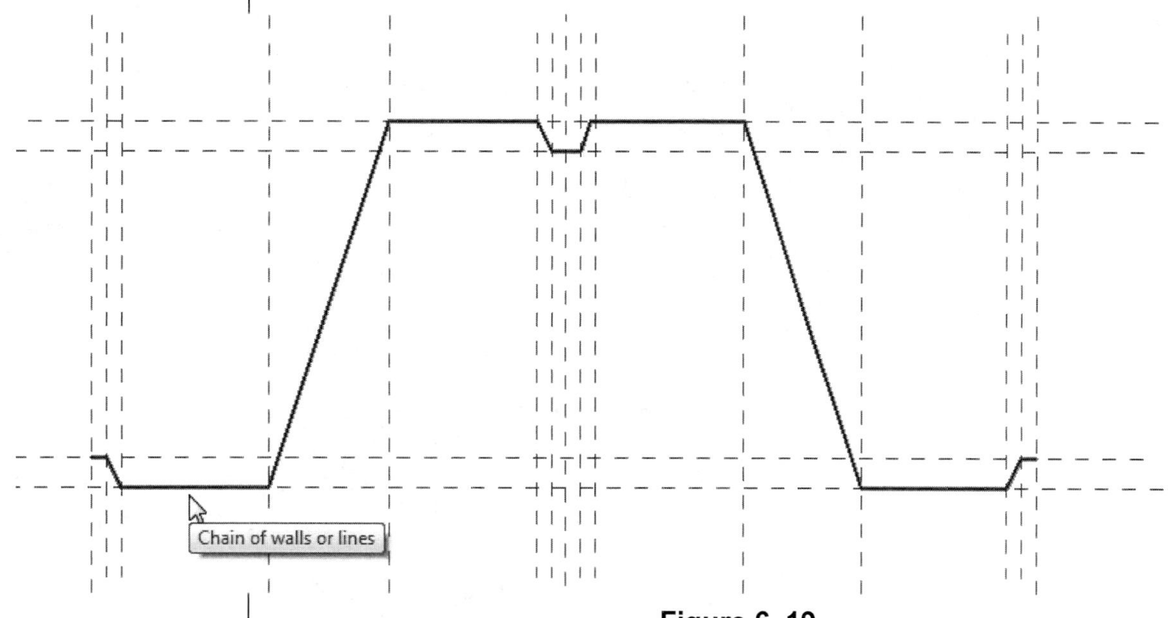

Figure 6–19

12. Click (Modify) or press <Esc> when the profile is complete.

13. In the *Create* tab>Properties panel, click (Family Types).

14. In the Family Types dialog box, in the *Family Types* area, click ***New...***.

15. In the Name dialog box, name it **3"**, and click **OK**.

16. Click **OK** again to close the dialog box.

17. Save the family without closing the file.

Task 2 - Load the composite Metal Deck profile.

1. Start a new project based on the default Structural template.

2. In the Quick Access Toolbar or *View* tab>Windows panel, expand (Switch Windows), and select **Composite-Metal-Deck.rfa** or press <Ctrl>+<Tab> to toggle back to the family file.

If more than one project is open, select the new one from the list that displays.

3. In the Family Editor panel, click (Load into Project) to insert the family into the new project. You should be in the Structural Plans>**Level 2** view of the new project.

4. In the *Structure* tab>Structure panel, click (Floor: Structural).

5. in the Type Selector, verify that **Floor: 3"LW Concrete on 2" Metal Deck** is selected.

6. In the *Modify | Create Floor Boundary* tab>Draw panel, use the **Line** or **Rectangle** tool to draw a slab of any type or size. Set the *Span Direction* to **Horizontal**.

7. Click (Finish Edit Mode).

8. In the Quick Access Toolbar or *View* tab>Create panel, click (Section) and place a section vertically through the slab from north to south.

9. Open the new section view and change the *Scale* to **3/4"=1'-0"** and the *Detail Level* to **Fine**. You might need to adjust the crop region and/or zoom in on the slab section. It should look similar to the one shown in Figure 6–20.

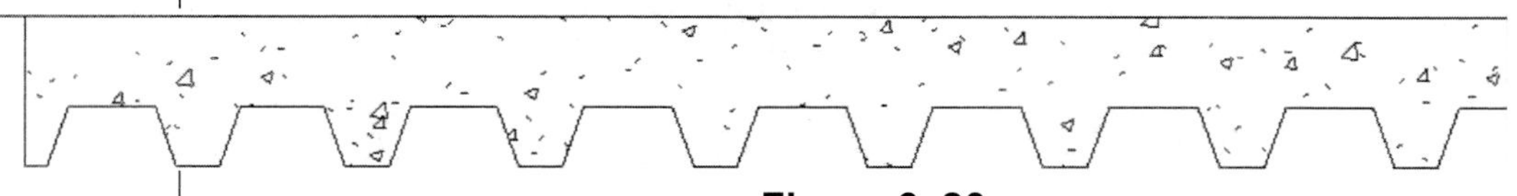

Figure 6–20

10. Select the slab.

11. In Properties, click (Edit Type).

12. Duplicate the existing type. Name the duplicated type **6" Concrete on 3" Comp Metal Deck**.

13. In the Type Properties dialog box, next to the **Structure** parameter, click **Edit...**.

14. Modify the **Structure [1], Concrete, Lightweight - 4 ksi** layer to a *Thickness* of **6"**, as shown in Figure 6–21.

Layers

	Function	Material	Thickness	Wraps	Structural Material	Variable
1	Core Boundary	Layers Above Wrap	0' 0"			
2	Structure [1]	Concrete, Lightweig	0' 6"	☐	☑	☐
3	Structural Dec	Metal Deck	0' 0"	☐	☐	☐
4	Core Boundary	Layers Below Wrap	0' 0"			

Figure 6–21

15. For the **Structural Deck** layer, select **Composite-Metal-Deck: 3"** in the Deck Profile drop-down list, as shown in Figure 6–22.

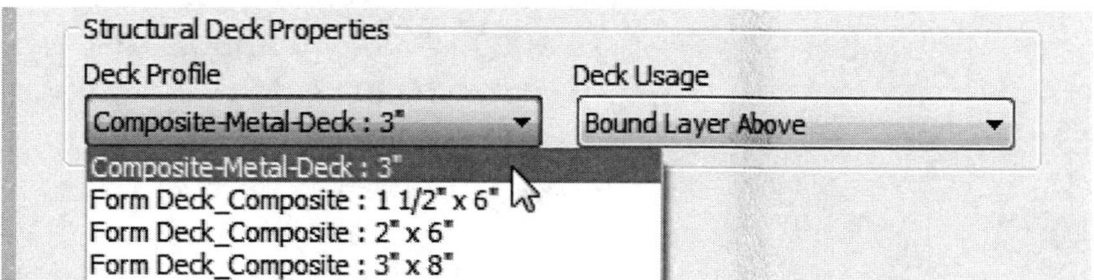

Figure 6–22

16. Click **OK** to close all of the dialog boxes. The section should display the metal deck profile that was previously created, as shown in Figure 6–23.

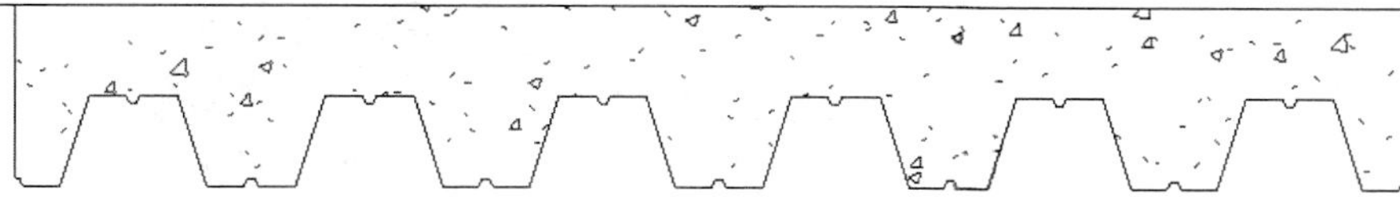

Figure 6–23

Task 3 - Change the Flute Direction of a comp metal deck.

1. Return to the **Level 2** structural plan view.
2. Select the **Span Direction Symbol**. Note that the **Direction Symbol** extends to the edges of the slab. The closed arrowheads represent the flute direction and the open arrowheads represent the extents of the slab in the perpendicular direction.
3. Start the **Rotate** command and rotate the span direction 90 degrees.
4. Return to the section view. The orientation of the composite metal deck has changed as shown in Figure 6–24.

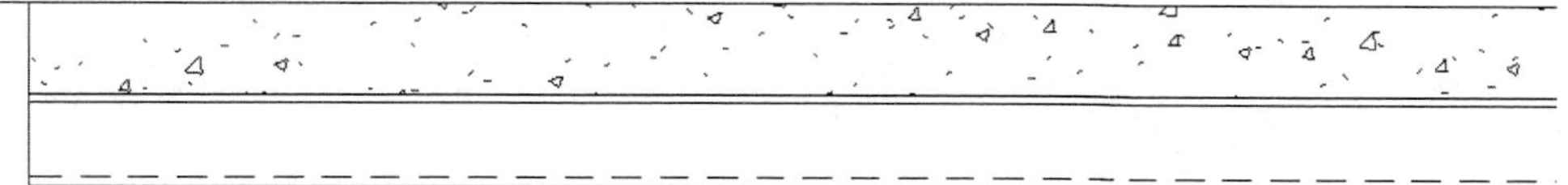

Figure 6–24

5. Save the project as **New Floor.rvt** and close the project.
6. Save and close the file **Composite-Metal-Deck.rfa**.

6.3 Creating Annotation Families

Annotation symbols are 2D elements that can be inserted into views or sheets to give additional information about a project. You can modify symbols that are supplied with the Autodesk Revit software to fit your company standards. You can also create new symbols using a variety of templates, as shown in Figure 6–25.

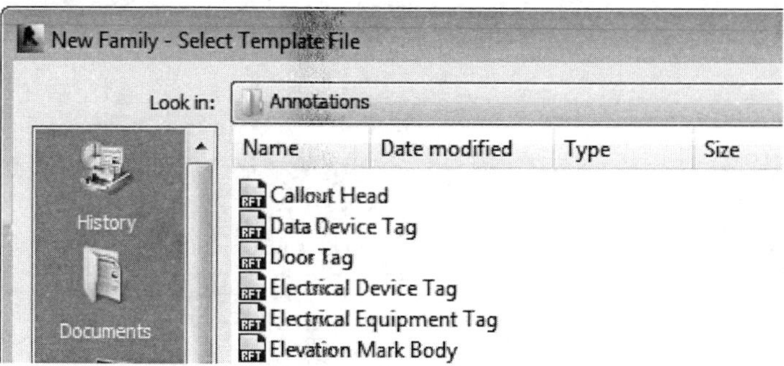

Annotation templates in Autodesk Revit

Figure 6–25

The drawing tools for annotation families include the standard line sketching tools, filled region, text, and label. You can also add existing symbols to the family.

- Most annotation symbols do not have family types. They are designed to change size according to the view scale.
- Some annotation family templates include notes to help you create the symbol. Delete them before using the family.

Hint: Modifying Labels

When you add labels, the length of the boundary box impacts the way the text word-wraps. For example, the Viewport: Title w Line tag that comes with the default metric template does not have enough room for a long title, as shown in Figure 6–26.

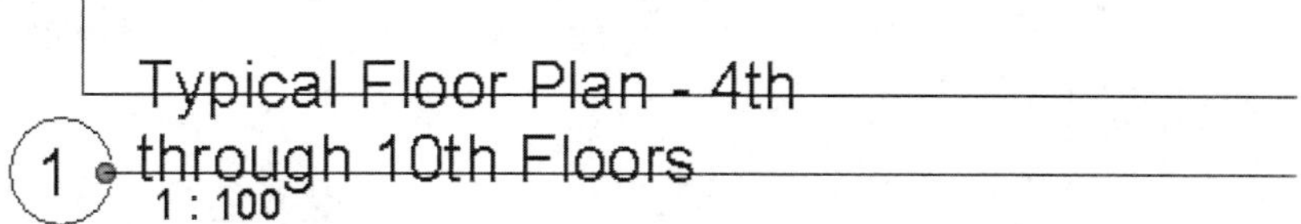

Figure 6–26

To modify it, open the related family file (in this case **M_View Title.rfa**) and stretch the length of the label boundary as shown in Figure 6–27.

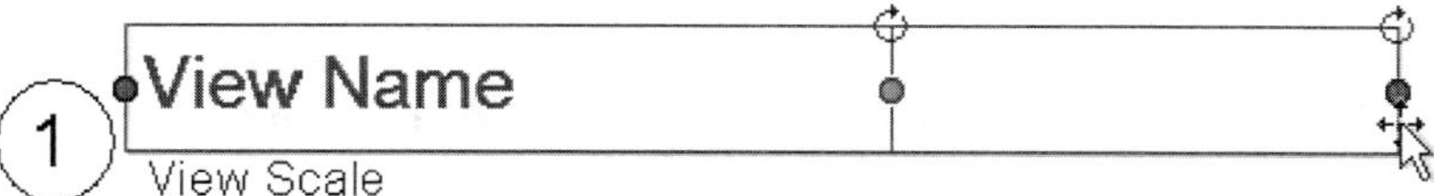

Figure 6–27

Save the annotation family and reload it into the project. The length of the title expands to fit the longer title as shown in Figure 6–28.

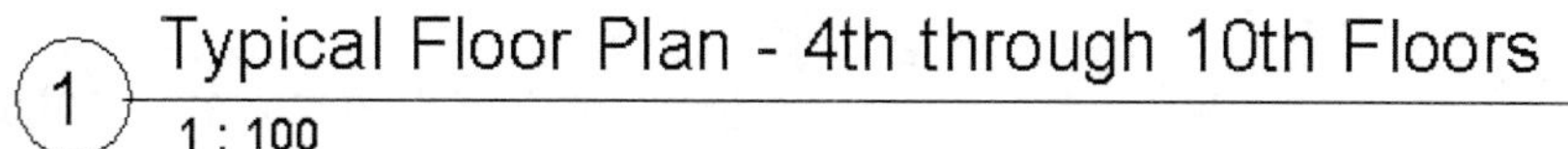

Figure 6–28

Practice 6d Create Annotation Families: Arrow Symbol (All Disciplines)

Practice Objective

- Create an arrowhead annotation symbol that includes a filled region and test it in a project to see the size change automatically with the scale change.

Estimated time for completion: 10 minutes.

In this practice you will create an annotation symbol and apply it to a project, as shown in Figure 6–29.

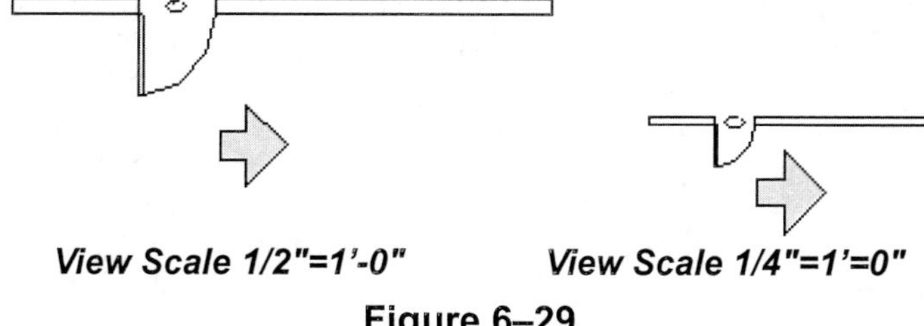

Figure 6–29

Task 1 - Create an annotation symbol.

1. Create a new annotation symbol family based on the **Generic Annotation.rft** template found in the Autodesk Revit Library in the *Annotations* folder.
2. Save the family to your practice folder as **Arrow.rfa**.
3. Draw the arrow as shown in Figure 6–30. The dimensions are for reference only; do not add them to the sketch.

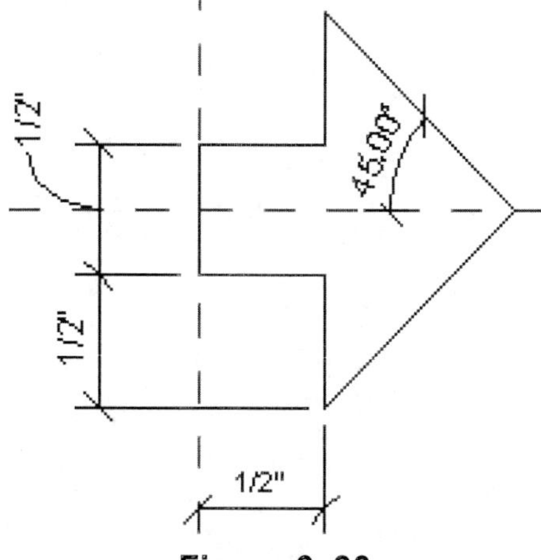

Figure 6–30

4. In the *Create* tab>Detail panel, click (Filled Region). Add a filled region to the arrow, as shown in Figure 6–31. You can use the default solid pattern or create a new pattern if time permits.

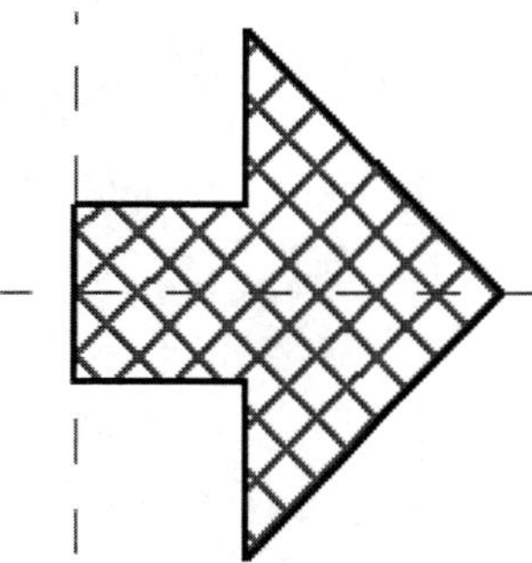

Figure 6–31

5. Delete the notes and save the family.

Task 2 - Add the symbol to a project.

1. Open a project file.
2. In the *Insert* tab>Load from Library panel, click (Load Family) and select the new file **Arrow.rfa**, found in the practice files folder.
3. In the *Annotate* tab>Symbol panel, click (Symbol).
4. Select the **Arrow** type.
5. Add some other full-scale elements near the arrow.
6. Change the *View Scale*, as shown in Figure 6–32. The arrow should resize according to the new view scale.

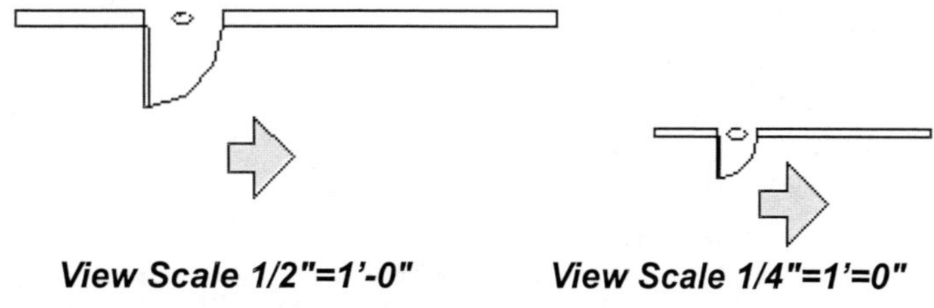

View Scale 1/2"=1'-0" ***View Scale 1/4"=1'=0"***

Figure 6–32

7. Close the project. You do not need to save it.

6.4 Working with Project and Shared Parameters

The Autodesk Revit software includes many parameters used in families and schedules.

- **Family parameters:** Are specific to a family that constrain or flex geometry and assign attributes, such as material.
- **Project parameters:** Can be assigned to schedules.
- **Shared parameters:** Can be referenced in multiple locations, projects, families, tags, and schedules, and shared between projects.

Project Parameters

Project parameters are created per project in the Parameter Properties dialog box, as shown in Figure 6–33. In the *Manage* tab>Settings panel, click (Project Parameters). Each parameter is assigned to one or more categories. Use *Filter list* to specify the discipline to limit the number of items displayed.

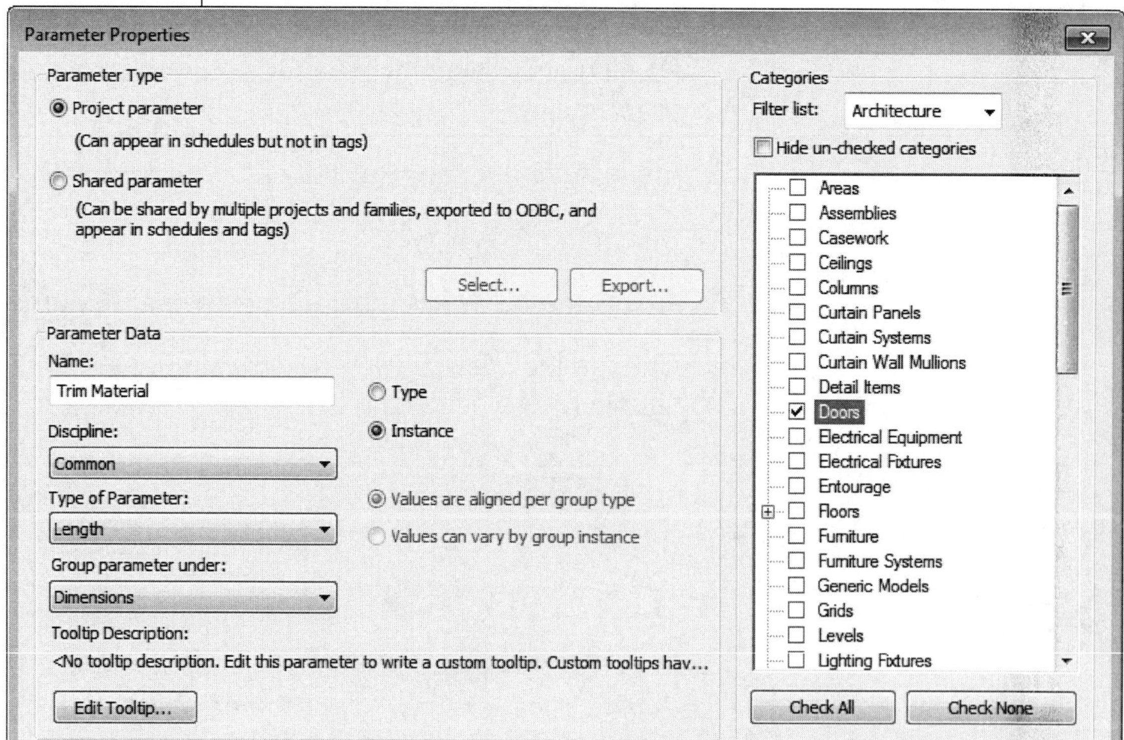

Figure 6–33

- If you want a parameter to be used in a schedule (such as the manufacturer of a desk family) it can be a project parameter.
- If you want to use a parameter in a tag and a schedule, create a shared parameter.

Shared Parameters

If you are creating a family to post to Autodesk Seek, use the Revit Master Shared Parameters file included in the Revit Model Content Style Guide available on the Autodesk Seek website.

Create shared parameters for custom information that needs to be displayed in both tags and schedules, used in multiple projects or families, or exported to ODBC to be put in a database.

- Shared parameters can be created wherever there is an option to create a parameter (as shown in Figure 6–34), including Schedule Properties (creating a Field) and the Family Types dialog box.

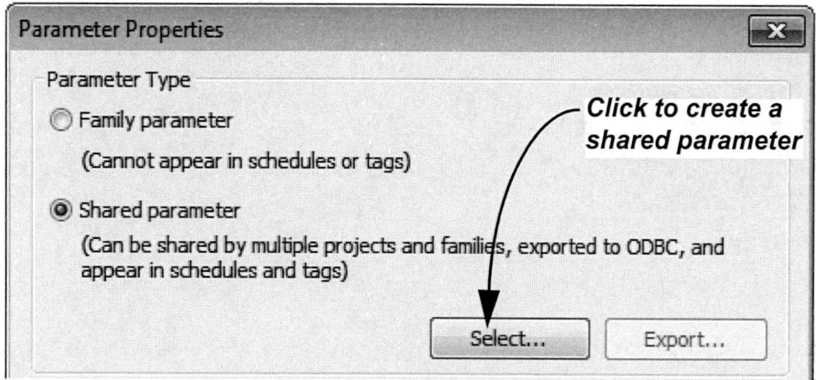

Figure 6–34

- You can also directly open the Edit Shared Parameters dialog box. In the *Manage* tab>Settings panel, click (Shared Parameters).
- If you are creating shared parameters in a family file, do it only for elements that constrain or flex the geometry and are used in a schedule.

How To: Create a Shared Parameter

1. Start the process of adding a parameter in Project Parameters, schedules, or family types.
2. Set the *Parameter Type* to **Shared parameter** in the Parameter Properties dialog box, as shown in Figure 6–35.

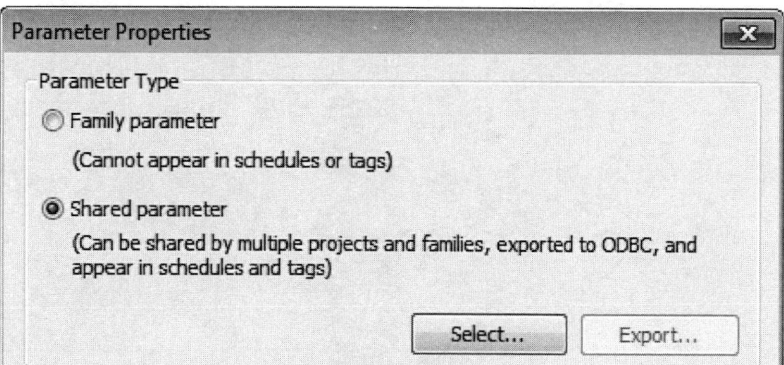

Figure 6–35

3. Click **Select...**.
4. If no shared parameter file has been specified, an alert box opens. Click **Yes** to select a shared parameter file.
5. In the Edit Shared Parameters dialog box, as shown in Figure 6–36, click **Create...** to create a shared parameter file or **Browse...** to select an existing one.

Shared parameters are stored in a text file that should be placed in a network location available to everyone. This ensures consistency between projects and team members by maintaining naming standards and limiting typical typing errors with spaces and case sensitivity.

Figure 6–36

6. Creating groups for related parameters makes working with numerous names much easier. In the *Groups* area, click **New** to create a new parameter group.
7. In the New Parameter Group dialog box, as shown in Figure 6–37, type a new name and click **OK**.

Only one shared parameter file can be active in a project at a time. Therefore, create one main file with multiple groups instead of creating many files.

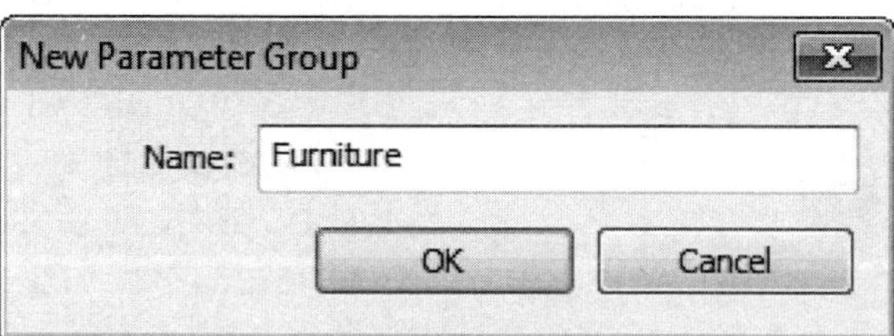

Figure 6–37

8. In the Edit Shared Parameters dialog box, in the *Parameters* area, click **New** to create a new shared parameter.

9. In the Parameter Properties dialog box, type a *Name* for the parameter, and set the *Discipline* (**Common**, **Structural**, or **Electrical**) and *Type of Parameter* in the drop-down lists. You can also add a Tooltip. The dialog box is shown in Figure 6–38.

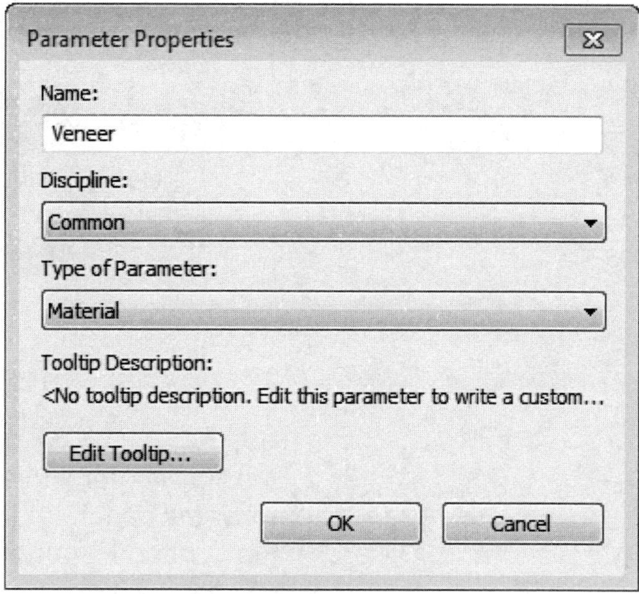

Figure 6–38

10. Click **OK**.
11. In the Edit Shared Parameters dialog box, once you have the parameters in place, you can:
    - Click **Properties...** to display the contents of the Parameter Properties. You cannot modify them.
    - Click **Move...** to switch a parameter from one parameter group to another.
    - Click **Delete** to remove a parameter. Be careful about using this because it might impact the current project and any other projects that use the parameter.
12. When you have finished creating the shared parameters, click **OK** to close the Edit Shared Parameters dialog box.
13. Continue with the other steps required to add the parameter to the project, family, or schedule.

How To: Add a Shared Project Parameter to a Project

1. In the *Manage* tab>Settings panel, click (Project Parameters).
2. In the Project Parameters dialog box, as shown in Figure 6–39, click **Add**.

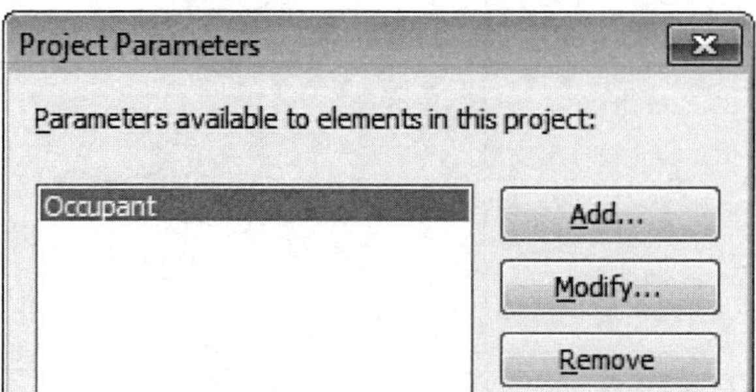

Figure 6–39

3. In the Parameter Properties dialog box, select the **Shared parameter** option and click **Select...**.
4. In the Shared Parameters dialog box, set the *Parameter group* and *Parameters* that you want to add to the project, as shown in Figure 6–40. Click **OK**.

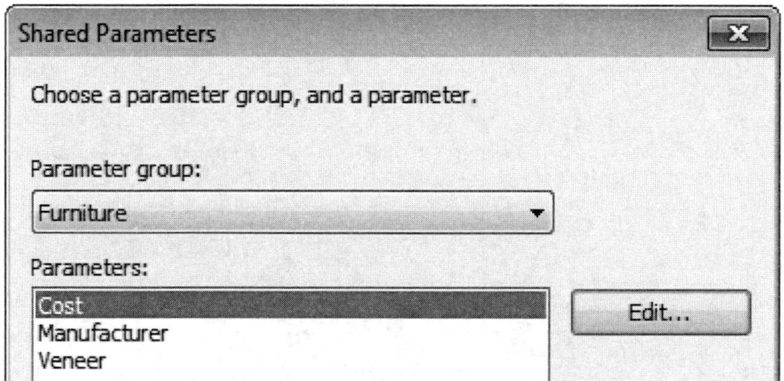

Figure 6–40

- If the parameter you want to use does not exist, click **Edit....** to create a new shared parameter.

5. In the Parameter Properties dialog box:

- In the *Categories* area, select the categories to which you want the parameter to apply, as shown in Figure 6–41.

Use the Filter List to limit the number of categories available.

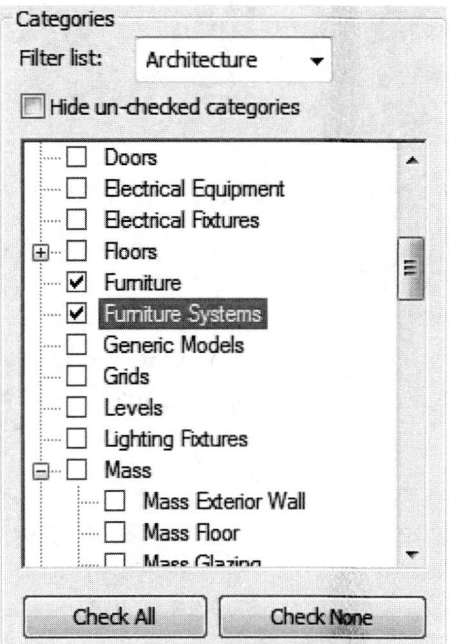

Figure 6–41

- In the *Parameter Data* area, the Name, Discipline, and Type of Parameter drop-down lists are grayed out, while the shared parameter you just selected is displayed.

6. Select from the Group parameter under drop-down list (this is where it displays in the Element Properties dialog box) and select the **Instance** or **Type** option, as shown in Figure 6–42. Add a Tooltip Description if needed.

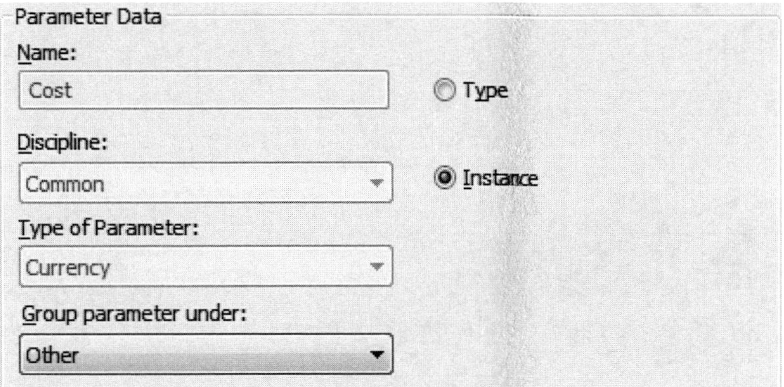

Figure 6–42

7. Click **OK** twice to add the parameter to the project.

Practice 6e

Work with Shared Parameters in Architectural Projects

Practice Objectives

- Create shared parameters in a project.
- Create an annotation tag family and link the shared parameters into the tag.
- Add the shared parameters to a family and insert them into the project along with the associated tags.
- Create a schedule using the shared parameters including the project, tag, and family.

Estimated time for completion: 20 minutes

In this practice you will create shared parameters through project parameters and use them in a family, tag, and schedule, as shown in Figure 6–43.

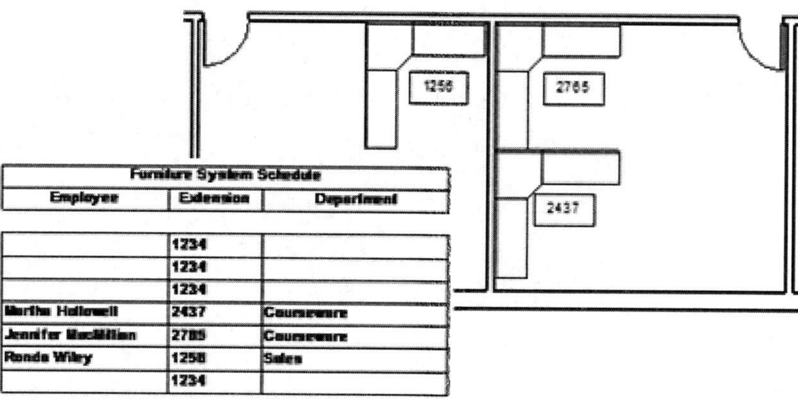

Figure 6–43

Task 1 - Create shared parameters.

1. In the ...*Architectural* folder, open **Office-Layout-A.rvt**.
2. In the *Manage* tab>Settings panel, click (Project Parameters).
3. In the Project Parameters dialog box, click **Add**.
4. In the Parameter Properties dialog box, select the **Shared parameter** option and click **Select...**.

5. If there is no existing shared parameter file, an alert box displays prompting you to select a shared parameter file. Click **Yes**. If the Shared Parameters dialog box opens, click **Edit...**.

6. In the Edit Shared Parameters dialog box, click **Create...**.

7. Open the practice folder and save the new file as **Office-Parameters.txt**.

8. In the *Groups* area, click **New**. Type the name **Office Information** and click **OK**. This becomes the current group, as shown in Figure 6–44.

Figure 6–44

9. In the *Parameters* area, click **New...** and create three new shared parameters using the following information:

Name	Discipline	Type of Parameter
Extension	Common	Integer
Employee	Common	Text
Department	Common	Text

10. Click **OK**.

11. In the Shared Parameters dialog box, as shown in Figure 6–45, select **Department** and click **OK**.

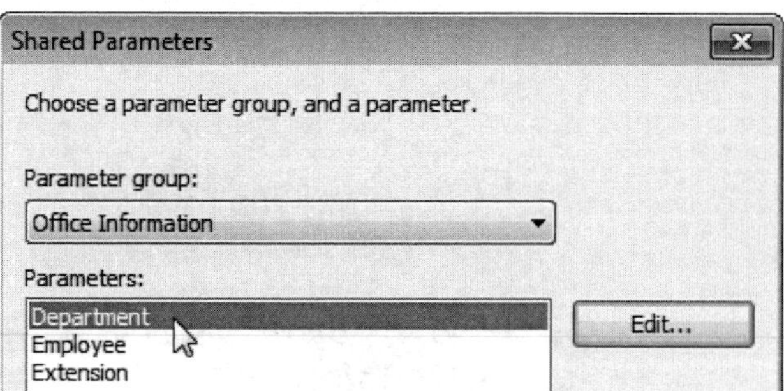

Figure 6–45

12. In the Parameter Properties dialog box, in the *Parameter Data* area, group the parameter under **Other** and make it an **Instance** parameter. In the *Categories* list, select **Furniture** and **Furniture Systems**, as shown in Figure 6–46.

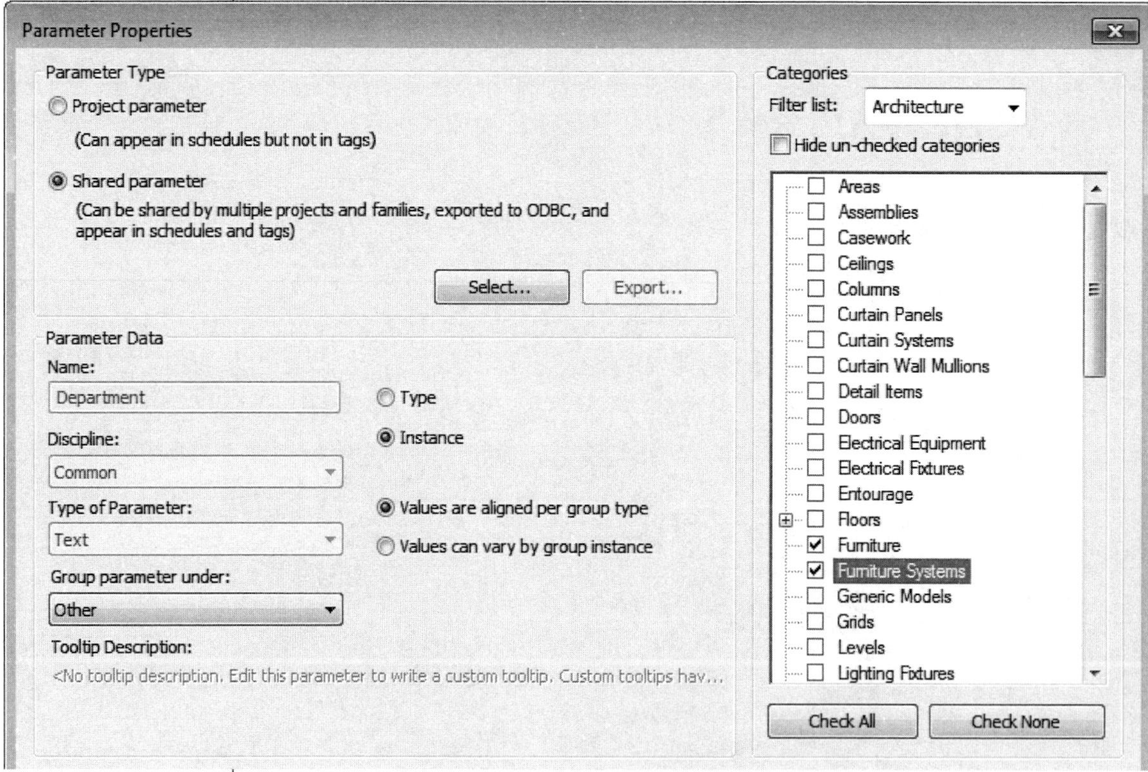

Figure 6–46

13. Click **OK**. **Department** is added to the shared parameters in the project.

14. Repeat the process for **Employee** and **Extension**.

15. Click **OK** to finish creating the project parameters.

16. Save the project.

Task 2 - Create a tag with shared parameters.

1. Create a new family based on the **Generic Tag.rft** template found in the *Annotation* folder of Autodesk Revit Library.
2. Save the family to the practice files folder as **Office-Information.rfa**.
3. Zoom in, read the information in red, and delete the note.
4. In the *Modify* tab>Properties panel, click (Family Category and Parameters).
5. In the Family Category and Parameters dialog box, set the *Family Category* to **Furniture Systems Tags**, as shown in Figure 6–47. Click **OK**.

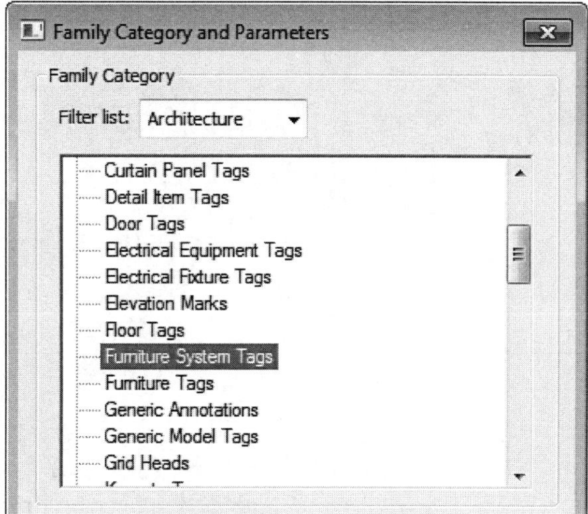

Figure 6–47

6. In the *Create* tab>Detail panel, click (Line). Draw a box, starting at the crossing of the two reference lines, as shown in Figure 6–48. Do not add the dimensions; they are for reference only.

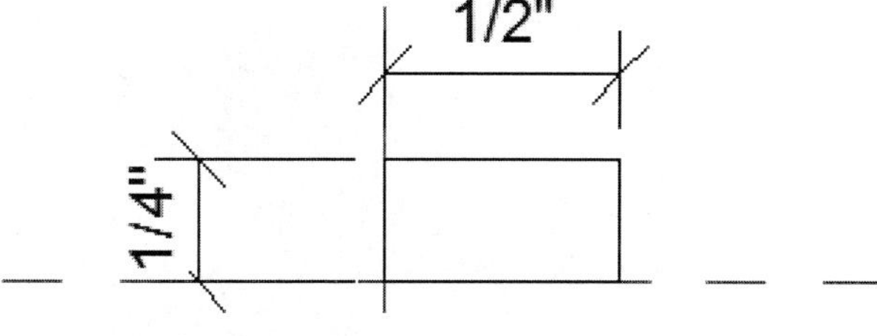

Figure 6–48

7. In the *Create* tab>Text panel, click (Label).
8. Use the default type, **Label: 3/32"**.
9. Click inside the box you just created.
10. In the Edit Label dialog box, which includes a list of the multi-category parameters supplied with the Autodesk Revit software, click (Add Parameter).

In the Parameter Properties dialog box, you can only create shared parameters.

11. In the Parameter Properties dialog box, click **Select...**.
12. The shared parameters you created earlier should be available by default in the Shared Parameters dialog box. Select **Extension**, as shown in Figure 6–49, and click **OK**.

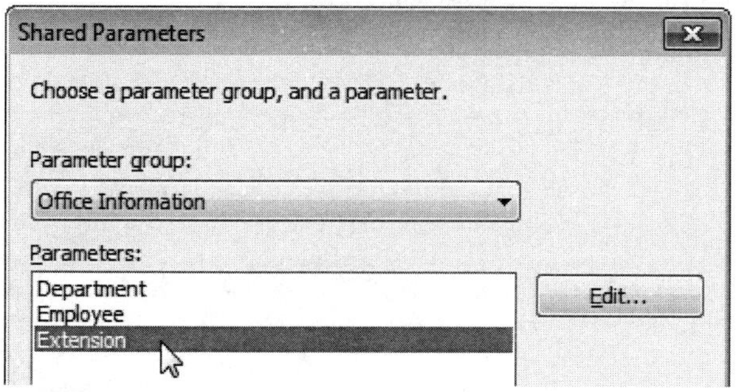

Figure 6–49

13. Click **OK** until you reach the Edit Label dialog box.
14. Select **Extension** and click (Add parameter(s) to label). Click **OK** to insert the label into the family.
15. Click (Modify) and select the new label. Resize the outline of the text box to fit inside the box. Move the label so that it is centered in the box, as shown in Figure 6–50.

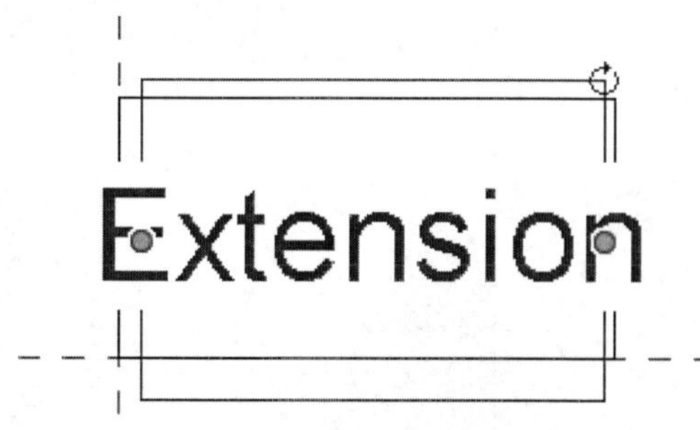

Figure 6–50

16. Save the family and load it into the project **Office-Layout-A.rvt**.

Task 3 - Add shared parameters to a furniture family.

1. Continue working in the project **Office-Layout-A.rvt**.
2. In the *Architecture* tab>Build panel, click (Component).
3. In the *Modify | Place Component* tab>Mode panel, click (Load Family).
4. In the ...*Architectural* folder, open **Work-Station-Cubicle-A.rfa**.
5. Place a workstation in one of the rooms in the wing to the left of the main building.
6. Click (Modify) and select the new workstation.
7. In the *Modify | Furniture Systems* tab>Mode panel, click (Edit Family).
8. In the Properties panel, click (Family Types).
9. In the Family Types dialog box, in the *Parameters* area, click **Add**.
10. Select the **Shared Parameter** option and click **Select...**.
11. The shared parameters you created earlier are also available in this family file. Add each of the parameters from the *Office-Information* category into the family as an Instance parameter in the *Other* category.
12. In the Family Types dialog box, set the *Extension (default)* to **1234**, as shown in Figure 6–51. The other values can be left blank.

Other			
Department (default)		=	
Employee (default)		=	
Extension (default)	1234	=	☐

Figure 6–51

13. Click **OK**.

14. Save the family file and load it into the **Office-Layout-A.rvt** project. Overwrite the existing version and its parameter values.

15. Add several other cubicles in various offices.

Task 4 - Add tags and create a schedule.

1. In the *Annotate* tab>Tag panel, click (Tag by Category).

2. Select the workstations and tag each of them. The default **1234** extension number should display on each tag, as shown in Figure 6–52.

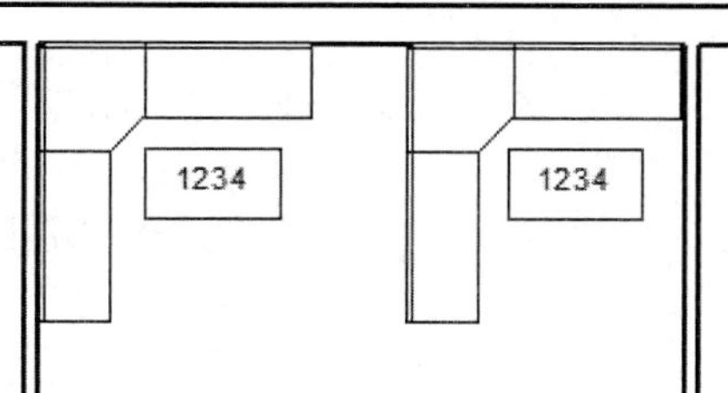

Figure 6–52

3. Select a workstation. In Properties, the **Extension**, **Employee**, and **Department** parameters should be available in the *Other* category, as shown in Figure 6–53.

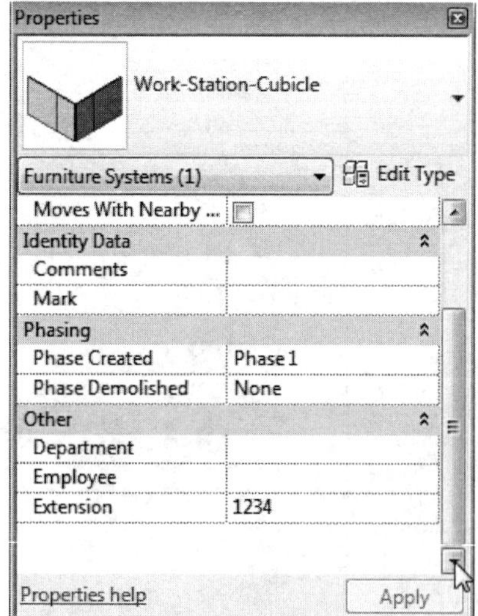

Figure 6–53

4. Change the *Extension*, and add an *Employee* name and *Department*.

5. Repeat with several other workstations.

6. In the *View* tab>Create panel, expand (Schedules) and click (Schedule/Quantities).

7. In the New Schedule dialog box, create a new Furniture Systems schedule and set the *Phase* is set to **Phase 1**, as shown in Figure 6–54.

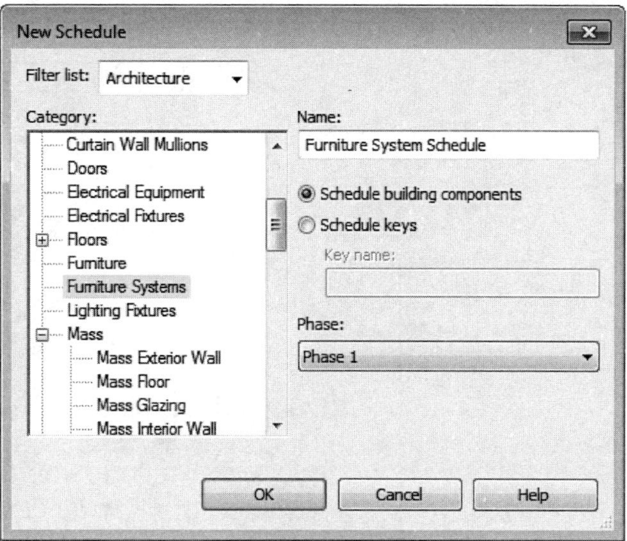

Figure 6–54

8. In the *Fields* tab, add the **Family and Type** parameter and then add **Extension**, **Employee**, and **Department**, as shown in Figure 6–55.

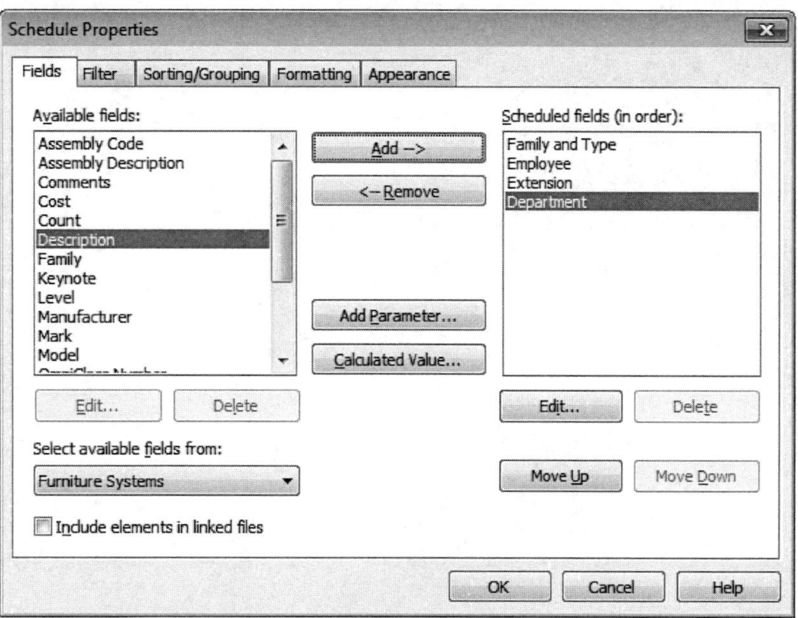

Figure 6–55

9. At this point, you can modify the other schedule properties as needed. Click **OK** when you are finished. The schedule should display the information you entered in the workstation's properties. Change the name of the schedule to **Office Information** and **Hide** the *Family and Type* column, as shown in Figure 6–56.

<Office Information>		
A	B	C
Employee	Extension	Department
Ronda	1234	Sales
Pual	1235	Engineering
Anne-Marie	1236	Sales
Kareen	1237	Marketing
Martha	1238	Engineering
Jen	1239	Enginering

Figure 6–56

10. Save the project.

11. If you have time, you can copy the Office Information tag and change the *Family Category* to **Furniture Tags**. Then you can use the new tag to tag the desks in the project. Furniture and Furniture Systems are different categories and cannot use the same tag.

Practice 6f

Work with Shared Parameters in MEP Projects

Estimated time for completion: 20 minutes

Practice Objectives

- Create a schedule that uses shared parameters.
- Add the shared parameters to a family file and note the impact on the schedule.

In this practice you will create a mechanical equipment schedule and add shared parameters to the project through the schedule. You will then modify and add the shared parameters in a family file and update the family in the project. The shared parameters then populate the schedule, as shown in Figure 6–57.

Mechanical Equipment Schedule						
		Dimensions			Gas Line	Liquid Line
Mark	Type	Height	Width	Depth	Diameter	Diameter
1	HPD60C	44 3/4"	38 7/8"	34 3/4"	1"	1/2"
2	HPD60C	44 3/4"	38 7/8"	34 3/4"	1"	1/2"
3	HPD60C	44 3/4"	38 7/8"	34 3/4"	1"	1/2"
4	HPD60C	44 3/4"	38 7/8"	34 3/4"	1"	1/2"

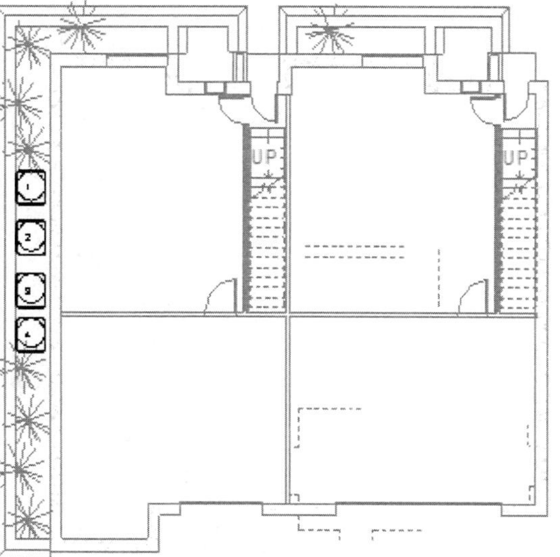

Figure 6–57

Task 1 - Create a schedule and add shared parameters.

1. In the ...*MEP* folder, open **Townhouse-Complex-MEP.rvt**. Note that four heat pumps have been added to serve the townhouses.

2. In the *View* tab>Create panel, expand (Schedules) and click (Schedule/Quantities).

3. In the New Schedule dialog box, set the *Filter list* to **Mechanical**, select the *Category* of **Mechanical Equipment**, and click **OK**.

4. In the Schedule Properties dialog box, in the *Fields* tab, select the following fields: **Mark** and **Type**. The rest of the parameters need to be shared parameters.

5. Click **Add Parameter...**.

6. In the Parameter Properties dialog box, select **Shared parameter** and click **Select...**.

7. If there is no existing shared parameter file, an alert box displays prompting you to select a shared parameter file. Click **Yes**. If the Shared Parameters dialog box opens, click **Edit...**.

8. In the Edit Shared Parameters dialog box, click **Create...**.

9. In the Create Shared Parameter File dialog box, navigate to the ...*MEP* folder and save the new file as **Townhouse-Parameters.txt**.

10. In the *Groups* area, click **New**. Set the name as **Heat Pump Information** and click **OK**. This becomes the current group, as shown in Figure 6–58.

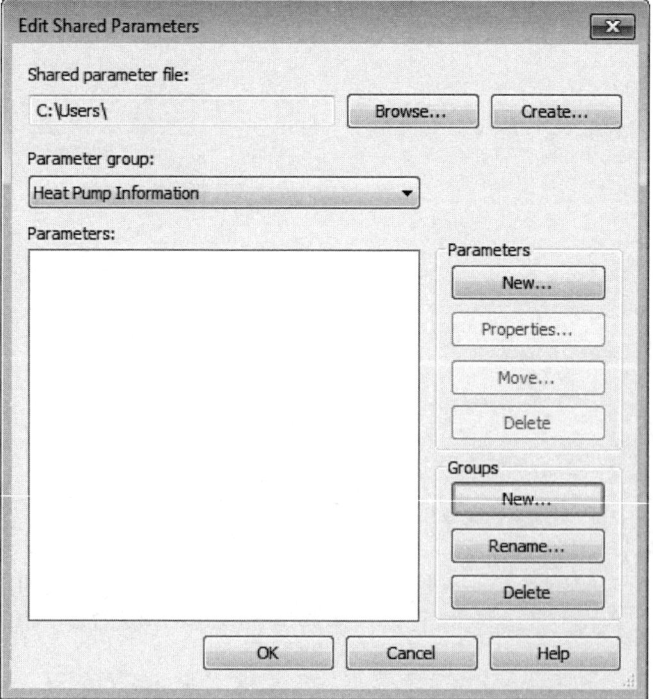

Figure 6–58

11. In the *Parameters* area, click **New**. Create new shared type parameters using the following information:

Name	Discipline	Type of Parameter
Height	Common	Length
Width	Common	Length
Depth	Common	Length
Gas Line Diameter	Common	Length
Liquid Line Diameter	Common	Length

12. Click **OK**.

13. In the Shared Parameters dialog box, select **Height** (as shown in Figure 6–59), and then click **OK**.

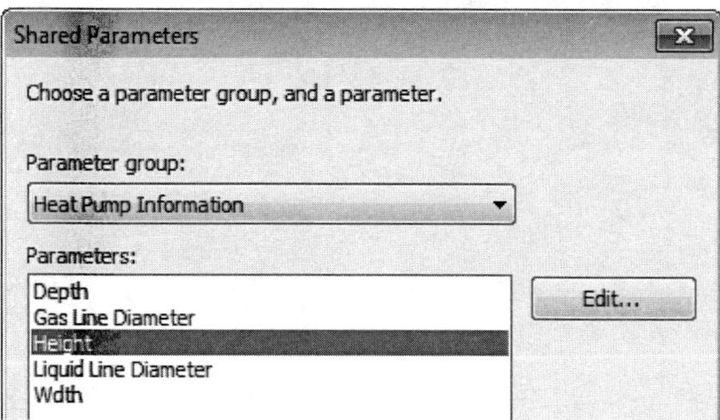

Figure 6–59

14. In the Parameter Properties dialog box, in the *Parameter Data* area, group the parameter under **Dimensions**, as shown in Figure 6–60.

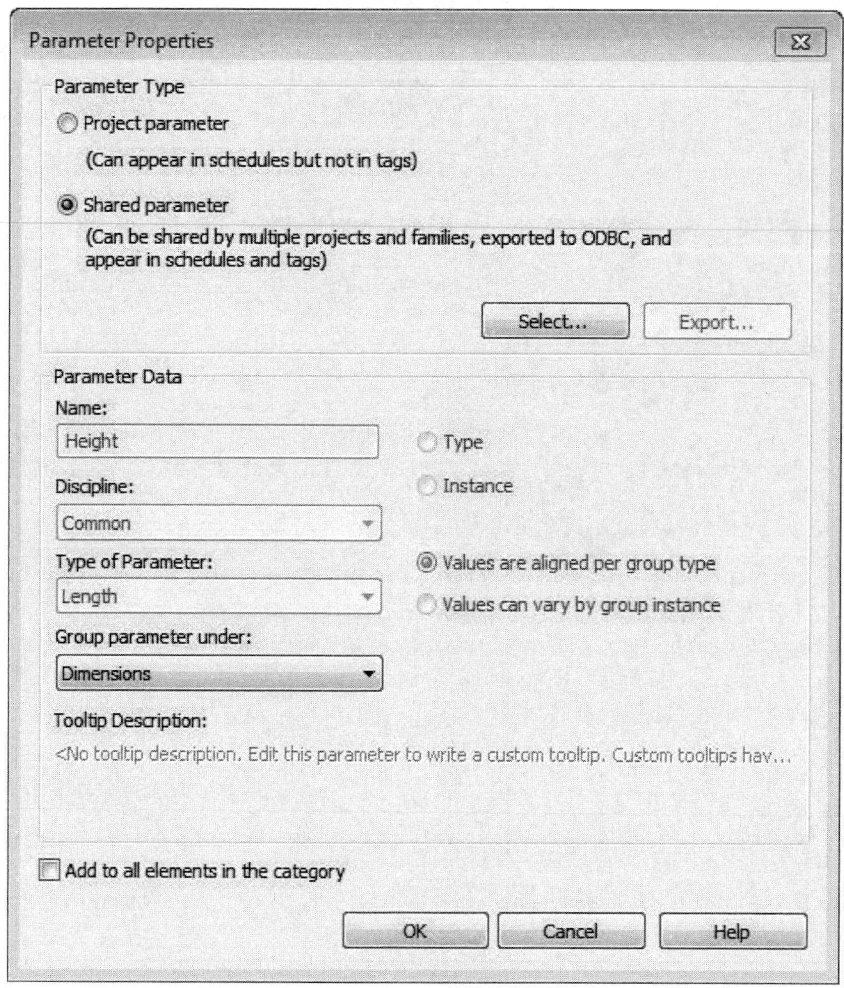

Figure 6–60

15. Click **OK**. Height is added to the scheduled fields.

16. Click **Add Parameter...** again.

17. In the Parameter Properties dialog box, select **Shared parameter** and click **Select...**.

18. This time you do not have to create a shared parameter file, as the files created in the previous steps are available. Select **Width** and click **OK** to apply the parameter to the scheduled fields.

19. Repeat the process and add the rest of the new shared parameters in the order shown in Figure 6–61.

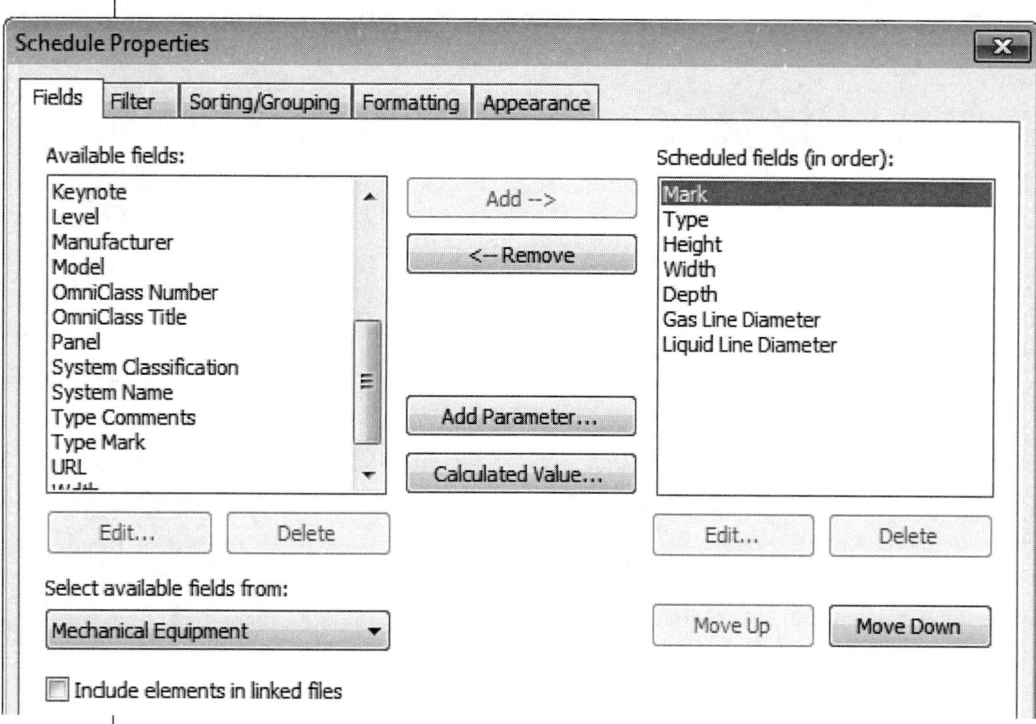

Figure 6–61

20. Click **OK**. The Mechanical Equipment Schedule displays but only includes values for the **Mark** and **Type**, as shown in Figure 6–62. The shared parameters are not yet identified in the Heat Pump family.

<Mechanical Equipment Schedule>

A	B	C	D	E	F	G
Mark	Type	Height	Width	Depth	Gas LIne Diameter	Liquid Line Diamete
1	HP060C					
2	HP060C					
3	HP060C					
4	HP060C					

Figure 6–62

21. Save the project.

Task 2 - Add shared parameters to a family.

1. Open the Mechanical>HVAC>Floor Plans>**Ground-Mech** view and select one of the heat pumps.
2. In the *Modify | Mechanical Equipment* tab>Mode panel, click (Edit Family). The family file opens in the Family Editor.
3. In the Properties panel, click (Family Types).

4. In the Family Types dialog box several of the parameters that already exist in the family are displayed, as shown in Figure 6–63, but they are not connected as shared parameters.

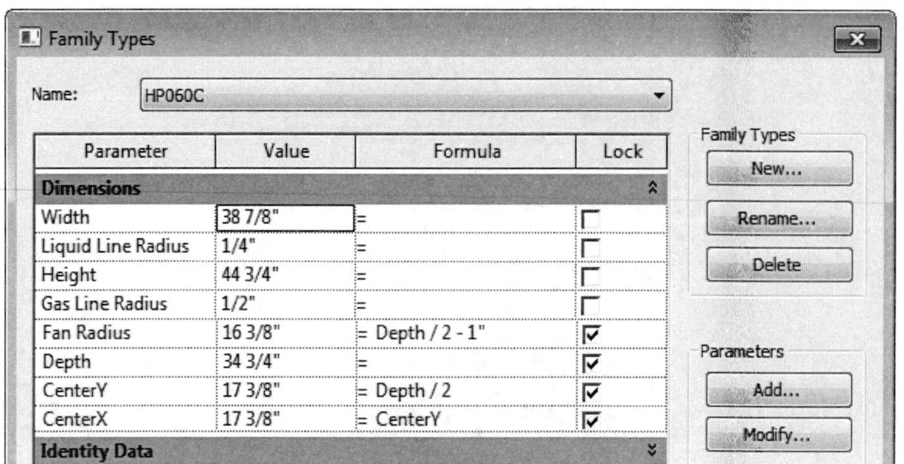

Figure 6–63

5. Select the **Width** parameter and in the *Parameters* area, click **Modify**.

6. In the Parameter properties dialog box, select the **Shared Parameter** option and click **Select...**.

7. The shared parameters you created earlier are also available in this family file, as shown in Figure 6–64.

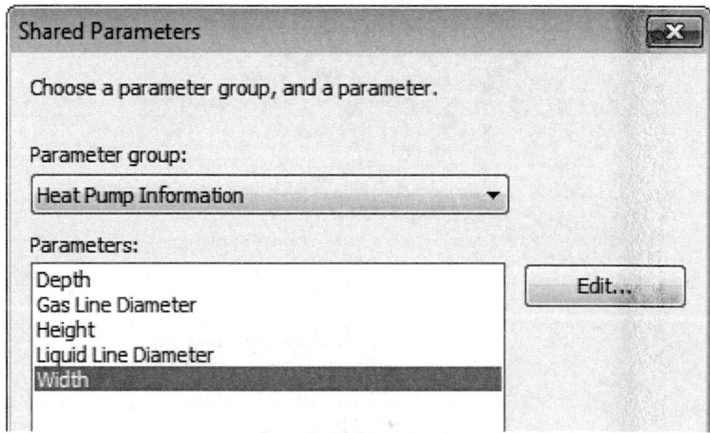

Figure 6–64

8. Select **Width** and click **OK** twice. The existing parameter is now connected with the shared parameter.

9. Repeat the process with the **Height** and **Depth** parameters.

10. The information stored in the Heat Pump family for the Liquid Line and Gas Line are set to a radius, because it is required by the connectors. Therefore, you have to add Parameters and formulas for the diameters.

11. In the *Parameters* area, click **Add...**.

12. In the Parameter Properties dialog box, select **Shared parameter** and click **Select...**.

13. In the Shared Parameters dialog box, select **Gas Line Diamete**r and click **OK**.

14. Set it as a **Type** parameter and group it with the **Dimensions** group. Click **OK**.

15. Repeat the process for **Liquid Line Diameter**.

16. Set the following formulas for the two diameter parameters, as shown in Figure 6–65.

- Gas Line Diameter = **Gas Line Radius * 2**
- Liquid Line Diameter = **Liquid Line Radius * 2**

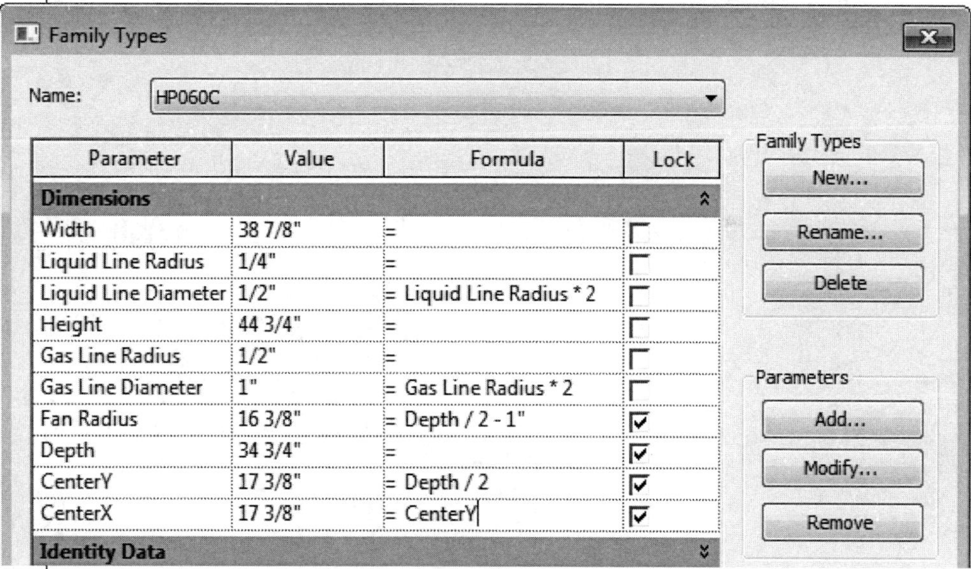

Figure 6–65

17. Click **OK** to close the Family Types dialog box.

18. Save the family to the practice folder.

Task 3 - Load the family and modify the schedule.

1. In the *Create* tab>Family Editor panel, click (Load into Project).
2. Overwrite the existing version and its parameter values.
3. Open the Mechanical Equipment Schedule. The values for the parameters are now included. Add headers for the dimensions, as shown in Figure 6–66.

<Mechanical Equipment Schedule>

A	B	C	D	E	F	G
		Overall Dimensions			Pipe Dimensions	
Mark	Type	Height	Width	Depth	Gas Line Diameter	Liquid Line Diamete
1	HP060C	3' - 8 3/4"	3' - 2 7/8"	2' - 10 3/4"	0' - 1"	0' - 0 1/2"
2	HP060C	3' - 8 3/4"	3' - 2 7/8"	2' - 10 3/4"	0' - 1"	0' - 0 1/2"
3	HP060C	3' - 8 3/4"	3' - 2 7/8"	2' - 10 3/4"	0' - 1"	0' - 0 1/2"
4	HP060C	3' - 8 3/4"	3' - 2 7/8"	2' - 10 3/4"	0' - 1"	0' - 0 1/2"

Figure 6–66

4. Typically the values for these parameters should only be in inches. To fix this, in Properties next to **Formatting** parameter, click **Edit...**.
5. The Schedule Properties dialog box opens with the *Formatting* tab selected.
6. In the *Fields* area, select **Height** and click **Field Format...**.
7. Clear **Use project settings** and change the *Units* to **Fractional inches**. Click **OK**.
8. Repeat this for the rest of the length based parameters so that they display in inches and not in feet and inches.
9. Click **OK** to close the dialog box. The schedule updates as shown in Figure 6–67.

<Mechanical Equipment Schedule>

A	B	C	D	E	F	G
		Overall Dimensions			Pipe Dimensions	
Mark	Type	Height	Width	Depth	Gas Line Diameter	Liquid Line Diamete
1	HP060C	44 3/4"	38 7/8"	34 3/4"	1"	1/2"
2	HP060C	44 3/4"	38 7/8"	34 3/4"	1"	1/2"
3	HP060C	44 3/4"	38 7/8"	34 3/4"	1"	1/2"
4	HP060C	44 3/4"	38 7/8"	34 3/4"	1"	1/2"

Figure 6–67

10. Save the project.

Practice 6g

Work with Shared Parameters in Structural Projects

Estimated time for completion: 20 minutes

Practice Objectives

- Create shared parameters in a project.
- Use the shared parameters in a family, associated tag, and a schedule.

In this practice you will create shared parameters through project parameters and use them in family, tag, and schedule, as shown in Figure 6–68. The context here is a public plaza with columns, some free-standing and some supporting canopies and roofs. Each column will have custom brackets for banners, flags, etc.

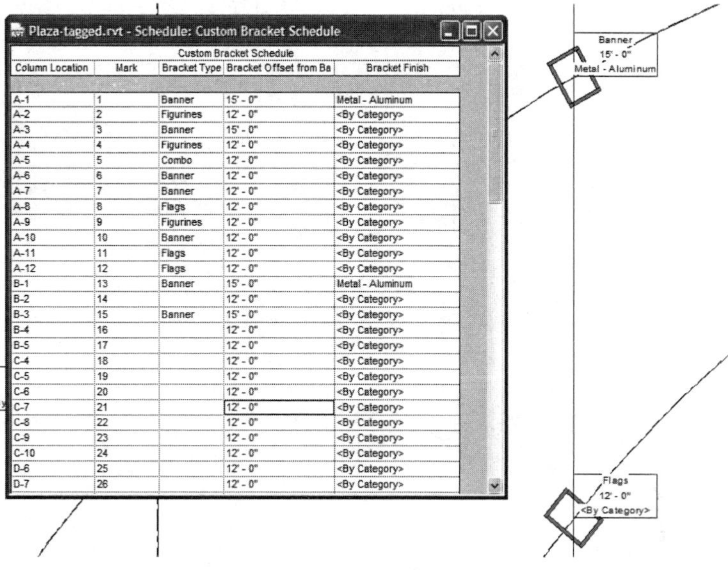

Custom Bracket Schedule				
Column Location	Mark	Bracket Type	Bracket Offset from Ba	Bracket Finish
A-1	1	Banner	15' - 0"	Metal - Aluminum
A-2	2	Figurines	12' - 0"	<By Category>
A-3	3	Banner	15' - 0"	<By Category>
A-4	4	Figurines	12' - 0"	<By Category>
A-5	5	Combo	12' - 0"	<By Category>
A-6	6	Banner	12' - 0"	<By Category>
A-7	7	Banner	12' - 0"	<By Category>
A-8	8	Flags	12' - 0"	<By Category>
A-9	9	Figurines	12' - 0"	<By Category>
A-10	10	Banner	12' - 0"	<By Category>
A-11	11	Flags	12' - 0"	<By Category>
A-12	12	Flags	12' - 0"	<By Category>
B-1	13	Banner	15' - 0"	Metal - Aluminum
B-2	14		12' - 0"	<By Category>
B-3	15	Banner	15' - 0"	<By Category>
B-4	16		12' - 0"	<By Category>
B-5	17		12' - 0"	<By Category>
C-4	18		12' - 0"	<By Category>
C-5	19		12' - 0"	<By Category>
C-6	20		12' - 0"	<By Category>
C-7	21		12' - 0"	<By Category>
C-8	22		12' - 0"	<By Category>
C-9	23		12' - 0"	<By Category>
C-10	24		12' - 0"	<By Category>
D-6	25		12' - 0"	<By Category>
D-7	26		12' - 0"	<By Category>

Figure 6–68

Task 1 - Create shared parameters.

1. In the ...*Structural* folder, open **Events-Plaza-S.rvt**.
2. In the *Manage* tab>Project Settings panel, click (Project Parameters).
3. In the Project Parameters dialog box, click **Add...**.
4. In the Parameter Properties dialog box, select the **Shared parameter** option and click **Select...**.

5. If there is no existing shared parameter file, an alert box displays prompting you to select a shared parameter file. Click **Yes**. If the Shared Parameters dialog box opens, click **Edit...**.

6. In the Edit Shared Parameters dialog box, click **Create...**.

7. Save the new file as **Plaza-Parameters.txt** in the practice files folder.

8. Create a new parameter group. In the *Groups* area, click **New...**. Set the name as **Columns** and click **OK**. This becomes the current group, as shown in Figure 6–69.

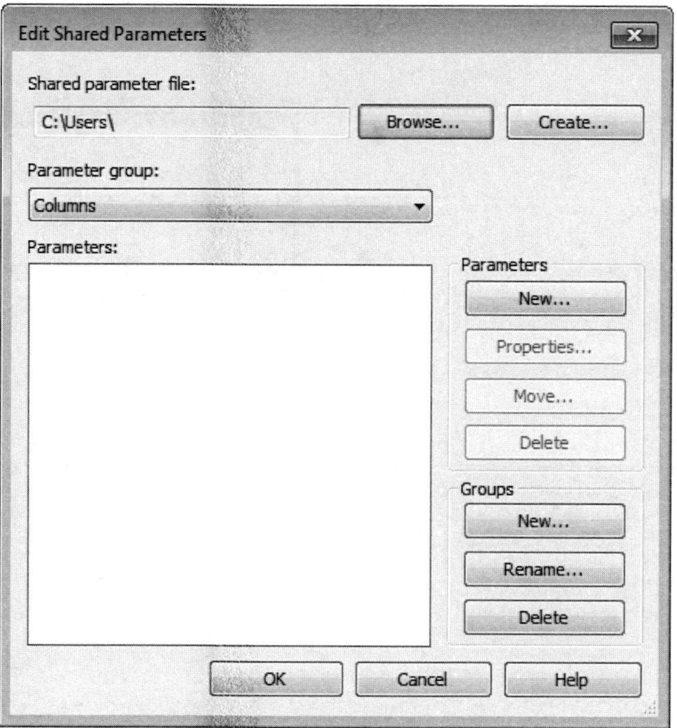

Figure 6–69

9. In the *Parameters* area, click **New...** and create three new shared parameters using the following information:

Name	Discipline	Type of Parameter
Bracket Type	Common	Text
Bracket Offset from Base	Common	Length
Bracket Finish	Common	Material

10. Click **OK**.

11. In the Shared Parameters dialog box, select **Bracket Type** as shown in Figure 6–70, and click **OK**.

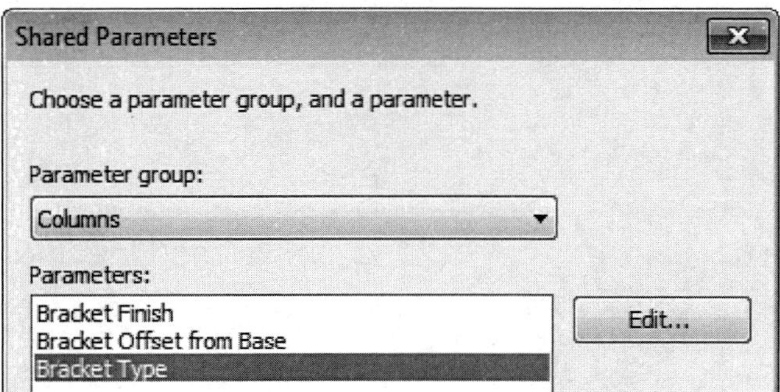

Figure 6–70

12. In the Parameter Properties dialog box, in the *Parameter Data* area, group the parameter under **Other** and make it an **Instance** parameter. In the *Categories* area, filter the list based on **Structure** and select **Structural Columns,** as shown in Figure 6–71.

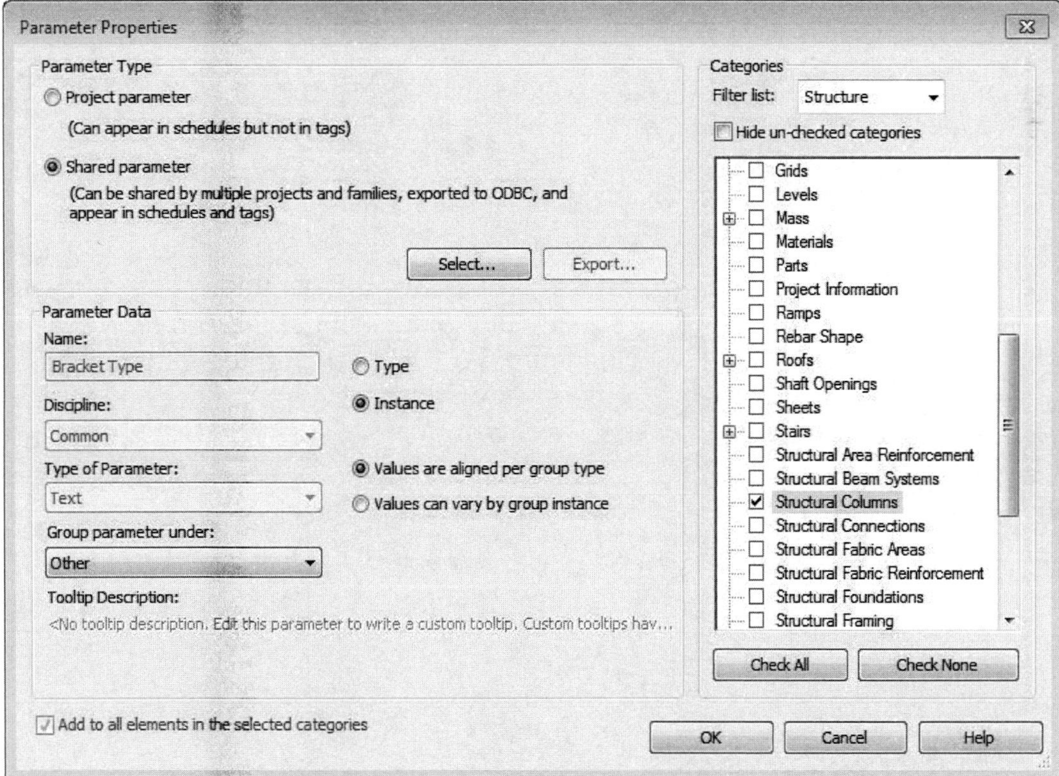

Figure 6–71

13. Click **OK**. **Bracket Type** is added to the shared parameters in the project.

14. Repeat the process to add **Bracket Offset from Base** and **Bracket Finish** to the Project Parameters list.

15. Click **OK** to finish creating the project parameters.

16. Save the project.

Task 2 - Create a Tag with Shared parameters.

1. Start a new family based on the **Generic Tag.rft** template found in the default library's *Annotations* folder.

2. Zoom in, read the information in red, and delete the note.

3. In the Properties panel, click (Family Category and Parameters).

4. In the Family Category and Parameters dialog box, set the *Family Category* to **Structural Column Tags**, as shown in Figure 6–72. Click **OK**.

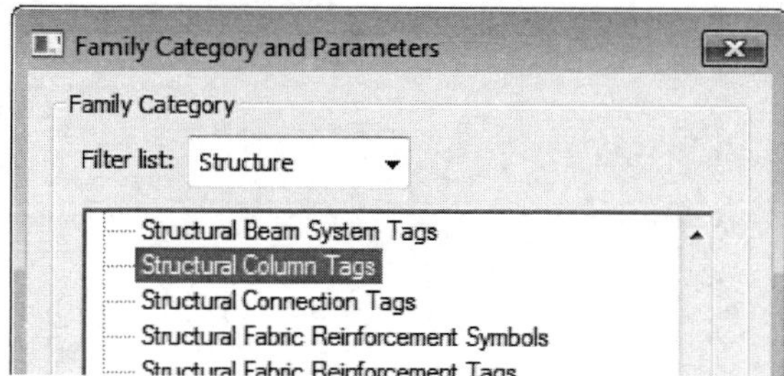

Figure 6–72

5. In the *Create* tab>Detail panel, click (Line). Draw a box, starting at the crossing of the two reference lines, as shown in Figure 6–73. Do not add the dimensions. They are for reference only.

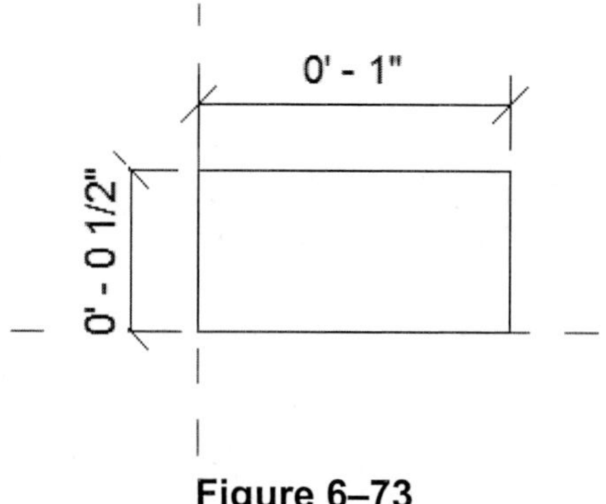

Figure 6–73

6. In the *Create* tab>Text panel, click (Label).

7. Use the default type, **Label: 3/32"**.

8. In the *Modify | Place Label* tab>Format panel, set *Justification* to (Align Center) and (Align Top).

9. Click inside the box you just created.

10. In the Edit Label dialog box, which includes a list of the multi-category parameters supplied with the Autodesk Revit software, click (Add Parameter).

In this dialog box, you can only create shared parameters.

11. In the Parameter Properties dialog box, click **Select...**.

12. In the Shared Parameters dialog box, the shared parameters you created earlier should be available by default. Select **Bracket Type**, as shown in Figure 6–74, and click **OK**.

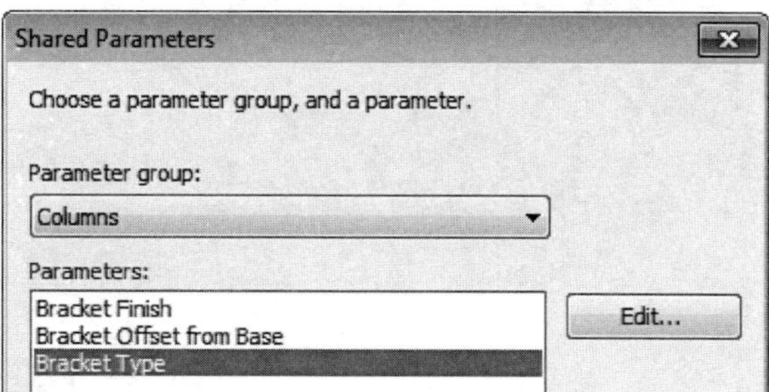

Figure 6–74

13. Click **OK** until you reach the Edit Label dialog box.

14. Select **Bracket Type** and click (Add parameter(s) to label).

15. Repeat with the **Bracket Offset from Base** and **Bracket Finish** parameters. In the *Break* column, select the option for the first two parameters, as shown in Figure 6–75. This forces each parameter to be on a separate line.

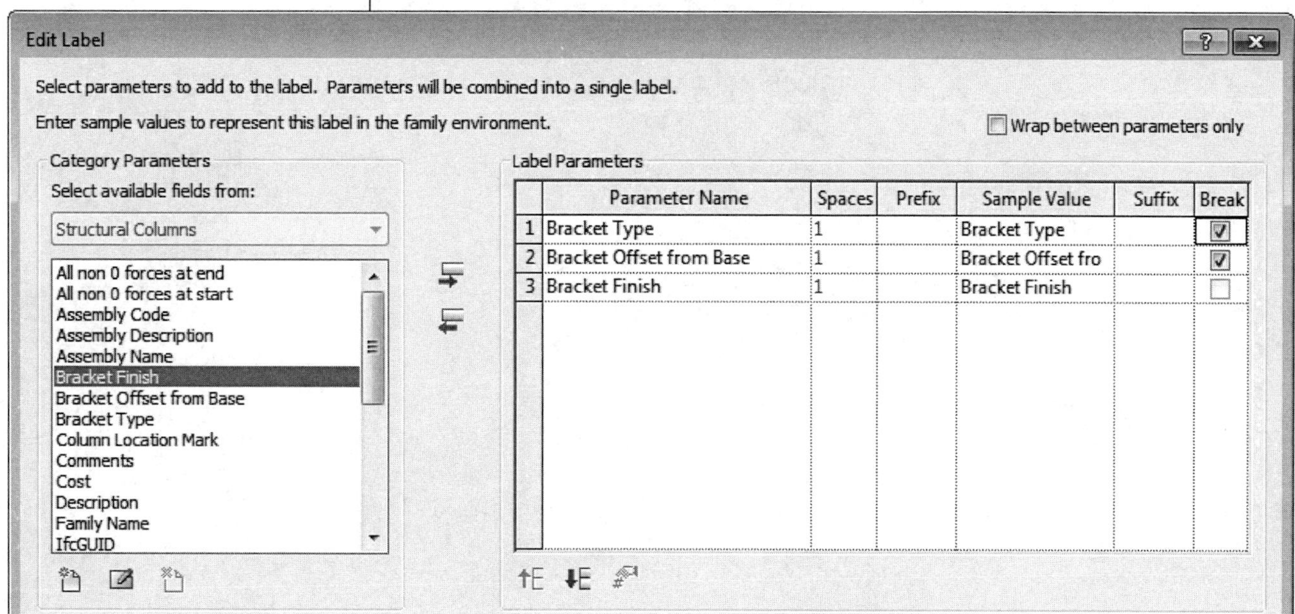

Figure 6–75

16. Click **OK** to insert the label into the family.

17. Click (Modify) and select the new label. Resize the outline of the text box so that it fits inside the box, and move the label so that it is top-centered in the box, as shown in Figure 6–76.

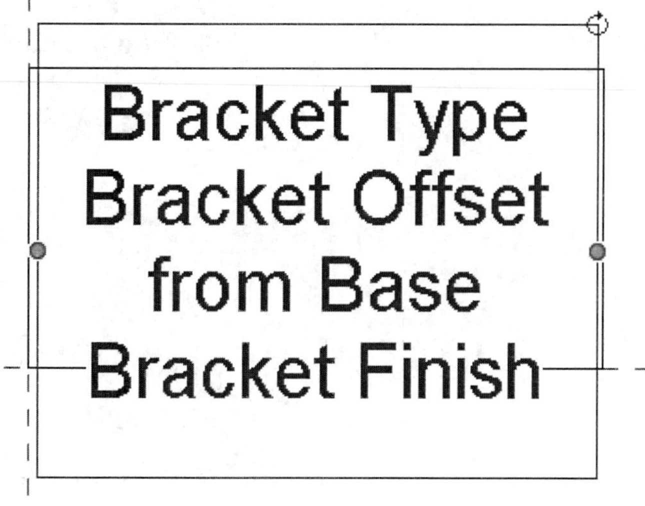

Figure 6–76

18. Save the new family with the name **Column-Bracket-Tag.rfa**.

19. Load the family into the project **Events-Plaza-S.rvt**.

20. The **Tag** command is automatically started. Tag one of the columns.

21. The tag is empty as shown in Figure 6–77, because no values have been added to the parameters.

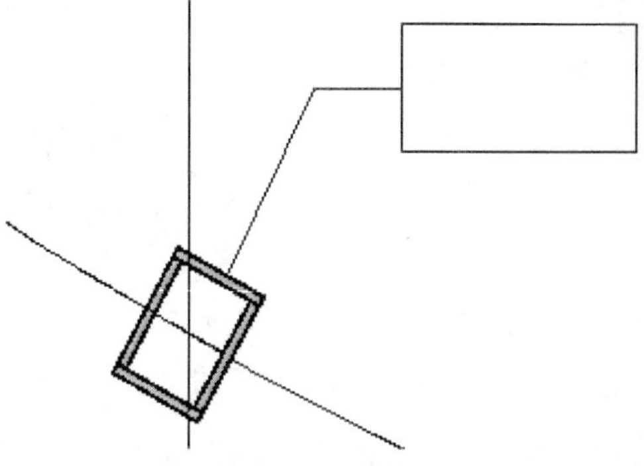

Figure 6–77

22. Save the project.

Task 3 - Add Tags and modify information.

1. In the *Annotate* tab>Tag panel, click (Tag All).
2. In the Tag All Not Tagged dialog box, select the *Category* **Structural Column Tags** with the *Loaded Tags* set to **Column-Bracket-Tag.** Select **Leader** and then click **OK**.
   - All of the columns are tagged. The tags are blank but display question marks when selected, as shown in Figure 6–78. This is because the corresponding information has not been entered into the properties.

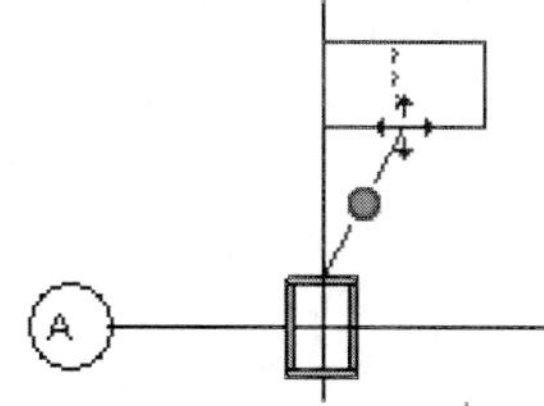
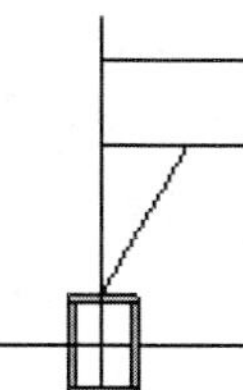
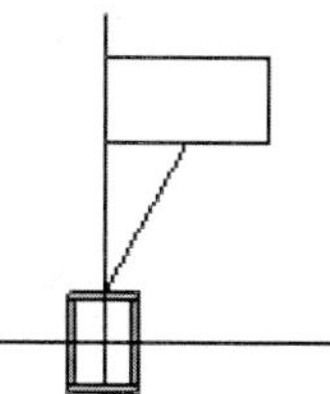

Figure 6–78

3. Select a column.
4. In Properties, scroll down to the *Other* area. The **Bracket Finish**, **Bracket Type**, and **Bracket Offset from base** parameters are available but have no information stored in them yet, as shown in Figure 6–79.

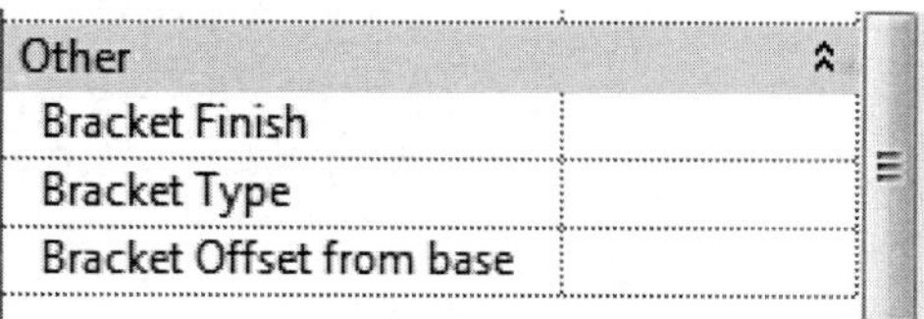

Figure 6–79

5. Right-click on the column and select **Select All Instances> Visible in View**.
   - In this case, you will set the default values for all columns once. Individual columns can be changed later.
6. In Properties, set the *Bracket Offset from Base* to **12'-0"**, set the *Bracket Type* to **Banner**, and set the *Bracket Finish* to **Metal**.

7. Select two or three individual columns, and in Properties change the *Bracket Type* to **Sculptural** and the *Bracket Offset from Base* to **15'-0"**. The tags update only for those columns, as shown in Figure 6–80.

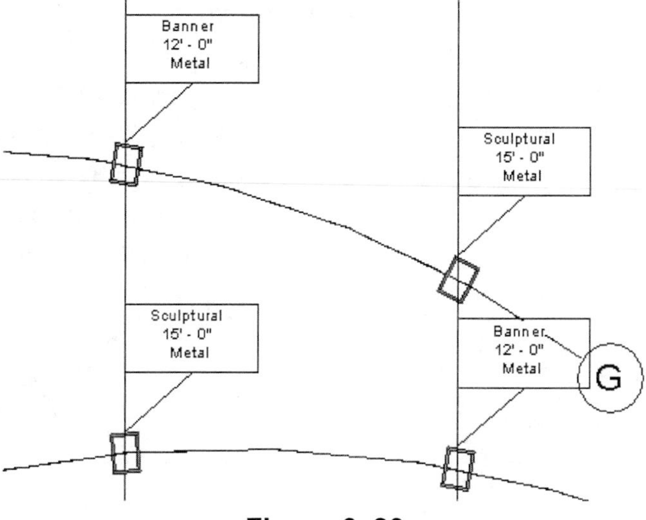

Figure 6–80

8. Save the project.

Task 4 - Create a Structural Column schedule.

1. Create a new building component schedule for structural columns.
2. In the Schedule Properties dialog box, in the *Fields* tab, add the **Column Location Mark** parameter and then add **Mark**, **Bracket Type**, **Bracket Offset from Base**, and **Bracket Finish**, as shown in Figure 6–81.

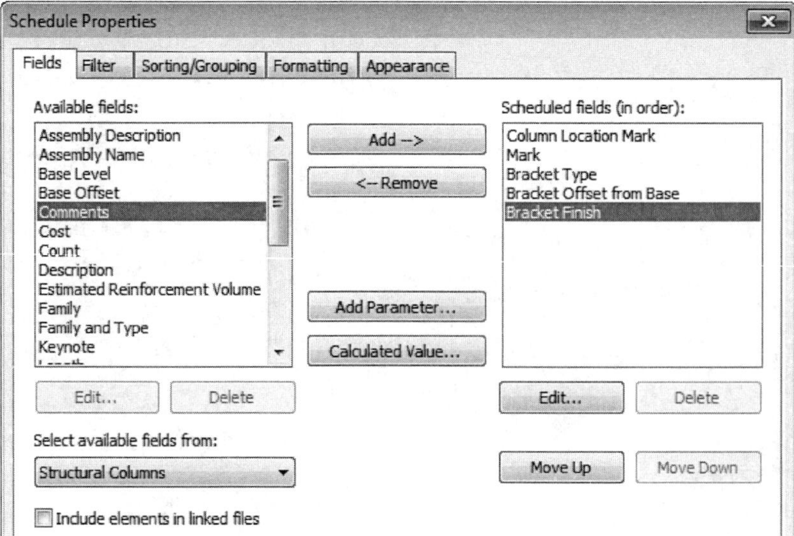

Figure 6–81

3. Sort by the Column Location Mark and modify the other schedule properties as required. Click **OK** when finished. The schedule should display the information you entered in the column's properties, as shown in Figure 6–82.

<Structural Column Schedule>				
A	B	C	D	E
Column Location M	Mark	Bracket Type	Bracket Offset fro	Bracket Finish
A-1	1	Banner	12' - 0"	Metal
A-2	2	Sculptural	15' - 0"	Metal
A-3	3	Banner	12' - 0"	Metal
A-4	4	Sculptural	15' - 0"	Metal
A-5	5	Banner	12' - 0"	Metal
A-6	6	Sculptural	15' - 0"	Metal
A-7	7	Banner	12' - 0"	Metal

Figure 6–82

4. Save the project.

Chapter Review Questions

1. An in-place family is created whenever you need a custom family that is:
   a. Not available in your library.
   b. Built using profiles.
   c. Directly connected to elements within the project.
   d. Requires shared parameters.
2. Which of the following are examples of elements for which you can create profiles? (Select all that apply.)
   a. Wall reveals.
   b. The top rail of railings.
   c. In-place Sweeps.
   d. Curtain Wall mullions.
3. At what size should you create the tag family elements, as shown in Figure 6–83?

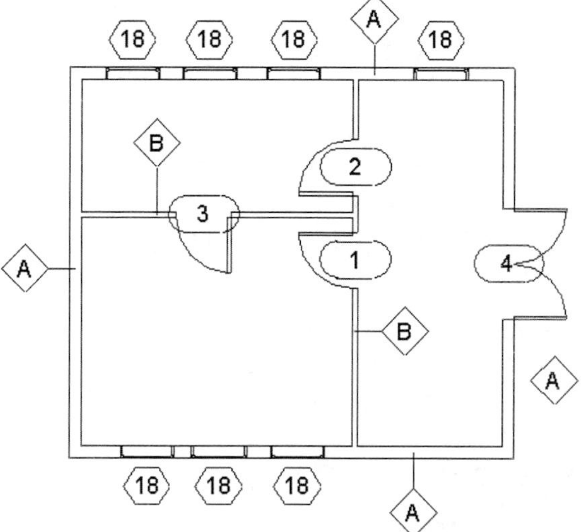

Figure 6–83

   a. Full scale to match the size of the tagged element in the project.
   b. Full scale to match the size of the annotation on the sheet.
   c. One size for each scale.
   d. It varies according to the type of annotation.

4. Where are shared parameters, as shown in Figure 6–84, stored?

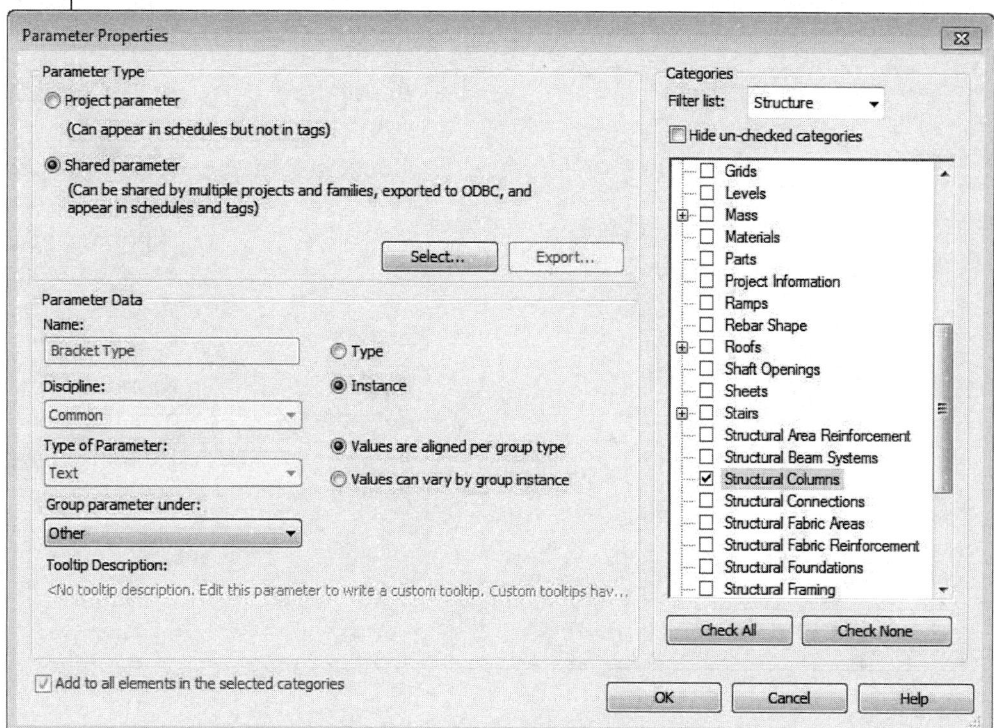

Figure 6–84

a. In a text file.

b. In the current project.

c. In a shared project.

d. In the family file.

Command Summary

Button	Command	Location
	Annotation Symbol	• **Application Menu:** expand New
	Model In-Place	• **Ribbon:** *Architecture* tab>Build panel> expand Component • **Ribbon:** *Structure* tab>Model panel> expand Component • **Ribbon:** *Systems* tab>Model panel> expand Component
	Project Parameters	• **Ribbon:** *Manage* tab>Project Settings panel
	Shared Parameters	• **Ribbon:** *Manage* tab>Project Settings panel

Creating Architectural Specific Families

Architectural families include elements that are created using standard component family methods (such as doors and windows). There are also ways to create other elements that use profiles and are best created directly in the project as in-place families (such as angled cornices and coping). Another important family is railings. These families are put together using other families, including rails, balusters, posts and panels.

Learning Objectives in this Chapter

- Create custom door and window families from existing families and templates.
- Use in-place families to create angled cornices, fascias, and copings.
- Create railing types, including the rails, balusters, posts, and panels.

7.1 Creating Custom Doors and Windows

You can create custom doors and windows by copying and modifying an existing family, enabling you to use all of the existing detailed elements and parameters. For example, in Figure 7–1 the full panel door family on the left can be transformed into the square panel on the right.

Figure 7–1

When you use a template, only the most basic elements are included. For example, in the Window with Trim template (shown in Figure 7–2), only the opening for the window, basic trim, and parameters are included. Everything else must be added.

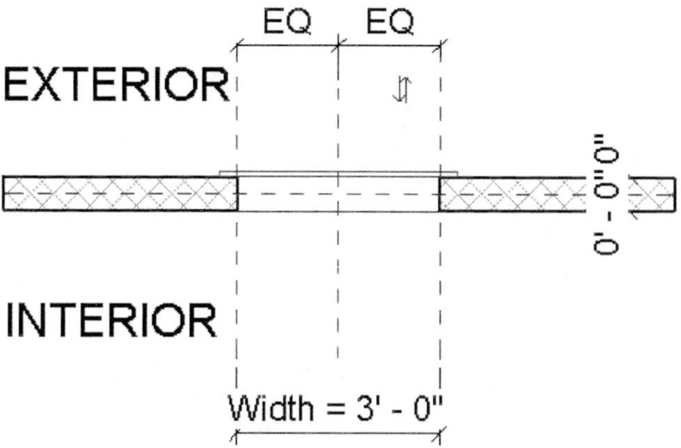

Figure 7–2

Practice 7a Create Custom Doors

Estimated time for completion: 20 minutes

Practice Objectives

- Create a custom family based on an existing family.
- Add reference planes and dimensions.
- Set formulas for parameters.

In this practice you will create new door family based on an existing door family. You will modify a nested door grill family so that it is square. Then you will insert the new square grill into the door and modify dimensions and labels so that the grill, glass, and panel all work together at different sizes, as shown in a project in Figure 7–3.

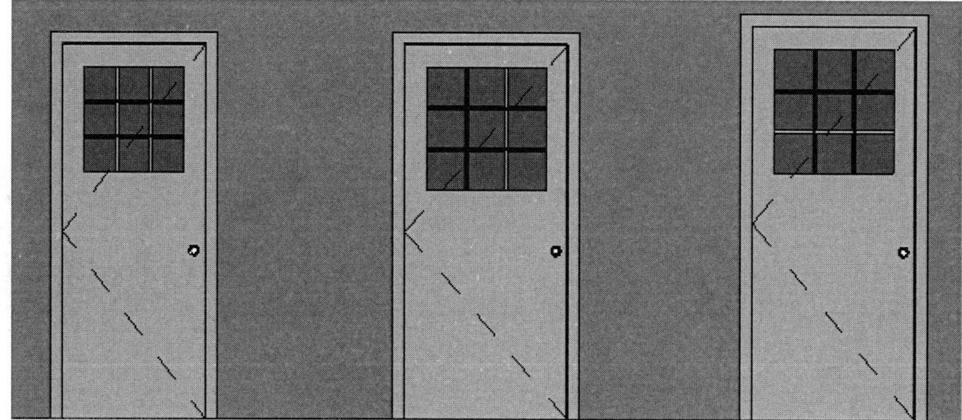

Figure 7–3

Task 1 - Create door types,

1. In the Autodesk Revit Library's ...*Doors**Residential* folder, open **Door-Interior-Single-Full Glass-Wood.rfa**.
    - The family is also available in the practice files...*Architectural* folder.
2. Save the file in the practice files folder with the name **Single-Glass-Square.rfa**.
3. Create three Family Types. Modify the *Width* and *Height* values to match the name.
    - 32" x 80"
    - 36" x 80"
    - 36" x 84"
4. Use the new types to flex the parameters.

5. Select the **Doors: Door Grill: Door Grill**, as shown in Figure 7–4. In the Type Selector, note that this is a nested family, along with the Door Swing and Handle families, as shown in Figure 7–5.

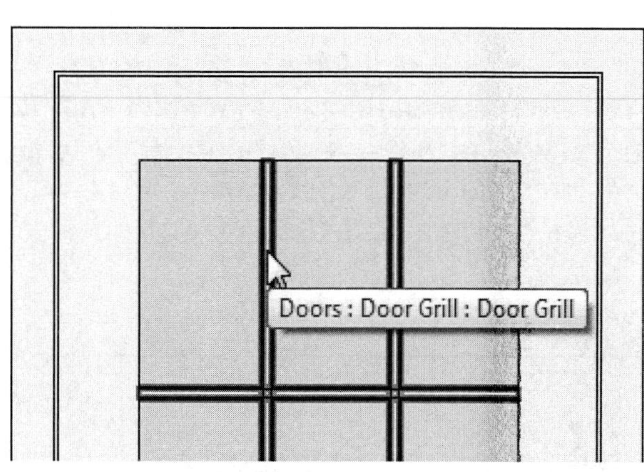

Figure 7–4

Figure 7–5

6. With the Door Grill family selected, in the *Modify | Doors* tab> Mode panel click (Edit Family). Note that it might take some time for this to open.

7. Once you are in the family editor, save the file as **Door Grill-Square.rfa** in the practice folder.

Task 2 - Modify the door grill.

1. Working in the **Door Grill-Square.rfa** family, open the **Front** elevation view. This is where the primary parameters display.

2. Delete the reference planes and associated elements shown in Figure 7–6, as well as the center vertical sweep and line (do not delete the center reference plane).

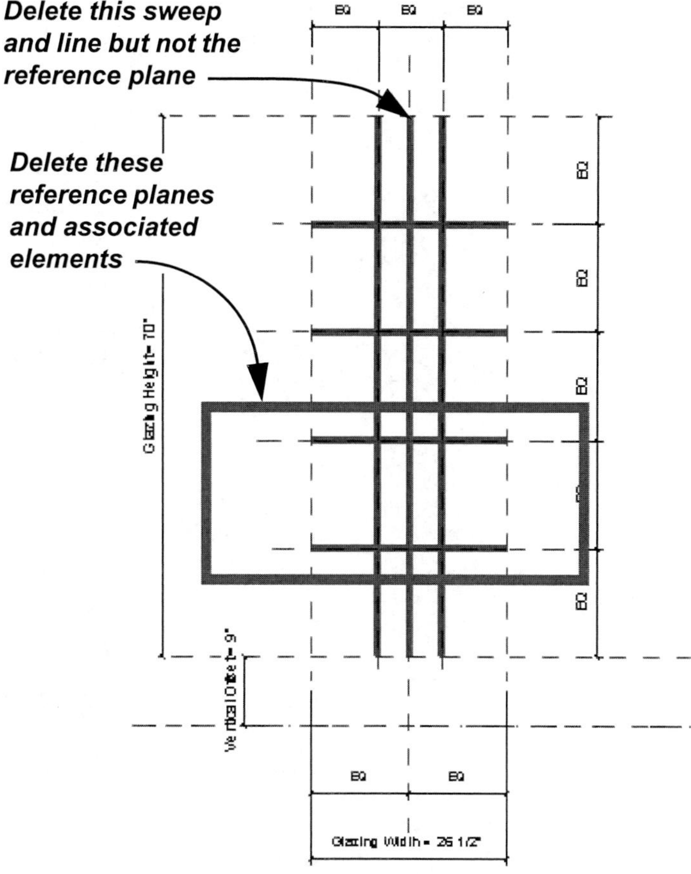

Figure 7–6

3. Select the dimension on the right that controls the placement of the remaining grids and click the **Toggle Dimension Equality** control. The muntin grids are spaced out equally again.

4. Open the Family Types dialog box.

5. Under the *Dimensions* heading, change the formula for the *Glazing Height* to **Glazing Width** and click **OK**.

6. Save the family.

Task 3 - Load and place the square grill in the door.

1. Load the square grill family into the door family and then click **Modify**.

2. Open the interior elevation of the door family. Select the existing grill and, in the Type Selector, change it to **Door Grill-Square**.

    - The existing grill is placed in the correct location within the door panel.

3. Align and lock the bottom of the square grill to the reference plane associated with the Bottom Rail as shown in Figure 7–7.

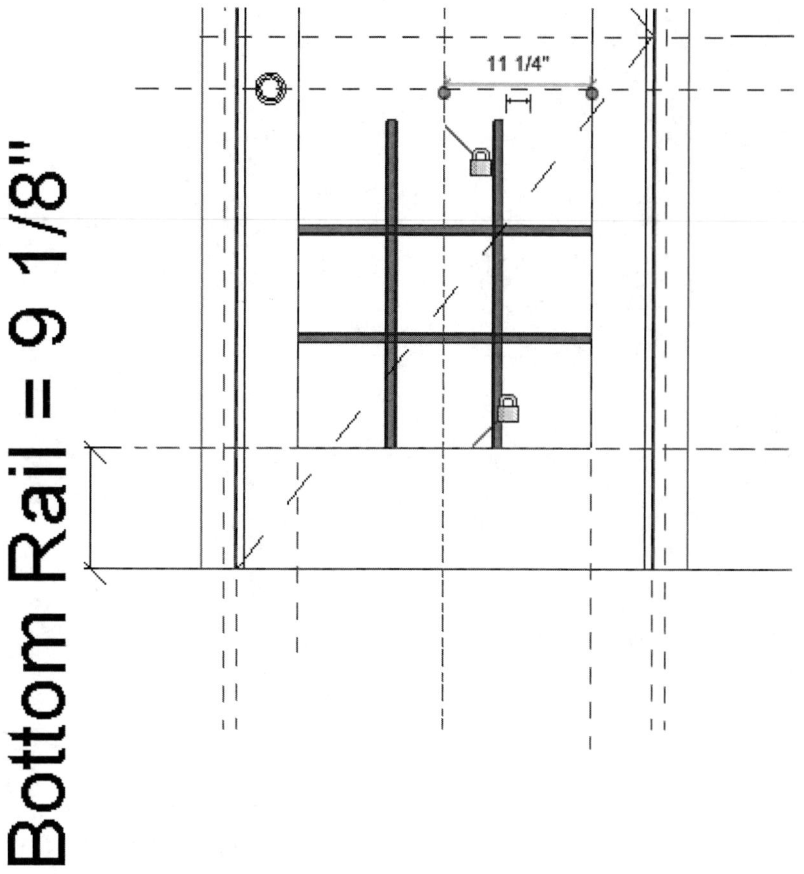

Figure 7–7

4. Delete the Bottom Rail dimension, but not the reference plane.

5. Create a new dimension from the Top Rail reference plane to the former Bottom Rail reference plane. Label it **Glass Height** (a **Type** parameter in the *Dimensions* category), as shown in Figure 7–8.

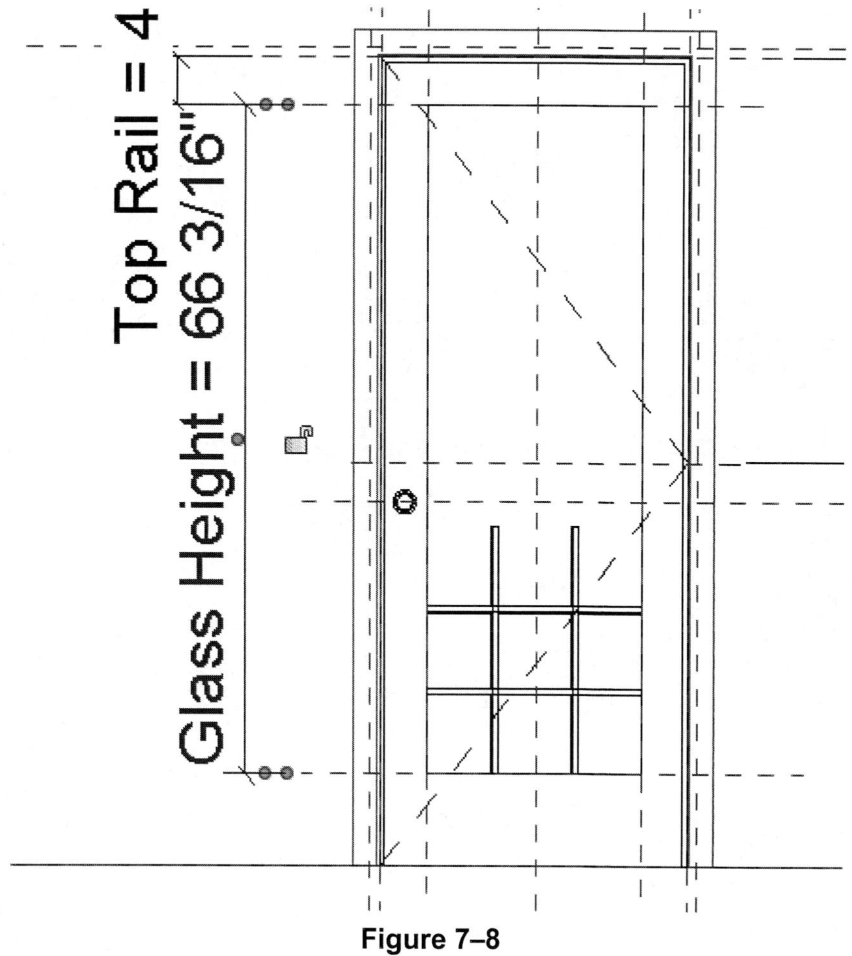

Figure 7–8

6. Select the square door grill. In Properties, click **Edit Type**.

7. Under the *Dimensions* heading, beside the *Width* parameter, click the **Associate Family Parameter** button, as shown in Figure 7–9.

8. In the Associate Family Parameter dialog box, select **Width**, (as shown in Figure 7–10) and then click **OK**.

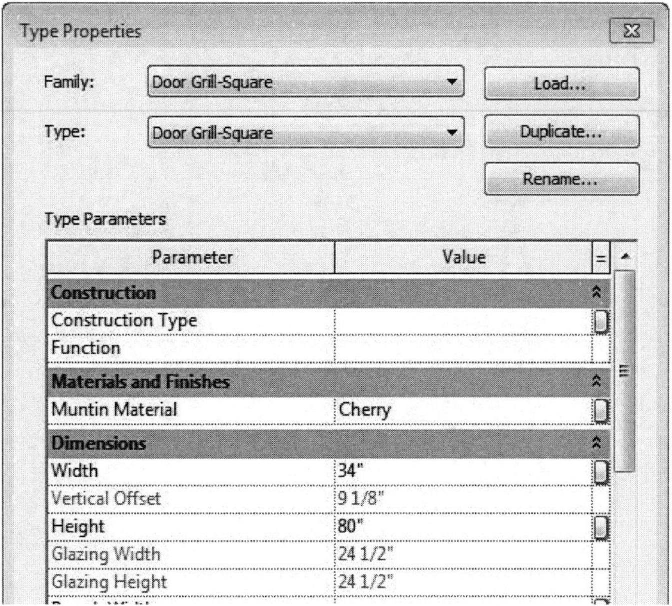

Figure 7–9

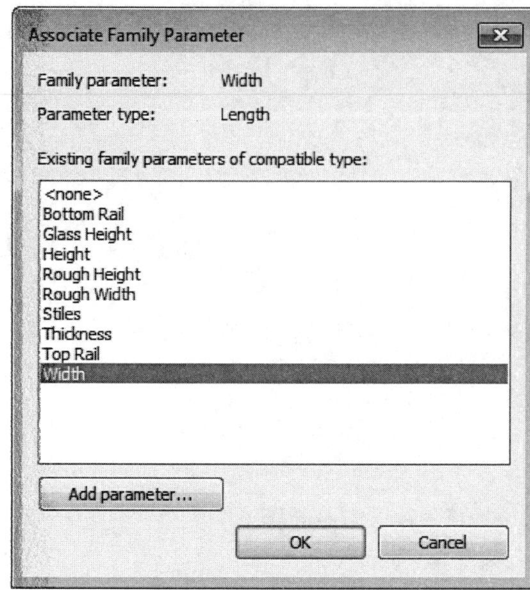

Figure 7–10

9. Repeat the process and associate the **Height** parameters.

10. Associate the **Muntin Material** with the **Trim Material** and then click **OK**.

11. Open the Family Types dialog box and set the formula for the *Glass Height* to **Width - (Stiles *2)**, as shown in Figure 7–11.

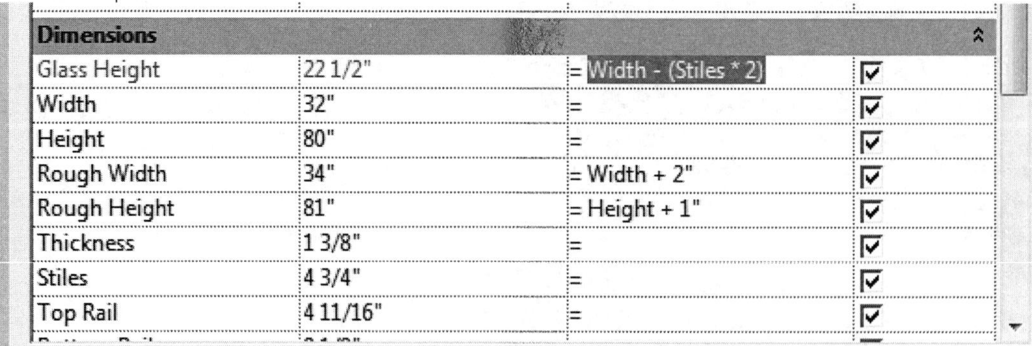

Figure 7–11

12. Click **Apply**. The grill should move to the top of the window. The panel and glass move up as well because they were locked to the former Bottom Rail reference plane.

13. Use the family types to flex the parameters and ensure that the square grill is working with th different door sizes.
14. Click **OK**.
15. Save the family.

Task 4 - Set the exterior grill placement.

1. Open the **Exterior** elevation view and select the grill. Note that it is still the full sized grill.
2. Change the grill to **Door Grill-Square**.
3. Change the *Visual Style* to **Wireframe** so you can see the square grill.
4. Align and lock the bottom of the grill to the reference plane, as shown in Figure 7–12.

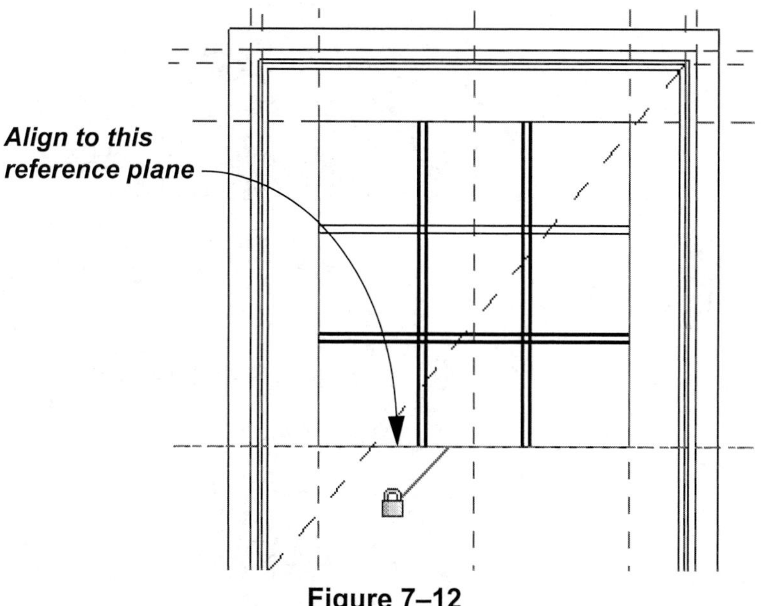

Figure 7–12

5. Flex the family to ensure that the grill is working correctly.
6. Return the *Visual Style* to Hidden Line.
7. Save the family.
8. Load and test the new door type in a project.

Practice 7b Create Custom Windows

Practice Objectives

- Create a custom family based on a template.
- Add trim and glass to the window.

Estimated time for completion: 20 minutes

In this practice you will create a new window family based on a family template. You will modify the existing opening, trim it, and add materials and glass to the window, as shown in Figure 7–13.

Figure 7–13

Task 1 - Create a custom window family.

1. Create a new family based on the **Window with Trim.rft** template found in the Autodesk Revit Library.
   - The family is also available in the practice files...*\Architectural* folder.
2. Save the family to the practice files folder as **Arched Window with Trim.rfa**.
3. Open the **Elevations: Exterior** view.
4. Edit the sketch of the **Opening Cut** element, as shown in Figure 7–14.

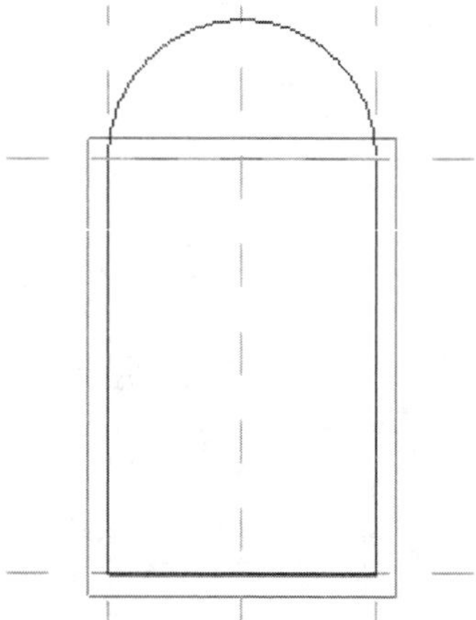

Figure 7–14

5. Click (Finish Edit Mode).

6. Add a dimension to the arc and label it **Radius** (a **Type** parameter in the *Dimensions* category).

7. Edit the **Trim: Extrusion** element and modify the sketch so that it matches the new arc opening, as shown in Figure 7–15.

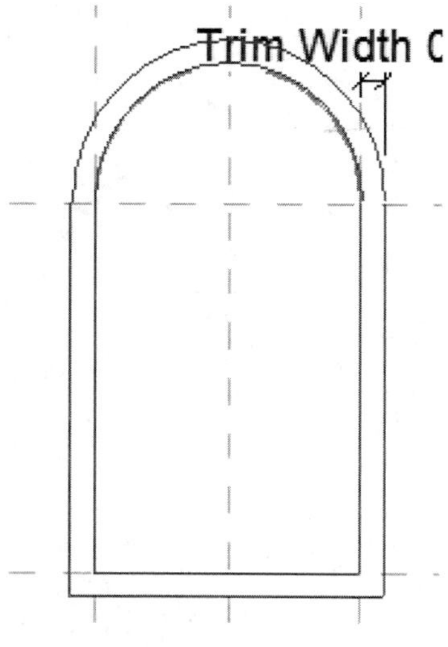

Figure 7–15

8. Click (Finish Edit Mode).

9. Open the Family Types dialog box and modify the **Radius** parameter to use the formula **Width/2**. Flex the model by trying different widths.

10. Set the final *Width* to **3'-0"** and click **OK**

11. Save the family.

12. Click (Load into Project) and test it in the same project as the door you just created.

Task 2 - Add a material to the window trim.

1. View the project with the door(s) and window(s) in a 3D view.
2. Set the *Visual Style* to (Shaded). The door has glass in the windows and materials applied to it, but the window does not have glass or materials, as shown in Figure 7–16.

Figure 7–16

3. Select one of the windows. In the *Modify* | *Windows* tab> Mode panel, click (Edit Family).
4. In the Family Editor, open the Elevations>**Exterior** view.
5. Set the *Visual Style* to (Shaded).
6. Select the **Trim Projection : Extrusion**.
7. In Properties, next to the **Materials** parameter, select **<By Category>**. Click (Browse) at the end of the row.
8. In the Materials Browser, under *Autodesk Materials*, select **Wood** and click (Add material to document), as shown in Figure 7–17.

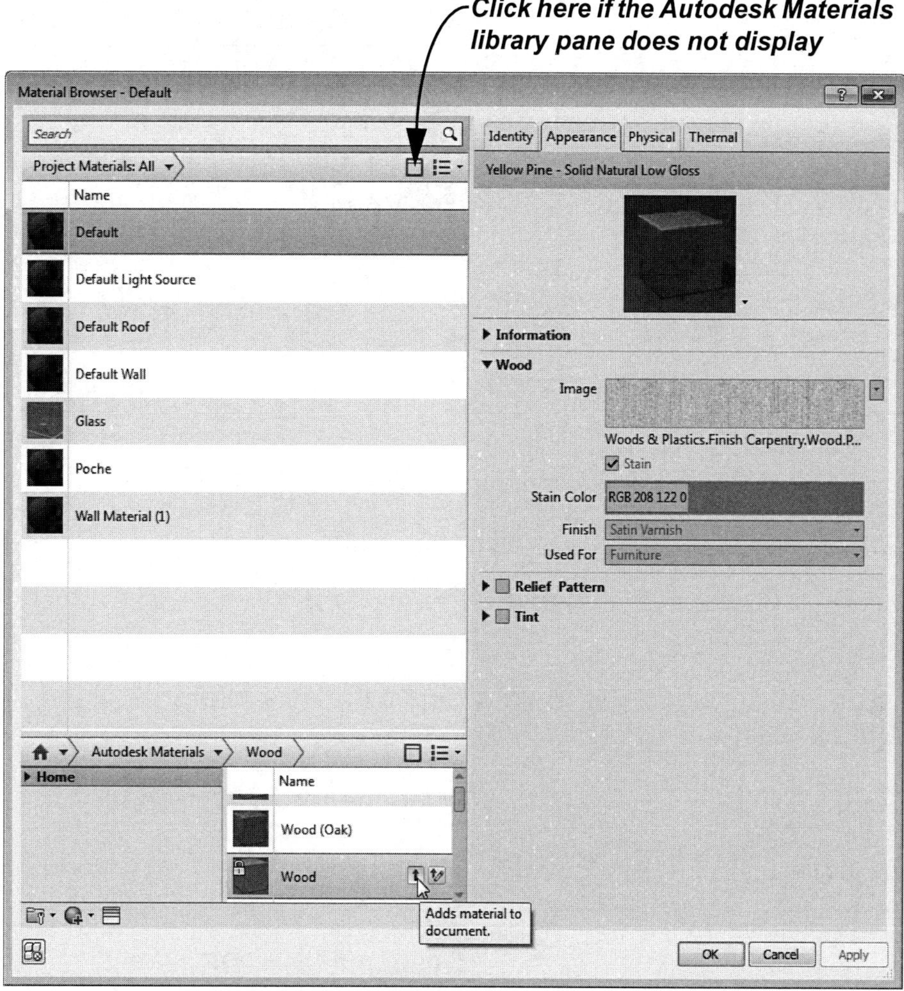

Figure 7–17

9. Select the new material. In the *Graphics* tab, in the *Shading* area, select **Use Render Appearance**, as shown in Figure 7–18.

By default this material only displays when the Visual Style is set to Realistic. By using the Render Appearance, the color displays when the Visual Style is set to Shaded or Consistent Colors.

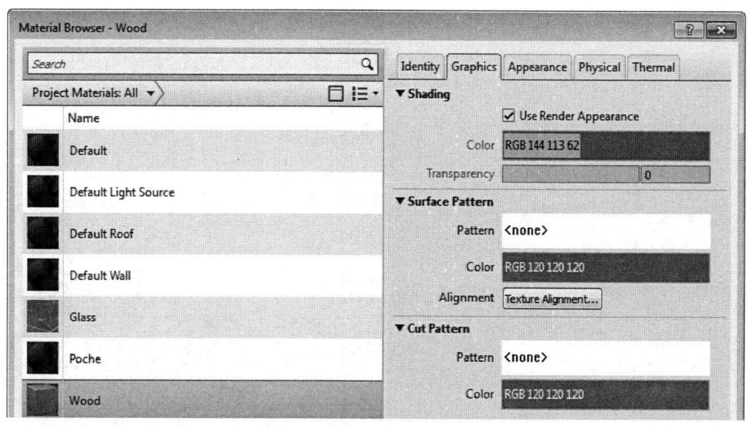

Figure 7–18

10. Close the Material Browser.

11. Save the window family file.

Task 3 - Add glass to the window.

1. Open the Floor Plans>**Ref. Level** view.
2. Draw a reference plane **1"** from the exterior face of the window and lock it in place, as shown in Figure 7–19.

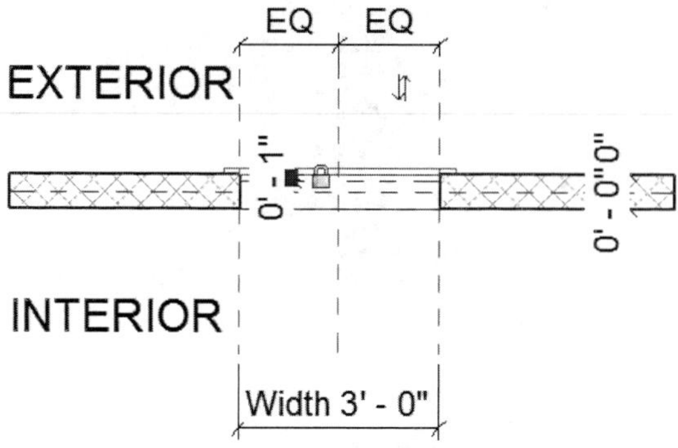

Figure 7–19

3. Select the new reference plane. In Properties, set the *Name* to **Glass**.
4. Open the Elevations>**Exterior** view.
5. In the *Create* tab>Work Plane panel, click (Set).
6. In the Work Plane dialog box, select the new reference plane **Glass** and click **OK**.
7. In the *Create* tab>Form panel, click (Extrusion).
8. Sketch the outline of the glass inside the window.
9. In Properties, set the *Extrusion End* to **1/2"**.
10. In the Material Browser, under *AEC Materials>Glass*, set the *Material* to **Glass, Clear Glazing**. Select **Use Render Appearance** so that it displays when the Visual Style is set to **Shaded**.
11. Close the Material Browser.
12. Click (Finish Edit Mode). The glass material displays.
13. Save the family file.
14. Click (Load into Project) and select **Overwrite the existing version and its parameter values**. The new glass and trim material display in the project.

7.2 Creating Angled Cornices and Copings

You might have tried to add a wall sweep to a wall at an angle, and found that it does not work. The **Wall Sweep** command only works horizontally or vertically. For example, to add a cornice to the top of a wall following the angled profile, as shown in Figure 7–20, you need to use a workaround, such as an in-place family using a solid sweep. A solid sweep can also be used to create a coping on the top of a short wall running beside steps.

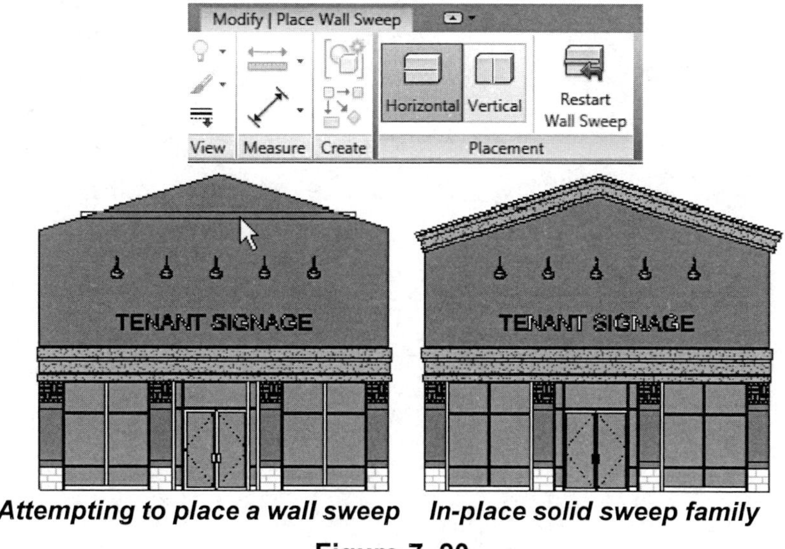

Attempting to place a wall sweep *In-place solid sweep family*

Figure 7–20

Creating Fascias

Another way to create profiles along an angle is to use a workaround to create a roof fascia on a model line, as shown in Figure 7–21. Typically, a fascia is applied to a roof edge but it can also be applied to a model line placed anywhere in a project.

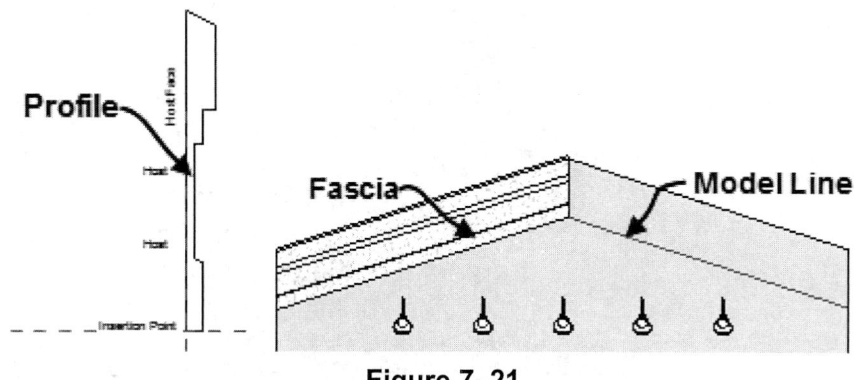

Figure 7–21

The process is similar for creating new Gutter types.

How To: Create a Fascia Type

1. Load the profile for the fascia. (In the *Insert* tab>Import panel, click (Load Family). In the Load Family dialog box, navigate to the appropriate folder in which your profile is stored. For example, in the Autodesk Revit library, in the *Profiles>Roofs* folder, select **Fascia Built-up.rfa** and click **Open**.)
2. In the *Architecture* tab>Build panel, expand (Roof) and click (Roof: Fascia).
3. In Properties, click (Edit Type).
4. In the Type Properties dialog box, click **Duplicate**.
5. In the Name dialog box, type a name and click **OK**.
6. In the Type Properties dialog box, select the appropriate profile, as shown in Figure 7–22.

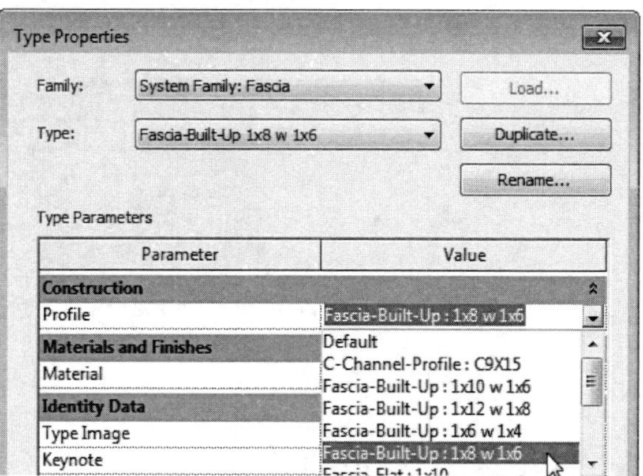

Figure 7–22

7. Click **OK** and the new fascia type is ready to be used.

How To: Add a Fascia Using Model Lines

1. In the *Architecture* tab>Work Plane panel, click (Set) and select the appropriate work plane.
2. In the *Architecture* tab>Model panel, click (Model Line).
3. Check the height of the fascia profile first so that you know where to put the model lines. Draw the model lines on the face of the wall where you want to add a fascia.
4. In the *Architecture* tab>Build panel, expand (Roof) and click (Roof: Fascia).
5. Select the model lines.

Hint: Object Style Subcategories

When creating an in-place family, you might want the family member to be modified in Visibility/Graphic Overrides separately from the main component. For example, if you are creating a cornice or coping for a wall and want to be able to turn it off separately from the wall, you can create an Object Style subcategory.

1. In the *Manage* tab>Settings panel, click (Object Styles).
2. In the *Modify Subcategories* area, click **New**.
3. In the New Subcategory dialog box, type a *Name* and select the category it is going to be a *Subcategory of*, as shown in Figure 7–23.

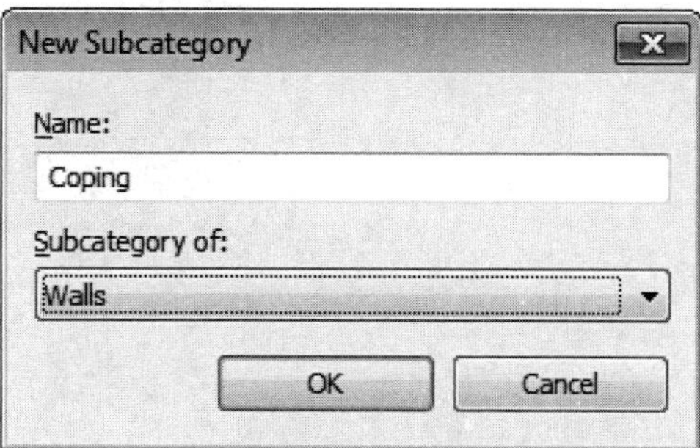

Figure 7–23

4. Click **OK** twice to finish the command.
5. The object style is now available in Visibility/Graphic Overrides, as shown in Figure 7–24.

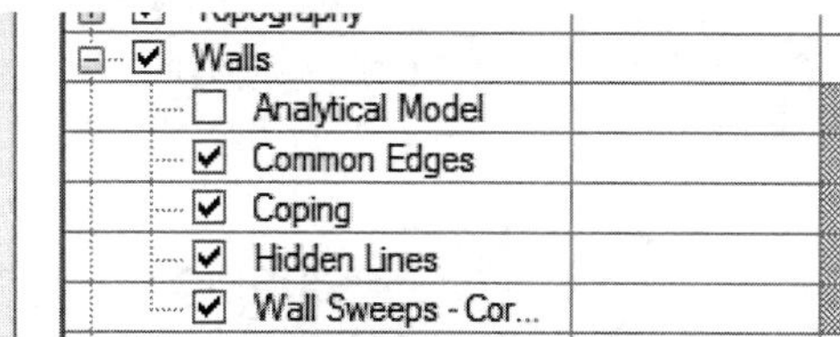

Figure 7–24

6. In the in-place family, select the object you want to add to the new subcategory. In Properties, under the *Identity Data* heading, select a *Subcategory* from the list (as shown in Figure 7–25), and then click **OK**.

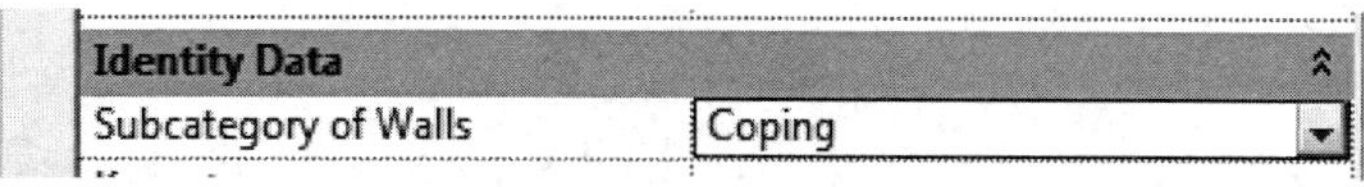

Figure 7–25

Practice 7c Create Angled Cornices and Copings

Estimated time for completion: 15 minutes

Practice Objectives

- Create a coping using an in-place family.
- Use Model Lines and Roof Fascia to add a Profile to a Sloped Wall.

In this practice you will create a coping using an in-place family with a solid sweep. You will then add a profile to a sloped wall using the **Roof: Fascia** command and a model line, as shown in Figure 7–26.

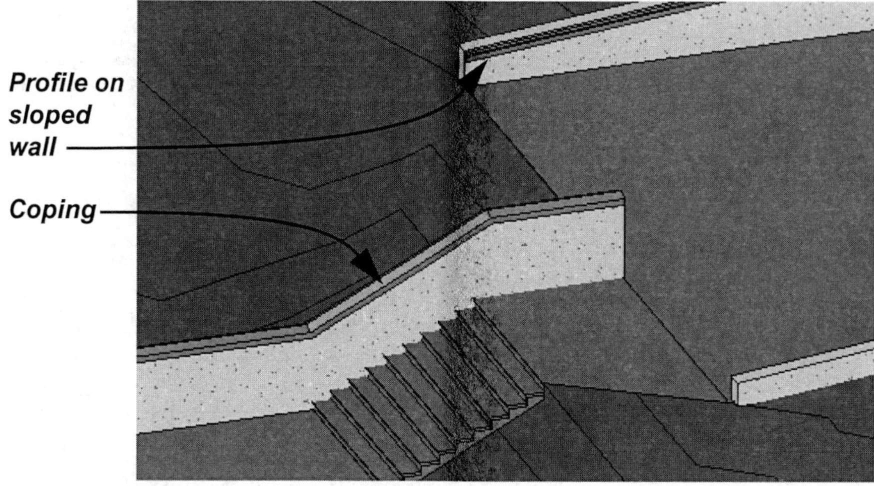

Figure 7–26

Task 1 - Create a coping using an in-place family.

1. In the ...*\Architectural* folder open **Plaza-A.rvt**. This is the beginning of a site design consisting of two slabs with retaining walls connected by a stair.
2. Open the **Floor Plans: Site** view. This project has three sections to help you draw additional elements.
3. Open the Sections>**East-West Site Section** view and zoom in on the lower wall, where you are going to add a stone coping to the top of the wall.
4. In the *Architecture* tab>Build panel, expand (Component) and click (Model In-Place).
5. In the Family Category and Parameters dialog box, select **Walls** and click **OK**.

You are placed in the Family Editor with the corresponding Ribbon tabs.

6. Name the family **Coping**.
7. In the *Create* tab>Forms panel, click (Sweep).
8. In the *Modify | Sweep* tab>Sweep panel, click (Sketch Path).
9. In the Work Plane dialog box, select **Pick a plane** and click **OK**. Select the face of the concrete retaining wall.
10. Draw the path, as shown in Figure 7–27.

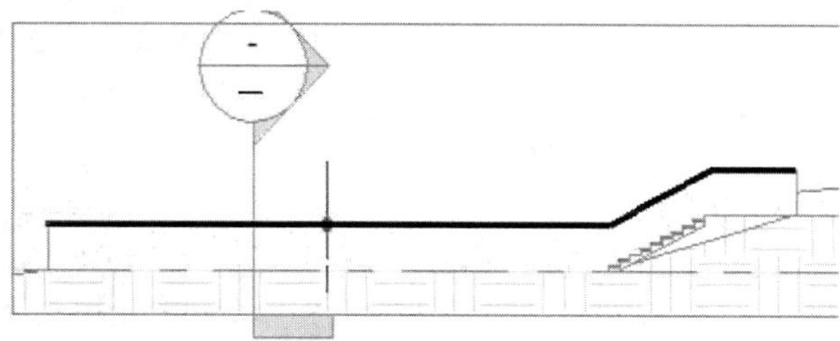

Figure 7–27

11. Click (Finish Edit Mode).
12. In the Sweep panel, click (Edit Profile).
13. In the Go To View dialog box, select **Section: Wall Section** and click **Open View**.
14. Zoom in and draw the profile, as shown in Figure 7–28.

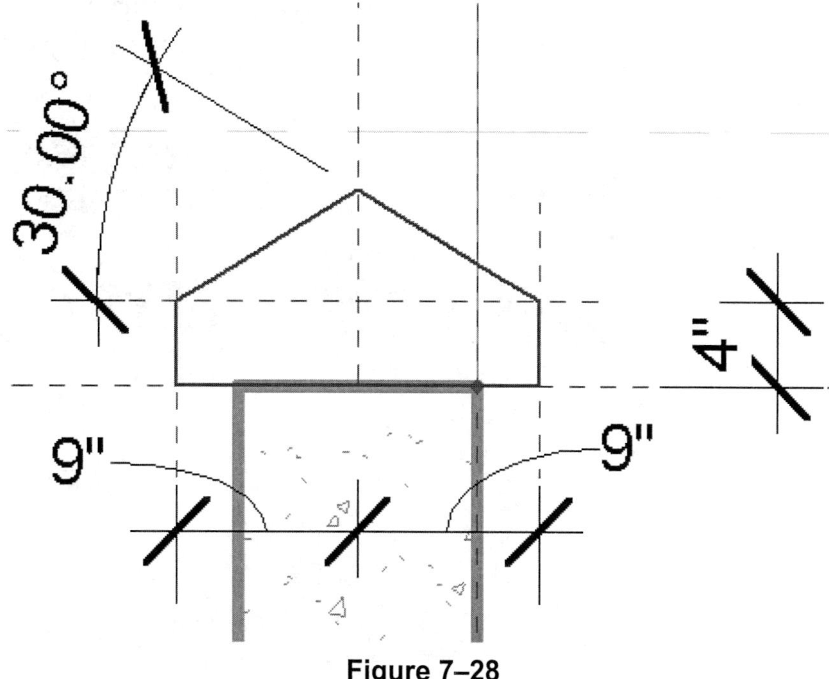

Figure 7–28

15. Finish the profile, sweep, and model.

16. Switch to a 3D view to see the coping in place.

Task 2 - Use Model Lines and Roof Fascia to add a profile to a Sloped Wall.

1. Open the **Sections>East-West Site Section** view and zoom in on the upper plaza.

2. In the *Insert* tab>Load from Library panel, click (Load Family). Navigate to the ...*Architectural* folder and load the **Wall-Profile-A.rfa.**

3. In the *Architecture* tab>Model panel, click (Model Line).

4. In the Work Plane dialog box, select **Pick a Plane** and click **OK** and select the face of the concrete retaining wall.

If you use Pick Lines or draw from the right to the left, the fascia faces the wrong direction when it is applied.

5. Draw a model line from the left to the right, **7"** down from the top of the retaining wall.

6. In the *Architecture* tab>Model panel, expand (Roof) and click (Roof: Fascia).

7. In Properties, click **Edit Type**.

8. In the Type Properties dialog box, create a new fascia type named **Wall Sweep** using the profile you loaded earlier, and set the *Material* to **Masonry-Stone**, as shown in Figure 7–29.

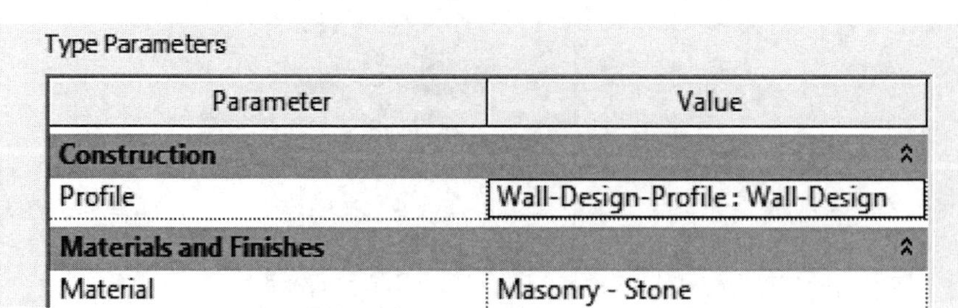
Type Parameters

Parameter	Value
Construction	
Profile	Wall-Design-Profile : Wall-Design
Materials and Finishes	
Material	Masonry - Stone

Figure 7–29

If you do not see the fascia, switch to a plan view, select it and click the Flip control.

9. Once you have finished creating the **Roof Fascia** type, select the model line to place it in the project, as shown in Figure 7–30.

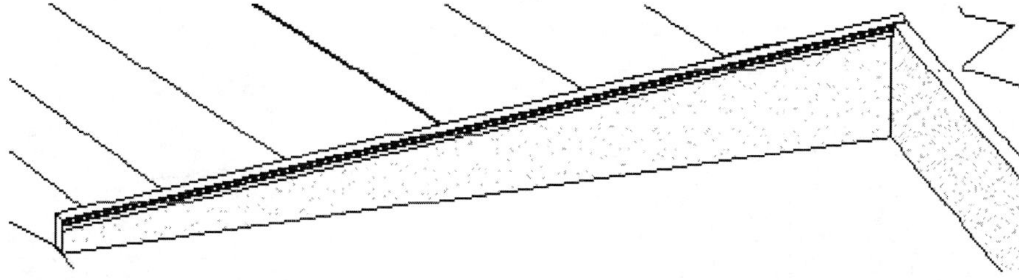

Figure 7–30

10. Save the project.

7.3 Creating Custom Railings

Creating a railing is more complex than creating a standard component family; it consists of several preset families for the rails, balusters, posts, and panels, as shown in Figure 7–31. Once you have these ready, you can create a Railing type.

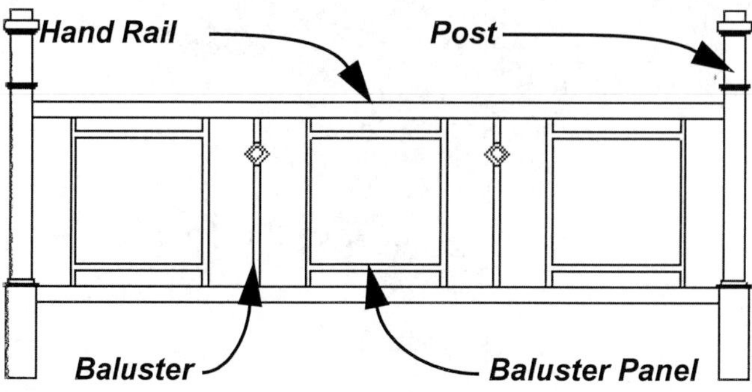

Figure 7–31

- Railings are system families made up of profiles and component families.
- Create all of the components for a Railing type (e.g., rails, handrails, top rails, posts, balusters, and panels) before starting to create the type.

How To: Create a Railing Type

1. Select an existing railing and open the Type Properties dialog box.
2. Create a duplicate type and set up the Type Parameters, as shown in part in Figure 7–32.

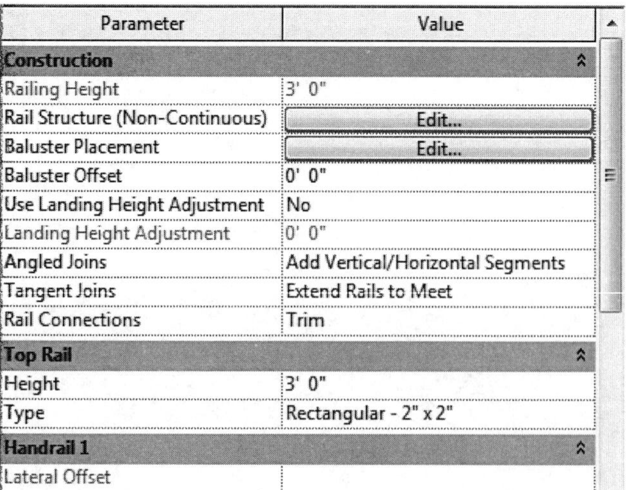

Parameter	Value
Construction	
Railing Height	3' 0"
Rail Structure (Non-Continuous)	Edit...
Baluster Placement	Edit...
Baluster Offset	0' 0"
Use Landing Height Adjustment	No
Landing Height Adjustment	0' 0"
Angled Joins	Add Vertical/Horizontal Segments
Tangent Joins	Extend Rails to Meet
Rail Connections	Trim
Top Rail	
Height	3' 0"
Type	Rectangular - 2" x 2"
Handrail 1	
Lateral Offset	

Figure 7–32

3. Click **OK** to close the dialog box.

How To: Set the Rail Structure (Non-Continuous)

1. In the Type Properties dialog box, next to the **Rail Structure (Non-Continuous)** parameter, click **Edit...**.
2. In the Edit Rails (Non-Continuous) dialog box, click **Insert** to add a new rail to the list. You can have a single rail or multiple rails in a Railing type, as shown in Figure 7–33.

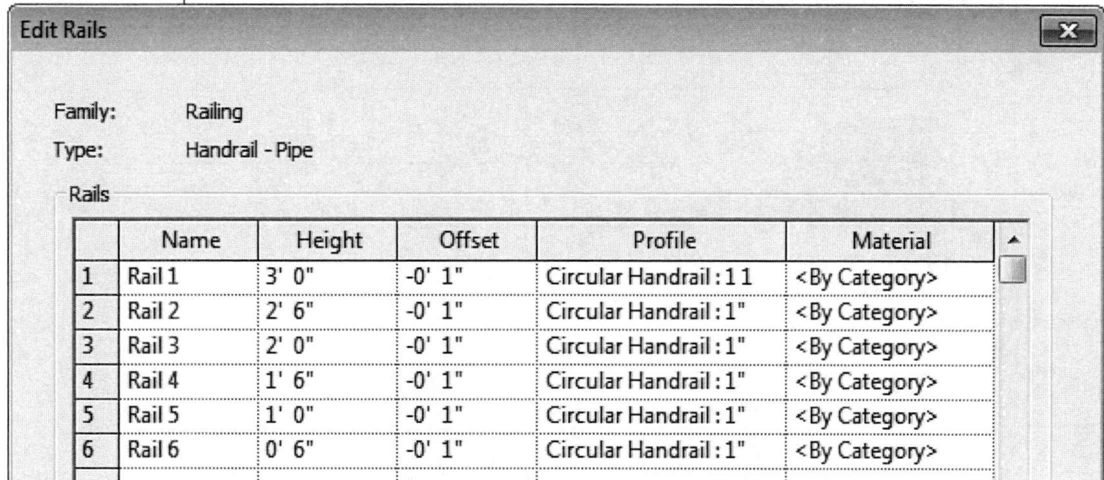

Edit Rails
Family: Railing
Type: Handrail - Pipe
Rails
Name | Height | Offset | Profile | Material
1 Rail 1 | 3' 0" | -0' 1" | Circular Handrail : 1" | <By Category>
2 Rail 2 | 2' 6" | -0' 1" | Circular Handrail : 1" | <By Category>
3 Rail 3 | 2' 0" | -0' 1" | Circular Handrail : 1" | <By Category>
4 Rail 4 | 1' 6" | -0' 1" | Circular Handrail : 1" | <By Category>
5 Rail 5 | 1' 0" | -0' 1" | Circular Handrail : 1" | <By Category>
6 Rail 6 | 0' 6" | -0' 1" | Circular Handrail : 1" | <By Category>

Figure 7–33

- Rails are horizontal members other than the Top Rail and Handrail (separate elements). If there are no other horizontal members, this dialog box remains empty.

3. Fill out the information about the rails as needed:

Name	The name of the individual railing component. Select the name to modify it.
Height	The height for each rail from the base height.
Offset	The distance from the rail sketch line.
Profile	Select a profile in the drop-down list. You need to have the profile loaded before you can use it. (This is just a profile family, not a rail type as is needed when you specify the top rails and handrails.)
Material	Specify a material for the rail. This is the only place where you can apply the material for rails.

4. Click **Up** and **Down** to move the rail to the correct location in the group.
5. Click **OK** to finish the command.

How To: Set up the Baluster Placement

1. In the Type Properties dialog box, next to the **Baluster Placement** parameter, click **Edit...**. The Edit Baluster Placement dialog box has two main areas: *Main pattern* and *Posts*, as shown in Figure 7–34.

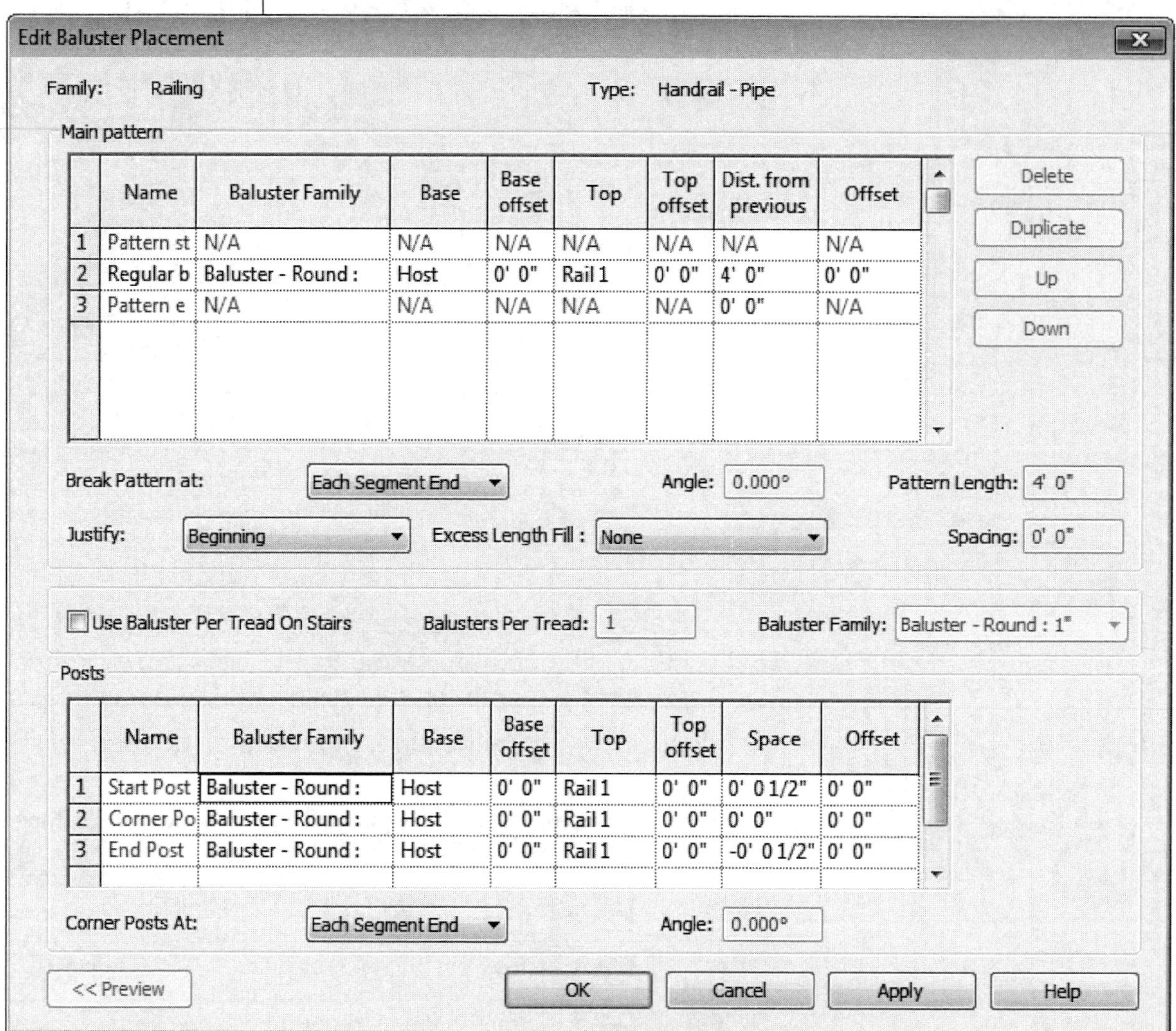

Figure 7–34

2. In the *Main pattern* area, add the baluster family and its sizes, such as the distance from the previous baluster. You can also add panels in this area.
3. If you are using a baluster on stairs and want a specific number per tread, select the **Use Baluster Per Tread On Stairs** option and specify the number per tread.
4. In the *Posts* area, specify the profiles for the *Start*, *Corner*, and *End* posts, and then specify their dimensions and locations.

How To: Set the Top Rail and/or Handrail

1. Select a railing type and open the Type Properties dialog box.
2. In the Top Rail area, specify the *Height* and *Type* as shown in Figure 7–35. The *Type* is a separate Top Rail type that needs to be set up before you get to this point in the Railing type creation.

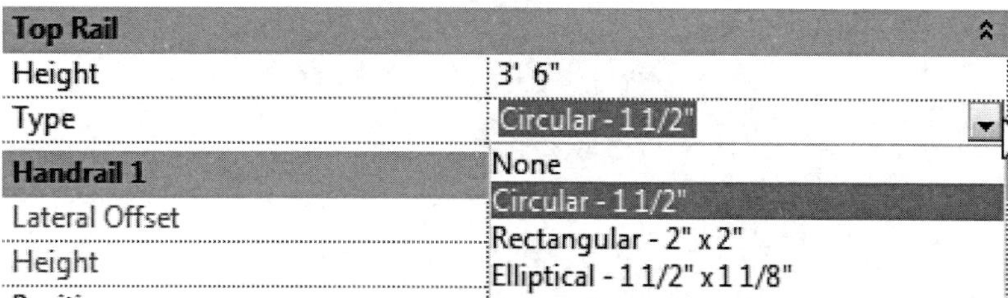

Figure 7–35

- If you are creating a guardrail railing type, this is the top of the guardrail and you would specify the handrail separately. If it is a handrail railing type, this is the top rail and you do not specify a separate handrail type.

3. If you are creating a guardrail, you also need to specify the *Type* and *Position* of the handrail. The handrail can be set to **Left**, **Right**, or **Left and Right** of the main railing as shown in Figure 7–36. The **Lateral Offset** and **Height** are read-only parameters.

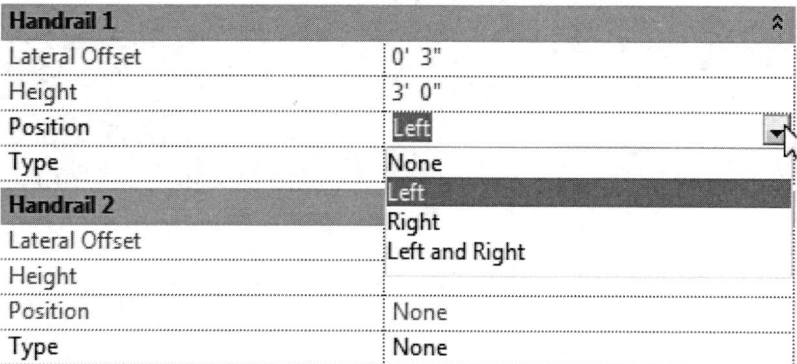

Figure 7–36

- If you do not want to include a separate handrail in the railing type, set the *Type* to **None**.

7.4 Families for Railings, Balusters, and Panels

To make a completely custom Railing type, you first need to create individual families for the rails, top rails, hand rails, posts, balusters, and panels, as shown in Figure 7–37.

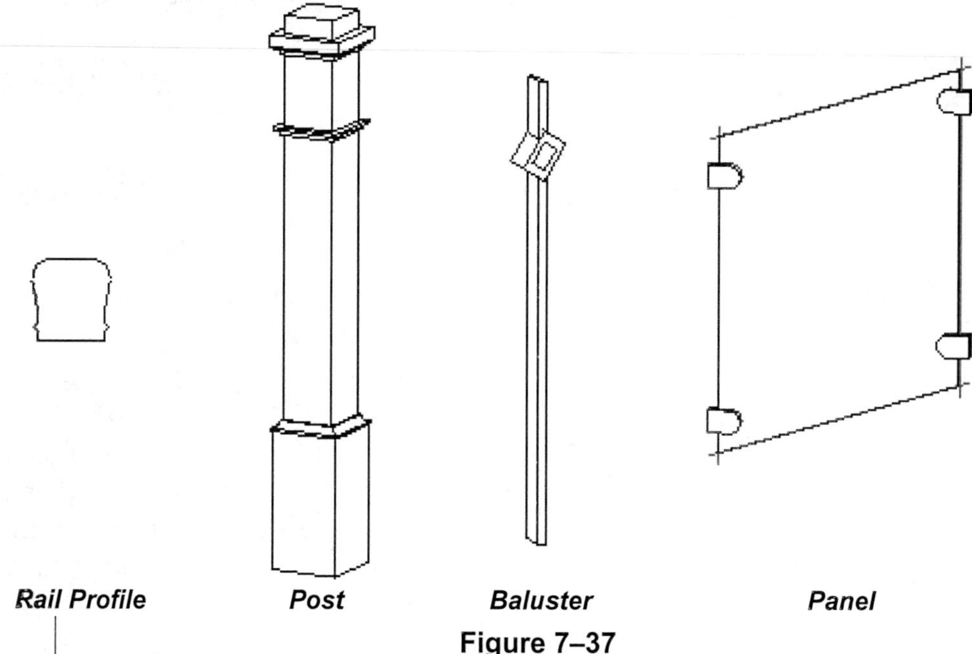

Figure 7–37

- Profiles for rails are found in the Autodesk Revit library's *...\Profiles\Railings* folders.
- Component families for other elements are found in the *...\Railings* folder in sub-folders for *Balusters*, *Supports*, and *Terminations*.
- To create the elements from scratch, use the following family templates:
  - **Baluster.rft**
  - **Baluster-Panel.rft**
  - **Baluster-Post.rft**
  - **Profile-Rail.rft**
  - **Railing Support.rft**
  - **Railing Termination.rft**

Creating Rails

Creating rails is a two part process:

1. Create a rail profile.
2. Apply the rail profile to a Top Rail or Handrail system family type.

Rail Profiles

A rail profile is made of 2D elements. The program uses the profile as the base of a sweep following the railing lines. The template **Profile-Rail.rft** is specifically created for railings. The template indicates where to start drawing the outline, as shown in Figure 7–38.

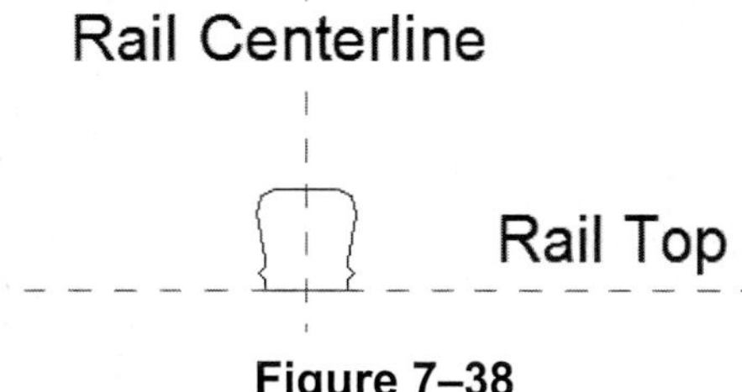

Figure 7–38

- Rail profiles must be closed shapes.

How To: Create a Top Rail or Handrail Type

1. In the Project Browser, expand **Families>Railings>Handrail Type** or **Top Rail Type** (as shown in Figure 7–39) and select one of the existing types.

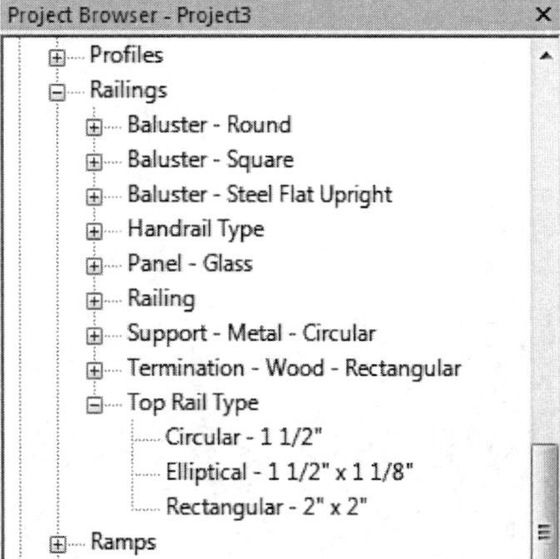

Figure 7–39

2. Right-click and select **Type Properties** or double-click on the name.
3. Duplicate and name a new type.

4. In the Type Properties dialog box, select a profile and set up the other **Construction** and **Material** parameters as shown in Figure 7–40.

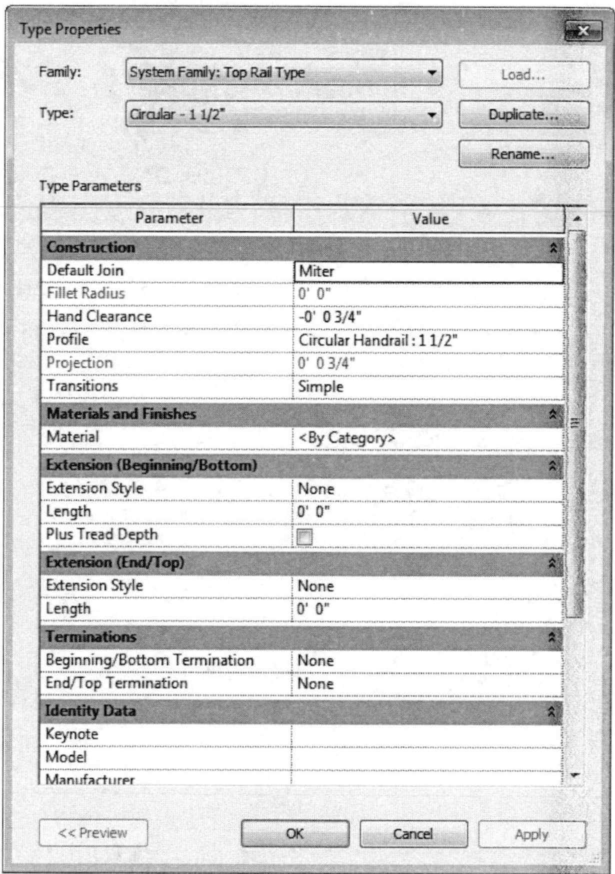

Figure 7–40

5. In the *Extension (Beginning/Bottom)* area, set the *Extension Style*. The options are **None**, **Wall**, **Floor**, and **Post** as shown in Figure 7–41.

Wall

Floor

Post

Figure 7–41

6. Set the *Length* and select **Plus Tread Depth** if required by the local codes.
7. Click **Apply** to check the addition.
8. Repeat the process for *Extension (End/Top)*.
9. A Termination can be added if needed. The default rectangular termination works best with the Floor Extension Style, but you can also create custom ones.
10. For Handrail types you can also specify *Supports*, as shown in Figure 7–42. These are typically used if you are creating a wall based railing type.

Supports	
Family	Support - Metal - Circular
Layout	Align With Posts
Spacing	None
Justification	Fixed Distance
Number	Align With Posts
	Fixed Number
Identity Data	Maximum Spacing
Keynote	Minimum Spacing

Figure 7–42

11. Make any other changes and click **OK** to finish.

- If you want to modify a continuous railing type after it has been added to the project, in a 3D view, hover the cursor over the top rail or handrail. Press <Tab> until top rail or handrail is highlighted and then select it as shown in Figure 7–43

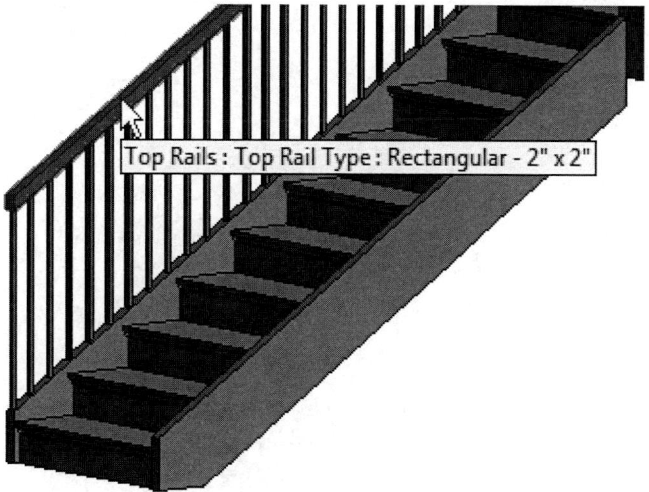

Figure 7–43

Creating Baluster, Post, and Panel Families

Balusters and posts are created from solid elements made with extrusions and/or revolves. Having the correct reference planes is important for flexibility. Standard balusters also require void elements to cut the angle on stair railings. There are three baluster templates: **Baluster.rft** (for standard balusters), **Baluster-Post.rft**, and **Baluster-Panel.rft**.

- The example in Figure 7–44 shows a custom baluster modified from an existing baluster family supplied with the Autodesk Revit software.

When you are creating a family, verify that you can modify an existing one to suit your needs. This is much faster than creating one from scratch and helps you learn how to construct the specific family.

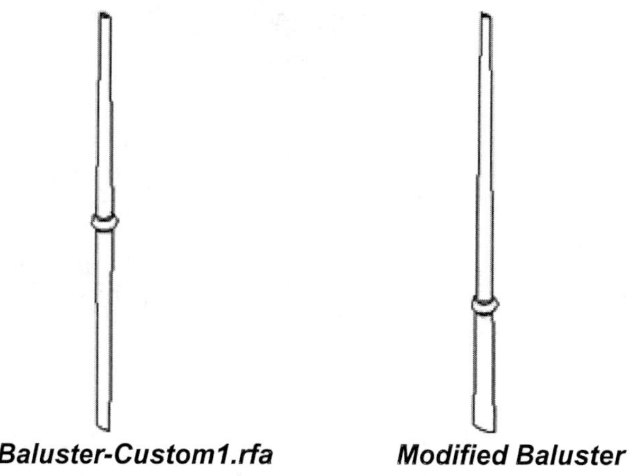

Figure 7–44

The templates for these families include reference planes for the slope angles, as shown for a baluster panel in Figure 7–45. Lock the solid elements to these angles for stair railings so that they respond to the slope of the stair.

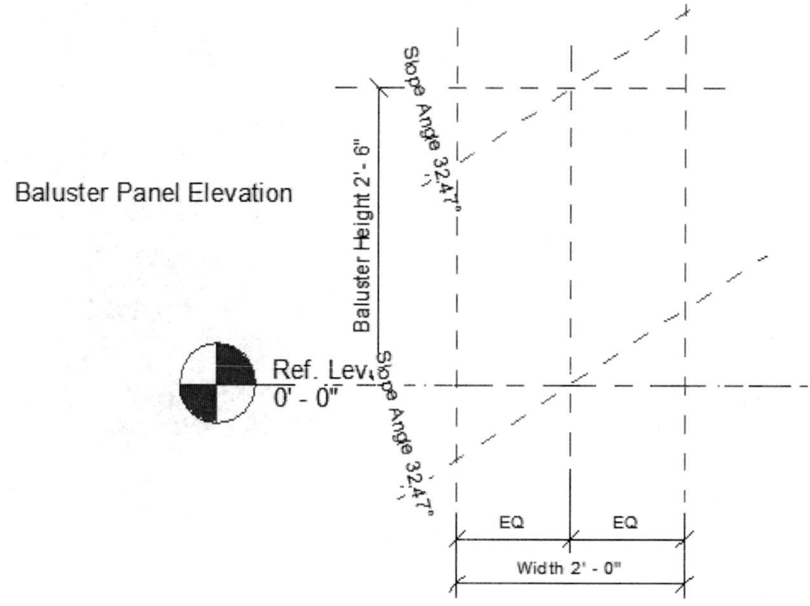

Figure 7–45

Adding Custom Posts

Posts are the point at which railings come together. In many cases, this is the area in which railings need to change pitch and extend into adjacent railings, creating an integral, non-interrupted system. Designing a post the first time is not easy. It is critical to have a good grasp of creating custom families and to want to experiment. An example is shown in Figure 7–46.

Figure 7–46

Post Categories vs. Generic Models

If you create a post as an actual post family, you cannot insert it as an independent component in the model. However, if you create a post as a generic model, you cannot use it in the Baluster dialog box and add it to a rail system. You should create the post family as an actual post and then save a copy as a generic model (recommended). This gives you the option of using the post in any situation.

Hint: Additional Cleanup Options

When railings are not cleaning up as expected, verify that the Stair Properties are correct. For example, the *Rail Connections* can be set to **Trim**, as shown in Figure 7–47, or **Weld**. The **Weld** option forces a miter joint rather than a butt joint.

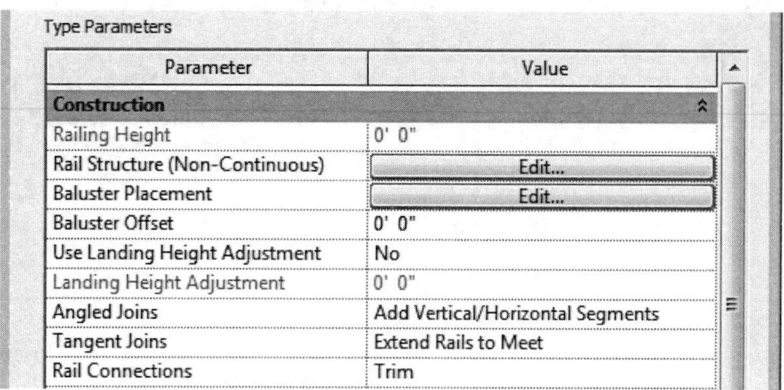
Type Parameters

Parameter	Value
Construction	
Railing Height	0' 0"
Rail Structure (Non-Continuous)	Edit...
Baluster Placement	Edit...
Baluster Offset	0' 0"
Use Landing Height Adjustment	No
Landing Height Adjustment	0' 0"
Angled Joins	Add Vertical/Horizontal Segments
Tangent Joins	Extend Rails to Meet
Rail Connections	Trim

Figure 7–47

When a straight rail joins an arc rail, the Autodesk Revit software tends to distort the geometry. Select the segment of a rail in the railing's Sketch mode and modify the *Height Correction* in the Options Bar, as shown in Figure 7–48.

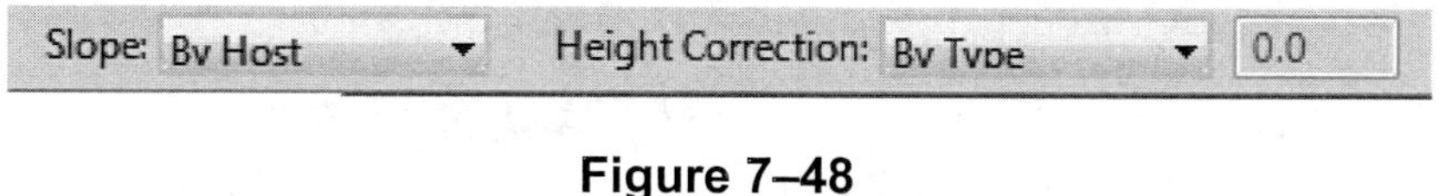

Figure 7–48

Practice 7d Create Custom Railings

Practice Objectives

- Duplicate and modify existing baluster and baluster panel families.
- Create a custom railing type using the baluster and baluster panel families.

Estimated time for completion: 25 minutes

In this practice you will make copies of existing baluster and baluster panel families and modify them slightly. You will also create new Top Rail and Handrail types. You will then use the various families and types to create a new Railing type, as shown in Figure 7–49.

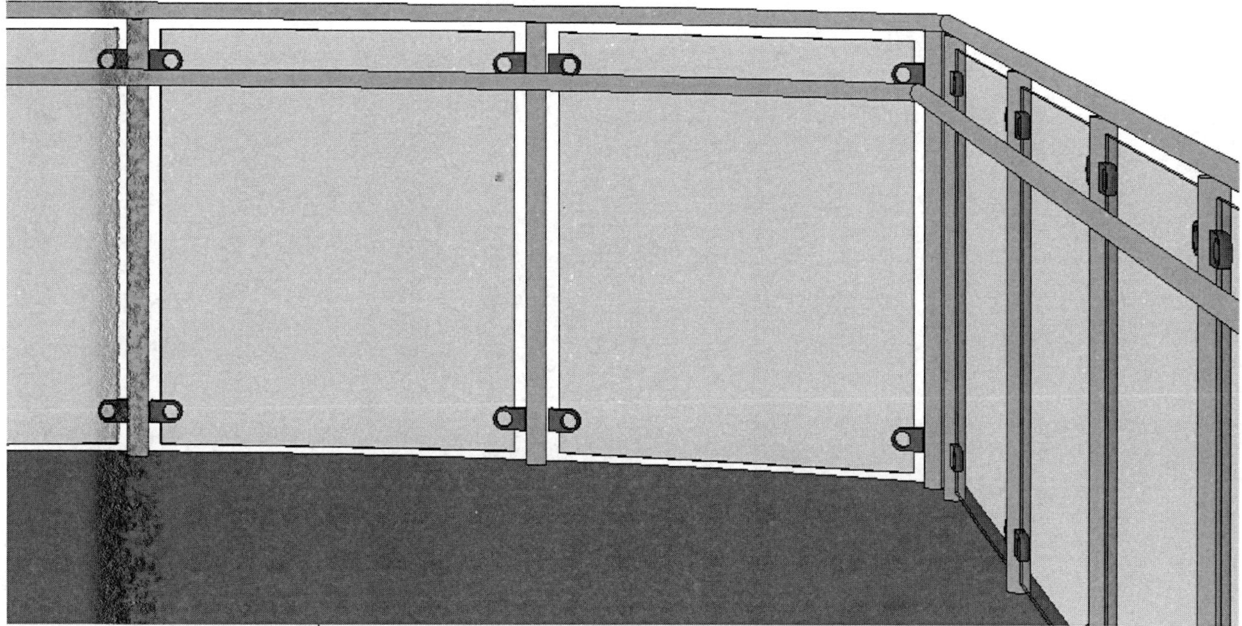

Figure 7–49

Task 1 - Set the material of a baluster family.

1. In the ...*Architectural* folder, open **Modern-Hotel-Railings-A.rvt**.
2. Open the 3D Views>**Hall Balcony** view. This is the railing style you are going to use to create a new railing style.
3. Switch to the Floor Plans>**Floor 2** view. Families cannot be opened when you are in a 3D camera view.
4. In the Application Menu, expand (Open) and click (Family).

5. In the Library, expand the *Railings>Balusters* folder and open **Baluster-Round.rfa**.
   - The family is also available in the practice files ...*Architectural* folder.
6. Save the file as **Baluster-Round-Wood.rfa** in your practice files folder.
7. In the Baluster family file, in the *Manage* tab>Settings panel, click (Materials). A short list of materials is available in the family and no wood materials are available.
8. In the Material Browser, expand Autodesk Materials and select **Wood** as shown in Figure 7–50.

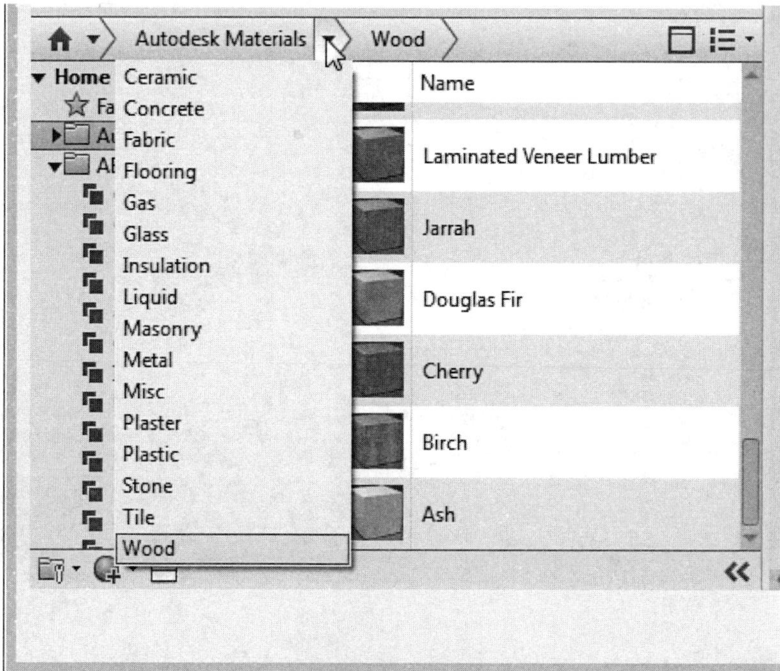

Figure 7–50

9. In the selection set next to **Birch**, select **Adds material to document** as shown in Figure 7–51.

Figure 7–51

10. In the *Project Materials* area, select the **Birch** material and in the *Graphics* tab, select **Use Render Appearance**, as shown in Figure 7–52.

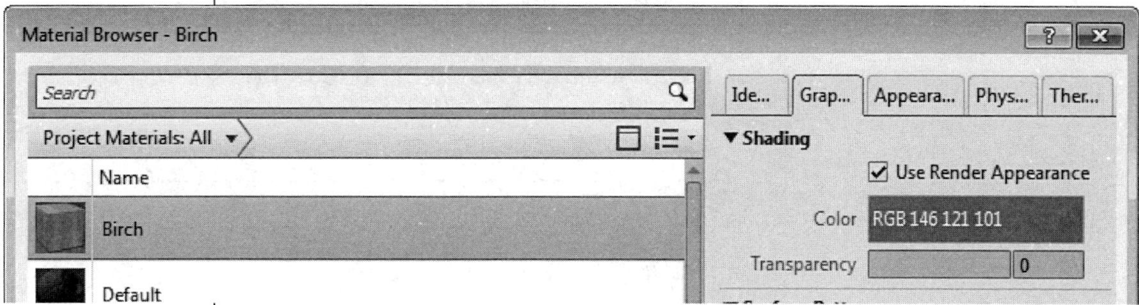

Figure 7–52

11. Click **OK**.

12. In the *Create* tab>Properties panel, click (Family Types).

13. In the Family Types dialog box, change the current type to **2"**.

14. In the *Family Types* area, click **New**... and name the new type **2" - Birch** and click **OK**.

15. Set the *Baluster Material* to **Birch**, as shown in Figure 7–53.

The material for balusters cannot be set in the railing style, only in the family.

Family Types

Name: 2" - Birch

Parameter	Value	Formula	Lock
Construction			
Post	☐	=	
Materials and Finishes			
Baluster Material	Birch	=	
Dimensions			
Top Cut Angle (defa	32.470°	=	☐
Slope Angle (default)	0.000°	=	☐
Diameter	0' 2"	=	☑
Bottom Cut Angle (d	32.470°	=	☐
Baluster Height (defa	2' 6"	=	☐

Family Types: New... Rename... Delete

Parameters: Add... Modify...

Figure 7–53

16. Click **OK**.

17. In the View Control Bar, set the *Visual Style* to (Consistent Colors) and then to (Realistic) to display the new wood shade.

18. Save the baluster family.

19. In the *Create* tab>Family Editor panel, click (Load into Project and Close).

20. If the Modern Hotel project is the only one you have open, the family automatically loads and switches to the project file. If you have others open, select the Modern Hotel project from the list and click **OK**.

Task 2 - Modify a baluster panel family.

1. In the Revit Library's *...\Railings\Balusters* sub-folder, open the family file **Baluster Panel - Glass w Brackets.rfa.**
    - The family is also available in the practice files*...\Architectural* folder.
2. Save it as **Baluster Panel - Glass with Open Brackets.rfa** in your practice files folder.
3. Open the **Elevations>Left** view.
4. Zoom in on one of the brackets and select it.
5. In the *Modify | Extrusion* tab>Mode panel, click (Edit Extrusion).
6. Sketch a **3/4"** radius circle inside the bracket, as shown in Figure 7–54. This creates a hole in the extrusion.

*To snap to the center of the arc, type the shortcut **SC** for Snap to Center.*

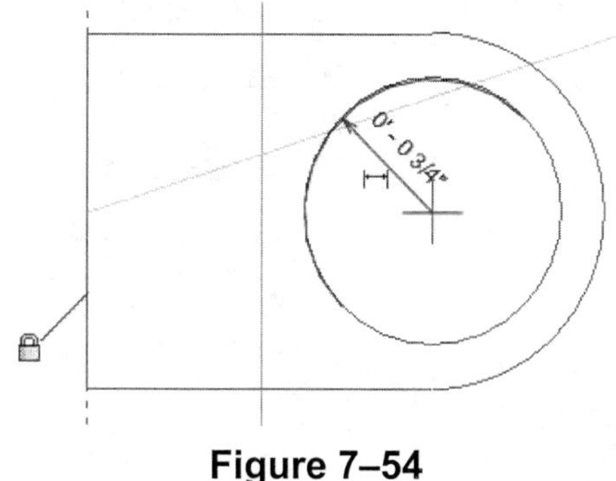

Figure 7–54

7. Click (Finish Edit Mode).
8. Repeat the process with the other three brackets.

9. Open the 3D Views>**View 1** view and set the *Visual Style* to (Consistent Colors). The panel displays with the opening in the brackets, as shown in Figure 7–55.

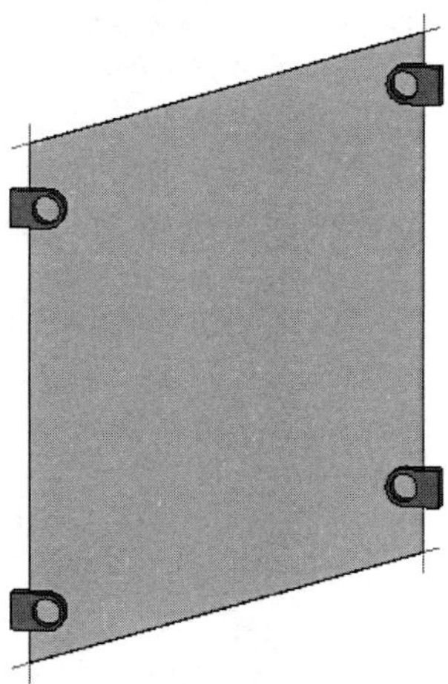

Figure 7–55

10. In the *Modify* tab>Properties panel, click (Family Types).
11. In the Family Types dialog box, in the *Family Types* area, click **New...**.
12. Name the new type **36" w 1" Gap**.
13. In the Family Types dialog box, change the *Width* to **3'-0"**.
14. Click **Apply**. The panel updates, as shown in Figure 7–56. Click **OK**.

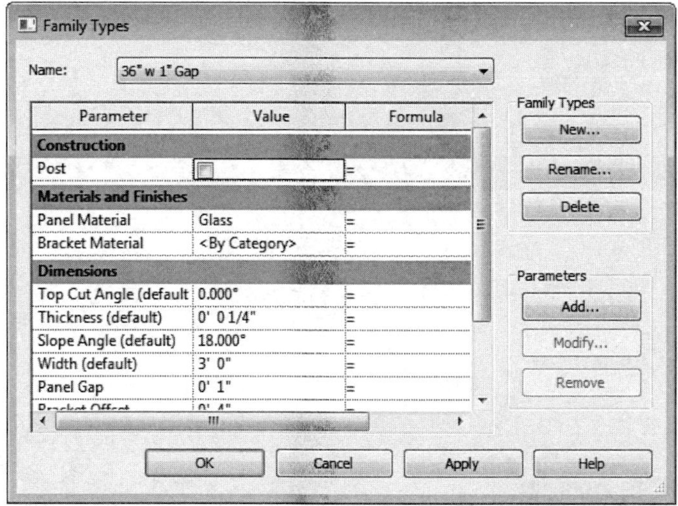

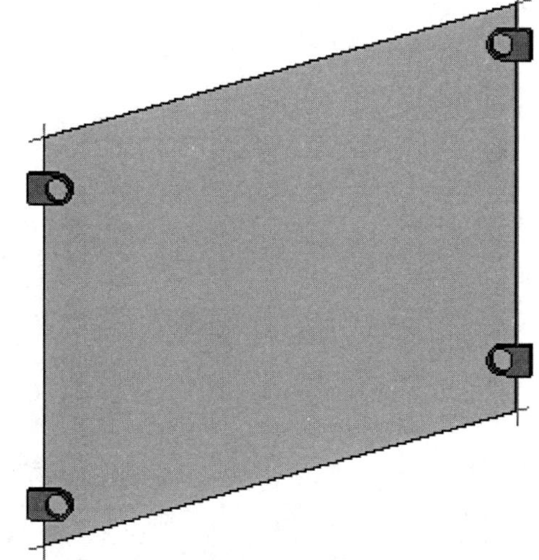

Figure 7–56

15. Save the baluster panel and load it into the Modern Hotel project.

Task 3 - Create Top Rail and Handrail types.

1. In the Modern Hotel project, in the Project Browser, expand **Families>Railings>Top Rail Type** node.
2. Right-click on **Circular - 1 1/2"** and select **Type Properties**.
3. In the Type Properties dialog box, duplicate the type, and name it **Circular - 1 1/2" - Birch**.
4. Change the *Material* to **Birch** and set the *Extension (Beginning/Bottom)* and *Extension (End/Top)* to **Wall**, as shown in Figure 7–57.

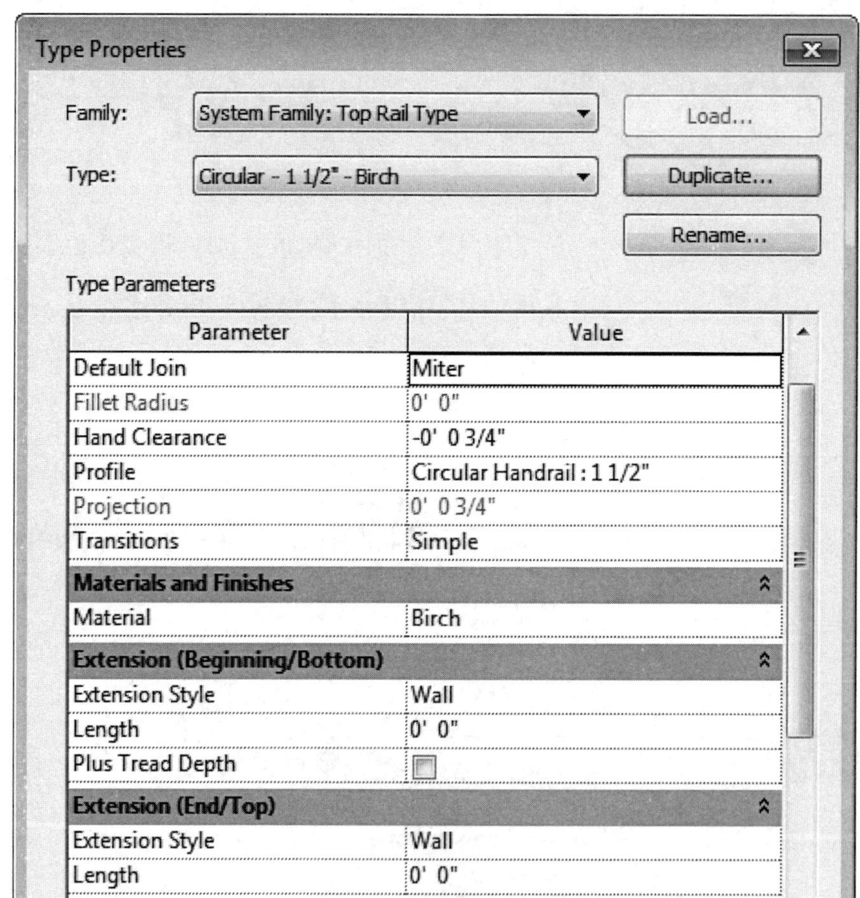

Figure 7–57

5. Click **OK**.
6. Repeat the process to create a new Birch Handrail type.
7. Save the project.

Task 4 - Create custom Railing types.

1. In the Modern Hotel project, open the 3D Views>**Hall Balcony** view.
2. Select the railing.
3. In Properties, click (Edit Type).
4. In the Type properties dialog box, click **Duplicate** and name the new railing type **Hotel Balcony Guardrail - Glass Panel**.
5. Next to the **Rail Structure (Non-Continuous)** parameter, click **Edit...**.
6. In the Edit Rails dialog box, delete all of the Rails and click **OK**.
7. Click **OK**. The new railing displays with wooden guardrail and handrail and existing posts, as shown in Figure 7–58.

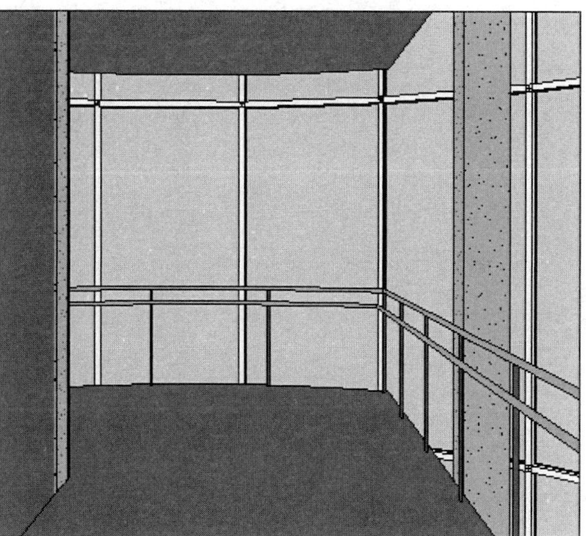

Figure 7–58

8. Select the railing and edit the Railing type again.
9. Next to the **Baluster Placement** parameter, click **Edit...**.

10. In the *Main pattern* area, next to *Regular Baluster* (in the *Baluster Family* column), change the type to **Baluster-Round-Wood: 2"-Birch**, as shown in Figure 7–59.

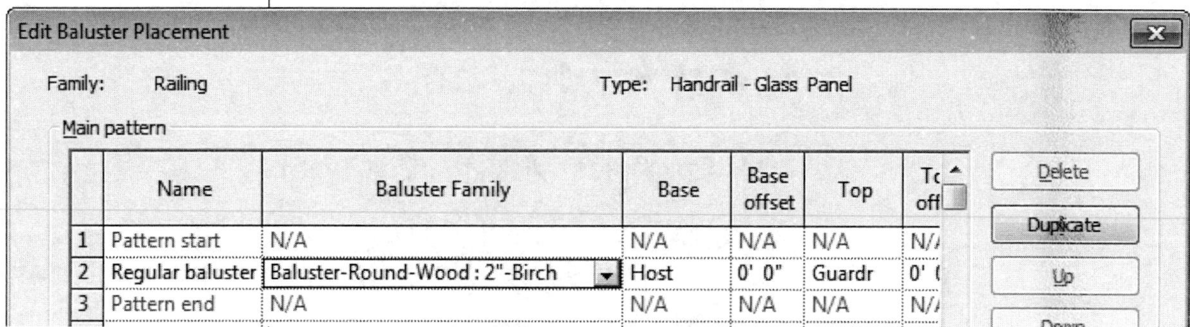

Figure 7–59

11. Click **Duplicate**.

12. In row *2*, change the *Name* to **Panel** and the *Baluster Family* to **Baluster Panel-Glass w Open Brackets: 36" w 1" Gap**.

13. Click **OK** twice and view the railings. The panel displays, but it is not in the correct location in relation to the posts, as shown in Figure 7–60.

Figure 7–60

Distances between balusters is measured from center to center.

14. Edit the type again and open the Edit Baluster Placement dialog box. In the *Dist. from Previous* column, you can see that the settings are now **4'-0"** for both the baluster and the panel. Therefore, the baluster and panel placement is spread out.

15. Set the *Dist. from Previous* to **1'-7 1/8"** for both the baluster and panel.

16. Set *Justify* to **Beginning** and *Excess Length Fill* to **None**, as shown in Figure 7–61.

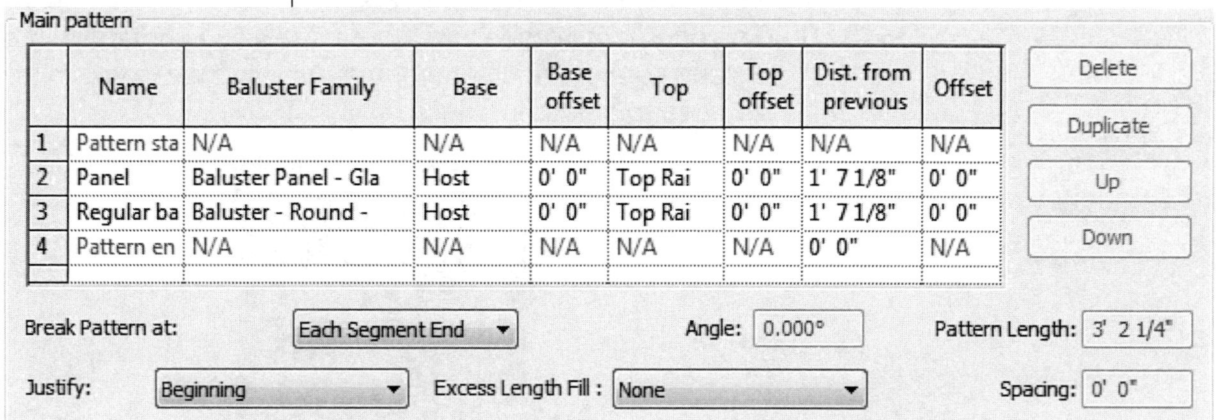

	Name	Baluster Family	Base	Base offset	Top	Top offset	Dist. from previous	Offset
1	Pattern sta	N/A	N/A	N/A	N/A	N/A	N/A	N/A
2	Panel	Baluster Panel - Gla	Host	0' 0"	Top Rai	0' 0"	1' 7 1/8"	0' 0"
3	Regular ba	Baluster - Round -	Host	0' 0"	Top Rai	0' 0"	1' 7 1/8"	0' 0"
4	Pattern en	N/A	N/A	N/A	N/A	N/A	0' 0"	N/A

Figure 7–61

17. While still in the Edit Baluster Placement dialog box, in the *Posts* area, change the *Baluster Family* for all of the posts to **Baluster - Round - Wood: 2" - Birch** and click **OK**.

18. In the Type Properties dialog box, change the *Top Rail Type* to the new **Circular - 1 1/2" - Birch** top rail type and *Handrail 1* to the new **Circular - 1 1/2" - Birch** handrail type.

19. Close the dialog boxes and view the new Railing type. This time, the baluster to panel fit is better, as shown in Figure 7–62.

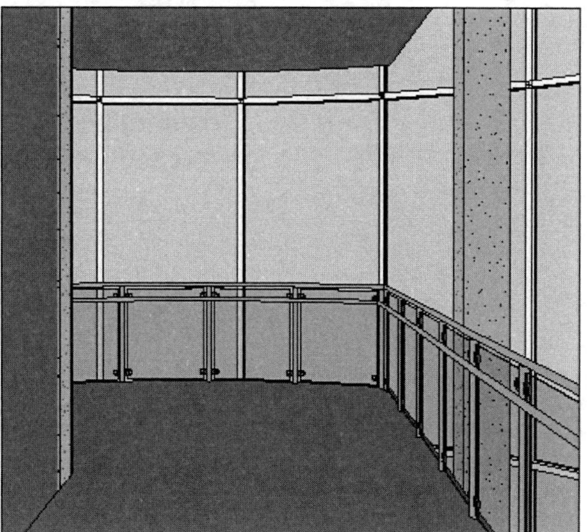

Figure 7–62

20. Save the project.

Chapter Review Questions

1. The door family shown in Figure 7–63 does not exist in the default library. Which one of the following methods should you use to create the door family?

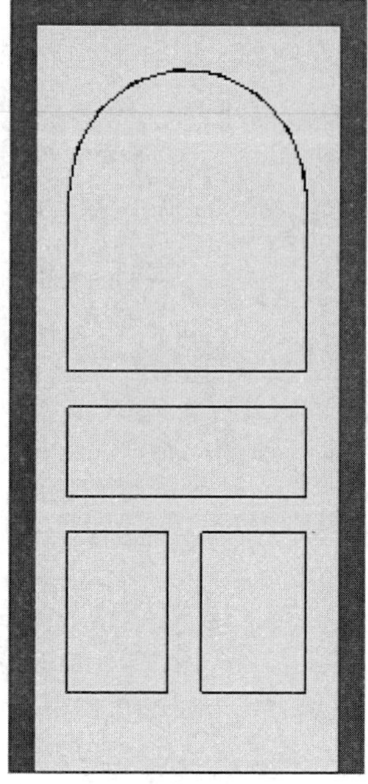

Figure 7–63

   a. Start a new door family based on the default door template and build it from there.
   b. Copy the existing **Single-Flush.rfa** family as the base and draw the panel outlines using detail lines.
   c. Copy an existing panel-style family and modify the panel extrusions.
   d. Start a new door family based on the **Single-Panel.rfa** template and modify the panel extrusions.

2. Based on the information in the Family Types dialog box (shown in Figure 7–64), if you changed the Trim Projection Ext value, would the Trim Projection Int also update?

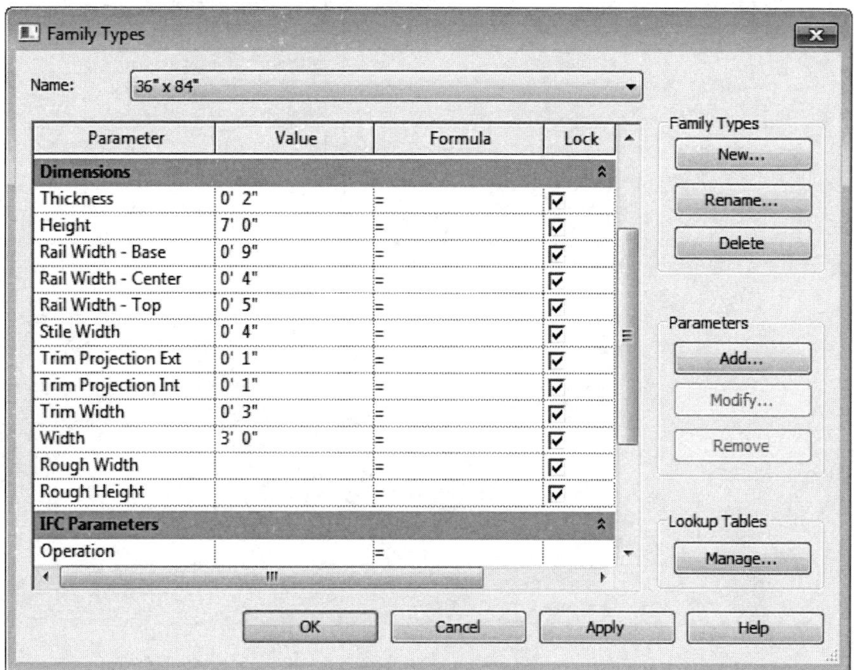

Family Types

Name: 36" x 84"

Parameter	Value	Formula	Lock
Dimensions			
Thickness	0' 2"	=	☑
Height	7' 0"	=	☑
Rail Width - Base	0' 9"	=	☑
Rail Width - Center	0' 4"	=	☑
Rail Width - Top	0' 5"	=	☑
Stile Width	0' 4"	=	☑
Trim Projection Ext	0' 1"	=	☑
Trim Projection Int	0' 1"	=	☑
Trim Width	0' 3"	=	☑
Width	3' 0"	=	☑
Rough Width		=	☑
Rough Height		=	☑
IFC Parameters			
Operation		=	

Figure 7–64

a. Yes

b. No

3. Which of the following must be in place before you can add a fascia to an angled wall? (Select all that apply.)

a. A profile.

b. A fascia type.

c. A wall sweep.

d. A model line.

4. Which of the following parts of a railing type (shown in Figure 7–65) is created by a profile?

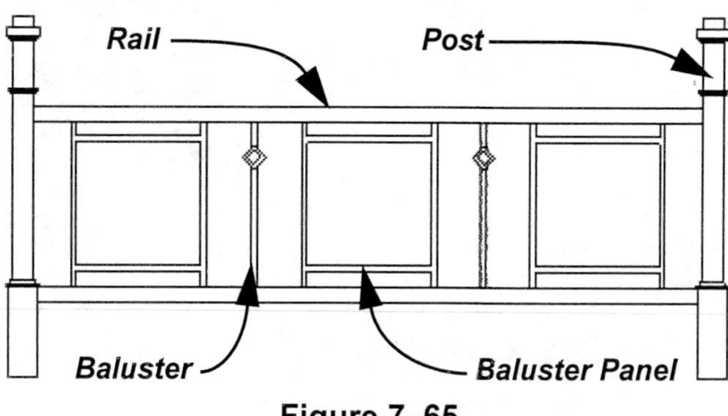

Figure 7–65

   a. Rail
   b. Post
   c. Baluster
   d. Baluster panel

5. What is the main difference between a Baluster family and a Post Family?
   a. A Baluster is created using a profile and a Post is created using solid forms.
   b. A Baluster has top and bottom angles programmed into the family, while a Post does not have any angles.
   c. Balusters can be placed on stair treads, but posts can only be placed on landings.
   d. Balusters are typically placed 4" apart, and Posts are placed 3'-0" apart.

Command Summary

Button	Command	Location
	Roof: Fascia	• **Ribbon:** *Architecture* tab>Build panel
	Object Styles	• **Ribbon:** *Manage* tab>Settings panel
	Model Line	• **Ribbon:** *Architecture* tab>Model panel

Creating MEP Specific Families

MEP families include elements such as plumbing fixtures, light fixtures, data devices, and pipe fittings. By working through practices, you can learn how to upgrade an architectural plumbing fixture to use connectors, create a custom light fixture, and create a data device with annotation options. For an advanced challenge you can also create a pipe fitting flange. Select the practices that you find most helpful and relevant to your work.

Learning Objectives in this Chapter

- Upgrade an architectural plumbing fixture with MEP connectors.
- Create a custom lighting fixture.
- Create a data device family that includes a tag within the family.
- Create a pipe fitting flange.

Practice 8a Upgrade an Architectural Plumbing Fixture to MEP

Practice Objective

- Upgrade an existing architectural toilet fixture with pipe sizes and connectors for use in a MEP project.

Estimated time for completion: 15 minutes

In this practice you will upgrade an existing Autodesk® Revit® Architectural toilet fixture to be used in the Autodesk Revit MEP software. You will create parameters for pipe sizes. You will also add connectors for cold water and Sanitary systems, as shown in Figure 8–1.

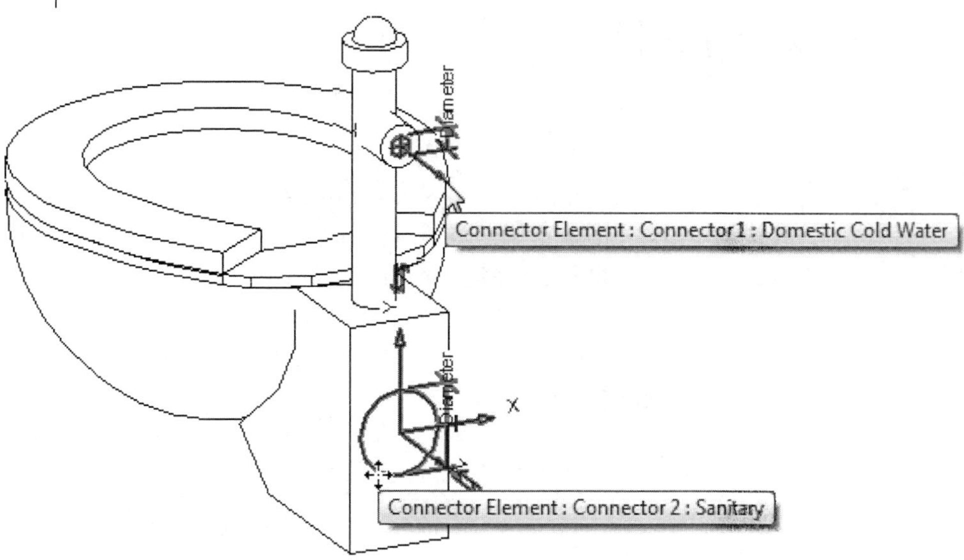

Figure 8–1

Task 1 - Create parameters for pipe sizes.

1. In the ...*MEP* folder, open **Toilet-Commercial-Wall-3D.rfa**.
2. Save the family as **Toilet-Commercial-Wall-3D-MEP.rfa**.
3. In the Family Types dialog box, add the following Type parameters:

Name	Discipline	Type of Parameter	Group
Sanitary Radius	Piping	Pipe Size	Dimensions
Cold Water Radius	Piping	Pipe Size	Dimensions

The related diameter parameters need to be Shared because they are often used in plumbing fixture schedules.

4. Set *Sanitary Radius* to **2"** and *Cold Water Radius* to **1/2"**. **Lock** the parameters.
5. Add shared type parameters selected from the **MEP.txt** parameter file:
   - Sanitary Diameter
   - Cold Water Diameter
6. Create formulas for *Sanitary Diameter* and *Cold Water Diameter* based on the corresponding radius parameters, as shown in Figure 8–2.

Dimensions			
Sanitary Radius	2"	=	☑
Sanitary Diameter	4"	= Sanitary Radius * 2	☑
Cold Water Radius	1/2"	=	☑
Cold Water Diameter	1"	= Cold Water Radius * 2	☑
Mounting Height	1' 3"	=	☑

Figure 8–2

7. Click **OK**.
8. Save the family.

Task 2 - Add connectors.

1. Rotate the view so that the back of the toilet is displayed. (If the wall displays, hide it temporarily or set the *Visual Style* to **Wireframe** while you are adding connectors.)
2. In the *Create* tab>Connectors panel, click (Pipe Connector).
3. In the Options Bar, select **Domestic Cold Water**.
4. Hover over the face of the pipe opening, as shown in Figure 8–3 and click to place the connector.

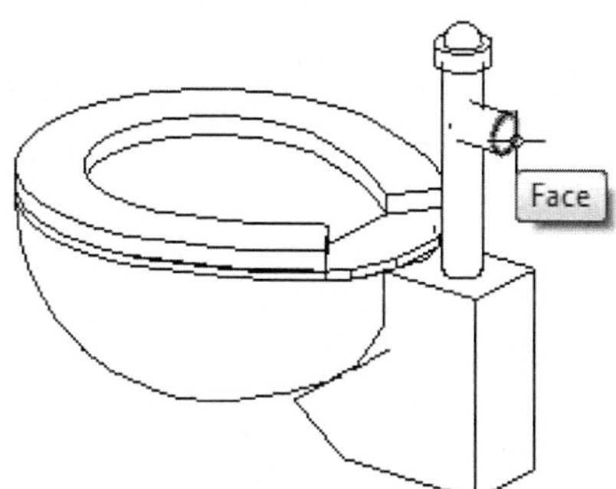

Figure 8–3

5. Click (Modify) and select the Connector.
6. In Properties, set the following values for parameters:
    - **Flow Configuration:** Fixture Units
    - **Flow Direction:** In
    - **Allow Slope Adjustments:** clear
    - **Fixture Units:** 2.5
7. Scroll down to the *Dimensions* area, and next to *Diameter*, click (Associate Parameter).
8. In the Associate Family Parameter dialog box, select **Cold Water Diameter** (as shown in Figure 8–4) and then click **OK**.

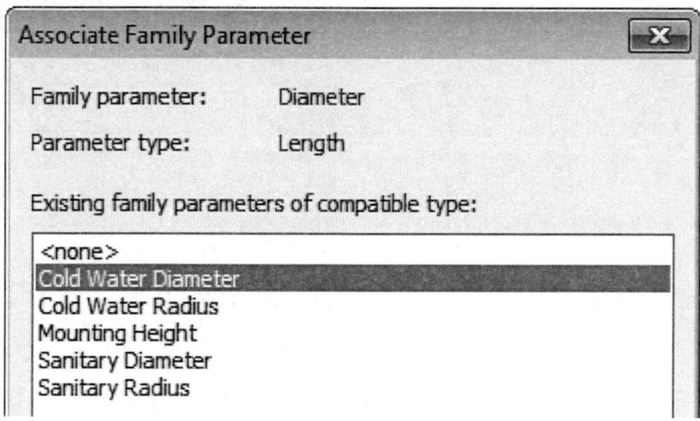

Figure 8–4

9. The connector resizes to fit the correct pipe size.
10. Add a Pipe Connector for a Sanitary system to the back of the fixture, as shown in Figure 8–5.

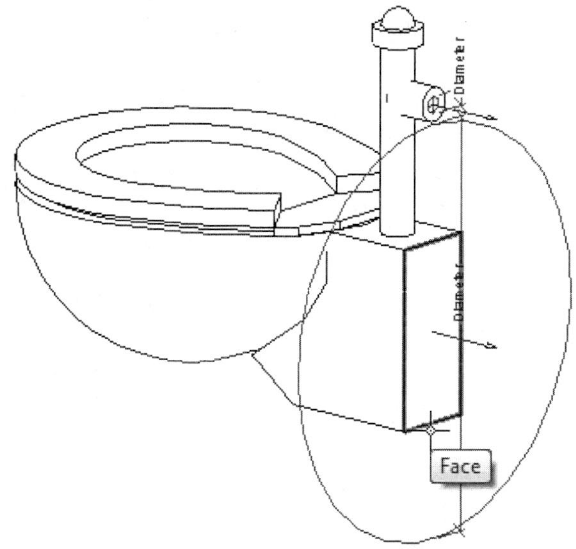

Figure 8–5

11. Select the connector and set the following properties.
    - **Flow Configuration:** Fixture Units
    - **Flow Direction:** Out
    - **Allow Slope Adjustments:** select
    - **System Classification:** Sanitary
    - **Fixture Units:** 4
    - **Diameter:** Associate with **Sanitary Diamete**r parameter.
12. Add any other parameters or values required to make this a more useful MEP fixture.
13. Save the family.
14. Test the fixture in a project.

Practice 8b Create a Custom Lighting Fixture Family

Practice Objectives

- Modify the light source.
- Add elements for pendant and glass shade.
- Apply a power connector.
- (Optional) Create family types with different materials for the pendant and glass shade.

Estimated time for completion: 20 minutes

In this practice you will create a new lighting fixture family, modify the light source, create geometry for a pendant and glass shade, and apply a connector, as shown in Figure 8–6. If you have time you can load materials and create family types using different materials.

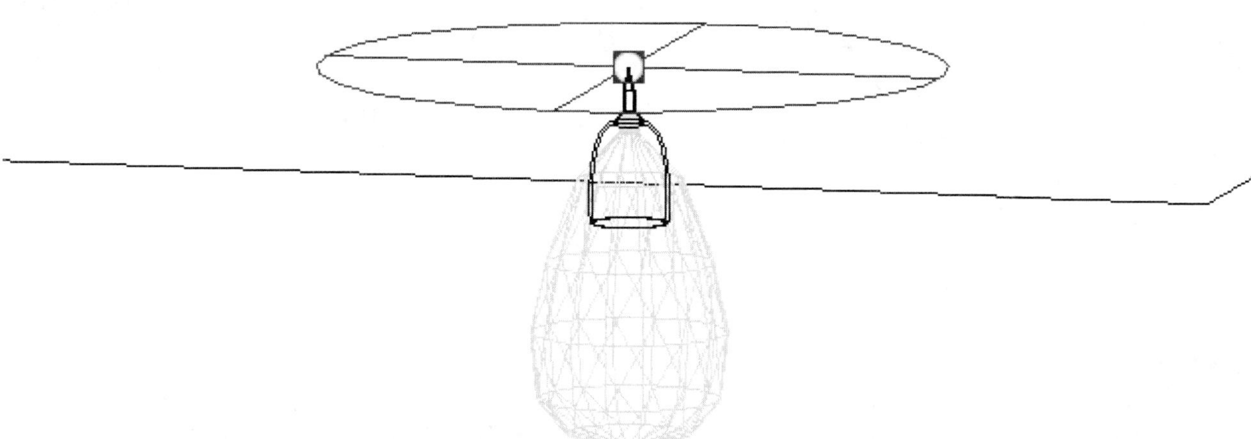

Figure 8–6

Task 1 - Set up a new light fixture family.

1. Create a new family based on the **Lighting Fixture ceiling based.rft** template found in the Autodesk Revit Library.
2. Save the family to your practice folder as **Pendant Light.rfa**.
3. Open the **Elevations: Front** view. It includes reference planes for the Ceiling and Light Source, a ceiling element, and the Ref. Level as shown in Figure 8–7.

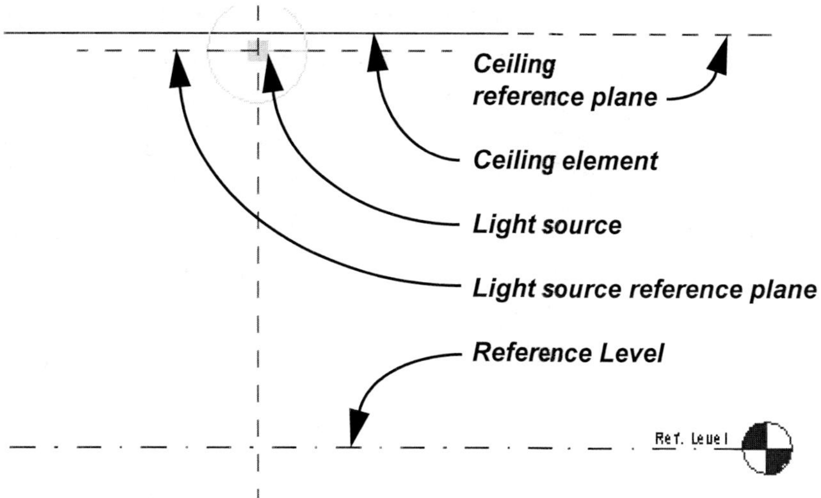

Figure 8–7

4. Select the light source. In the *Modify | Light Source* tab> Lighting panel, click (Light Source Definition).
5. In the Light Source Definition dialog box, change the Light distribution to (Photometric Web), as shown in Figure 8–8 and click **OK**.

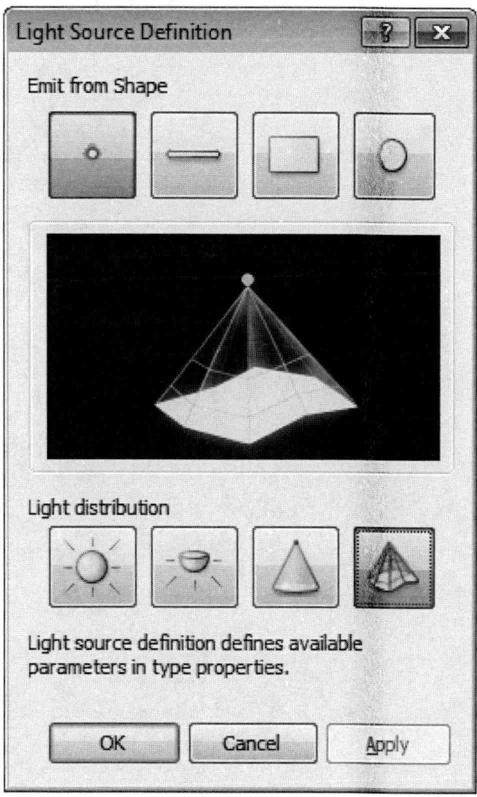

Figure 8–8

6. Open the Family Types dialog box and set the following values for the parameters:
   - **Lamp:** CF42TRT
   - **Apparent Load:** 44.21 VA
   - **Tilt Angle:** 90
   - **Photometric Web File:** Click [...] and select **spot_ideal.ies**.
   - **Light Loss Factor:** 0.85
   - **Initial Intensity:** Luminous Flux, 3200
7. In the Family Types dialog box, add two material Type parameters for **Pendant Material** and **Glass Material**, grouped under **Materials and Finishes.**
8. Save the family.

Task 2 - Create the fixture geometry.

1. Zoom in on the area near the light source.

The light source is temporarily hidden in these graphics.

2. Add reference planes as needed to help you create two revolves for a pendant and glass shade, similar to that shown in Figure 8–9.

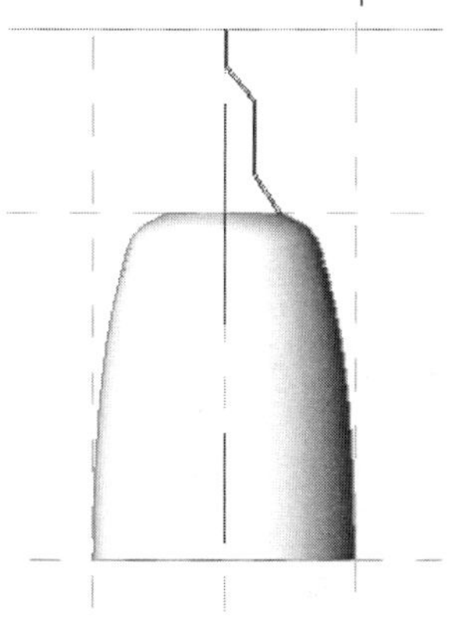

Sketch for Pendant

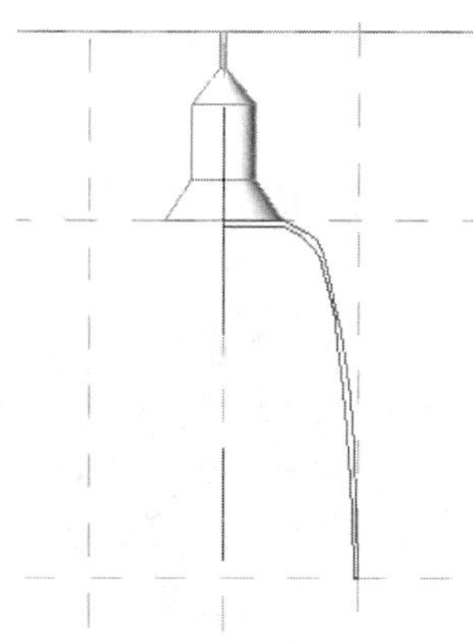

Sketch for Glass

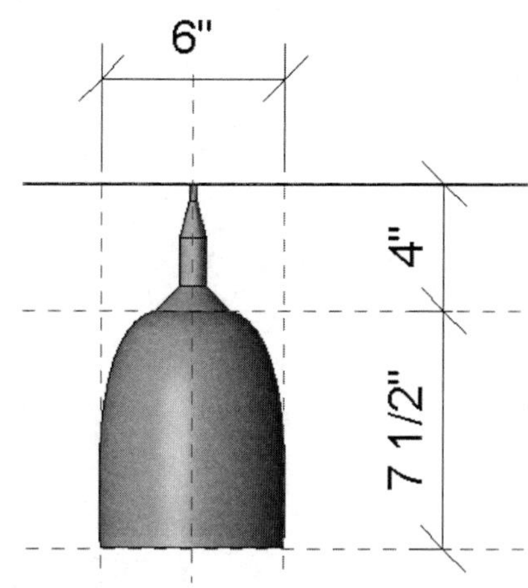

Figure 8–9

3. Select the pendant revolve. In Properties, in the *Materials and Finishes* area, for **Materials** parameter, click (Associate Family Parameter).
4. In the Associate Family Parameter dialog box, select **Pendant Material** and click **OK**.
5. Repeat this with the glass shade revolve.
6. Save the project.

Task 3 - Add a connector.

1. Open a 3D view and set the *Visual Style* to **Wireframe**.
2. In the *Create* tab>Connectors panel, click (Electrical Connector).
3. In the Options Bar, select **Power - Balanced**.
4. Attach the connector to the face of the pendant at the point where the pendant touches the ceiling, as shown in Figure 8–10.

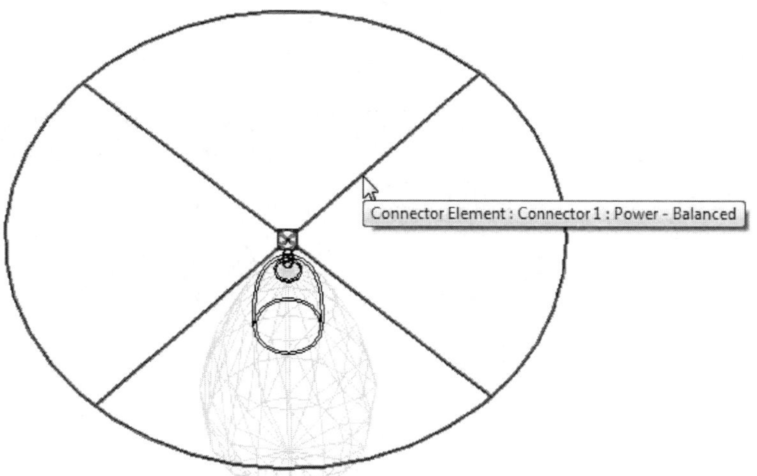

Figure 8–10

5. Click (Modify) and select the connector.
6. In Properties specify the values for the following parameters:
   - **System Type:** Power - Balanced (verify)
   - **Load Classification:** Lighting
   - **Voltage:** 277
   - **Power Factor:** 0.95
7. While still in Properties, beside the *Apparent Load* parameter, click (Associate Family Parameter).
8. In the Associate Family Parameter dialog box, select **Apparent Load**. Click **OK**.
9. Save the family.

Task 4 - (Optional) Add materials and create Family types.

1. In the *Manage* tab>Settings Panel, click (Materials). Note that in the Materials Browser there are not many materials to select from. Close the dialog box.
2. In the ...*MEP* folder open **Light-Fixture-Sample-Materials.rfa**. This is a finished version of the light fixture with several materials added.
3. Switch back to your new light fixture family.
4. In the *Manage* tab>Settings panel, click (Transfer Project Standards).
5. In the Select Items to Copy dialog box, verify that *Copy from* is set to **Light-Fixture-Sample-Materials.rfa** and clear everything but **Materials**, as shown in Figure 8–11.

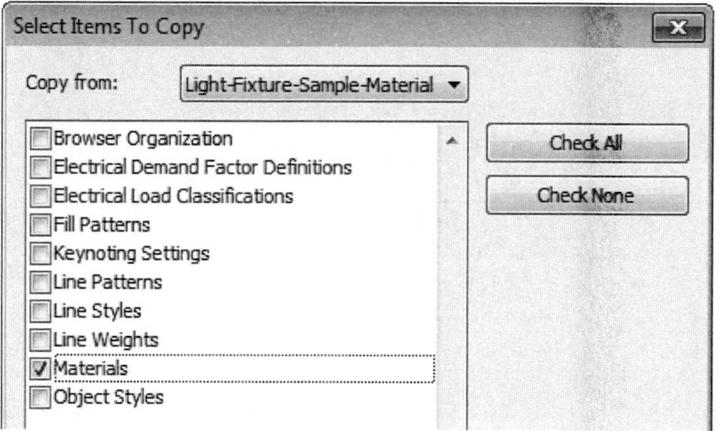

Figure 8–11

6. Click **OK**. The materials are copied into the current family.
7. In the Family Types dialog box, assign materials to the **Glass Material** and **Pendant Material** parameters.
8. If you have time, create several Family Types that use different colors of glass and different pendant materials.
9. Save the family.

Practice 8c Create an Data Device with Annotation Parameters

Practice Objective

- Create a data device with both solid elements and annotation elements.

Estimated time for completion: 20 minutes

In this practice you will create a data device using reference planes, extrusions, and connectors. You will then add an annotation symbol to the family with its related parameters, as shown in Figure 8–12. You will also test the family in a project.

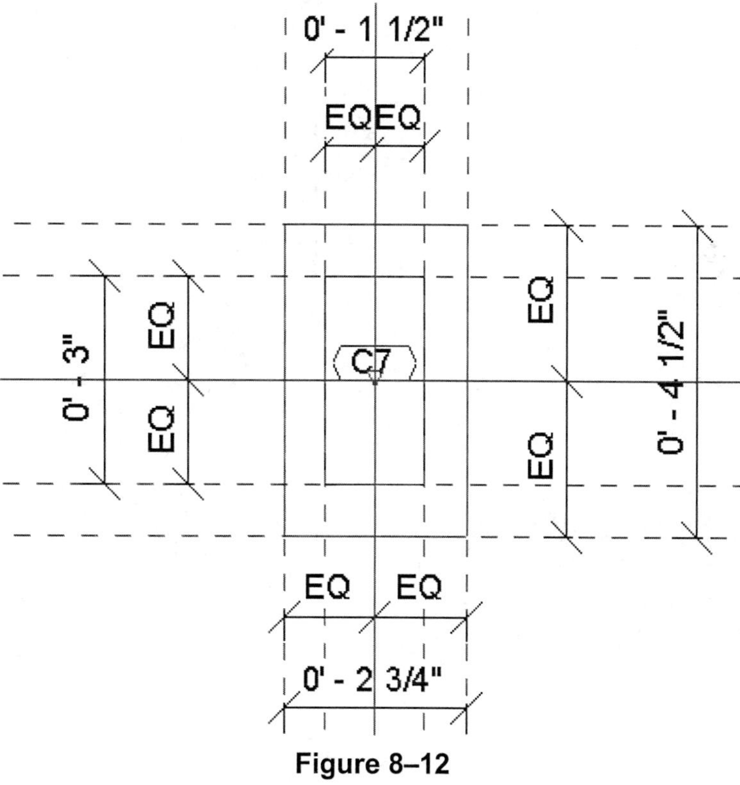

Figure 8–12

Task 1 - Set up reference planes.

1. Create a new family based on the **Data Device Hosted.rft** template found in the Autodesk Revit Library.
2. Save the family to your practice folder as **Data Receptacle Type C.rfa.**
3. In the **Floor Plans: Ref. Level** view, set the *View Scale* to **3"=1'-0"**.
4. Zoom in closer to the intersection of the reference planes.

5. Create and dimension the reference planes, as shown in Figure 8–13. You do not need to label these dimensions as all of the data receptacles are the same size.

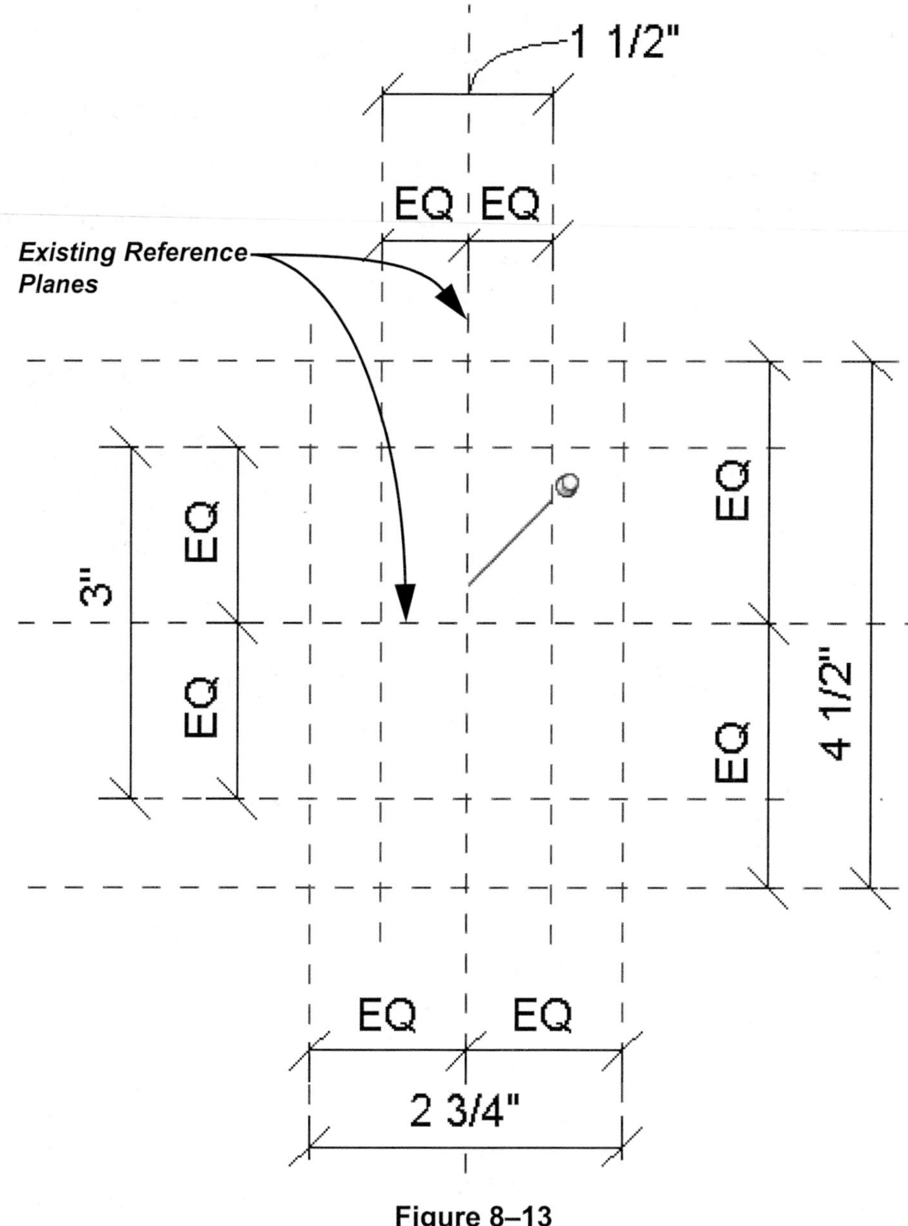

Figure 8–13

6. Save the family.

Task 2 - Create the model extrusions.

1. Create the two extrusions shown in Figure 8–14. Set the *Depth* of the extrusions as follows:
   - Extrusion 1 = (negative) **-2 1/8"**
   - Extrusion 2 = **3/16"**

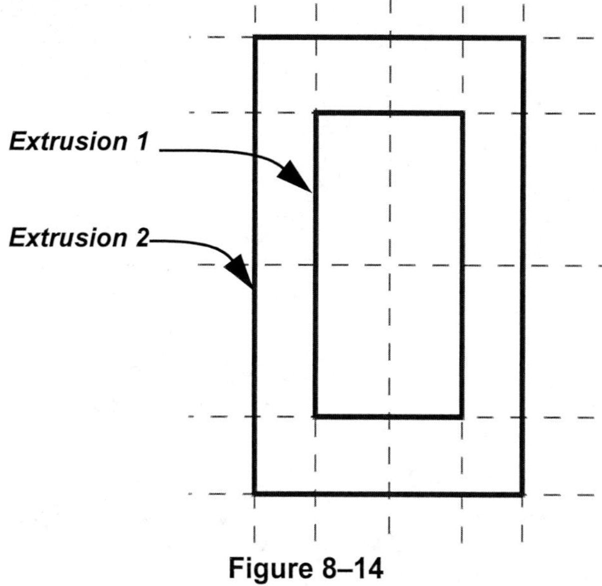

Figure 8–14

2. Open a 3D view and set the *Visual Style* to **Wireframe**. The finished view displays as shown in Figure 8–15.

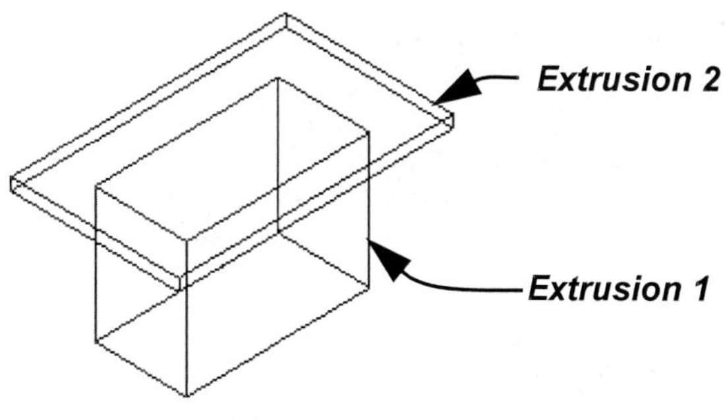

Figure 8–15

3. In the *Create* tab>Connectors panel, click (Electrical Connector).

4. In the Options Bar, select **Data**.

5. Select the face of the top extrusion.

6. Click (Modify) and select both extrusions.

7. In the *Modify | Extrusion* tab>Mode panel, click (Visibility Settings).

8. In the Family Element Visibility Settings dialog box, clear **Front/Back**, **Coarse**, and **Medium**. Click **OK**.
   - This causes the 3D elements to display only at the Fine detail level where it is often used to coordinate the placement of items by their real dimensions. At the Coarse and Medium detail levels the tag is more useful.

9. Return to the **Ref. Level** view.
10. Save the family.

Task 3 - Create an Annotation symbol.

1. Create a new family based on the **Generic Annotation.rft** template found in the *Annotations* folder of the Autodesk Revit Library.
2. Save the family to your practice folder as **Data Device Annotation.rfa.**
3. Read and delete the red note. You will not modify the Family Category in this case because you need it to remain a generic annotation.
4. Draw the sketch as shown in Figure 8–16.

Do not add the dimensions, they are for information only.

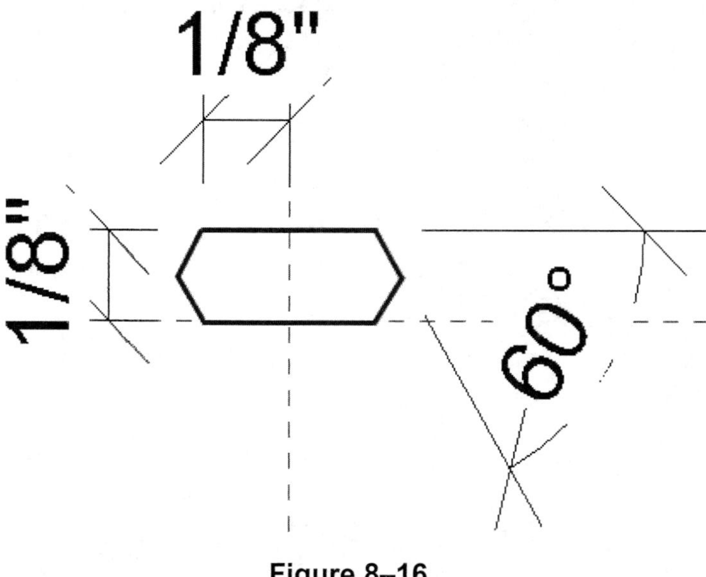

Figure 8–16

5. In the *Create* tab>Text panel, click (Label).
6. Click inside the new symbol sketch.
7. In the Edit Label dialog box there are no parameters listed. Click (Add Parameter).
8. Create an Instance parameter named **Label**, set *Type of Parameter* as **Text**, and set *Group parameter under* as **Identity Data**. Click **OK**.

9. Back in the Edit Label dialog box, add the parameter to the label and set the *Sample Value* to **A1**, as shown in Figure 8–17 and click **OK**.

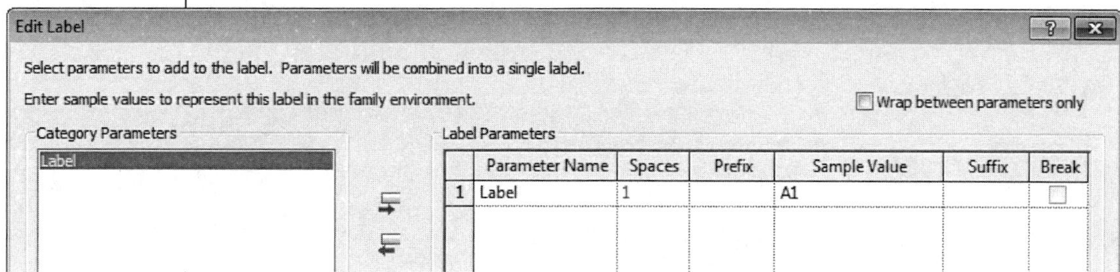

Figure 8–17

10. Select the new label and, in Properties, click (Edit Type). In the Type Properties dialog box, set the *Background* to **Transparent** and click **OK**.

11. Relocate the label to fit inside the symbol sketch.

12. With nothing selected, in Properties, select **Rotate with component** and **Keep text readable.**

13. Save the annotation family and load it into the data receptacle family.

Task 4 - Add the Label Family to the Data Receptacle.

1. In the data receptacle family, place a copy of the Data Device Annotation family at the center of the elements, as shown in Figure 8–18.

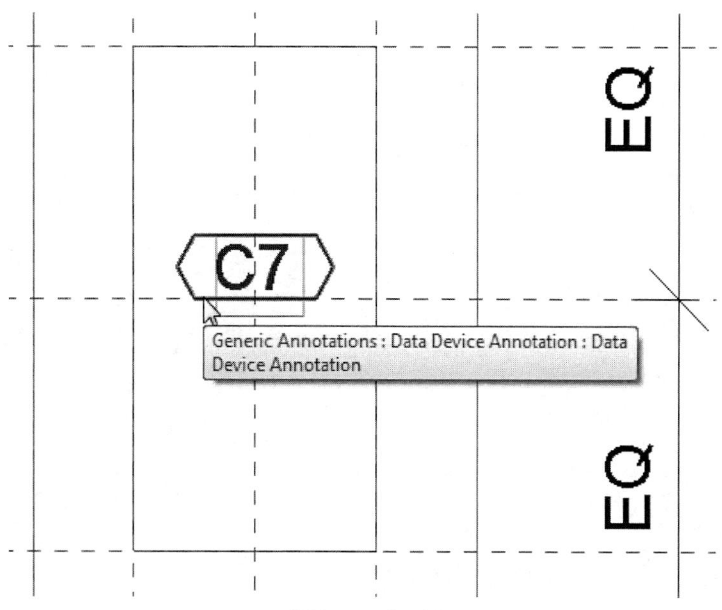

Figure 8–18

2. Open the Family Types dialog box and create an Instance parameter named **Label**. Set *Type of Parameter* as **Text** and *Group parameter under* as **Identity Data**. Click **OK**.

3. Expand the *Identity Data* heading (if required) and set *Label* to **C7**, as shown in Figure 8–19.

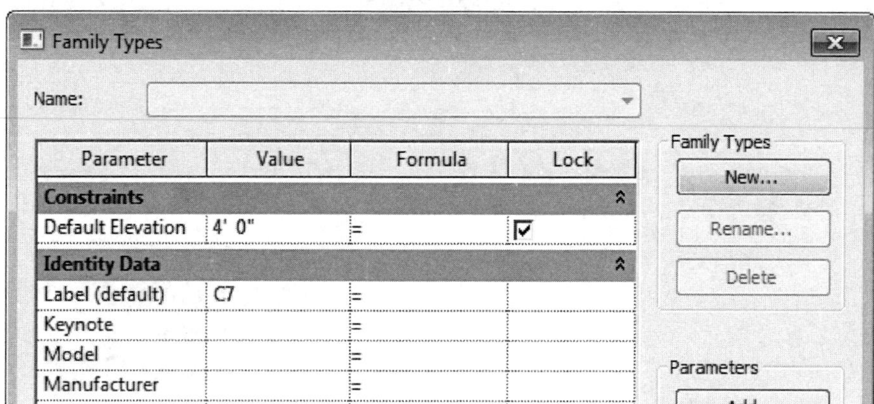

Figure 8–19

4. Close the dialog box.

5. Select the Data Device Annotation element.

6. In Properties, under the *Identity Data* heading, beside the *Label* parameter, click (Associate Family Parameter).

7. In the Associate Family Parameter dialog box, select **Label** and then click **OK**.

8. When you finish, the parameters are associated together and the label displays as C7, as shown in Figure 8–20.

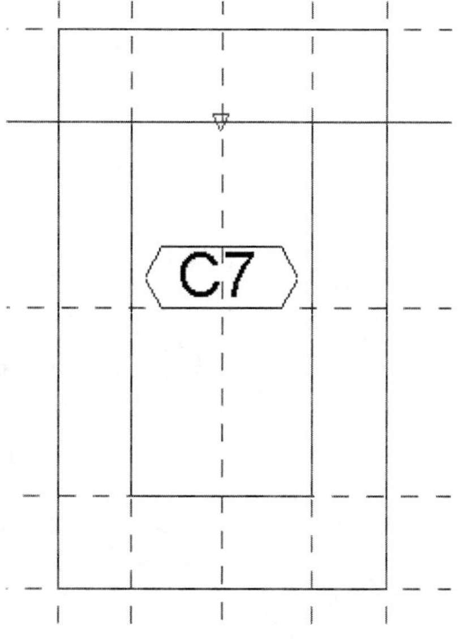

Figure 8–20

9. If you have time you can create Family Types for C1, C2, C3, C4, C5, C6, and C7. Change the value of the **Label** parameter for each of these as required.

10. Save the family.

Task 5 - Test the family.

1. Create a new project using the **Electrical-Default.rte** or **Systems-Default.rte** template.
2. Open the **Floor Plans: 1- Power** view and close the **1-Mech** view. (Data devices do not display in Mechanical views.)
3. Add a couple of walls.
4. Switch to the data receptacle family and load it into the project.
5. Add it several times using different types if you have created them.
6. Save the project and the family and close the files.

Practice 8d Create a Pipe Fitting Flange (Advanced)

Practice Objectives

- Create reference planes, dimensions, and labels.
- Draw single-line representations for Coarse and Medium detail levels.
- Create solids for the Fine detail level.
- Create and apply Family Types.

Estimated time for completion: 90-120 minutes

In this practice you will create a family file based on a template and add parameters, some of which include formulas. You will then add reference planes, dimensions, and labels and create family types to help you flex the framework. You will then create the flange components: a single-line representation for Coarse and Medium Detail Levels, and an extrusion and blend solid elements for Fine Detail Level, as shown in Figure 8–21. Finally, you will test the flanges in a project and draw pipe between them.

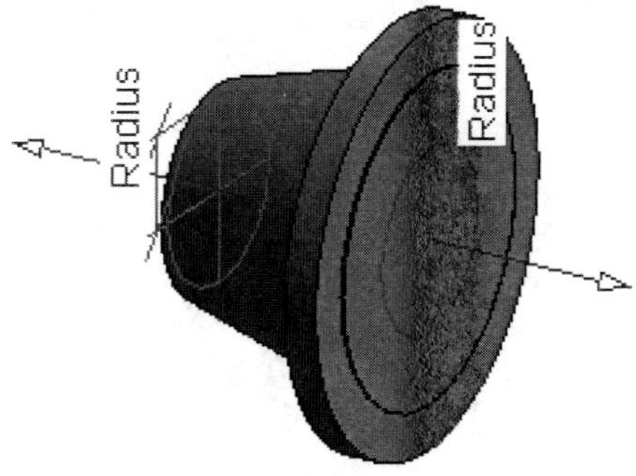

Figure 8–21

Task 1 - Open a family template.

1. Create a new family based on the **Plumbing Fixture.rft** template found in the Autodesk Revit Library.
2. Save the family to your practice folder as **Pipe Fitting Flange.rfa**.
3. In the *Create* tab>Properties panel, click (Family Category and Parameters).
4. In the Family Category and parameters dialog box, set the *Filter list* to **Piping** and in the *Family Category* area, select **Pipe Fittings**.

5. In the *Family Parameters* area, set the following options:
   - Clear the **Work Plane-Based** option.
   - Select **Always Vertical**.
   - For the *Part Type*, select **Transition**.
   - Clear the **Shared** option.
6. Click **OK**.
7. Type **UN** to open the Project Units dialog box.
8. Change the format of *Length* to **Fractional inches**, and set it to the nearest **1/16"**
9. Set the *Discipline* to **Piping** and change the format of *Pipe Size* to the nearest **1/16"**.
10. Click **OK**.
11. Set the *Scale* to **1" = 1'-0"**.
12. Save the family.

Task 2 - Create parameters.

1. In the *Create* tab>Properties panel, click (Family Types).
2. In the *Parameters* area, click **Add**.
3. In the Parameter Properties dialog box, in the *Parameter Type* area, select **Family Parameter** (if it is not already selected) and specify the following:
   - **Name:** NomRad
   - **Discipline:** Piping
   - **Type of Parameter:** Pipe Size
   - **Group Parameter Under:** Dimensions
   - Select the **Type** option.
4. Click **OK**.
5. Repeat the steps, creating parameters named **NomDia**, **FlgRad**, **FlgDia**, **PipeOD**, **PipeOR**, **RFD**, **RFR**, **RFT**, **FT**, **LTH**, **HT**, and **HubRad**. Define the parameters as follows:
   - **Parameter Type:** Family Parameter
   - **Discipline:** Piping
   - **Type:** Pipe Size
   - **Group Parameter Under:** Dimensions
   - Select the **Instance** option.

By default, the parameters are listed in alphabetical order.

6. In the Family Types dialog box, specify the following values:
   - **FT:** 3/4"
   - **FlgDia:** 9"
   - **LTH:** 5"
   - **NomRad:** 2"
   - **PipeOD:** 4 1/2"
   - **RFD:** 7"
   - **RFT:** 1/16"
7. In the Family Types dialog box, enter the following formulas for the parameters:
   - **FlgRad:** FlgDia / 2
   - **HT:** LTH - (FT + RFT)
   - **HubRad:** PipeOR * 1.5
   - **NomDia:** NomRad * 2
   - **PipeOR:** PipeOD / 2
   - **RFD:** FlgDia - 2"
   - **RFR:** RFD / 2

   The final information displays as shown in Figure 8–22.

Parameter	Value	Formula	Lock
Graphics			
Use Annotation Scale (defa	☐	=	
Mechanical			
K Coefficient		=	
K Coefficient Table		=	
Loss Method		=	
Dimensions			
FT (default)	3/4"	=	☐
FlgDia (default)	9"	=	☐
FlgRad (default)	4 1/2"	= FlgDia / 2	☐
HT (default)	4 3/16"	= LTH - (FT + RFT)	☐
HubRad (default)	3 3/8"	= PipeOR * 1.5	☐
LTH (default)	5"	=	☐
NomDia (default)	4"	= NomRad * 2	☐
NomRad	2"	=	☐
OffsetHeight (default)		=	
OffsetWidth (default)		=	
PipeOD (default)	4 1/2"	=	☐
PipeOR (default)	2 1/4"	= PipeOD / 2	☐
RFD (default)	7"	= FlgDia - 2"	☐
RFR (default)	3 1/2"	= RFD / 2	☐
RFT (default)	1/16"	=	☐
Other			

Figure 8–22

8. Click **OK**.

9. Save the family.

Task 3 - Create plan view reference planes.

1. Place three vertical reference planes 2" apart, with one to the left of the **Center (Left/Right)** reference plane, and two on the right, as shown in Figure 8–23.

2. Name the new reference planes as shown in Figure 8–23, with the *Is Reference* parameter set to **Not a Reference**.

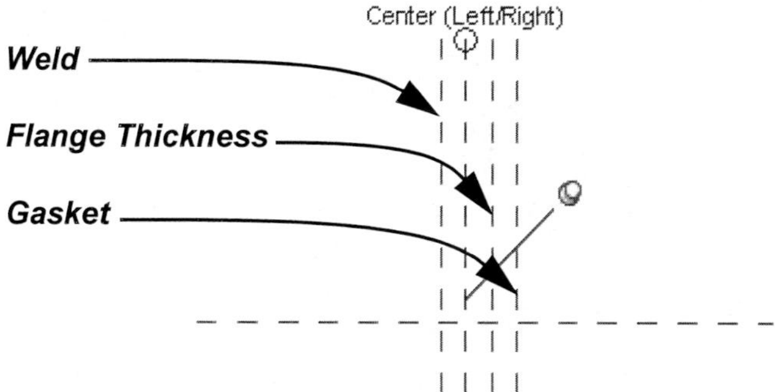

Figure 8–23

3. Dimension the reference planes as shown in Figure 8–24. Apply the labels shown in Figure 8–25 to the dimensions.

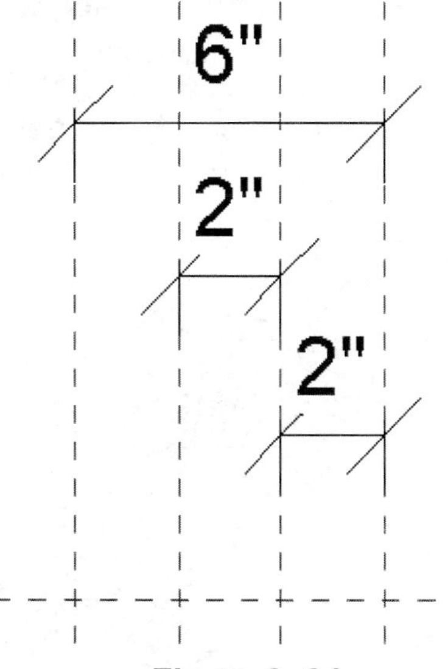

Figure 8–24

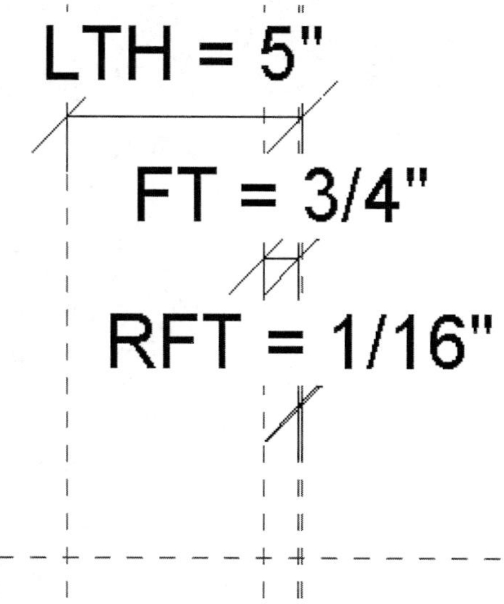

Figure 8–25

4. Add horizontal reference planes, dimensions, and labels on either side of the **Center (Front/Back)** reference plane, as shown in Figure 8–26.

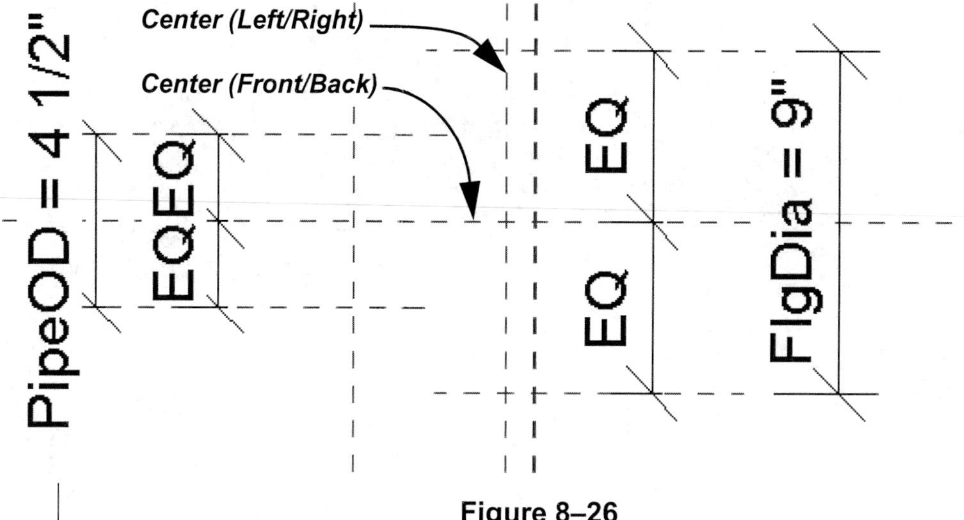

Figure 8–26

5. Save the family.

Task 4 - Create family types so that you can test the framework.

1. In the Properties panel, click (Family Types).
2. In the Family Types dialog box, in the *Family Types* area, click **New**.
3. In the Name dialog box, type **4"**, and then click **OK**. Note that the existing values are applied to this type by default.
4. Click **New** again. In the Name dialog box, type **6"**.
5. In the *Dimensions* area, change the parameter values as follows:
    - **FT:** 15/16"
    - **FlgDia:** 11"
    - **LTH:** 3 1/2"
    - **NomRad:** 3"
    - **PipeOD:** 6 5/8"
6. Move the Family Types dialog box so that the reference planes are displayed.

7. Click **Apply**. The reference plane locations change according to the new parameter values, as shown in Figure 8–27.

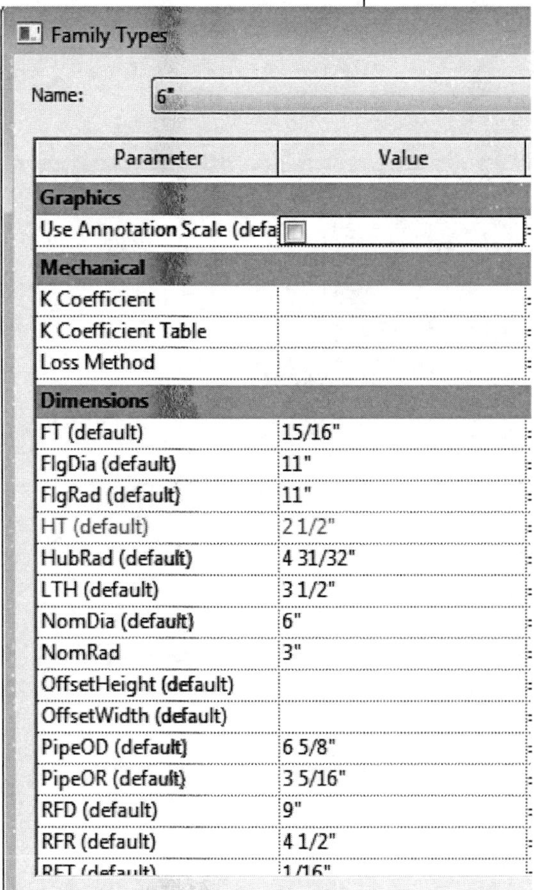

Family Types

Name: 6"

Parameter	Value
Graphics	
Use Annotation Scale (defa	☐
Mechanical	
K Coefficient	
K Coefficient Table	
Loss Method	
Dimensions	
FT (default)	15/16"
FlgDia (default)	11"
FlgRad (default)	11"
HT (default)	2 1/2"
HubRad (default)	4 31/32"
LTH (default)	3 1/2"
NomDia (default)	6"
NomRad	3"
OffsetHeight (default)	
OffsetWidth (default)	
PipeOD (default)	6 5/8"
PipeOR (default)	3 5/16"
RFD (default)	9"
RFR (default)	4 1/2"
RFT (default)	1/16"

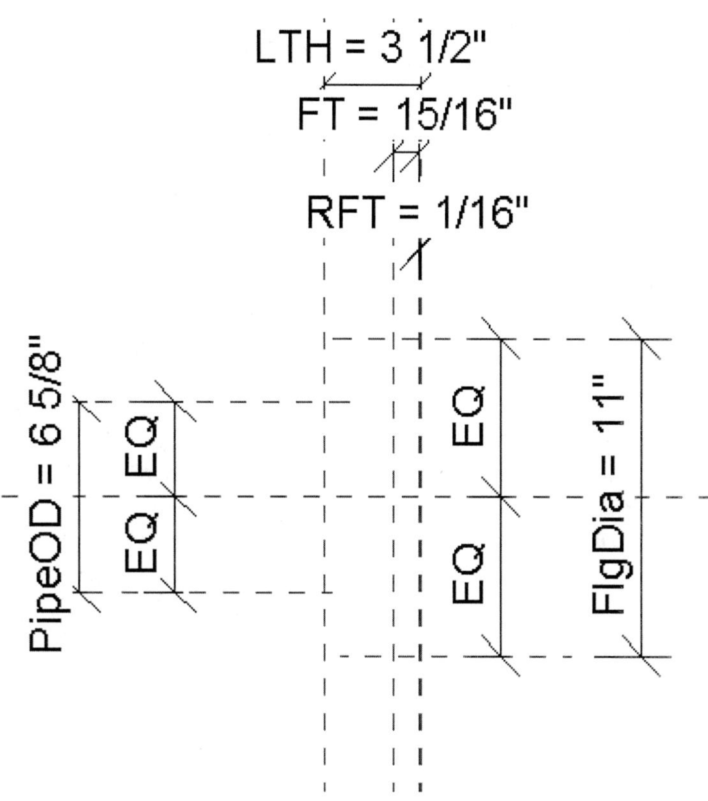

Figure 8–27

8. In the Name drop-down list, select **4"**.

9. Click **Apply**. Note that the flange returns to its original size.

10. Click **OK** to close the dialog box.

11. Save the family.

Task 5 - Create plan view graphics.

1. Continue working in the **Floor Plans: Rev. Level** view.

2. Open the Visibility Graphics dialog box. In the *Model Categories* tab, click **Object Styles**.

3. In the Object Styles dialog box, click **Select All**. Change the *Projection Line Weight* to **5** for all objects, as shown in Figure 8–28.

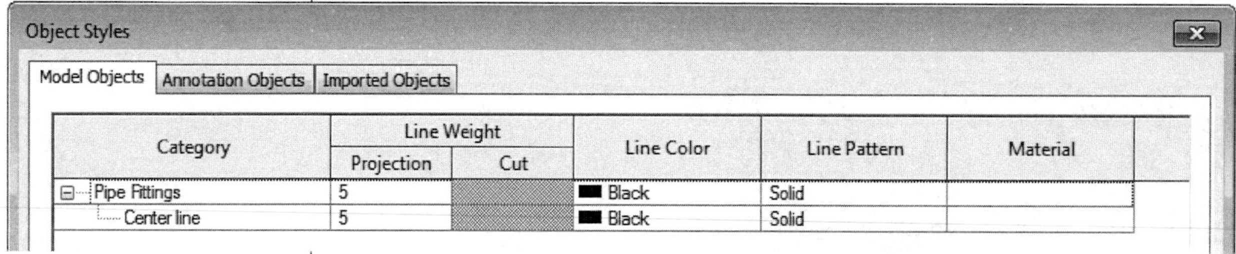

Figure 8–28

4. Click **OK** twice to close the dialog boxes.
5. In the *Create* tab>Model panel, click (Model Line).
6. Draw the lines shown in Figure 8–29. If you do not see the thickness, toggle (Thin Lines). Ensure that you draw from the intersections of the reference planes so that the lines will flex with the reference planes.

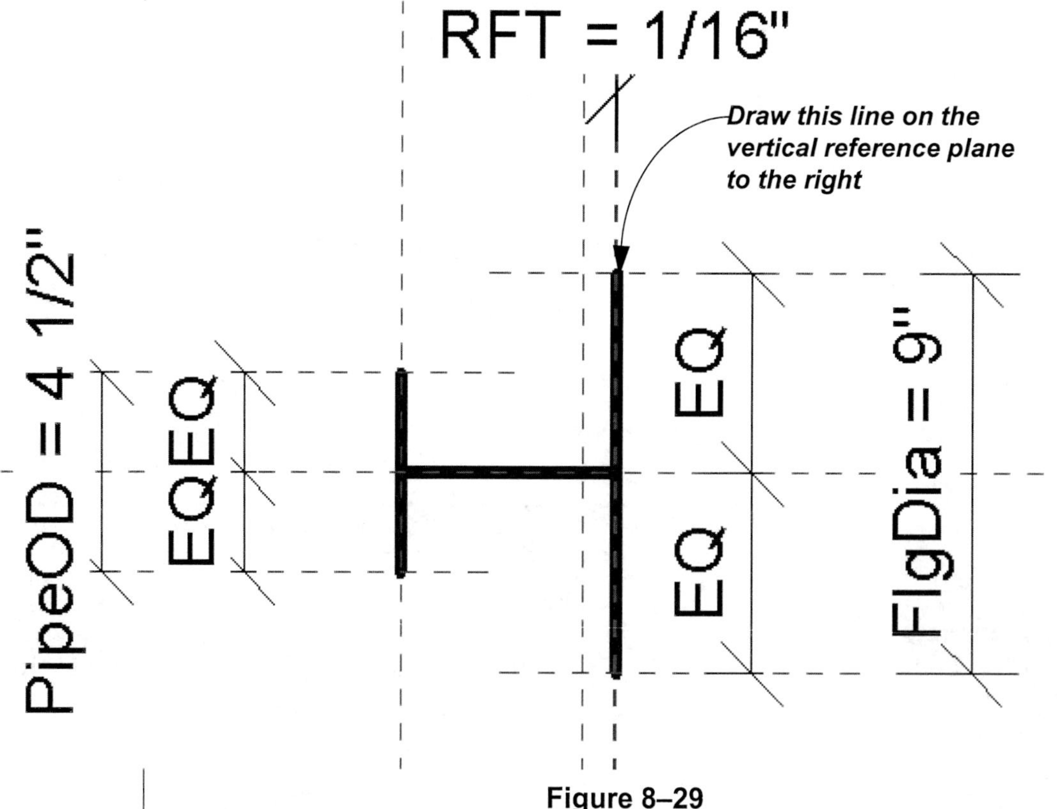

Figure 8–29

7. Open the Family Types dialog box. Use the new Types to verify that the lines work with the changes.

8. Reset *Name* to **4"** and then click **OK** to close the dialog box.

9. Select the three lines.

10. In the *Modify | Lines* tab>Visibility panel, click (Visibility Settings).

11. In the Family Element Visibility Settings dialog box, in the *Detail Levels* area, clear the **Fine** option. Leave all of the others selected, as shown in Figure 8–30.
    - This enables the selected elements to be visible only at the Coarse and Medium detail levels.

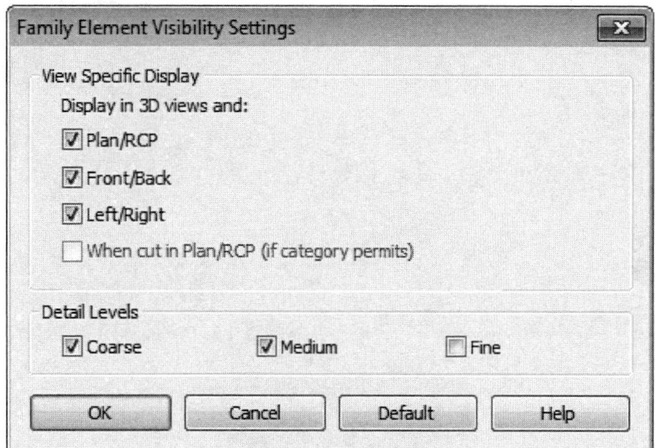

Figure 8–30

12. Click **OK**.

13. Test the *Detail Levels.* When you select **Fine**, the line elements should fade out. End on **Medium.**

14. Save the family.

Task 6 - Create the Raised Face and Flange extrusions.

1. In the Project Browser, open the **Elevations: Left** view.

2. Select the model lines and temporarily hide the elements.

3. In the *Create* tab>Forms panel, click (Extrusion).

4. In the *Modify | Create Extrusion* tab>Work Plane panel, click (Set).

5. In the Work Plane dialog box, in the Name drop-down list, select **Reference Plane: Flange Thickness**, as shown in Figure 8–31 and click **OK**

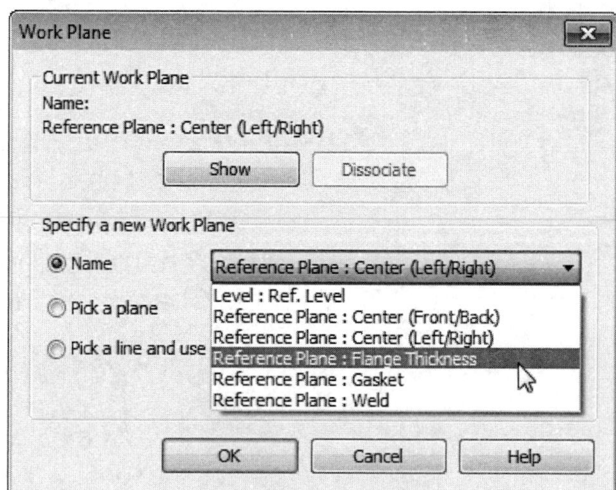

Figure 8–31

6. Draw a 3 1/2" circle at the intersection of the **Ref Level** and **Center (Front/Back)** reference planes. Dimension and label it using the **RFR** parameter, as shown in Figure 8–32.

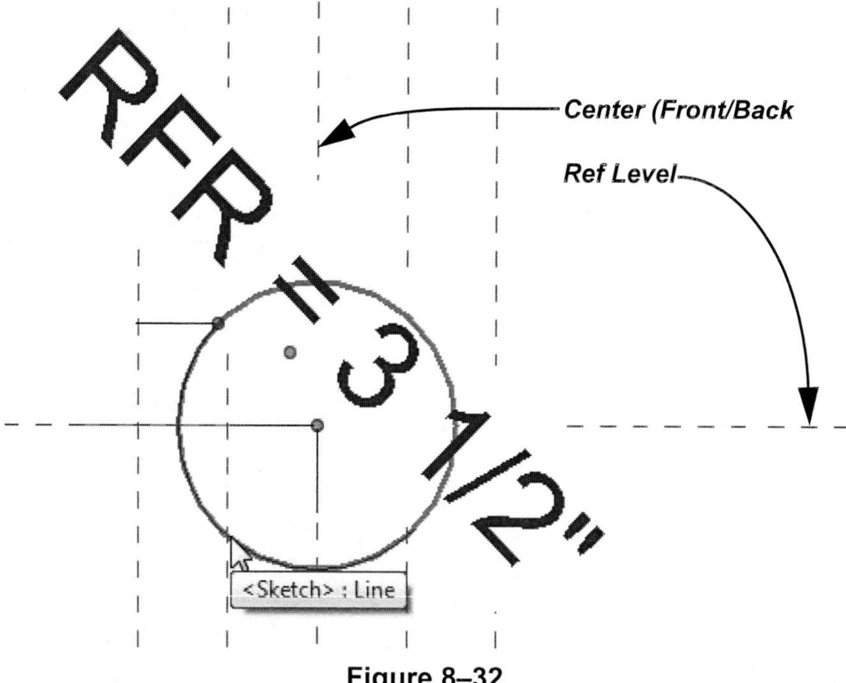

Figure 8–32

7. Click (Modify).

8. In Properties, in the *Extrusion End* row, click the Associate Family Parameter button, as shown in Figure 8–33.

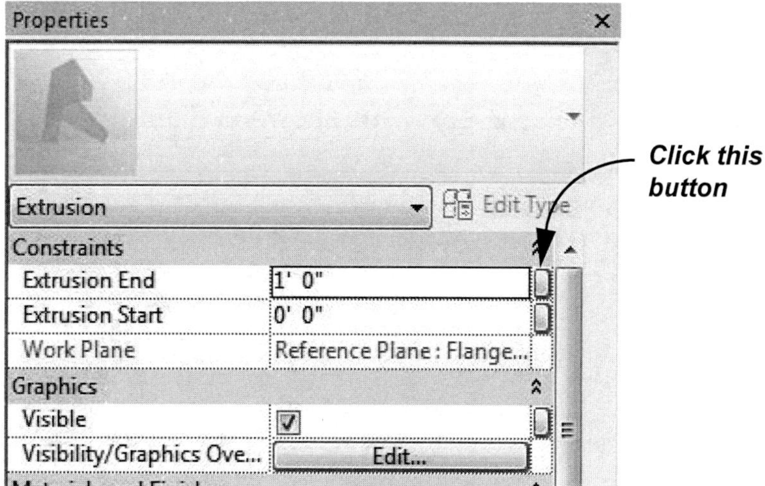

Figure 8–33

9. In the Associate Family Parameter dialog box, select **RFT** (as shown in Figure 8–34), and then click **OK**.

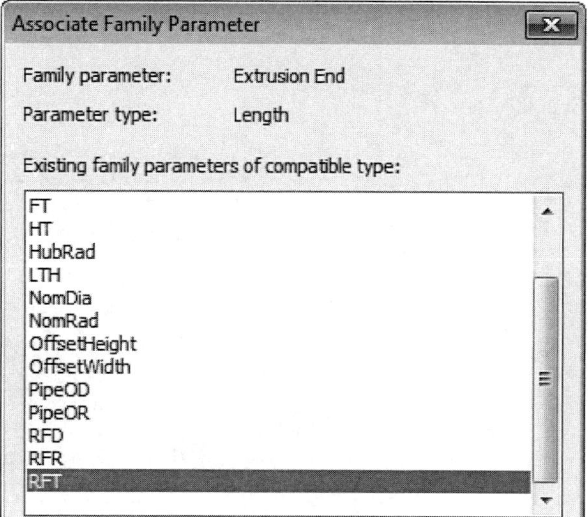

Figure 8–34

10. Click (Finish Edit Mode).

11. Click (Extrusion).

12. Set the Work Plane to **Reference Plane: Center (Left/Right)**.

13. Draw a circle at the same intersection using a radius of **4 1/2"**.

14. Make the temporary dimension permanent and label it as **FlgRad**.

15. In Properties, in the *Extrusion End* row, click the Associate Family Parameter button. In the Associate Family Parameter dialog box, select **FT** and click **OK**. The new radius displays as shown in Figure 8–35.

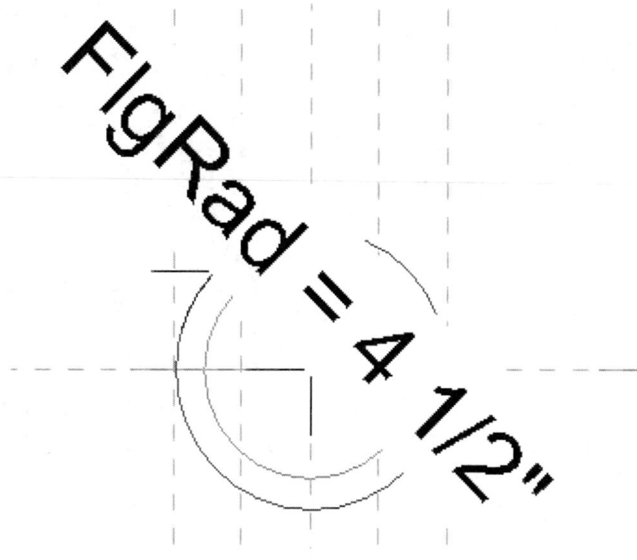

Figure 8–35

16. Click (Finish Edit Mode).

17. Open the **3D views: View 1** view. The two circular extrusions and the model lines should display. Shade and rotate the element as required, as shown in Figure 8–36. Flex it using the family types.

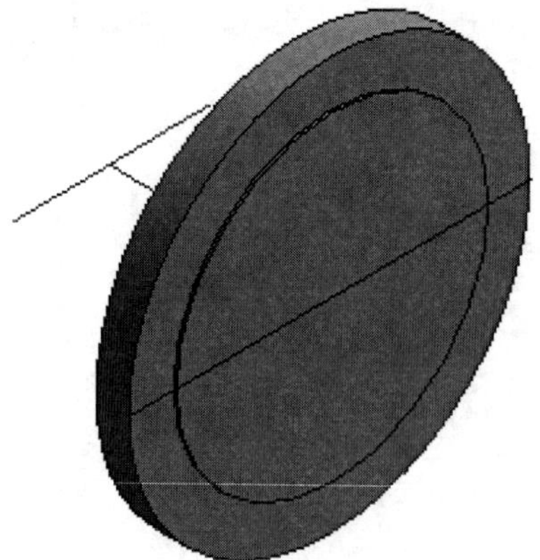

Figure 8–36

18. Save the family.

Task 7 - Create the pipe connection blend.

1. Return to the Elevations**: Left** view.
2. In the *Create* tab>Forms panel, click (Blend).
3. Draw a 3 3/8" circle at the same intersection. Dimension and label it **HubRad**.
4. In the *Modify | Create Blend Base Boundary* tab>Mode panel, click (Edit Top).
5. Draw a 2 1/4" circle at the same intersection as the other element. Dimension and label it **PipeOR**.
6. In Properties, set the *Second End* to (negative) **-4"**.
7. Click (Finish Edit Mode).
8. Open the **Elevations: Front** view. Align and lock the small end of the blend to the Weld reference plane, as shown in Figure 8–37.

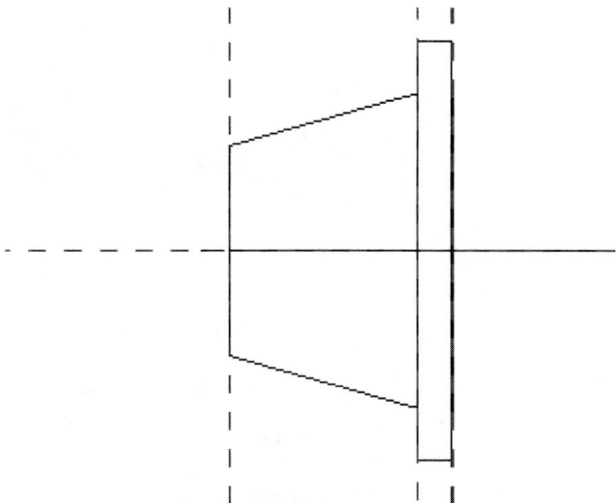

Figure 8–37

9. Open the **3D Views: View 1** view and flex the types to ensure that everything is working, as shown for two possible flexes in Figure 8–38 and Figure 8–39.

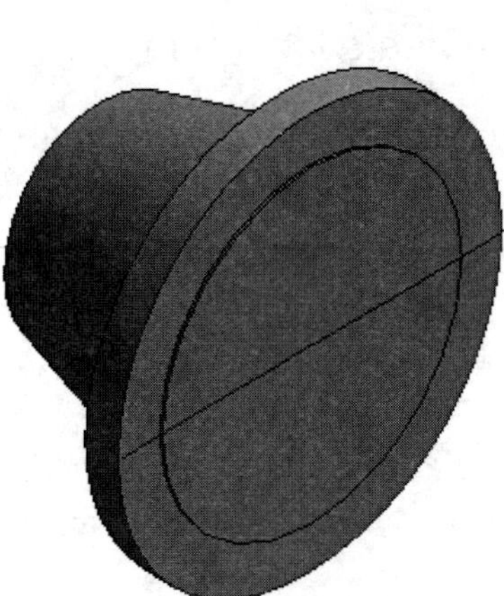

Figure 8–38

Figure 8–39

10. Temporarily hide the model lines.

11. Save the family.

Task 8 - Add connectors.

1. Continue working in the **3D Views: View 1** view.
2. In the *Create* tab>Connectors panel, click (Pipe Connector).
3. In the *Modify | Place Pipe Connector* tab>Placement panel, verify that (Face) is selected.
4. In the Options Bar, in the System type drop-down list, select **Hydronic Supply**.
5. In Properties, set the *Round Connector Dimension* parameter to **Use Radius**.
6. Rotate the view and press <Tab> as required to highlight the small face of the blend and click to add the connector.
7. The arrow of the connector should be pointing outward from the flange object, as shown in Figure 8–40. If the arrow is pointing inward, click the **Flip** control.

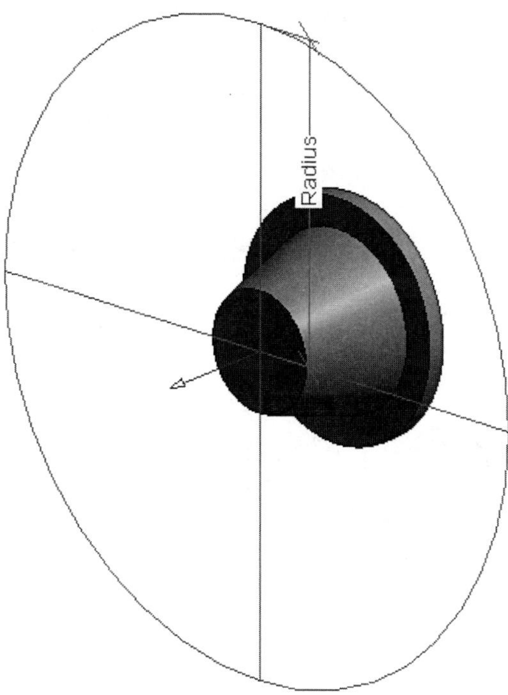

Figure 8–40

- The connector arrow does not indicate flow direction. Instead, it points in the direction from which the pipe is going to connect into it.

8. Spin the view so that the raised face is visible and apply another connector to that face as shown in Figure 8–41.

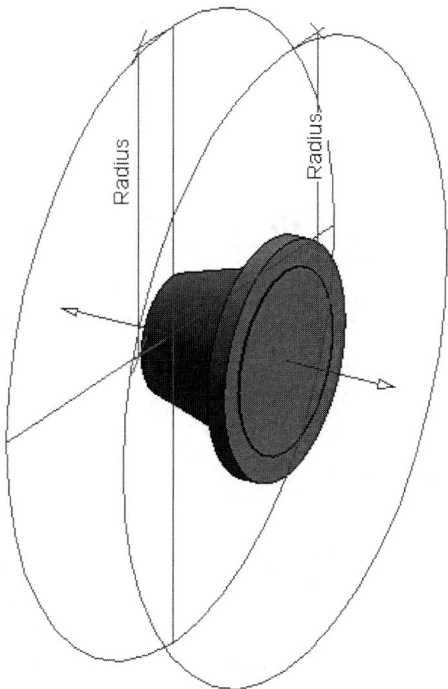

Figure 8–41

9. Click (Modify).
10. Select the first connector. In the *Modify | Connector Element* tab>Connector Links panel, click (Link Connectors).
11. Click the second connector to link it to the first.
12. Click (Modify).
13. Select both connectors.
14. In Properties, for *Radius,* associate it with the **NomRad** parameter and click **OK**.
15. The connectors are resized as shown in Figure 8–42.

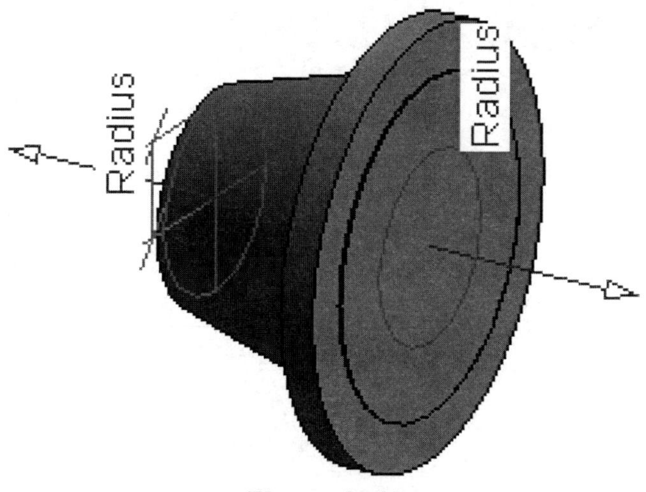

Figure 8–42

16. Save the family.

Task 9 - Test the flange in a project.

1. Start a new project and switch back to the family.
2. In the Family Editor panel, click (Load into Project).
   - If you have one project open, the new family loads directly into it. If you have more than one project open, select the right one(s) in the dialog box.
3. You are returned to the project.

4. Place an instance of each type into the drawing area of the project.
   - If a command is not automatically started, in the *Systems* tab>Model panel, click (Component).
5. Zoom in to display the plan view of the flanges at a Coarse Detail level. Ensure that Thin Lines is toggled off so that the line weight is displayed, as shown in Figure 8–43.
6. With Thin Lines on, change the *Detail Level* to **Fine**. The flanges display in detail at full size, as shown in Figure 8–44.

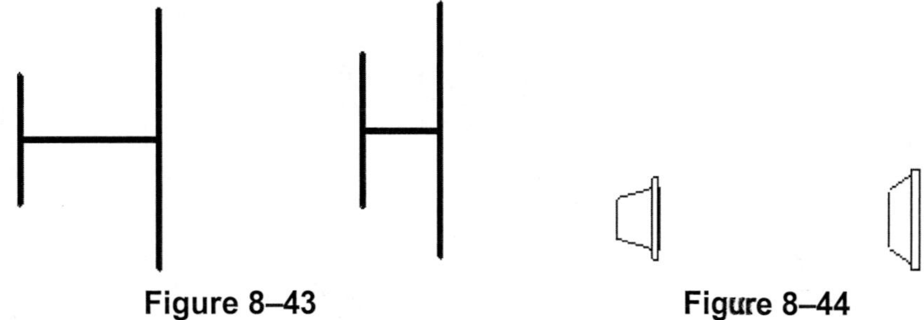

Figure 8–43 **Figure 8–44**

7. Ensure that the flanges are aligned with each other, and then draw pipe between them.
8. Open a 3D view.
9. Zoom in on the flanges and turn on (Thin Lines) as required.
10. Change Detail Level to **Fine** and type **SD** to shade the view.
11. Spin the view to examine the flanges from all sides.
12. Save the project file.
13. Close the project and family files. Save any changes you want to keep.

Chapter Review Questions

1. The architect has requested that you use a vanity design that is not designed to work with MEP systems, such as the one shown in Figure 8–45. What do you need to do to upgrade it?

Figure 8–45

   a. Add pipes to the family where you want the project pipes to tie in.
   b. Add connectors to the family where you want the project pipes to tie in.
   c. Create a separate MEP specific vanity and use **Copy/Monitor** to add it to the project.
   d. Import an existing sink designed to work with MEP systems into the family.

2. Which of the following Light Source Definition distribution options enables you to specify an IES file?

   a.  (Spherical)

   b. (Hemispherical)

   c. (Spot)

   d. (Photometric Web)

3. If you want to include an annotation element, such as a label that references a parameter in a family, you can only do so by creating a separate annotation tag.

   a. True

   b. False

4. How do you have a family display different elements when you change the *Detail Level* from *Coarse* to **Fine**?

   a. In the project file, select the family and in Properties change the detail level.

   b. In the family file, select the elements and modify the Visibility Settings.

   c. In the family file, in the View Properties, set the detail level.

   d. In the project file, set the Visibility Settings for the view.

5. In the Family Types dialog box shown in Figure 8–46, if the *Pump Length* changed to **2'-0",** what would the value of *Radius 2* be?

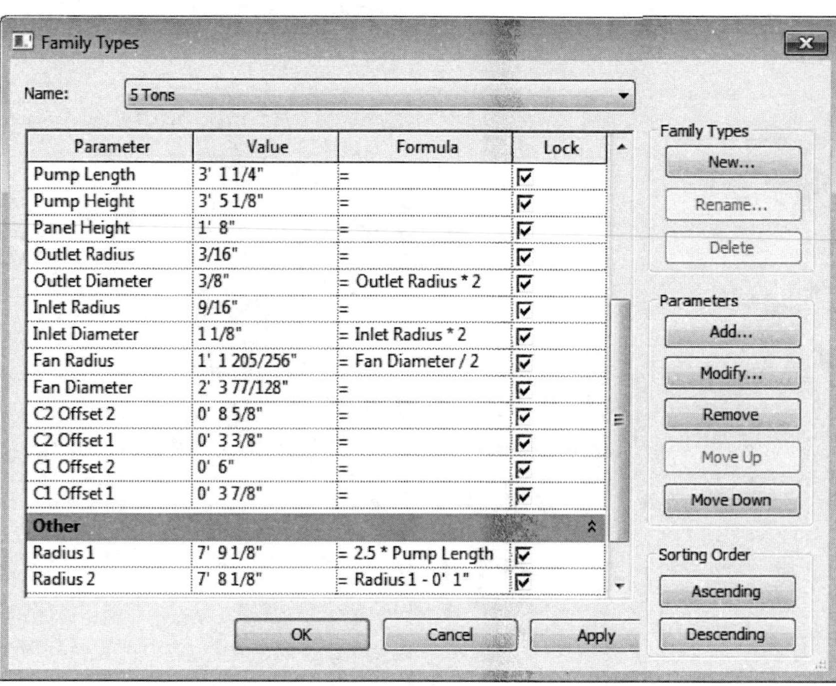

Figure 8–46

a. 5'-0"

b. 4'-0"

c. 4'-11"

d. 5'-1"

Creating Structural Specific Families

Structural specific families include everything from column and beam design, to detail items such as gusset plates. By working through a series of practices, you learn about creating a parametric gusset plate, an in-place stiffener plate, an in-place slab depression, a built-up column, a tapered concrete column, a truss family, a precast hollow core slab and a tapered moment frame. Select the ones that are most helpful in your work.

Learning Objectives in this Chapter

- Create a gusset plate with parametric properties.
- Create an in-place column stiffener family.
- Create a floor-based family with a void that cuts out a depression in a floor when inserted in a project.
- Create a built-up column family made of steel plates.
- Create a tapered concrete column family.
- Create a truss family.
- Create a precast hollow core slab family that uses an imported CAD file for the profile.
- Create a tapered moment frame family that includes two columns with a connecting beam.

Practice 9a Parametric Gusset Plate

Practice Objectives

- Create reference planes, dimensions, and parameters.
- Create solid geometry for the plates.
- Modify Visibility Settings and create Family Types.

Estimated time for completion: 30 minutes

In this exercise you will create a family file based on a template and create reference planes. You will then create gussets as a mass, create material parameters, and apply Family Types. Finally, you will turn off the visibility of the gusset plate from the coarse detail to all views, as shown in Figure 9–1.

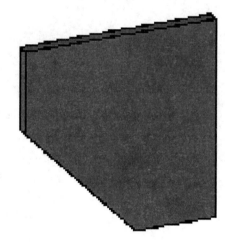

Figure 9–1

Task 1 - Start a new family using a template and add reference planes.

1. Create a new family based on the **Structural Stiffener.rft**. template found in the Autodesk® Revit® Library.
2. Save the family to your practice folder as **Parametric Gusset Plate.rfa**.
3. Open the Project Units dialog box. Change the **Length Format** to **Fractional inches**. This makes the dimensions easier to read.
4. Open the **Elevations: Left** view.

5. Create and dimension the new reference planes, as shown in Figure 9–2.
   - Ensure that the diagonal reference plane being added crosses the intersection of the **Center (Front/Back)** and **Ref. Level** reference planes.

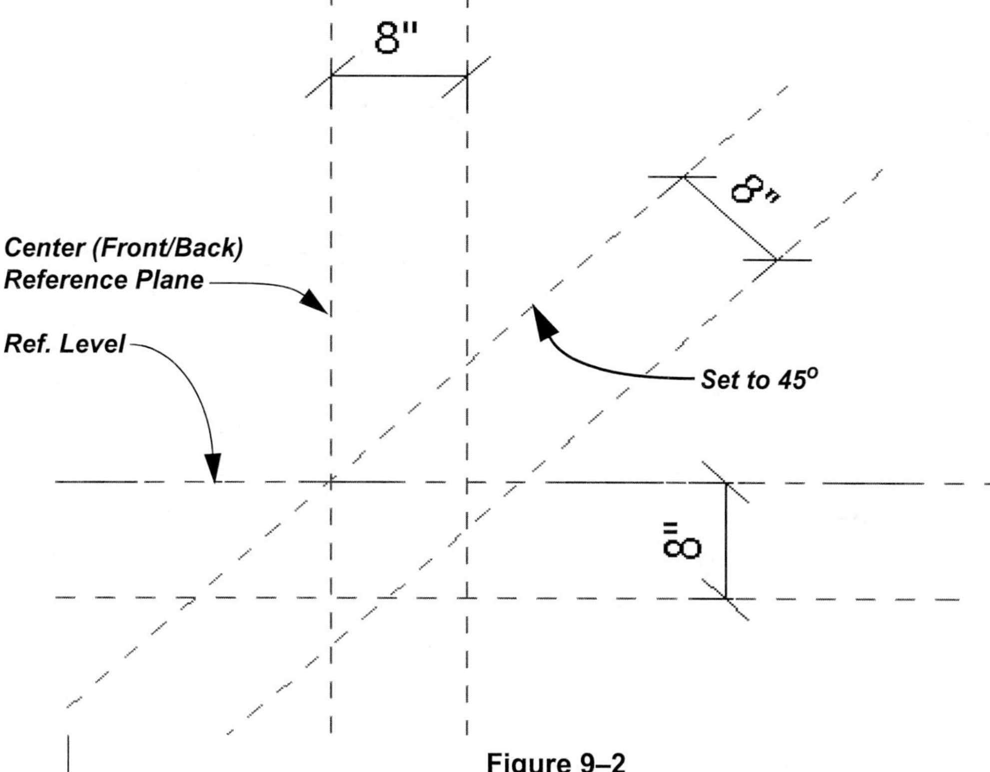

Figure 9–2

6. Label one of the dimensions with a new parameter named **Width**.
7. Select the other two dimensions and apply the same parameter to them.
8. Flex the parameter to test the stability of the framework.
9. Set the **Width** to **8"**
10. Save the family.

Task 2 - Create Front View reference planes.

1. Switch to the **Elevations: Front** view and create two reference planes, **1/4"** on either side of the Center (Left/Right) reference plane and add dimensions. Label the overall dimension as **Thickness**, as shown in Figure 9–3.

Modify the scale so that the dimensions are clearly displayed.

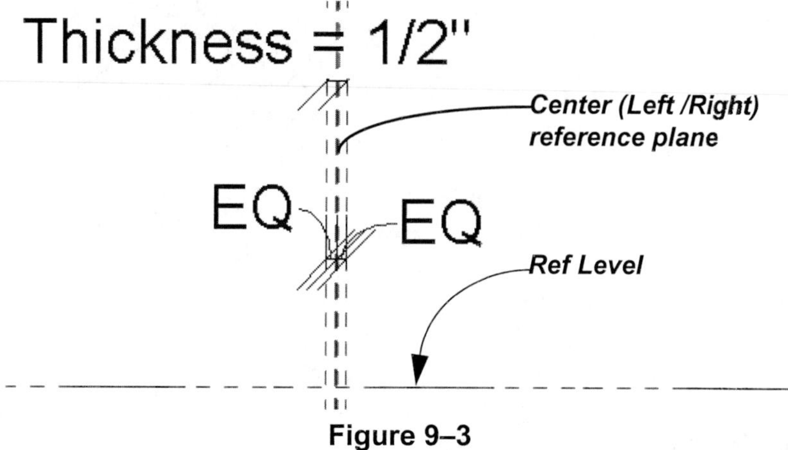

Figure 9–3

2. Flex the parameter to test the stability of the framework.
3. Set the **Thickness** to **1/2"**.
4. Save the family.

Task 3 - Creating family geometry.

1. Switch to **Elevations: Left** view.
2. In the *Create* tab>Forms panel, click (Extrusion). In the Options Bar, set the *Depth* to **1/2"**.
3. Draw individual lines across the outside intersections of the reference planes, and lock all five lines to the reference planes, as shown in Figure 9–4.

*Use the **Align** command to help lock the lines to the reference plane.*

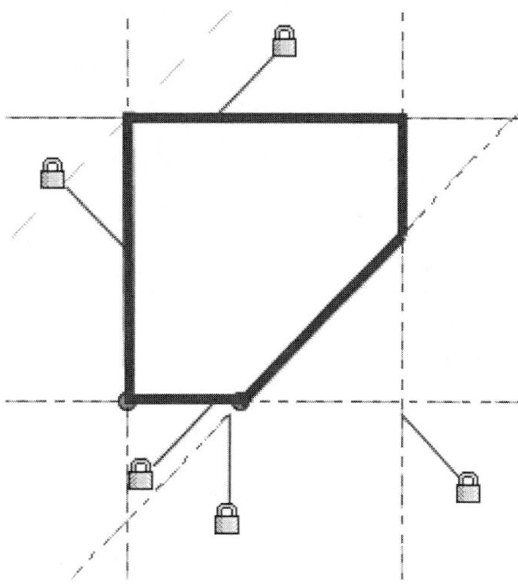

Figure 9–4

4. Click (Finish Edit Mode).
5. Switch to the **Elevations: Front** view. Align the left and right sides of the extrusion with the reference planes you created earlier and lock them.
6. Switch to a 3D view and shade the model.
7. In the *Create* tab>Properties panel, click (Family Types) to flex the model to ensure that everything is working together. The model displays as shown in Figure 9–5.

Figure 9–5

8. Save the family.

Task 4 - Add material parameters.

1. Open the Family Types dialog box.
2. In the *Parameters* area, click **Add...**.

*Setting the parameter as an **Instance** makes it easier to modify it in the project file.*

3. Create an Instance parameter named **Material** and set the *Type of Parameter* to **Material**. The Group automatically updates to Materials and Finishes, as shown in Figure 9–6. Click **OK**.

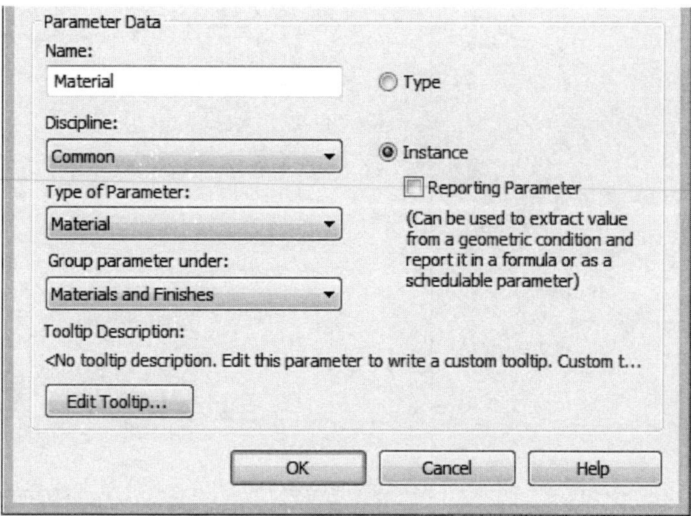

Figure 9–6

4. In the 3D view, select the plate.
5. In Properties, in the *Materials and Finishes* area, in *Material* row, click (Associate Family Parameter).
6. In the Associate Family Parameter dialog box, select **Material** as shown in Figure 9–7 and click **OK**.

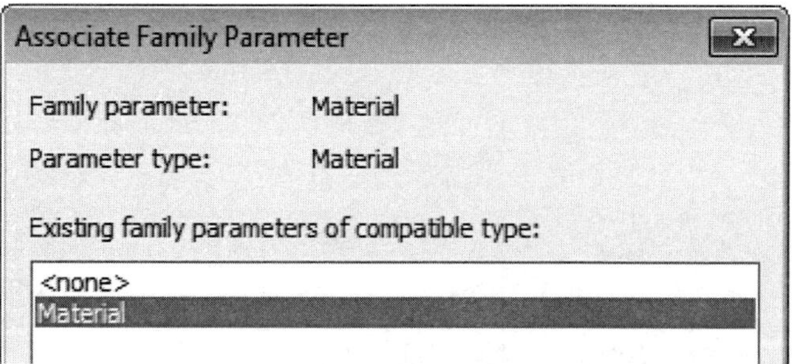

Figure 9–7

7. In Properties, the **Material** parameter is now marked with the **=** sign, as shown in Figure 9–8. Therefore, when the family is loaded into a project, you can control its material using Properties.

Figure 9–8

Task 5 - Change the visibility settings in a family.

1. Select the extrusion.
2. In the *Modify | Extrusion* tab>Mode panel, click (Visibility Settings).
3. In the Family Element Visibility Settings dialog box, clear the **Coarse** option as shown in Figure 9–9, so that the gusset plate does not display at a Coarse detail level.

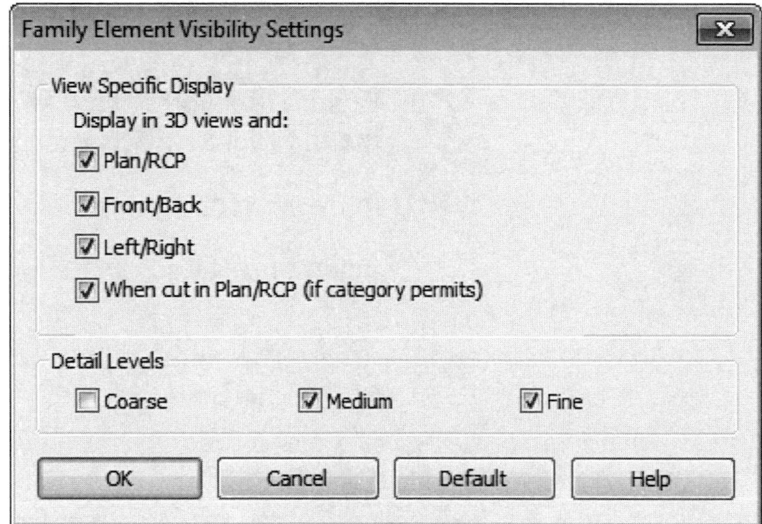

Figure 9–9

4. Click **OK**.
5. Press <Esc> to clear the selection.
6. Change the *Detail Level* to **Coarse**, **Medium**, and **Fine**. The outline of the Gusset Plate remains dark in the Medium and Fine Detail Levels but turns gray in the **Coarse** Detail Level. This means that it will not display in a project when inserted in a view set to the Coarse Detail Level.
7. Save the family.

Task 6 - Create family types.

1. Open the Family Types dialog box.
2. Create three Family Types. Set the *Name* and the *Width* parameter of each as follows:
   - **8"**
   - **10"**
   - **12"**
3. Use the new types to flex the gusset plate family.
4. Save the family.
5. Ensure that you have a structural project open and then load the family into the project.
6. In the project, open a 3D View.
7. Insert one of each Plate type into the project.
8. Switch to a 3D view to see the differences.
9. Shade the view, using the shortcut **SD** as required.
10. Select one of the Plates. In Properties, apply a material to it. Repeat with the others using different materials.
11. Save the family and project.

Practice 9b Column Stiffeners (In-Place Family)

Estimated time for completion: 15 minutes

Practice Objective

- Create an in-place family in a detail view

In this practice you will create a stiffener plate to strengthen a column's flange, as shown in Figure 9–10.

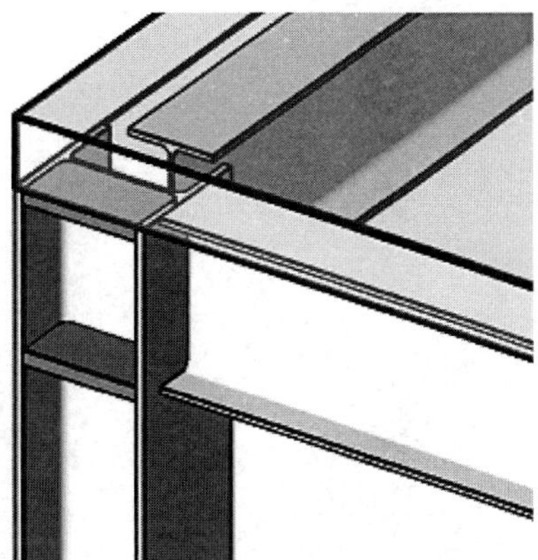

Figure 9–10

Task 1 - Create an in-place family in a detail view.

1. In the ...*Structural* folder, open **Column-Stiffeners.rvt**. The file opens in the **Structural Plans: Level 2** view.
2. In the Project Browser, open **Detail Views: Detail 0**. It displays a column supporting a beam and concrete on metal deck slab, as shown in Figure 9–11.

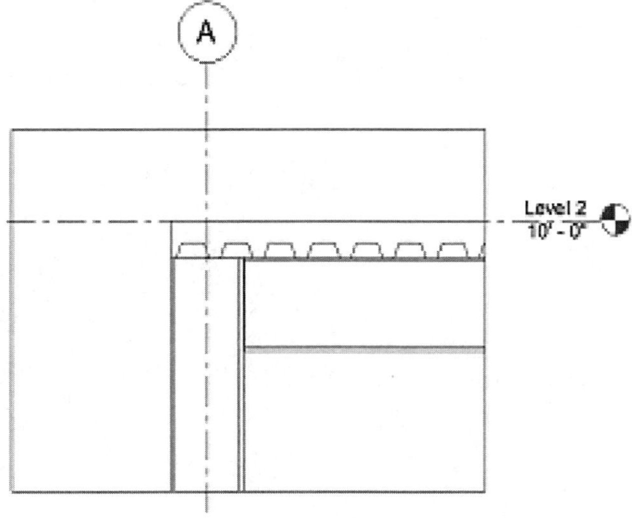

Figure 9–11

3. The column requires stiffeners to strengthen its flanges. In the *Structure* tab>Model panel, expand (Component) and click (Model In-Place).
4. In the Family Category and Parameters dialog box, set the *Filter list* to **Structure** and select **Structural Stiffeners**, as shown in Figure 9–12. Click **OK**.

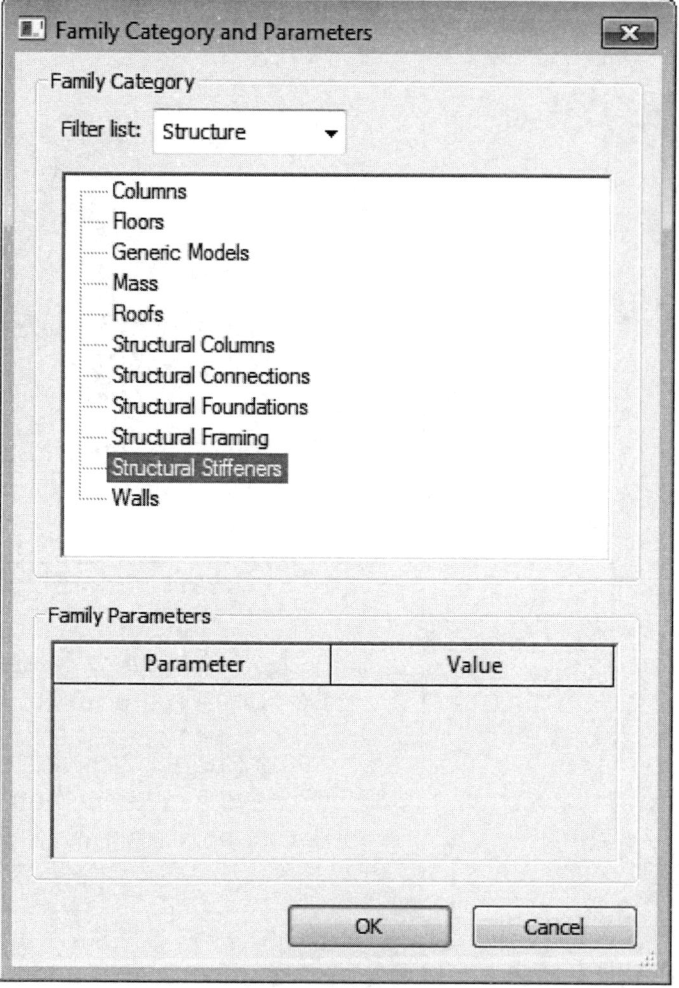

Figure 9–12

5. Enter a name for the stiffener plate, as shown in Figure 9–13.

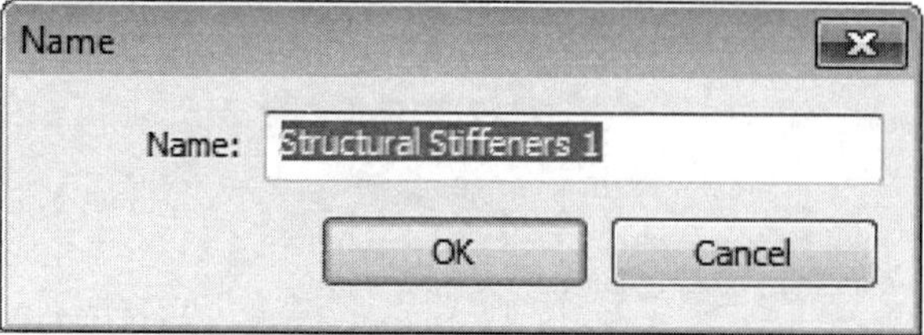

Figure 9–13

6. You are now in the In-Place Editor. In the **Detail 0** view, draw two horizontal reference planes and name them, as shown in Figure 9–14. These reference planes assist in locating the stiffener plates.

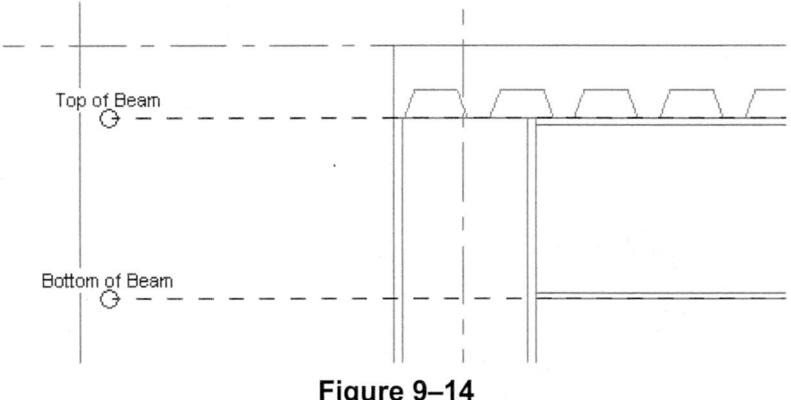

Figure 9–14

7. In the *Create* tab>Forms panel, click (Extrusion).

8. In the Work Plane dialog box, in the drop-down list, select **Reference Plane: Top of Beam**, and then click **OK**.

9. In the Go To View dialog box, select **Structural Plan: Level 2**. This view is used to draw the stiffener plate.

10. Click **Open View**.

11. In the **Structural Plans: Level 2** view, zoom into the grid intersection **1/A**.

12. In the *Modify | Create Extrusion* tab>Draw panel, click (Pick Lines). In the Options Bar, select **Lock**. This ensures that the stiffener edges adjust to any changes to the column size.

13. Select and trim the perimeter of the inside face of the column, as shown in Figure 9–15.

You might need to zoom in very close to display the faint background image.

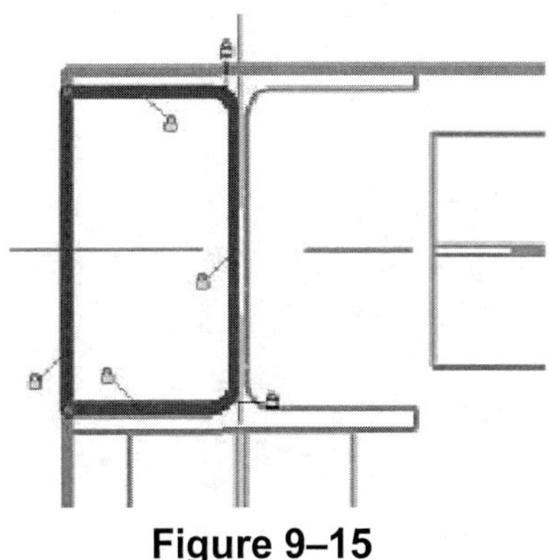

Figure 9–15

14. In Properties, set the *Extrusion End* value to (negative) **-1"**, as shown in Figure 9–16.

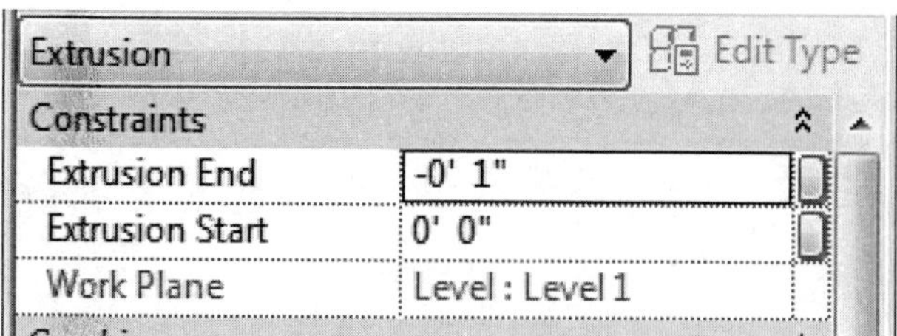

Figure 9–16

15. Click (Finish Edit Mode).

16. Switch to the **Detail 0** view. Note that the stiffener plate is located under the **Top of Beam** reference planes.

17. Select the stiffener plate and click (Copy). In the Options Bar, clear the **Constrain** option. Select the starting point as the bottom of the stiffener plate and copy it halfway down between the two reference planes.

18. Select the copied stiffener plate.

19. In the *Modify* | *Extrusion* tab>Work Plane panel, click **Edit Work Plane**.

20. In the Work Plane dialog box, select **Reference Plane: Bottom of Beam** and click **OK**. The plate moves into a new position.

21. The stiffener plate thickness must be adjusted upward. Select the new stiffener plate and change the *Extrusion End* to **1"**.

22. The new plates are now in place, as shown in Figure 9–17.

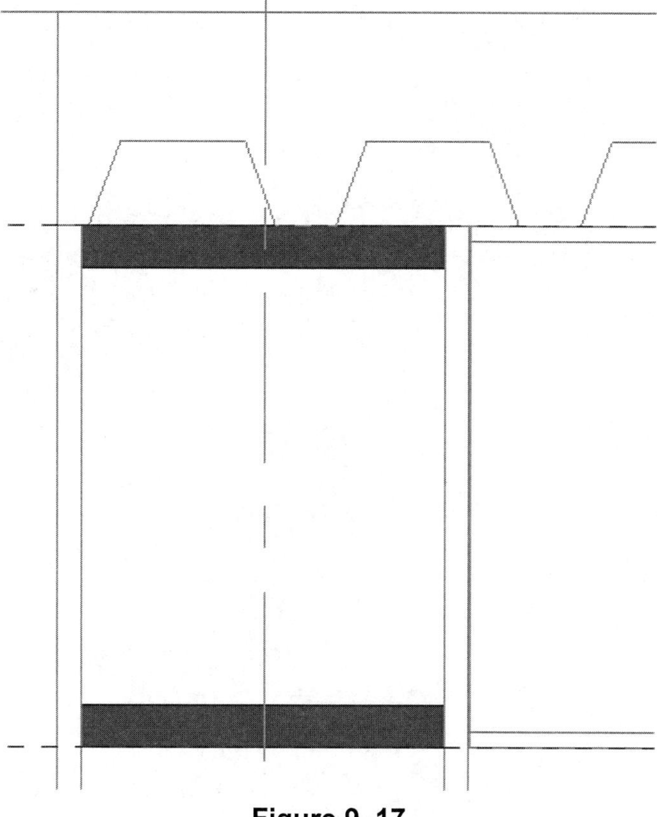

Figure 9–17

23. In the In-Place Editor panel, click ✔ (Finish Model).

24. In the project file, navigate through the different views to see the stiffener plates.

25. Save the project.

Practice 9c In-Place Slab Depression

Practice Objectives

- Create a floor-based family by setting up the parametric framework.
- Create a void form for a depression in the floor.
- Apply the family in a project.

Estimated time for completion: 15 minutes.

Slab depression family is an important host-based family that you need to modify for your projects. In this practice, you will create a slab depression using a template file and apply it in a project, as shown in Figure 9–18.

Figure 9–18

Task 1 - Create a family and add reference planes, insertion points, and dimensions.

1. Create a new family based on the **Generic Model floor based.rft** template found in the Autodesk Revit Library.
2. Save the family to your practice folder as **Slab Depression Component.rfa**.
3. In the *Create* tab>Datum panel, click (Reference Plane).
4. Add the horizontal and vertical reference planes and dimensions, as shown in Figure 9–19.

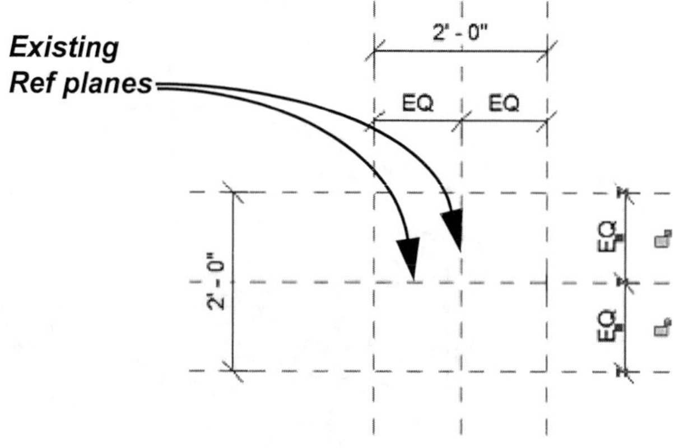

Figure 9–19

5. Label the horizontal dimension as **Slab Depression Length** and the Vertical dimension as **Slab Depression Width**. In both cases set the parameter to **Instance**. The parameters display as shown in Figure 9–20.

*Ensure that you set these parameters and the one below to **Instance** parameters. This ensures that they can be set in the project.*

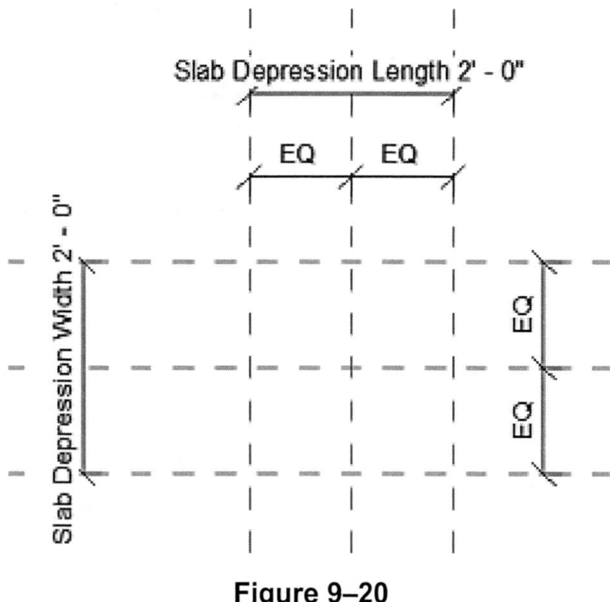

Figure 9–20

6. Open the **Elevations: Front** view where you create the depth of the depressed slab.

7. Add another reference plane **3"** down from the reference level. Add a dimension between the reference level and the new reference plane. Label the dimension as **Slab Depression Depth**, as shown in Figure 9–21.

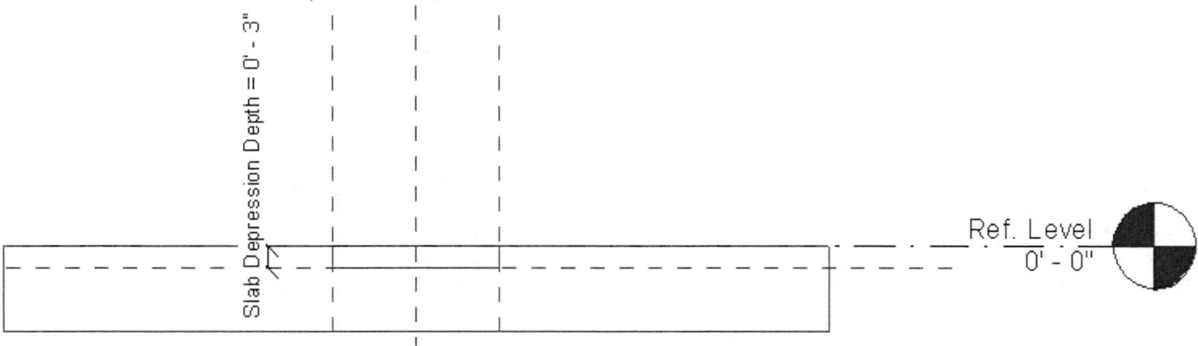

Figure 9–21

8. Save the family.

Task 2 - Create the depressed area for the slab.

1. Now you will create the depressed area for the slab by creating a void form and locking it to the reference planes. Switch to the **Floor Plans: Ref. Level** view.

2. In the *Create* tab>Forms panel, expand (Void Forms) and click (Void Extrusion).
3. In the Draw panel, click (Rectangle).
4. Add the rectangle by clicking at the top left corner of the intersecting reference planes and then the bottom right corner. Click the lock controls to constrain the void form to the reference planes, as shown in Figure 9–22.

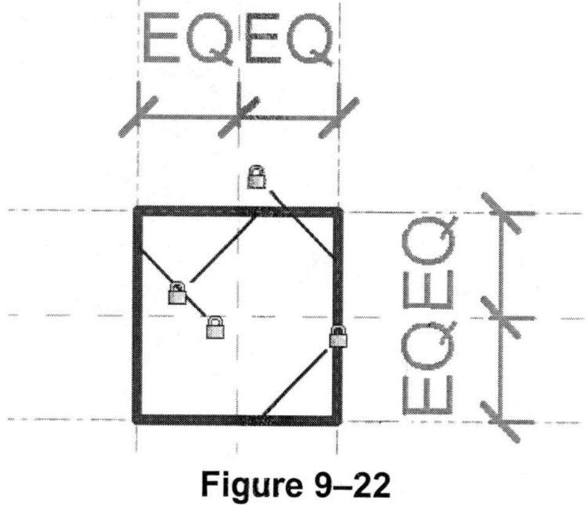

Figure 9–22

5. Click (Finish Edit Mode).
6. Press <Esc> to clear the extrusion selection.
7. Open the **Elevations: Front** view.
8. In the *Modify* tab>Modify panel, click (Align).
9. Select the lower reference plane first and then pick the bottom of the void form, as shown in Figure 9–23. Select the lock control to constrain the bottom of the void form to the reference plane.

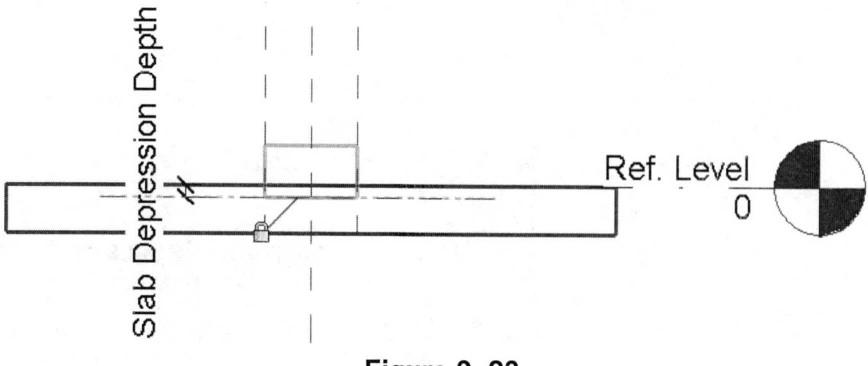

Figure 9–23

10. Still in the **Align** command, select the reference level and pick the top of the void form. Select the lock control to constrain the top of the void form to the reference level.

11. In the *Modify* tab>Geometry panel, click (Cut Geometry) and pick the void form and the slab object. (The order of the selection does not matter.)

12. Open a 3D view to see the depression within the slab object. Type **SD** to shade the view, as shown in Figure 9–24.

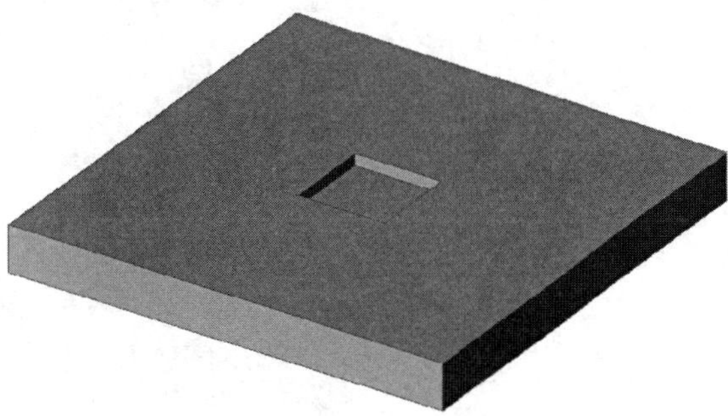

Figure 9–24

13. Use the Family Types dialog box to flex the depression.

14. Close the dialog box.

15. Save the Depressed Slab component family file. Do not close the family file.

Task 3 - Test the component family file.

1. Create a new project and save the file as **Depressed Slab Example.rvt**.

2. In the *Structure* tab>Foundation panel, click (Structural Foundation: Slab). Create an **8" Concrete Slab** in the new project.

3. When the slab has been created in the project file, return to the **Slab Depression Component.rfa** family file.

4. In the Family Editor panel, click (Load into Project).

5. If it is not automatically started, start the **Component** command.

6. Pick a location for the slab depression. The insertion is based on the origin within the family file. In the family file you based the insertion point at the center of the recessed slab.

7. Select different views to display the depression. Cut some sections through it.

8. To change the size of slab depression, return to the **Floor Plans: Level 2** view, and select the component.

9. In Properties, change the values to any required size, as shown in Figure 9–25.

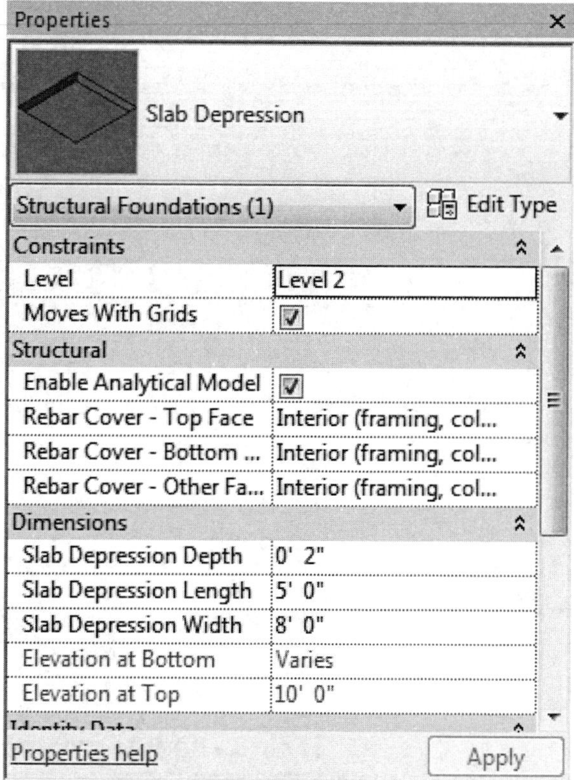

Figure 9–25

10. Return to the 3D view and Section view to see the new shape of the depressed slab.

11. Save and close the project and family files.

Practice 9d Built-Up Column

Estimated time for completion: 15 minutes.

Practice Objectives

- Set up the parametric framework for a column.
- Use extrusions to create the sides of the column.

The Autodesk Revit Structure software has an extensive structural columns library to meet many standard types and sizes in the industry. However, there is still a need for built-up (custom) made columns. In this practice you will create a built-up column from steel plates, as shown in Figure 9–26.

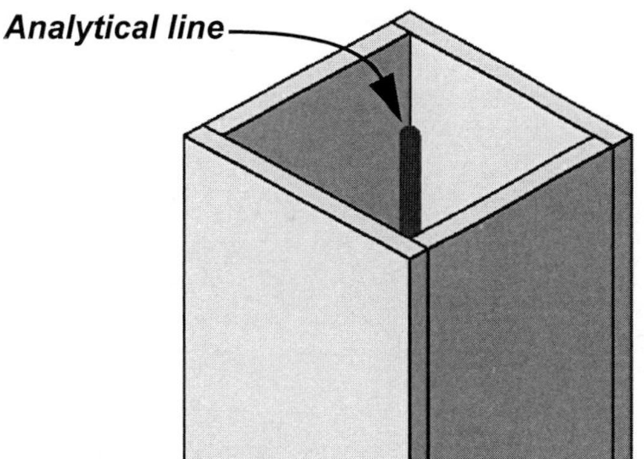

Figure 9–26

- The analytical line is already incorporated into the family in the structural system family template. Therefore, it is not required to be created in the new family.

Task 4 - Set up reference planes and dimensions.

1. Create a new family based on the **Structural Column.rft** template found in the Autodesk Revit Library.
2. Save the family to your practice folder as **Built-Up Column.rfa.**

3. In the **Structural Plans: Lower Ref. Level** view, add reference planes to the inside of the four exterior reference planes using (Pick Lines) with an *Offset* of **2"**, as shown in Figure 9–27.

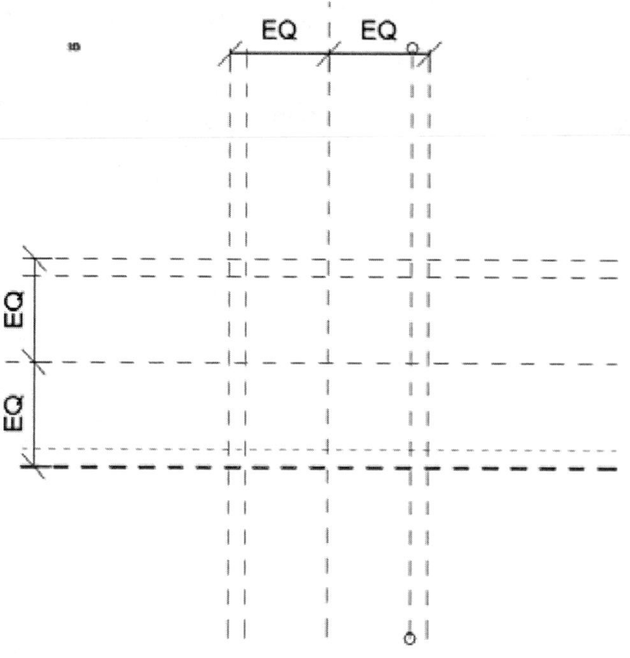

Figure 9–27

4. Delete the vertical **EQ** dimensions to prevent over-constraining and re-insert new vertical EQ dimensions at the inner reference planes, as shown in Figure 9–28.

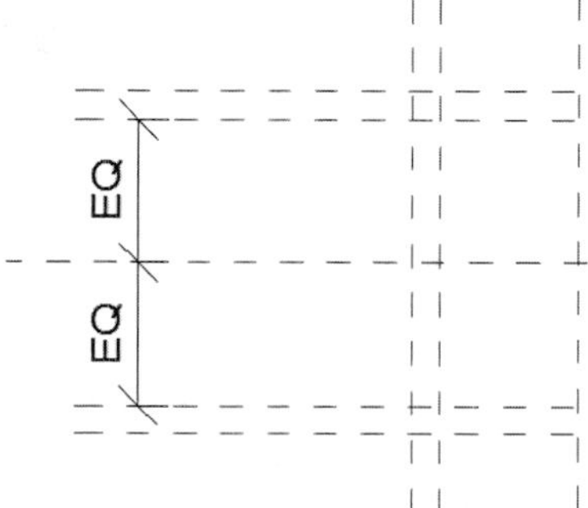

Figure 9–28

5. Place single dimensions horizontally and vertically above the EQ dimensions, as shown in Figure 9–29.

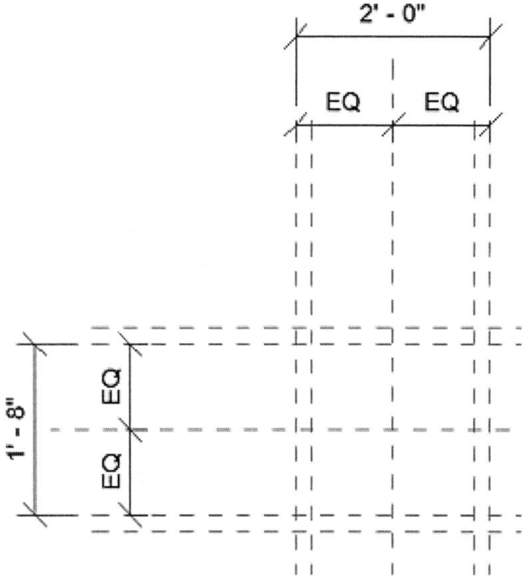

Figure 9–29

6. Select the horizontal dimension and label it. Set the *Name* to **Primary Plate Length** and select **Instance**, as shown in Figure 9–30.

*The **Instance** option permits the value to be modified per element.*

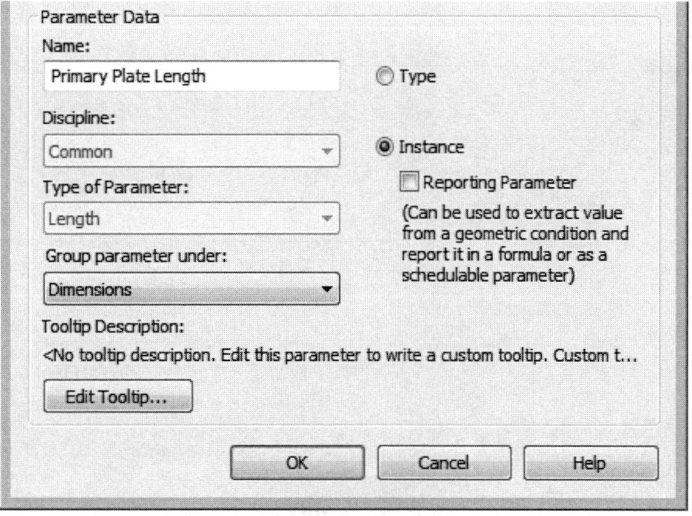

Figure 9–30

7. Follow the same procedures for the vertical dimension, naming it **Secondary Plate Length**.

8. Add four more dimensions, as shown in Figure 9–31. Set the parameters for these new dimensions to determine the thickness of the primary and secondary plates.

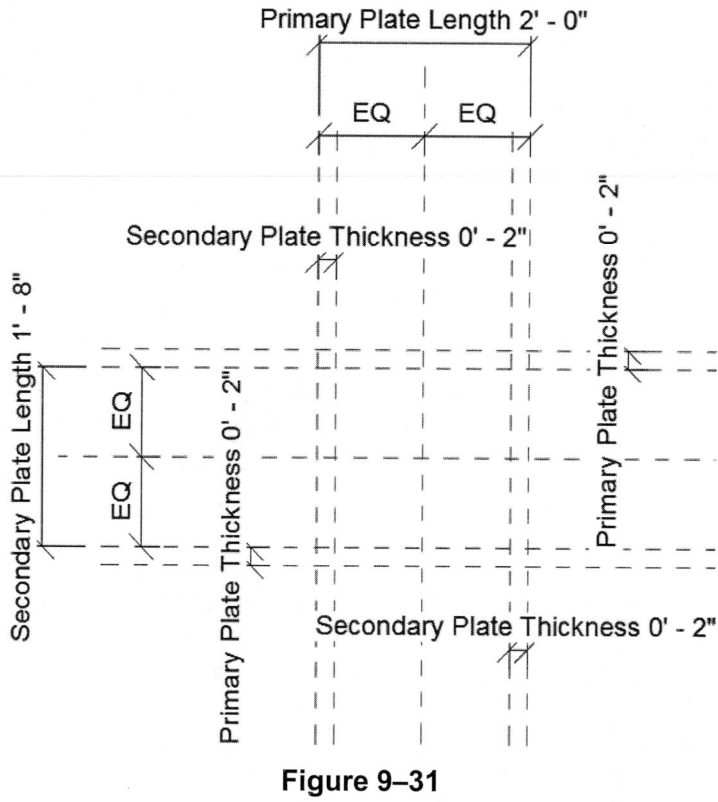

Figure 9–31

9. Flex the sizes to ensure that they are moving together.

10. Save the family.

Task 5 - Create the plates.

1. In the *Create* tab>Forms panel, click (Extrusion) to create the plates. Only one plate can be created at a time because each extrusion must be a closed loop.

2. Sketch the rectangle, based on the start and end points shown in Figure 9–32.
   - Lock the lines to the reference planes.
   - Leave the **Extrusion Start** and **Extrusion End** parameters as default. The column height is adjusted later by aligning with other reference planes.

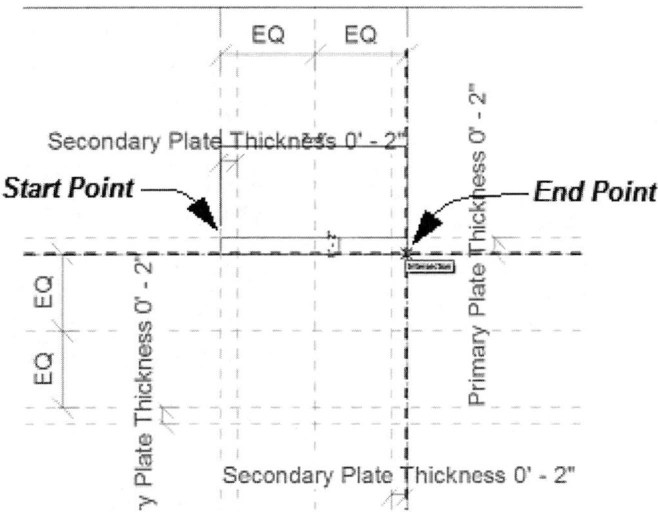

Figure 9–32

3. Click (Finish Edit Mode).

4. Select the primary plate created and mirror it using the center line as the axis of reflection.

5. Repeat the steps to create the secondary plate shown in Figure 9–33.

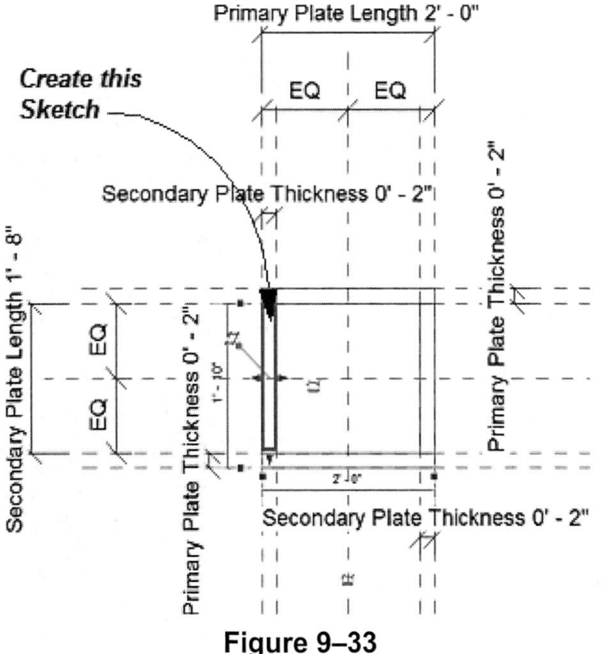

Figure 9–33

6. Flex the sizes to ensure that the extrusions move as expected.

7. Open the **Elevations: Front** view to establish the height of the column.

8. Use (Align) to align the four plates to the Upper Ref. Level. Select the **Upper Ref. Level** and then the top of a plate and lock it in place. Repeat this procedure until all four plates are locked to the **Upper Ref. Level** and **Lower Ref. Level**, as shown in Figure 9–34.

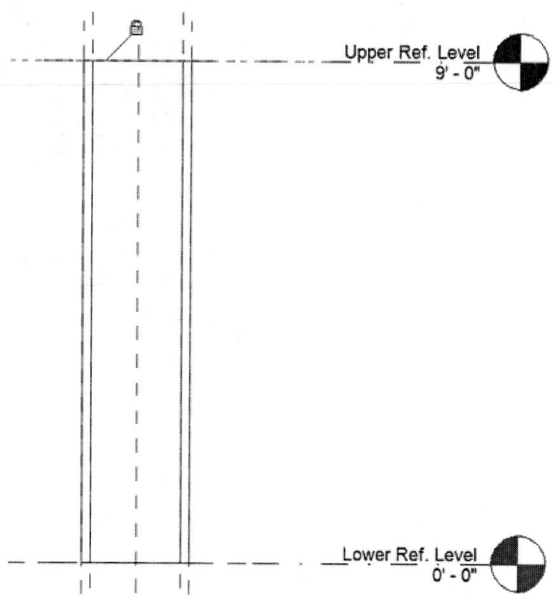

Figure 9–34

9. Set the *Visual Style* of the view to **Hidden Line**. Note that the secondary plates behind the primary plates disappear.

10. Use Symbolic Lines to display the secondary plates with hidden lines. In the *Annotate* tab>Detail panel, click  (Symbolic Line).

11. In the *Modify | Place Symbolic Lines* tab>Subcategory panel Type Selector, select **Hidden Lines (projection)**.

12. In the Options Bar, verify that the **Chain** option is cleared. Draw a line from the intersection of the Upper Ref. Level and the inside reference planes on both sides of the column, as shown in Figure 9–35. Lock these lines to both the Upper and Lower Ref. Levels.

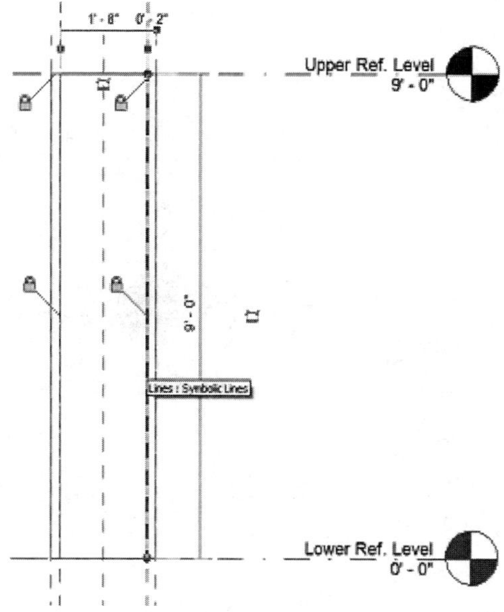

Figure 9–35

13. Flex the sizes and Upper Ref Level Height.

14. Save the family.

15. To test the Built-Up Column, open a new project and load the **Built-Up Column** family into the new project. Insert a column (as shown in Figure 9–36), and navigate from one view to another to see the new column. Add other columns and change the size and offset from the level.

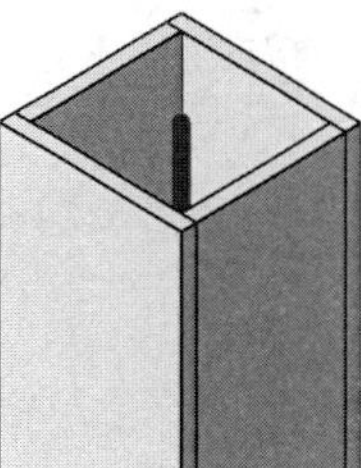

Figure 9–36

Practice 9e Tapered Concrete Column

Practice Objectives

- Create Reference Planes and Parameters as the framework of the column family.
- Create geometry using a solid blend.

Estimated time for completion: 15 minutes

In this practice you will create a tapered concrete column using a column family template. You will create its geometry using a solid blend and add a taper parameter that can be modified as shown in Figure 9–37.

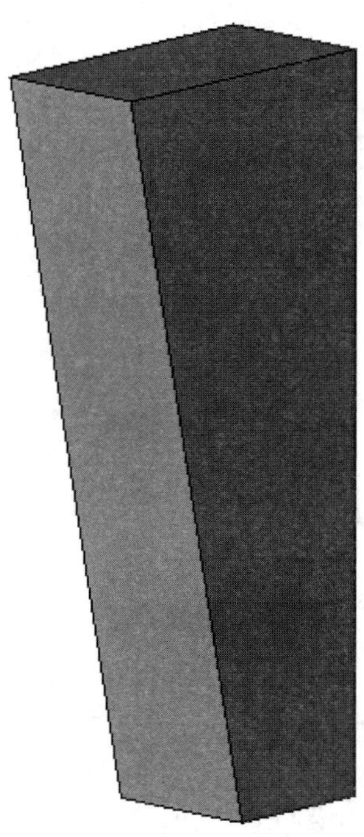

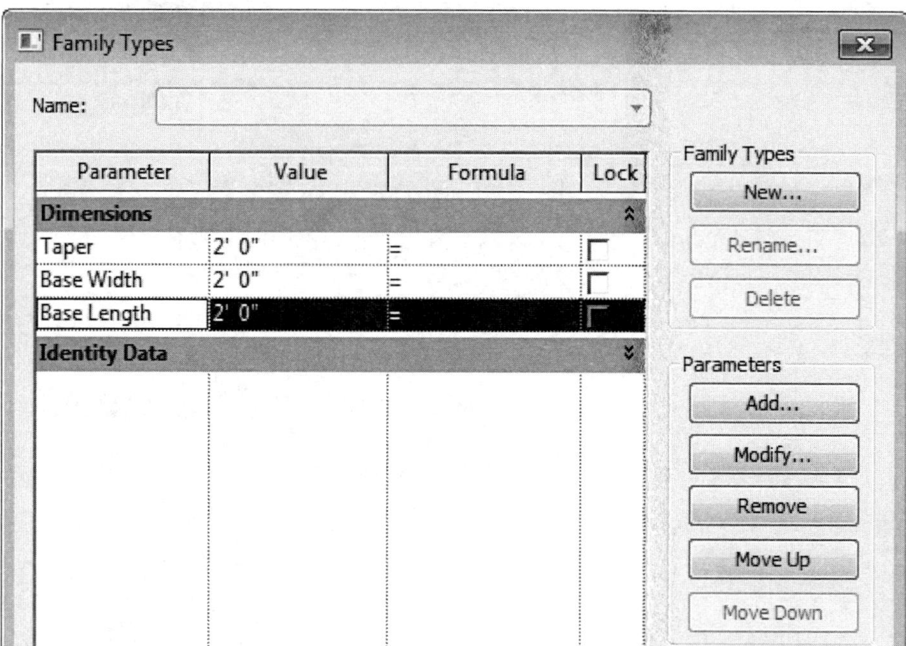

Figure 9–37

Task 1 - Create Reference Planes and Parameters.

1. Create a new family based on the **Structural Column.rft** template found in the Autodesk Revit Library.

2. Save the family to your practice folder as **Tapered-Concrete-Column.rfa**

3. In the *Create* tab>Datum panel, click (Reference Plane) and add one horizontal reference plane offset **2'-0"** from the lower existing horizontal plane. Add a **2'-0"** dimension as shown in Figure 9–38.

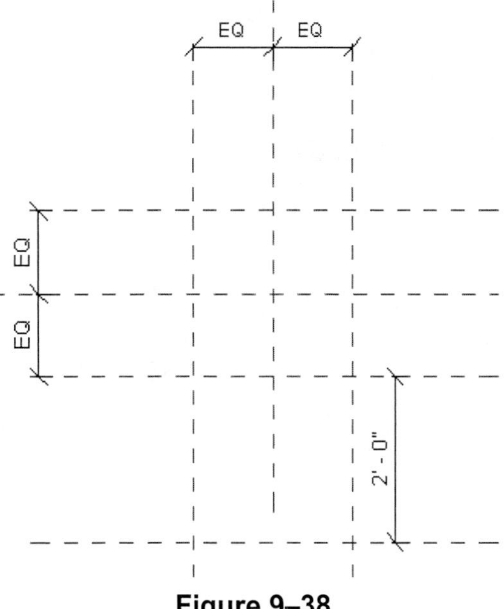

Figure 9–38

4. Select the new reference plane. In Properties, name the reference plane **Taper** as shown in Figure 9–39.

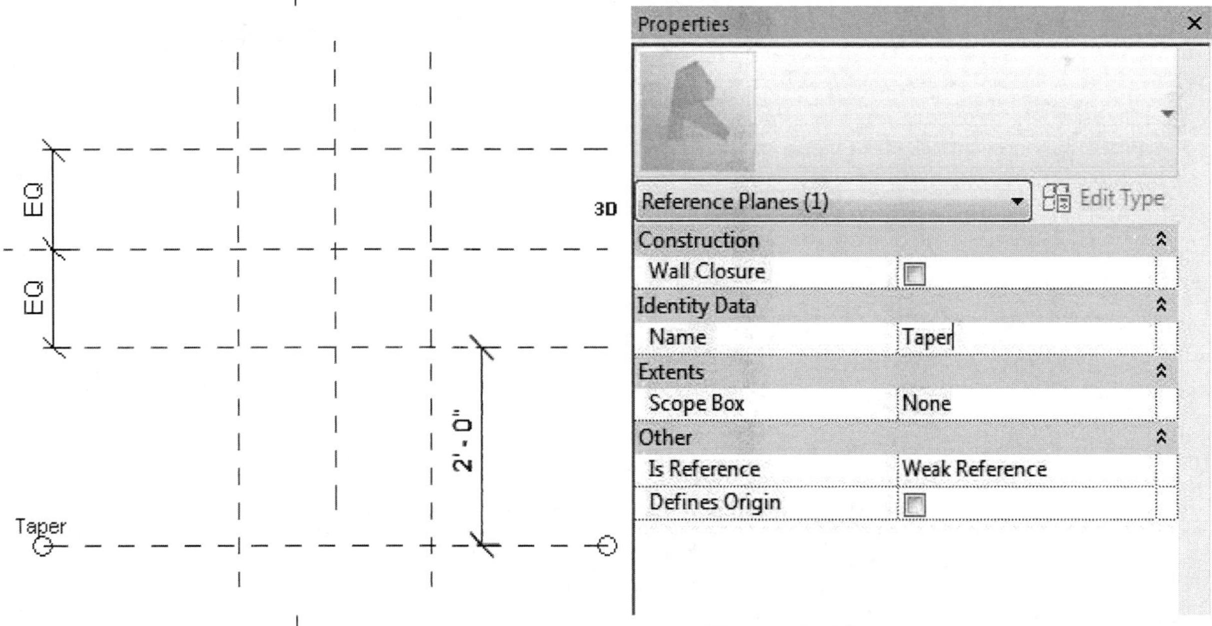

Figure 9–39

5. Select the **2'-0"** dimension. In the Options Bar, in the Label drop-down list, select **<Add parameter...>**.

6. In the Parameter Properties dialog box, set the *Name* to **Taper**, select the **Dimensions** group and **Type** parameter, as shown in Figure 9–40, and click **OK**.

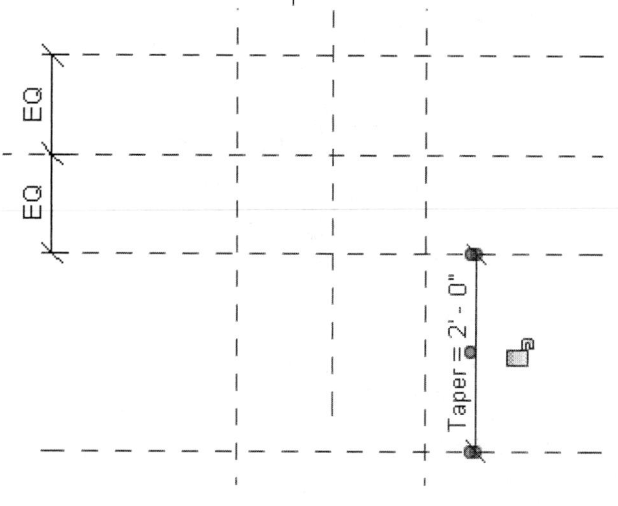

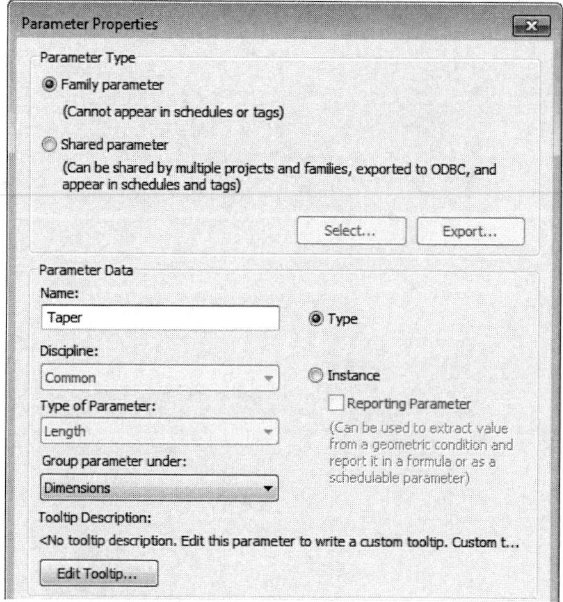

Figure 9–40

7. Add dimensions to the outer horizontal and vertical reference planes representing the base of the column. Label the horizontal and vertical dimensions as **Base Width** and **Base Length**, respectively, as shown in Figure 9–41. Add the parameters to the **Dimensions** group and set them as **Type** parameters.

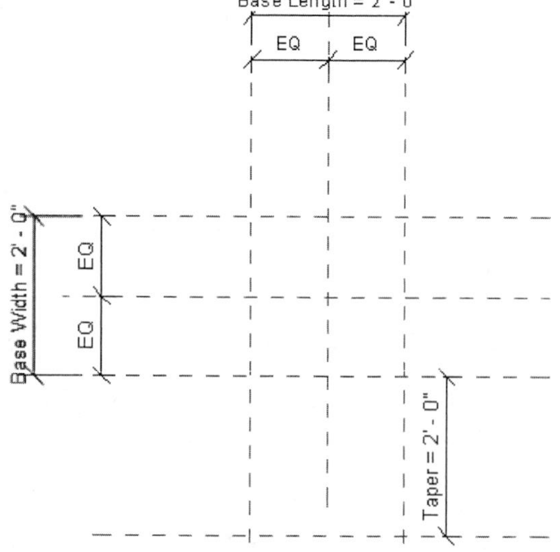

Figure 9–41

8. Flex the sizes to test the framework.

9. Save the family.

Task 2 - Create geometry.

1. In the *Create* tab>Forms panel, click (Blend).
2. The Base profile needs to be established first. In the *Modify | Create Blend Base Boundary* tab>Draw panel, click (Rectangle). Draw a square along the Base reference planes. Lock the lines to the reference planes, as shown in Figure 9–42.

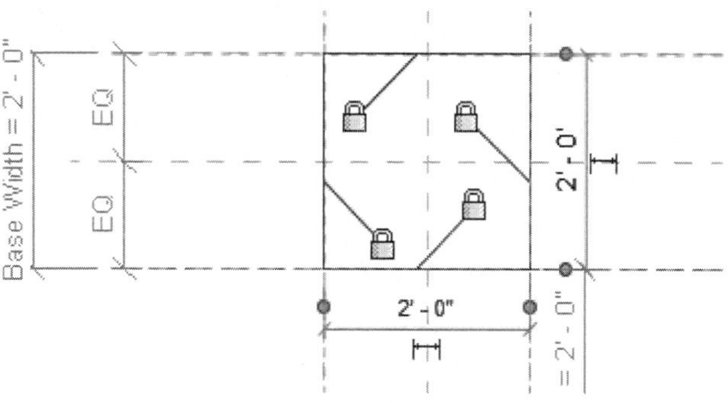

Figure 9–42

3. After completing the outline of the base of the column, the Top of the profile needs to be established. In the *Modify | Create Blend Base Boundary* tab>Mode panel, click (Edit Top).
4. In the Draw panel, click (Rectangle). Draw a **2'x 4'** rectangle along the full extents of the reference planes as shown in Figure 9–43, and lock them.

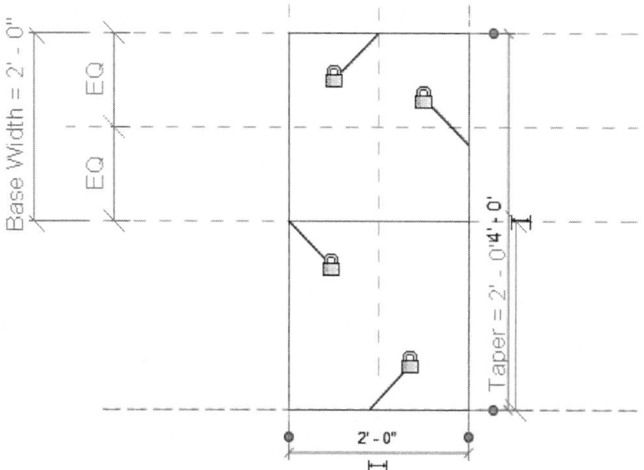

Figure 9–43

5. In the Mode panel, click (Finish Edit Mode).

6. Open the **Elevations: Front** view.

7. In the *Modify* tab>Modify panel, click (Align). Select the **Upper Ref. Level**, pick the top of the column, and lock it into place with the reference level, as shown in Figure 9–44. This is to ensure that the top of the column is always constrained to the Upper Ref. Level.

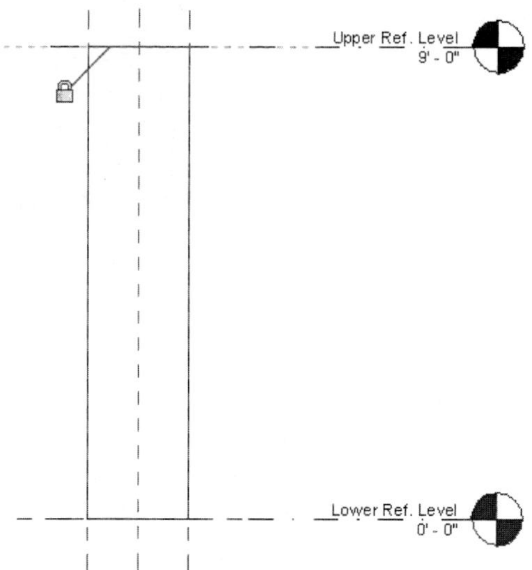

Figure 9–44

8. Open the **3D Views: View 1** view and rotate it to display the constructed version of the tapered concrete column. You can also shade it as shown in Figure 9–45.

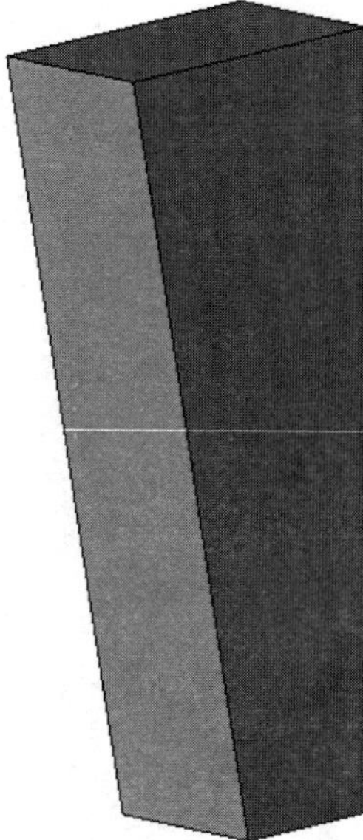

Figure 9–45

9. In the Properties panel, click (Family Types) and test the parameters by changing the values as shown in Figure 9–46.

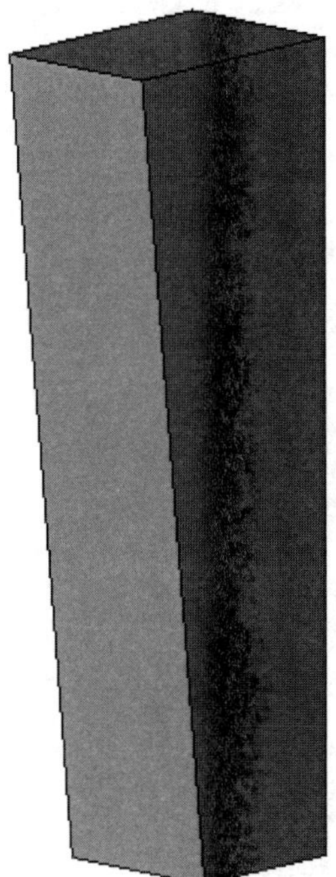

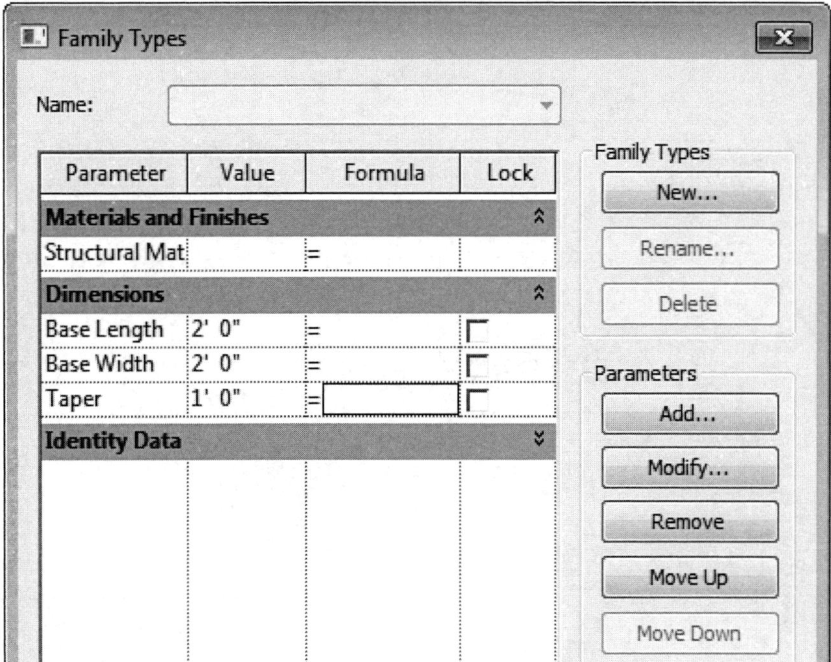

Figure 9–46

10. Reset the default values and close the Family Types dialog box.

11. Save the family.

12. If you have time you can create Family Types using various dimensions and test them in a project.

Practice 9f Truss Family

Practice Objectives

- Start a truss family and add reference planes and labels.
- Add Top Chord, Bottom Chord, and Web layout lines.

Estimated time for completion: 40 minutes

In this practice you will create a truss using a truss family template. You will add reference planes, create parameters, and add chords and webs to form the structure of the trusses, as shown in Figure 9–47. You will then test it in a project.

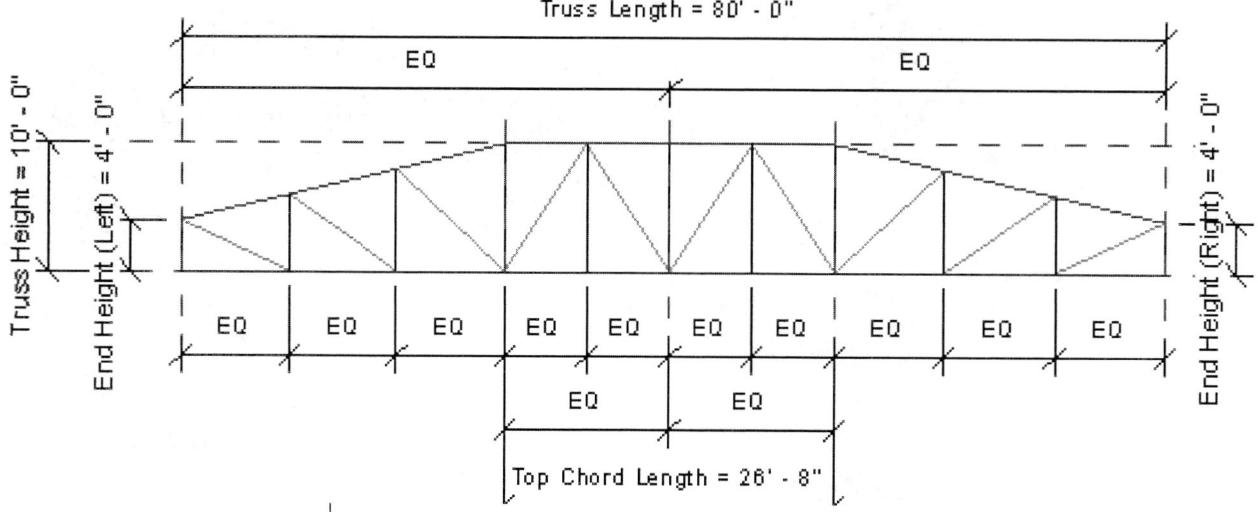

Figure 9–47

Task 1 - Add reference planes and labels.

1. Create a new family based on the **Structural Trusses.rft** template found in the Autodesk Revit Library.
1. Save the family to your practice folder as **Custom Truss.rfa**
2. In the *Create* tab>Datum panel, click (Reference Plane).
3. In the Draw panel, click (Pick Lines).
4. In the Options Bar, set the *Offset* to **10'-0"**.

5. Add a vertical reference plane to each side of the center vertical reference plane as shown in Figure 9–48.

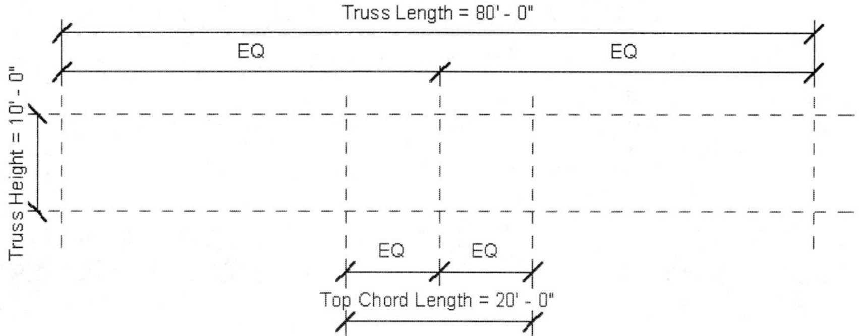

Figure 9–48

6. Dimension as shown previously in Figure 9–48. Label the overall dimension as **Top Chord Length**. When you create the parameter, create it as an **Instance** parameter as shown in Figure 9–49.

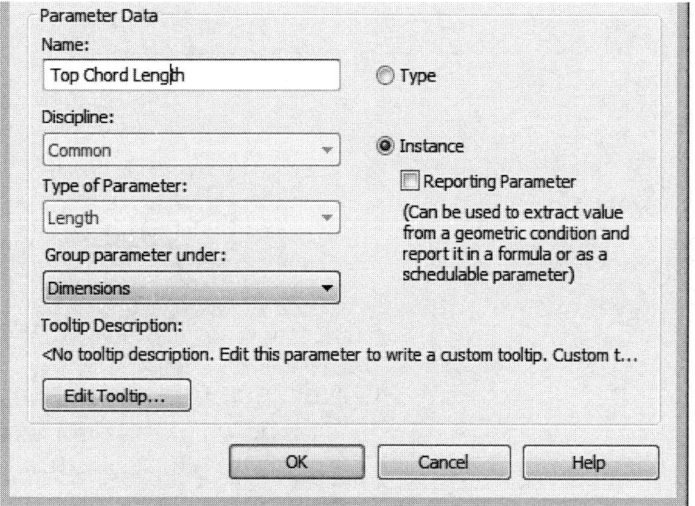

Figure 9–49

7. Open the Family Types dialog box and scroll down to the *Dimensions* area. Add the formula *Top Chord Length* = **Truss Length / 3**, as shown in Figure 9–50.

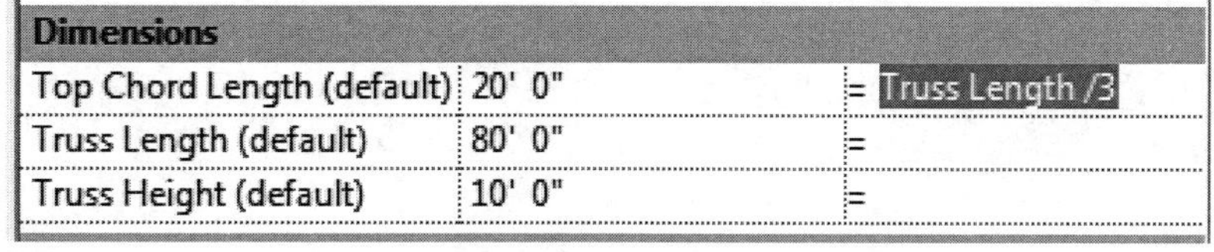

Dimensions		
Top Chord Length (default)	20' 0"	= Truss Length /3
Truss Length (default)	80' 0"	=
Truss Height (default)	10' 0"	=

Figure 9–50

8. Click **OK**.

Task 2 - Add the Top Chord, Bottom Chord, and Web layout lines.

1. In the *Create* tab>Detail panel, click (Top Chord).
2. In the Options Bar, verify that the **Chain** option is cleared so that you can lock each line as it is created.

*Use temporary dimensions to locate the start point or draw an additional reference plane **4'-0"** above the bottom reference plane.*

3. Starting from the left vertical reference plane, draw the top chord from **4'-0"** above the bottom reference plane to the intersection of the next vertical and top horizontal reference planes. Lock the sketch to the reference planes as shown in Figure 9–51.

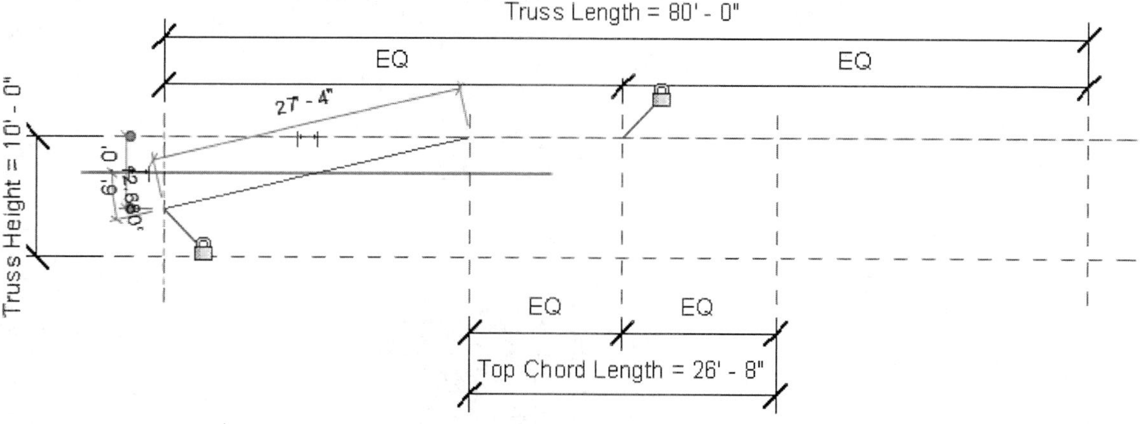

Figure 9–51

4. Continue drawing the chord across the top of the truss and then down to the right vertical reference plane, locking as you go, shown in Figure 9–52. The exact angle for the last web will be fixed later.

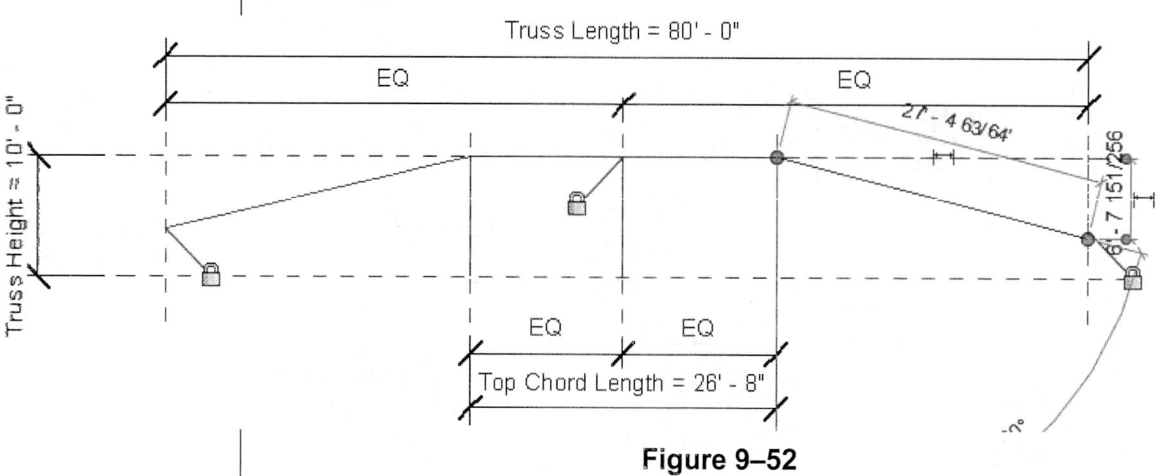

Figure 9–52

5. While you are still in the contextual tab in the Draw panel, click (Bottom Chord). (Alternatively, in the *Create* tab> Detail panel, click (Bottom Chord).)
6. Draw the bottom chord, beginning from the far left reference plane along the bottom reference plane to the far right reference plane. Remember to lock the bottom chord sketch line in place, as shown in Figure 9–53.

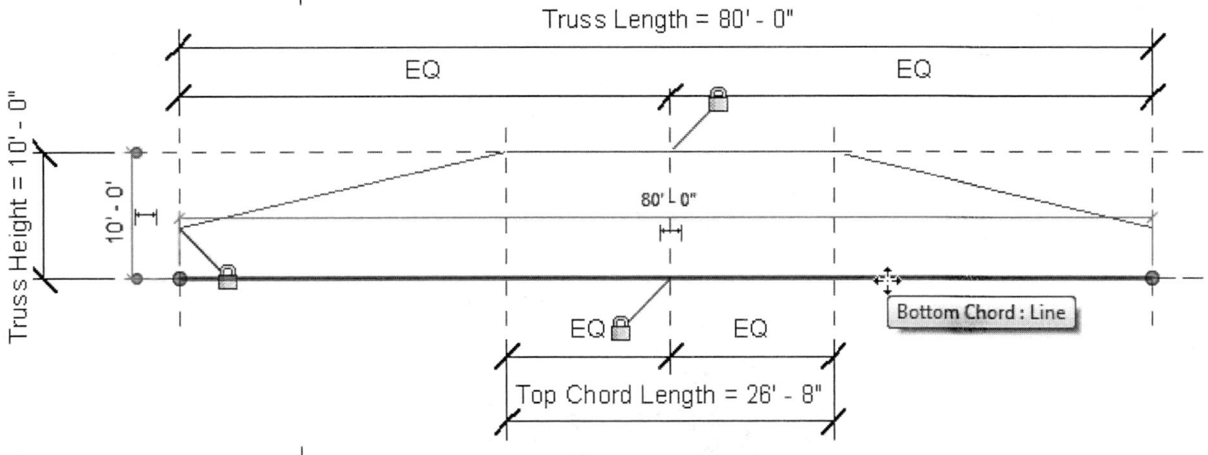

Figure 9–53

7. Add dimensions from the bottom reference plane to the intersection of the far left reference plane and the starting point of the top chord line and repeat on the right. Then create new labels for *End Height (Left)* and *End Height (Right)*, as shown Figure 9–54. When you create the parameter, set it as an **Instance** parameter. Adjust the *End Height (Right)* to **4'-0"** in the Family Types dialog box.

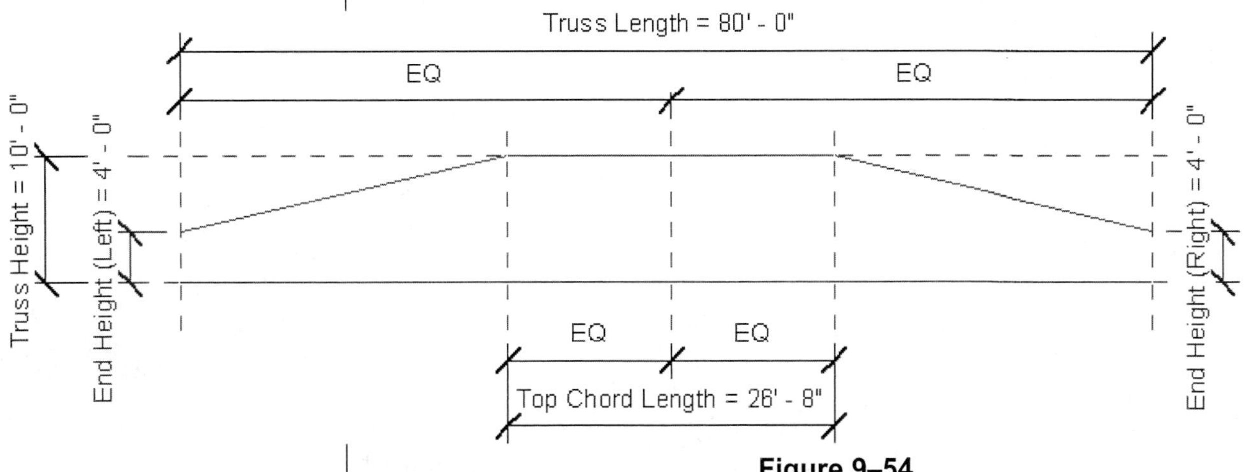

Figure 9–54

8. In the *Create* tab>Detail panel, click (Web). In the Options Bar, verify that the **Chain** option is not selected.

9. Draw vertical web lines along each vertical reference plane, as shown in Figure 9–55.

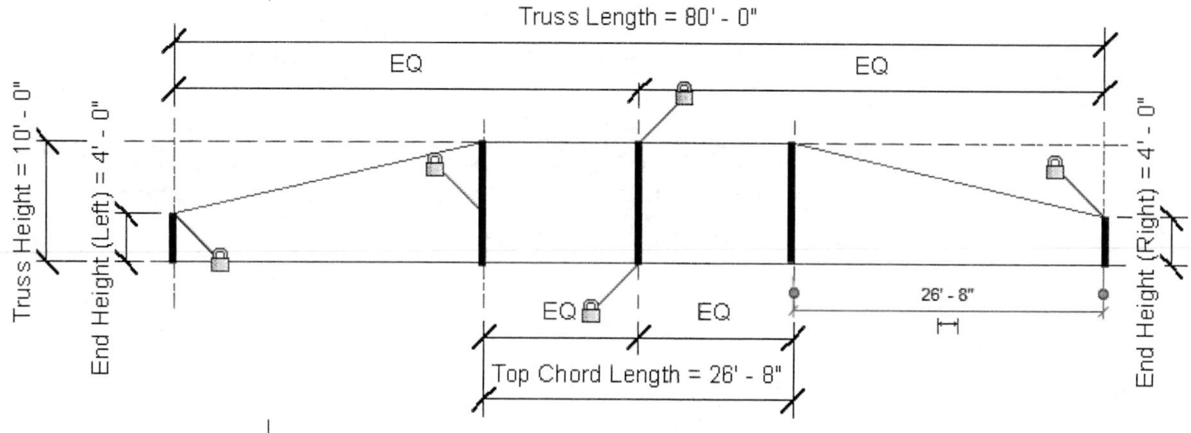

Figure 9–55

- Start at the far left reference plane and work over to the right.
- The webs at the two far ends of the truss should go from the bottom reference plane to the intersection of the top chord line.
- The three web lines along the middle three vertical reference planes should go from the top reference plane to the bottom reference plane.
- Lock all of the web lines to their respective reference planes.

10. Draw two new vertical web lines and dimension them so that they are equal, as shown in Figure 9–56.

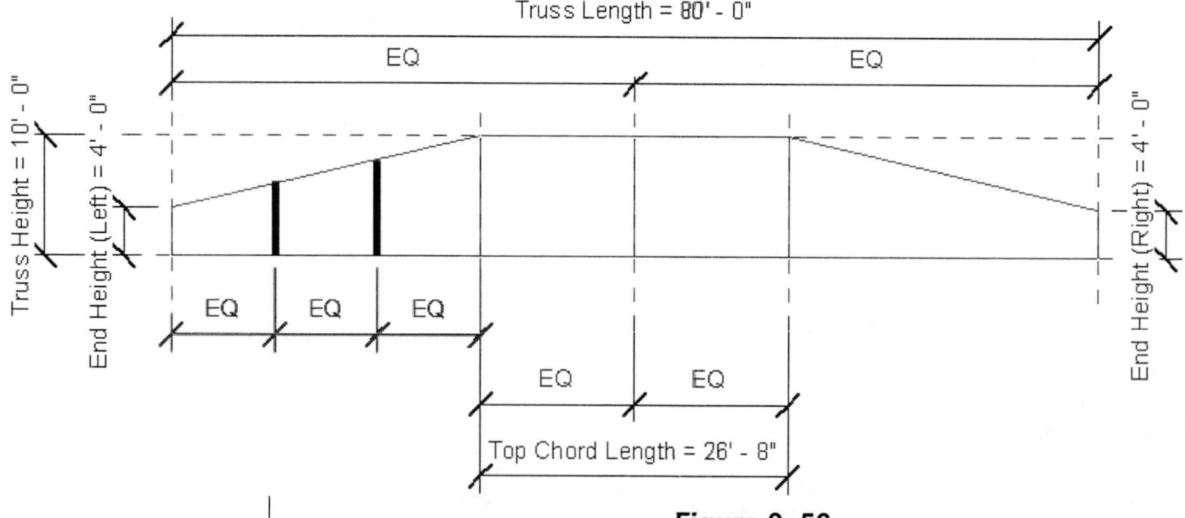

Figure 9–56

- Ensure that you dimension to the vertical reference planes where available or the truss will not flex correctly.

11. Draw one vertical web from the top to the bottom reference planes between the left **Truss Chord Length** reference plane and the center reference plane and dimension it to be equal as shown in Figure 9–57.

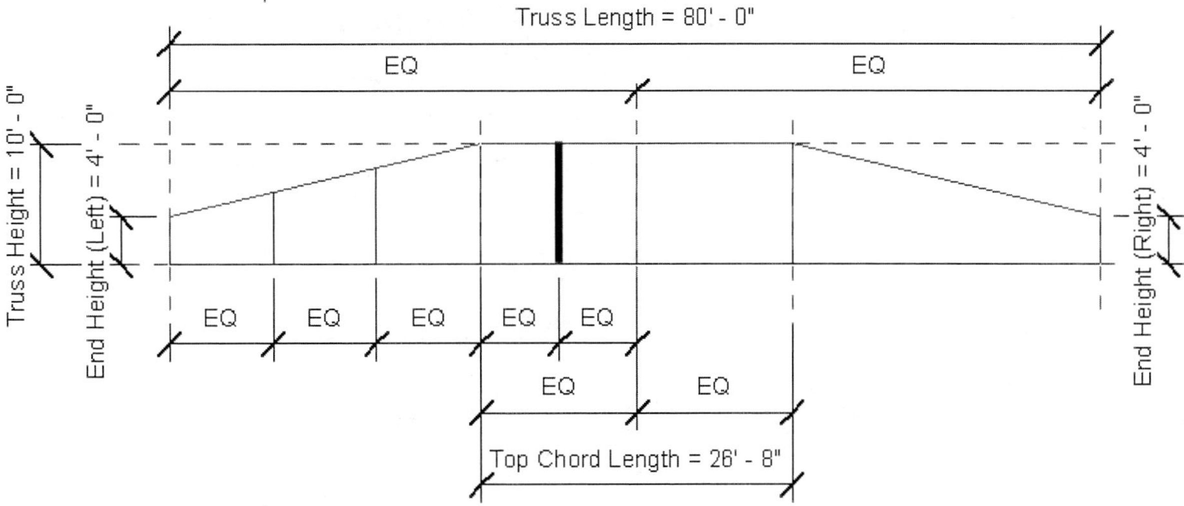

Figure 9–57

12. Repeat the steps to create vertical webs for the right side of the truss. (DO NOT mirror the Web lines as this might cause undesirable results or errors in the truss family.) The final outcome should be as shown in Figure 9–58.

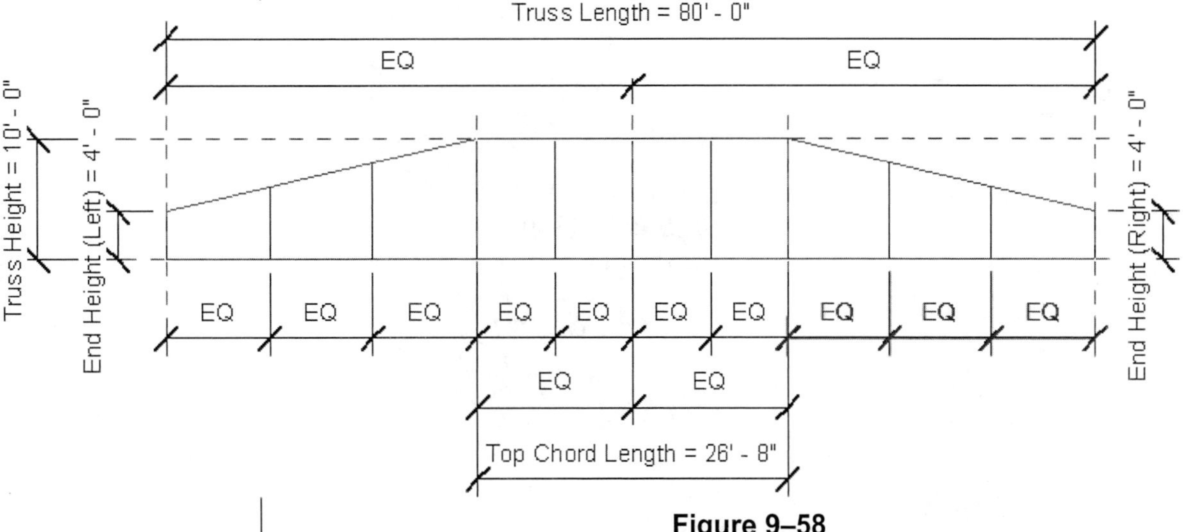

Figure 9–58

13. Add diagonal web members to the left of the truss, from the top chord to the bottom chord. Start from the far left side working toward the center of the truss. Repeat the process from the other direction. The final outcome is shown in Figure 9–59. DO NOT mirror the web lines, as this might cause errors when the truss is loaded into a project.

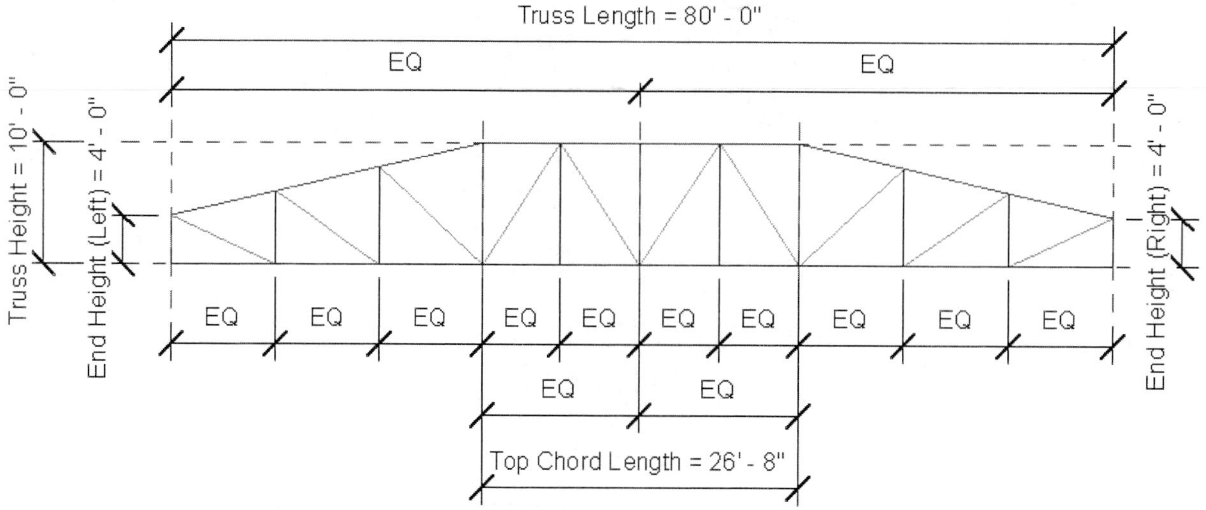

Figure 9–59

14. Using the Family Types dialog box, flex the geometry using the parameters found in the *Dimensions* area.

15. Reset the values to the dimensions shown in Figure 9–59.

16. Save the family.

17. Load the truss into a project and test it.
    - Note that the default framing sizes might be too large. In the truss family (or in the truss type in the project), set appropriate framing members for the chords and webs.

18. Save and close the project and the family.

Practice 9g Precast Hollow Core Slab

Practice Objectives

- Create a family using the Structural Framing - Beams and Braces template.
- Use a CAD file to modify the sketch of an extrusion.
- Assign a material to the extrusion.
- Set visibility settings for Medium and Fine detail levels along with simple geometry for the Coarse detail level.

Estimated time for completion: 20 minutes

In this practice you will use the **Structural Framing - Beams and Braces.rft** file as the base of the hollow core slab. You will modify an existing extrusion and import a CAD drawing to assist in setting up the new profile. You will assign a material to the extrusion and establish the visibility settings for Medium and Fine Detail Levels. You will then draw a second extrusion with a simple rectangular profile, assign the material and set the visibility settings so that it only displays at a Coarse Detail Level as shown in Figure 9–60.

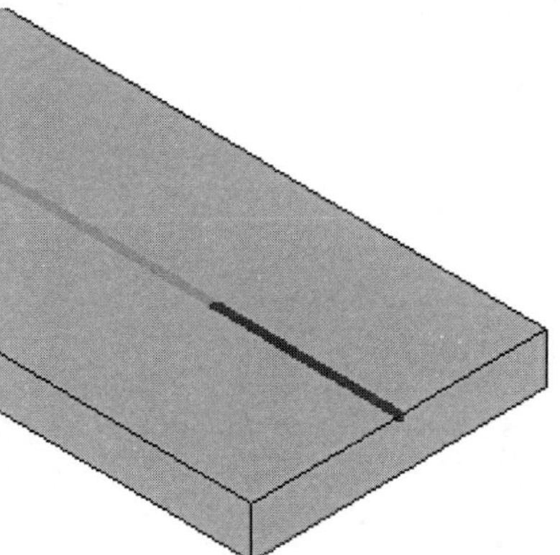

Detail Level: Coarse

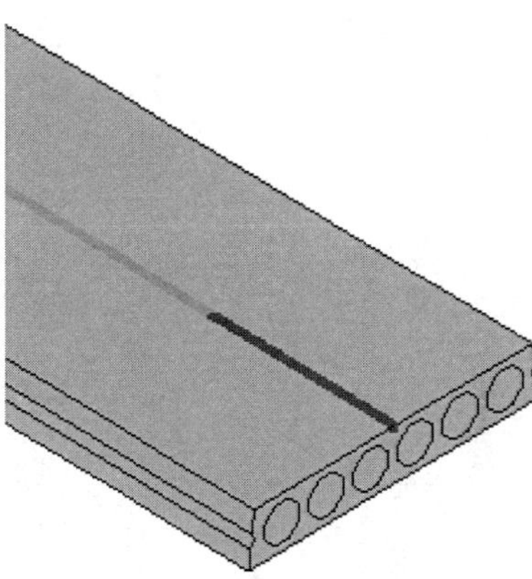

Detail Level: Medium and Fine

Figure 9–60

Task 1 - Import a CAD file into a family template to create the hollow core geometry.

1. Create a new family based on the **Structural Framing - Beams and Braces.rft** template found in the Autodesk Revit Library.
2. Save the family to your practice folder as **Precast-Hollow-Core-Slab.rfa**.
3. Close any other projects or family files that you might have open.

4. Open the following views:
   - **Floor Plans: Ref Level**
   - **Elevations: Front**
   - **Elevations: Left**
   - **3D Views: View 1**

5. Type **WT** (Tile Windows) and then **ZA** (Zoom All to Fit). The layout should be similar to that shown in Figure 9–61. This helps you to develop the precast hollow core slab.

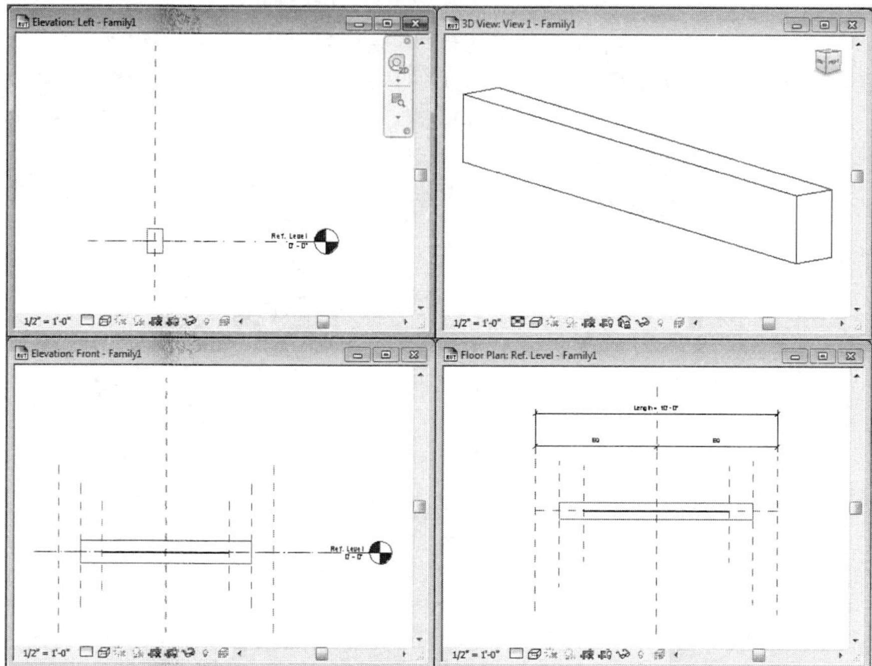

Figure 9–61

6. In the **Elevations: Left** view, zoom in on the geometry.

7. Select the extrusion. In the *Modify | Extrusion* tab>Mode panel, click (Edit Extrusion).

8. Select the outline geometry (as shown in Figure 9–62) and the reference planes that are part of the extrusion and delete them from the view.

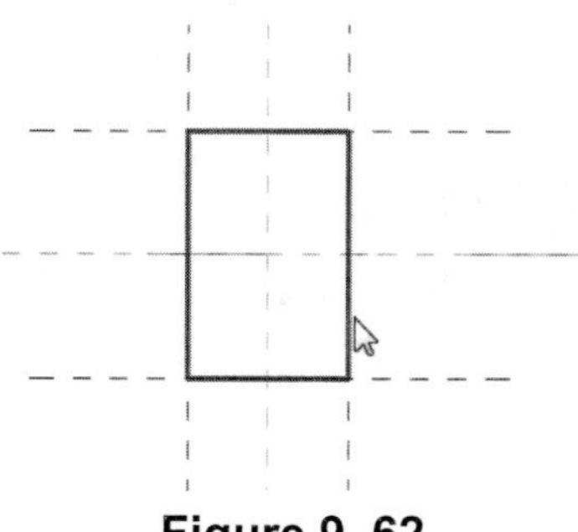

Figure 9–62

In this example, a CAD drawing is used to create the geometry of the hollow core slab.

9. In the *Insert* tab>Import panel, click (Import CAD).
10. In the Import CAD Formats dialog box, navigate to the ...*Structural* folder. Select **Precast-Hollow-Core-Slab.dwg** and then and set the following options, as shown in Figure 9–63:
    - **Select** Current View Only
    - **Colors:** Black and White
    - **Layers:** All
    - **Import units:** Auto-Detect
    - **Positioning:** Auto - Center to Center

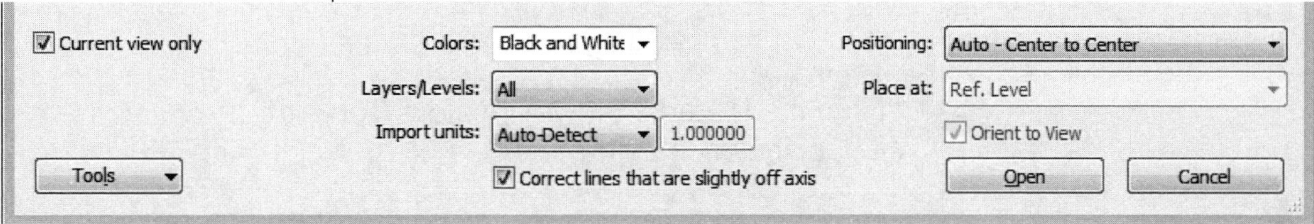

Figure 9–63

11. Click **Open**.
12. A message box opens prompting you about exploding import instances. Click **Yes** to continue. The family now includes the sketch lines from the AutoCAD drawing as shown in Figure 9–64.

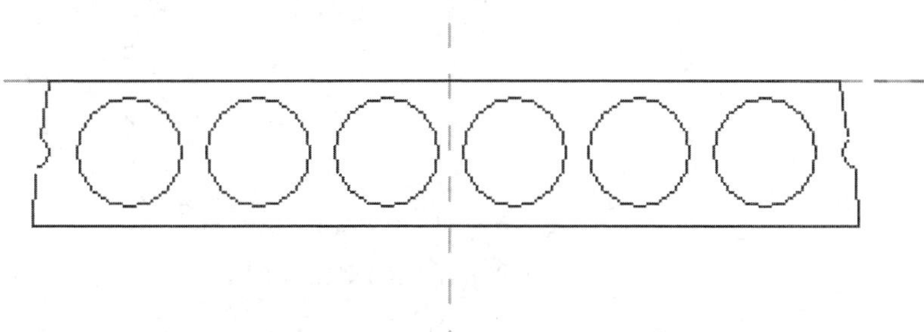

Figure 9–64

13. To select the material for the hollow core slab, in Properties, in the *Materials and Finishes* area, select **<By Category>**, and then click (Browse), as shown Figure 9–65.

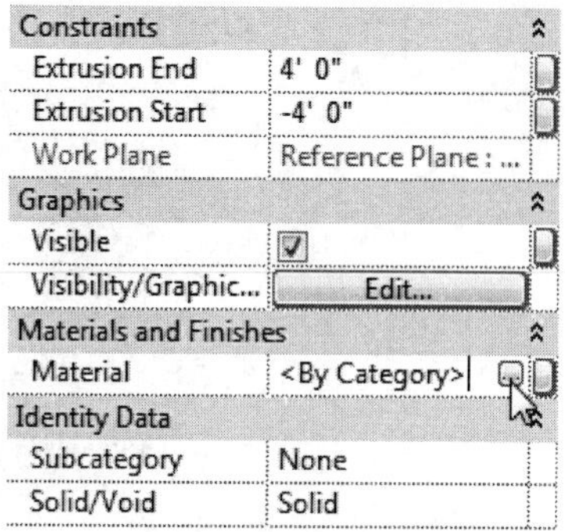

Figure 9–65

14. In the Material Browser, select **Concrete-Precast Concrete** as shown in Figure 9–66.

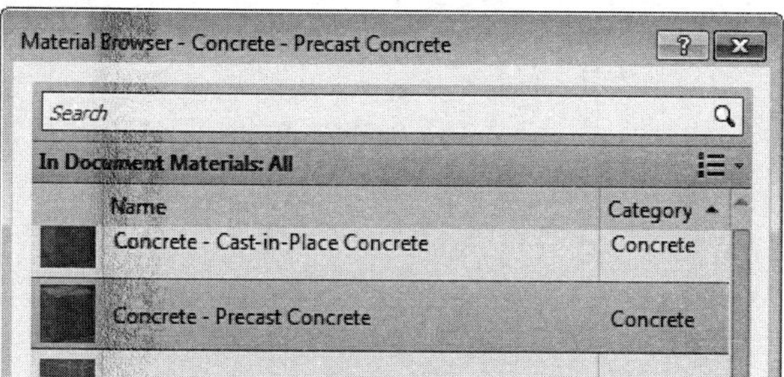

Figure 9–66

15. Click **OK** to close the dialog boxes. The Material is assigned.

16. Click (Finish Edit Mode).

17. Click in empty space to release the selection.

18. In the 3D view, zoom out as needed to display the slab. Set the *Visual Style* to (Shaded) to display the new slab with the concrete material applied, as shown in Figure 9–67.

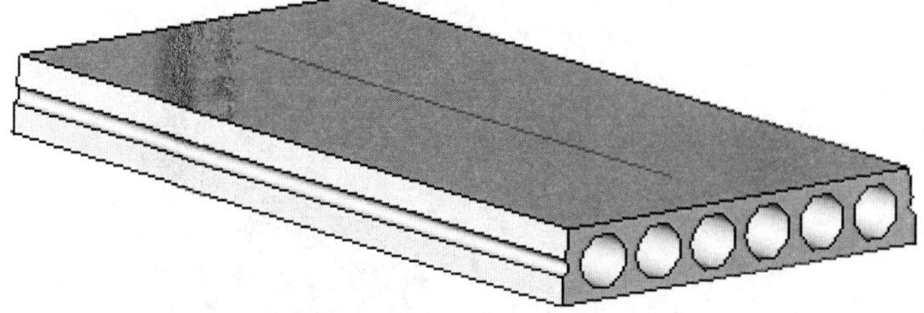

Figure 9–67

19. Save the file.

Task 2 - Establish visibility settings.

1. Select the new extrusion.
2. In the *Modify | Extrusion* tab>Mode panel, click (Visibility Settings).
3. In the Family Element Visibility Settings dialog box, in the *Detail Levels* area, verify that **Coarse** is cleared as shown in Figure 9–68.

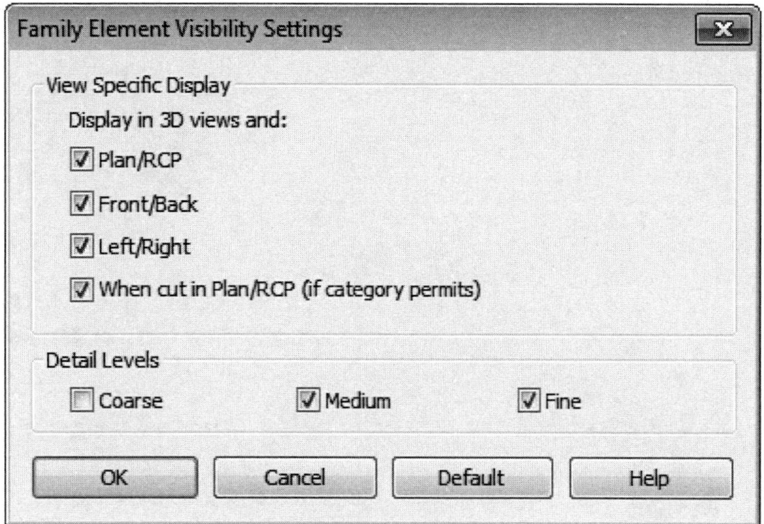

Figure 9–68

4. Click **OK**.
5. Clear the extrusion and test the Detail Levels. In the Coarse detail level, the extrusion should display in gray. This is only true in the family file. When inserted into a project this extrusion will not display at a Coarse detail level.
6. Activate the **Elevations: Left** view.
7. In the *Create* tab>Forms panel, click (Extrusion).
   - If a Work Plane dialog box opens, verify that **Pick a plane** is selected, and click **OK**. Click the face of the existing extrusion to select it as the work plane.

8. In the *Modify | Create Extrusion* tab>Draw panel, click (Rectangle) and draw a rectangle around the perimeter of the hollow core slab starting at the lower left corner. (When you are near the upper right corner, press <Tab> once to force the 4'-0" temporary dimension.) Lock all of the lines in place with the padlock icons. (There might only be three, as shown in Figure 9–69.)

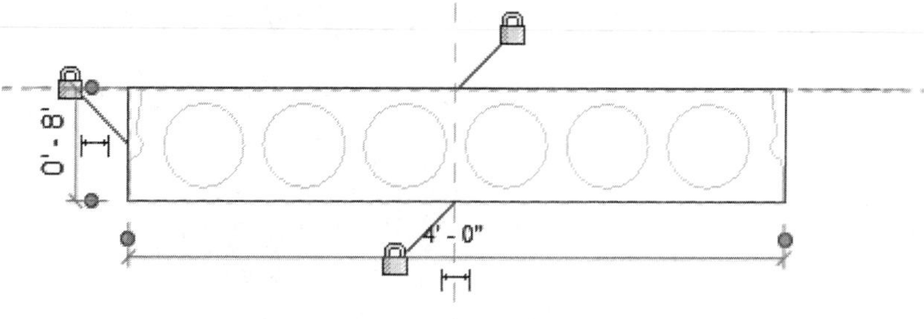

Figure 9–69

9. In Properties, set the *Extrusion End* to (negative) **-8'-0"** and the *Material* to **Concrete - Precast Concrete**.

10. Click (Finish Edit Mode).

11. Activate the **Floor Plans: Ref. Level** view.

12. Use the **Align** command to align the new extrusion to the **Member Left** reference plane as shown in Figure 9–70.

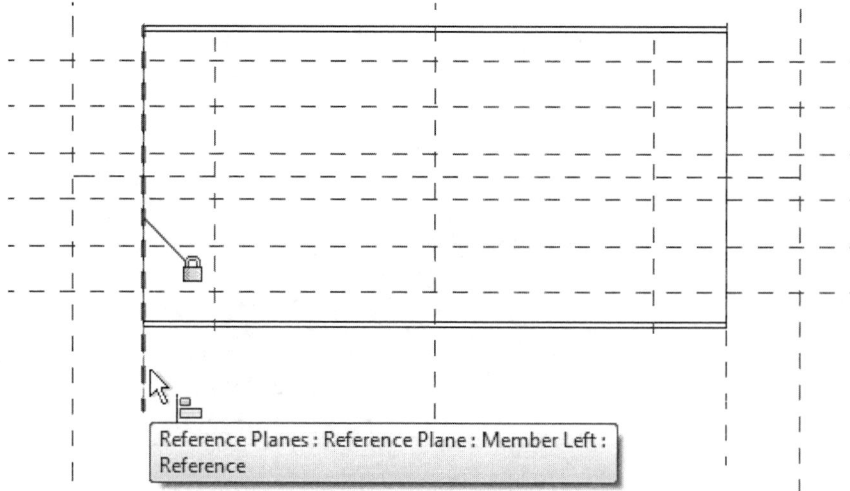

Figure 9–70

13. Repeat and align the other end to the **Member Right** reference plane.

14. Select the new rectangular extrusion.

15. In the *Modify | Extrusion* tab>Mode panel, click (Visibility Settings).

16. In the Family Element Visibility Settings dialog box, in the *Detail Levels* area, clear **Medium** and **Fine** as shown in Figure 9–71.

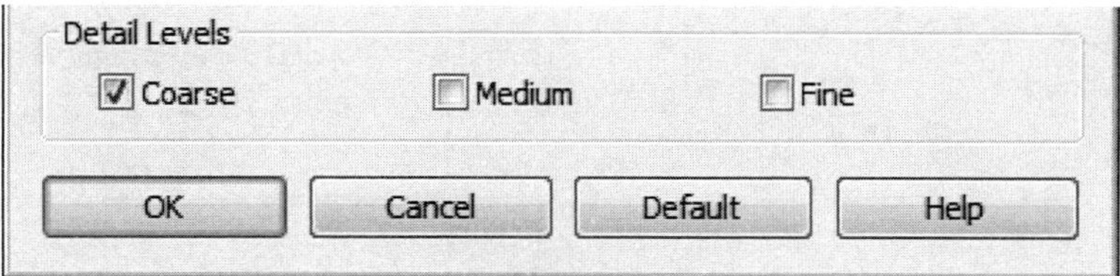

Figure 9–71

17. Click **OK**.

18. Release the selection and try the different Detail Levels.

Task 3 - Add hidden lines.

1. Activate the **Floor Plans: Ref. Level** view.

2. In the View Control Bar, set the *Visual Style* to (Hidden Line).

3. Select the heavy model line in the middle of the slab as shown in Figure 9–72 and delete it.

This line graphically represents the beam on coarse detail plan views. Because you are creating a slab system rather than a beam, the model line is not required.

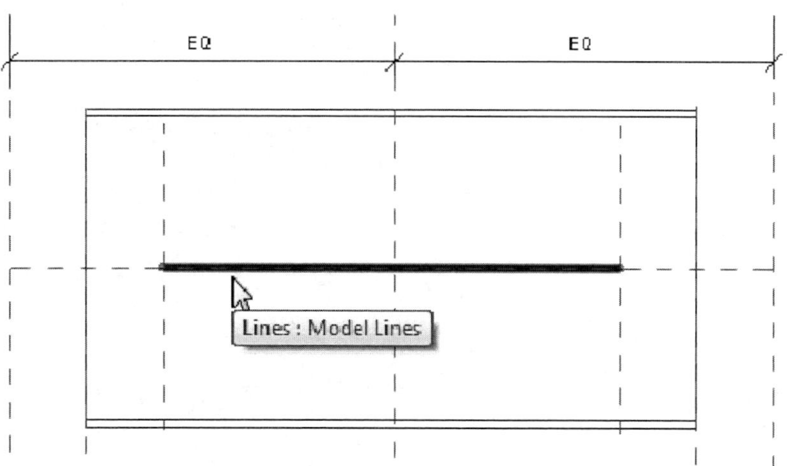

Figure 9–72

4. In the View Control Bar, set the *Visual Style* to (Wireframe).

5. In the *Annotate* tab>Detail panel, click (Symbolic Line).

6. In the *Modify | Place Symbolic Lines* tab>Subcategory panel, set the *Subcategory* to **Hidden Lines [projection]**.

7. In the *Modify | Place Symbolic Lines* tab>Draw panel, click (Line). In the Options Bar, clear the **Chain** option.

8. Draw hidden lines horizontally along the edges as shown in Figure 9–73. You cannot see them as you draw because they are in the same location as the elements displayed through the wireframe view.

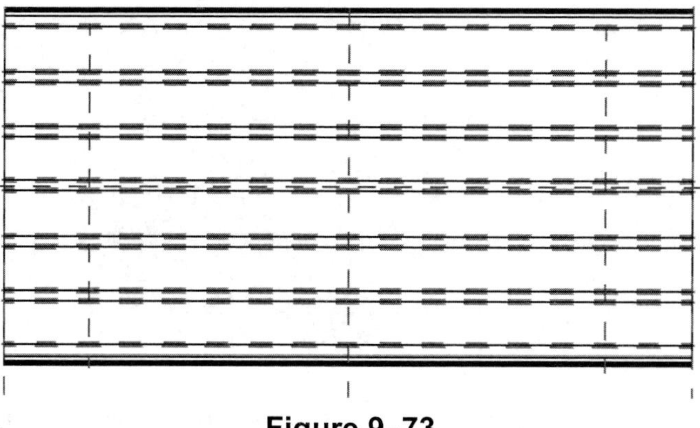

Figure 9–73

9. In the View Control Bar, set the *Visual Style* to (Hidden Line). The hidden lines should be displayed.

10. Click (Modify).

11. Select the 12 hidden lines that you just created.

12. In the *Modify | Lines* tab>Visibility panel, click (Visibility Settings).

13. Set the visibility of only the lines to the **Medium** and **Fine** *Detail Levels*.

14. Save the family in your practice files folder.

15. Load the family into a project.

16. Using (Beam) draw a beam using the new **Precast-Hollow-Core-Slab** family.

17. Test the visibility settings in the plan and elevation views.

18. Close but do not save the project.

19. Save and close the family.

Practice 9h Tapered Moment Frame

Practice Objectives

- Set up reference planes, reference lines, and parameters as the framework of the family.
- Create geometry for the columns and beams of the family.

Estimated time for completion: 40 minutes

In this practice you will use the family template file **Structural Column.rft** as the basis of the Tapered Moment Frame. You will create reference planes, reference lines, and parameters and then create the geometry for the columns and tapered beam using both solid sweeps and extrusions, as shown in Figure 9–74.

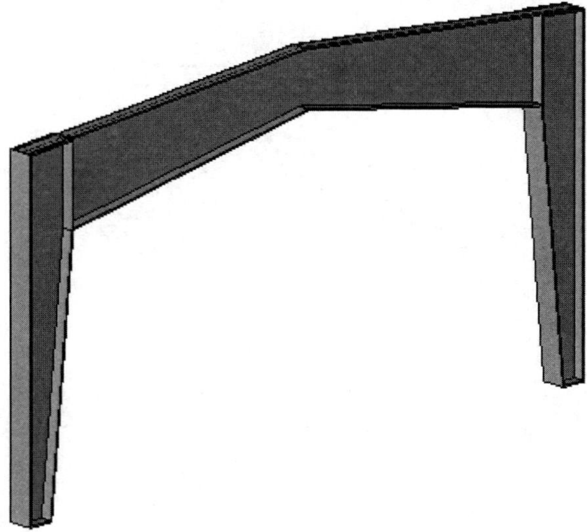

Figure 9–74

Task 1 - Set up the reference planes and parameters.

1. Create a new family based on the **Structural Column.rft** template found in the Autodesk Revit Library.
2. Save the family to your practice folder as **Tapered-Moment-Frame.rfa**.

3. Draw dimensions and add labels for **Overall Length** and **Column Width** to the existing reference planes as shown in Figure 9–75. Create them as **Type** parameters.

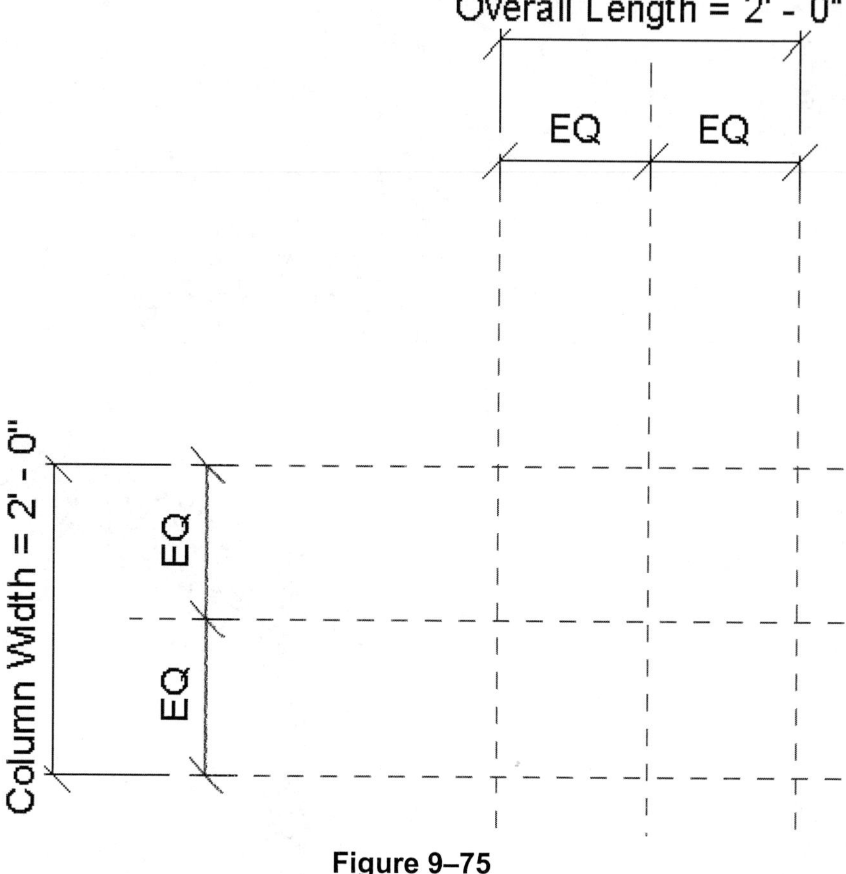

Figure 9–75

4. Open the Family Types dialog box

5. In the Family Types dialog box, change the values of *Overall Length* to **26'-0"** and *Column Width* to **1'-0"** (as shown in Figure 9–76), and then click **OK**.

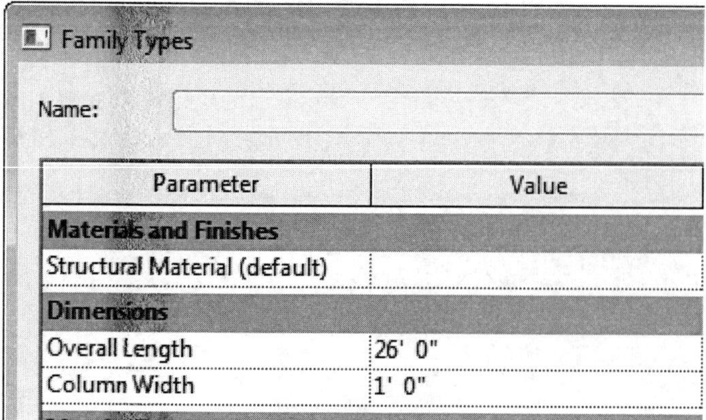

Figure 9–76

6. Zoom out and extend the horizontal reference planes beyond the new vertical reference planes, as shown in Figure 9–77.
   - Unpin the **Front/Back** reference plane to lengthen it. Click on the **Pin** icon to pin the reference plane back in place so that it will not move when you add dimensions and labels.

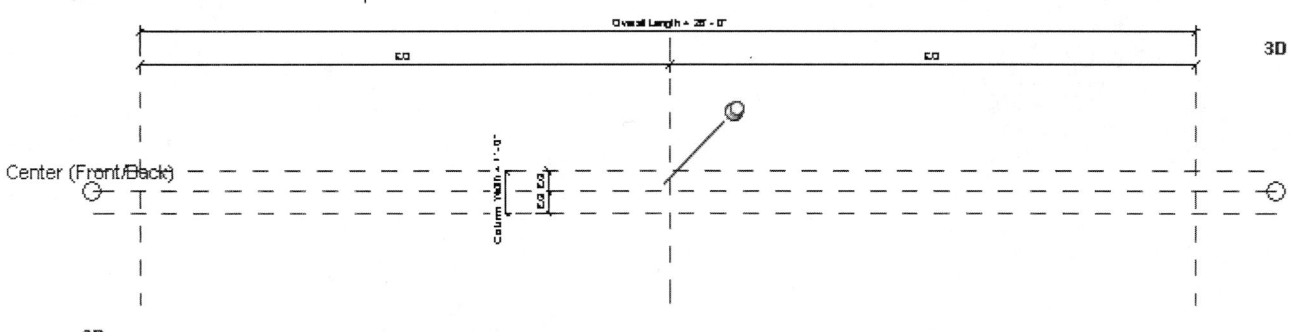

Figure 9–77

7. Add two new horizontal reference planes with associated dimensions and label (type parameter) to represent the **Beam Width**, as shown in Figure 9–78.
   - Hint: Use **Pick Lines** with an *Offset* of **2"** and place them inside the existing outer reference planes.

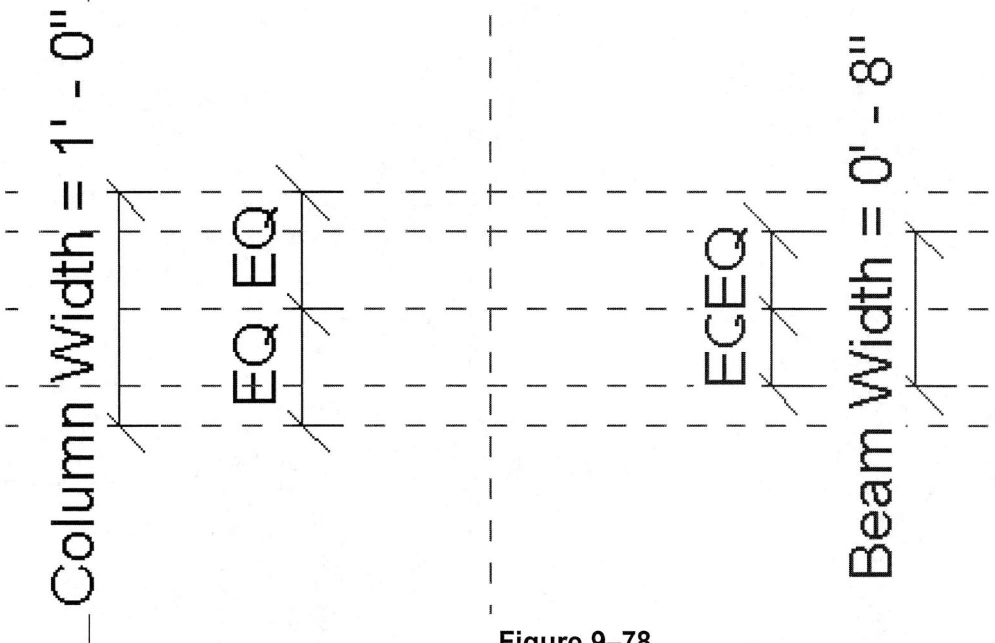

Figure 9–78

8. Open the **Elevations: Front** view.
9. Change the *elevation* of the Upper Ref. Level to **13'-0"**.
10. Extend the reference planes and levels so that they all overlap.

11. Add reference planes and labels (type parameters), as shown in Figure 9–79. Also add a reference plane on top of the Upper Ref. Level and lock it into the level.

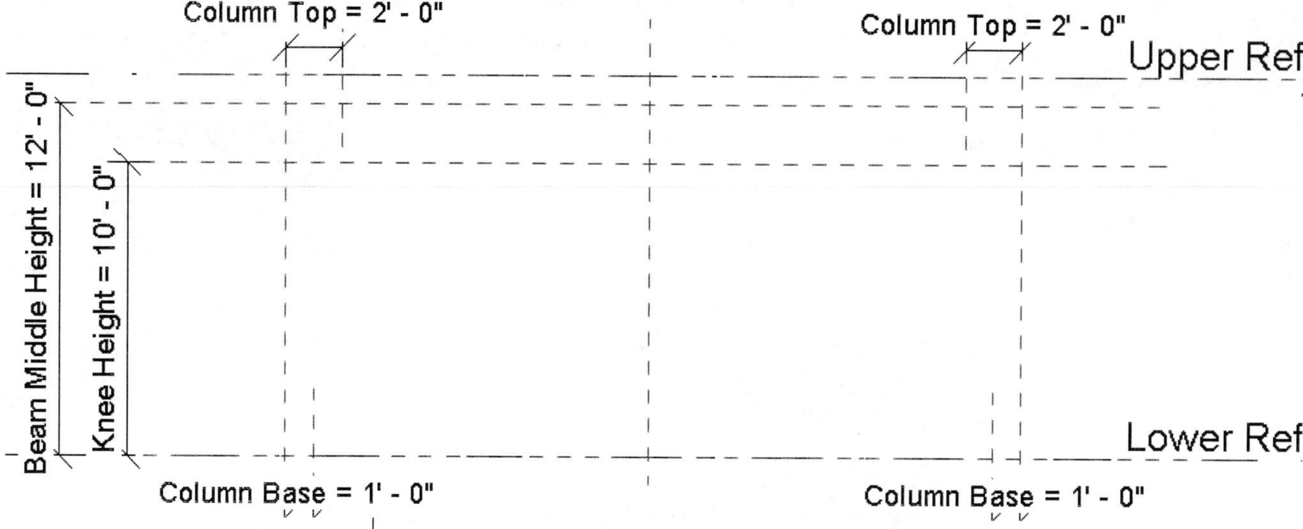

Figure 9–79

12. Temporarily hide the Upper Ref. Level, but not the associated reference plane.

13. Draw two reference lines from the intersection of the top and outside reference planes to the center vertical reference plane at an angle of about 5 degrees.
    - As you draw the reference lines, ensure that you lock the ends of the lines to the related reference planes.

14. Dimension and label the reference lines as **Beam Angle** (type parameter), as shown in Figure 9–80.

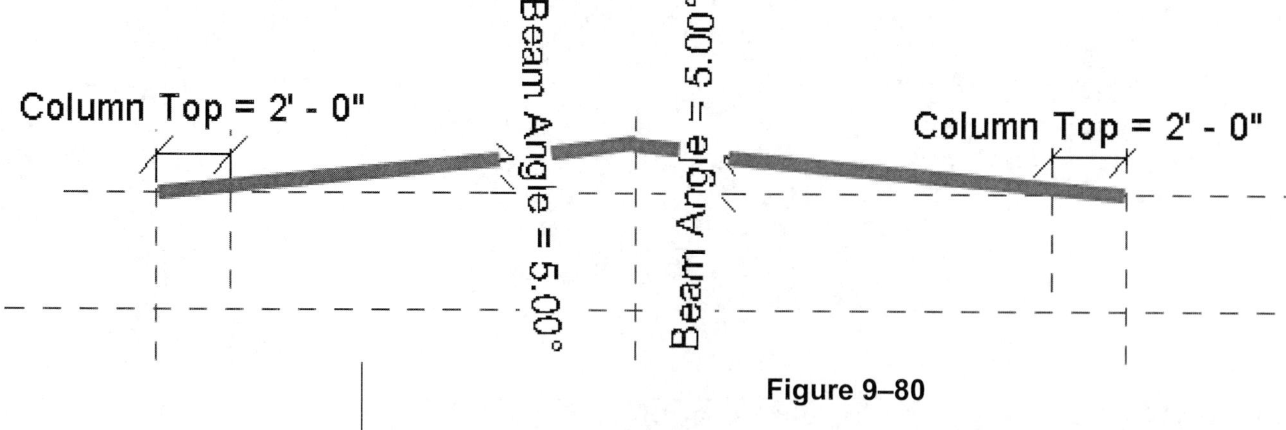

Figure 9–80

15. Flex the framework by changing the label dimensions.

16. Return to the dimensions shown in Figure 9–79 and Figure 9–80. Save the family file.

Task 2 - Create the geometry for the Tapered Columns.

1. In the *Create* tab>Forms panel, click (Sweep).

2. In the *Create* tab>Work Plane panel, click (Set).

3. In the Work Plane dialog box, select **Reference Plane: Center (Front/Back)** as shown in Figure 9–81. Click **OK**.

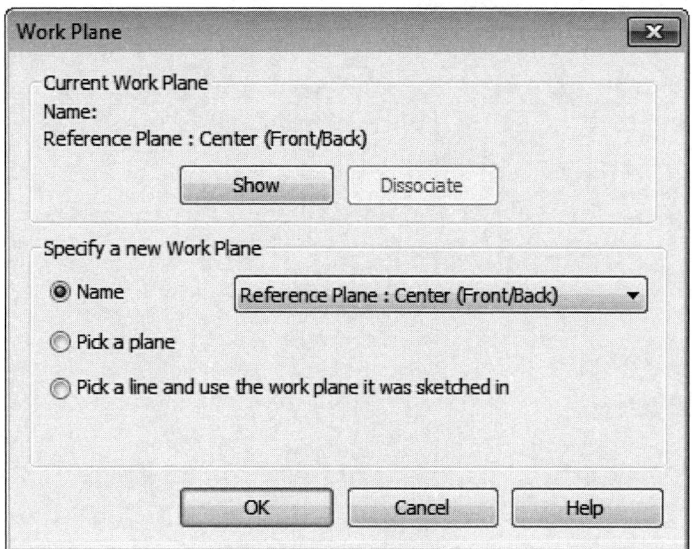

Figure 9–81

4. Click **Modify**.

5. In the *Modify | Sweep* tab>Sweep panel, click (Sketch Path).

6. In the Draw panel, (Line) is selected by default. In the Options Bar, verify that the **Chain** option is selected.

7. Sketch the path for the column of the moment frame. Align the two vertical sketch lines to their parallel reference planes.

8. Align and lock the angular sketch line to the reference line, as shown in Figure 9–82.

Draw the left vertical sketch first. The profile is drawn in this plane.

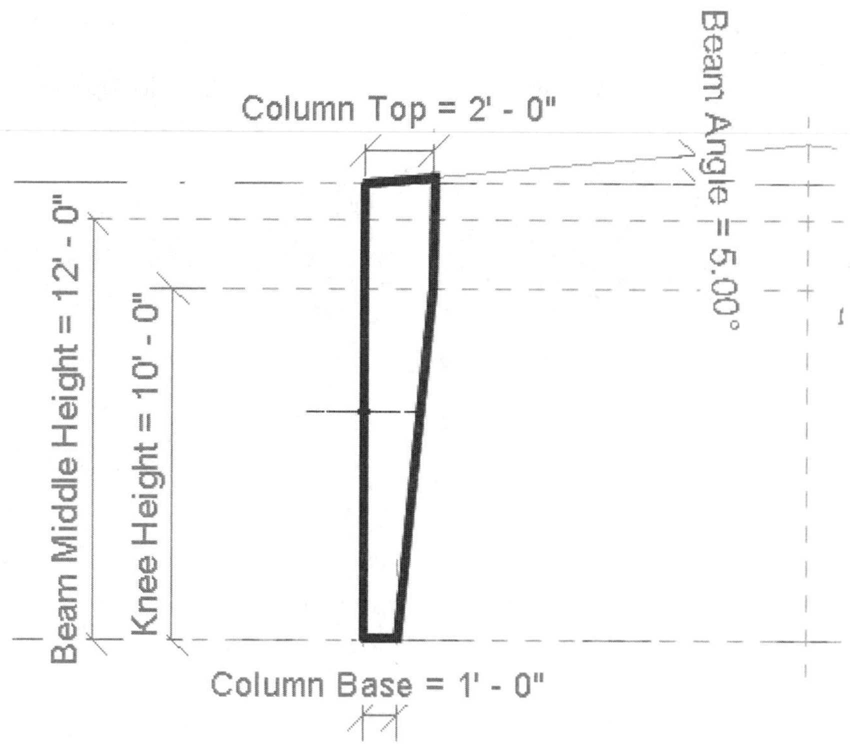

Figure 9–82

9. In the Mode panel, click (Finish Edit Mode).

10. In the *Modify* | *Sweep* tab>Sweep panel, click (Edit Profile).

11. When the Go To View dialog box opens, select **Floor Plan: Lower Ref. Level**. Click **Open View**.

12. If you have not already done so, zoom out to see the full framework and expand the Reference lines to the ends.

13. Zoom in on the Profile target and draw a rectangle **3/4" x 1'-0"**. Lock it to the reference planes, as shown in Figure 9–83.

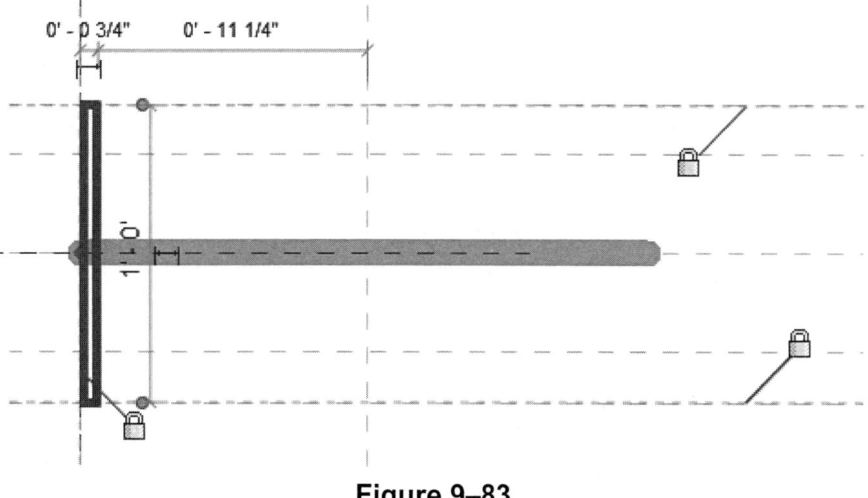

Figure 9–83

14. Click (Finish Edit Mode) twice to finish the Profile and the Sweep.

15. Open the **3D Views: View 1** view. Rotate and shade it to display the column up to this point as shown in Figure 9–84.

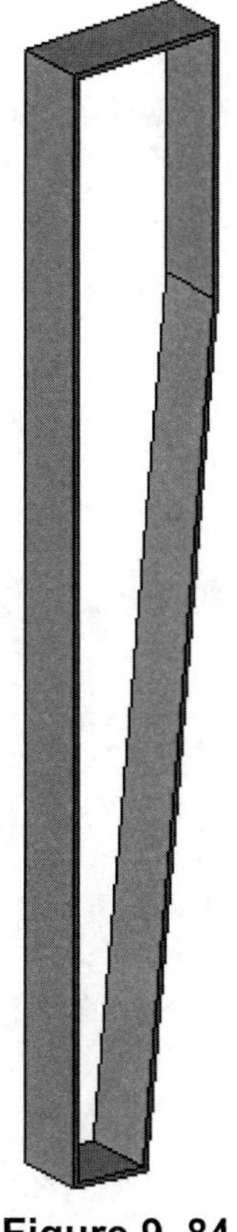

Figure 9–84

16. Return to the **Elevations: Front** view.

17. Click (Extrusion) and set the *Work Plane* to **Reference Plane: Center (Front/Back)**.

18. In the *Modify | Create Extrusion* tab>Draw panel, click (Pick Lines). In the Options Bar, select the **Lock** option.

19. Zoom into the sweep that was created earlier and select all of the interior lines, as shown in Figure 9–85.

*The lines are automatically locked to the selected lines because you selected the **Lock** option.*

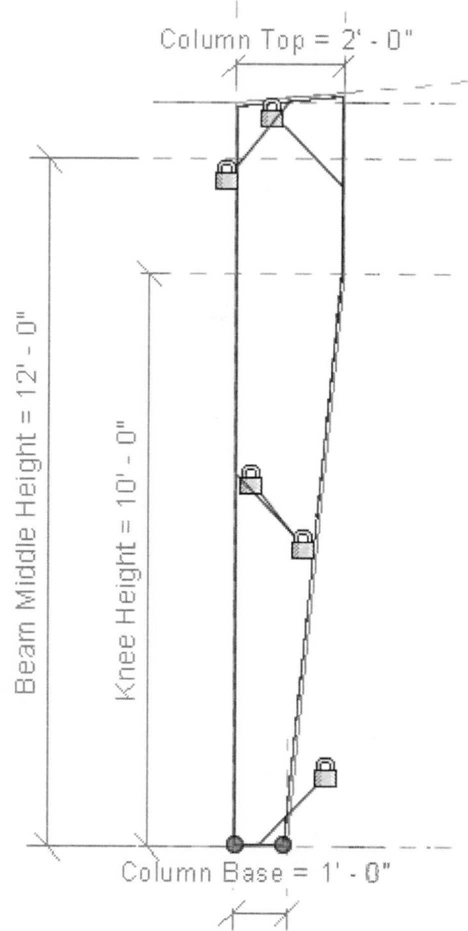

Figure 9–85

20. In Properties, set the values of the *Extrusion End* to **3/8"** and *Extrusion Start* to (negative) **-3/8"**, as shown in Figure 9–86.

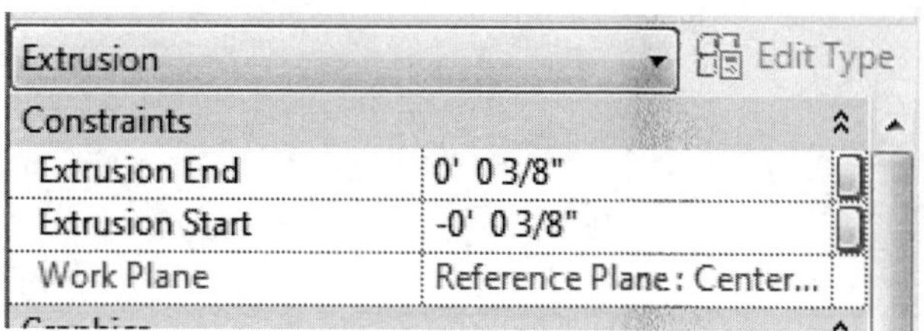

Figure 9–86

21. Click (Finish Edit Mode)

22. Display the column profile in a shaded 3D View and return to the **Elevations: Front** view. (Hint: Press <Ctrl>+<Tab>.)

23. Mirror the new tapered column (both extrusion and sweep) along the center vertical reference plane, as shown in Figure 9–87. Edit the sketch of the mirrored sweep and align and lock the sketch to the reference lines.

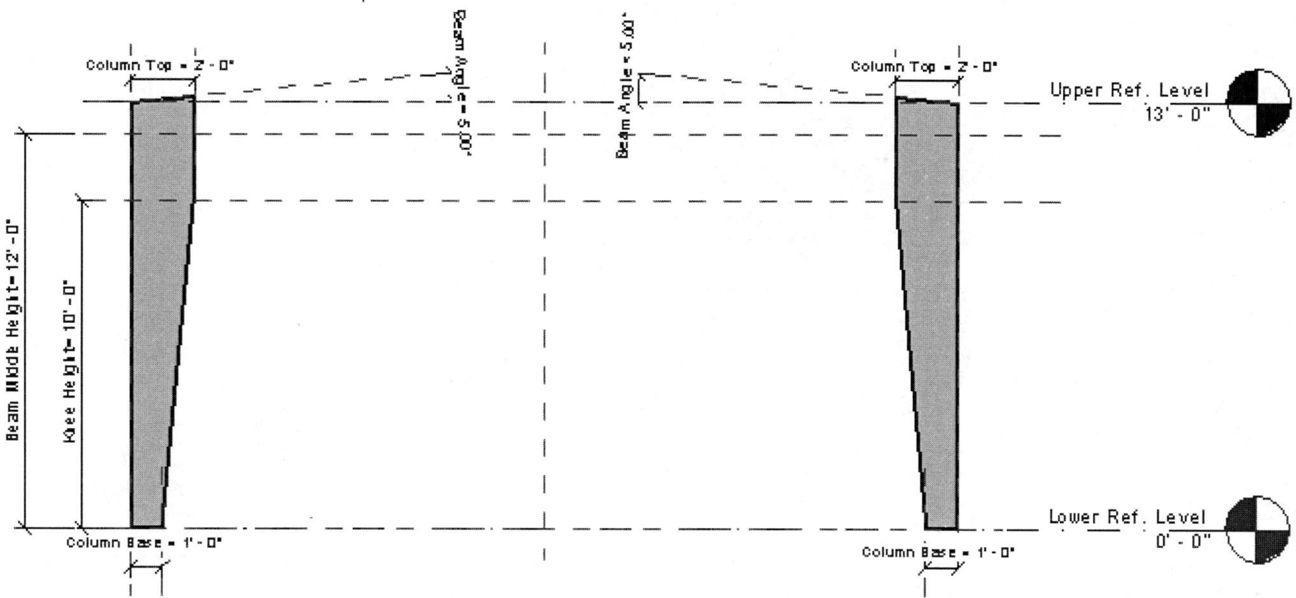

Figure 9–87

24. Flex the framework to ensure that everything is working as expected.

25. Save the family file.

Task 3 - Create the geometry for the Tapered Beam.

Ensure that you are aligning to the reference planes and not other elements. Press <Tab> to cycle through the options. The name of the selected element is displayed in the Status Bar.

1. Start the **Sweep** command and draw a path for the beam's top flange of the moment frame as shown in Figure 9–88. Align and lock the lines to the reference planes.

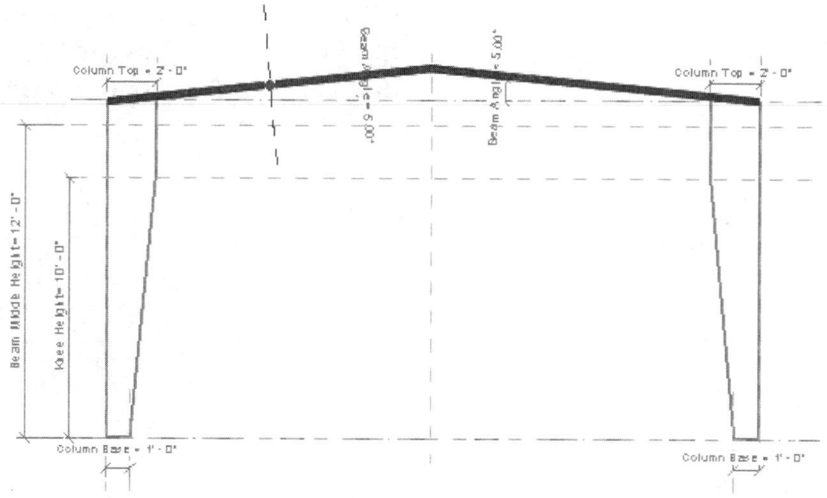

Figure 9–88

2. Click (Finish Edit Mode).

3. In the *Modify | Sweep* tab>Sweep panel, click (Load Profile).

4. In the Load Family dialog box, navigate to the ...*Structural* folder. Select **Tapered-Frame-Profile.rfa** and click **Open**.

5. In the *Profile:* drop-down list, select **Tapered-Frame-Profile**, as shown in Figure 9–89.

Figure 9–89

6. Click (Finish Edit Mode).

7. Display the top plate of the beam profile in a 3D View, as shown in Figure 9–90.

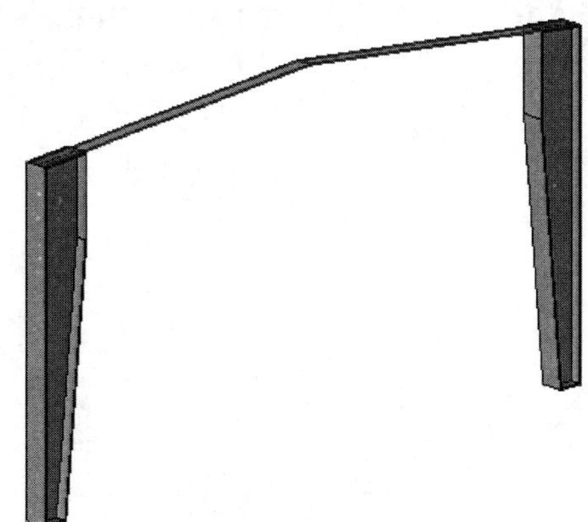

Figure 9–90

8. Return to the **Elevations: Front** view and create another sweep for the beam's bottom flange using the sketch for the path shown in Figure 9–91 and the same profile family.

Hint: You can temporarily hide existing solid elements so that you can attach the sketch to the reference planes.

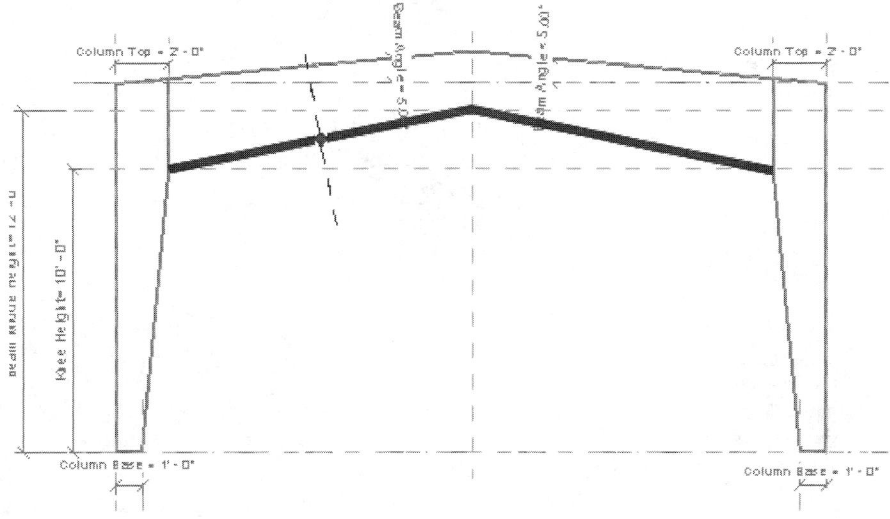

Figure 9–91

9. The final 3D view should display as shown in Figure 9–92.

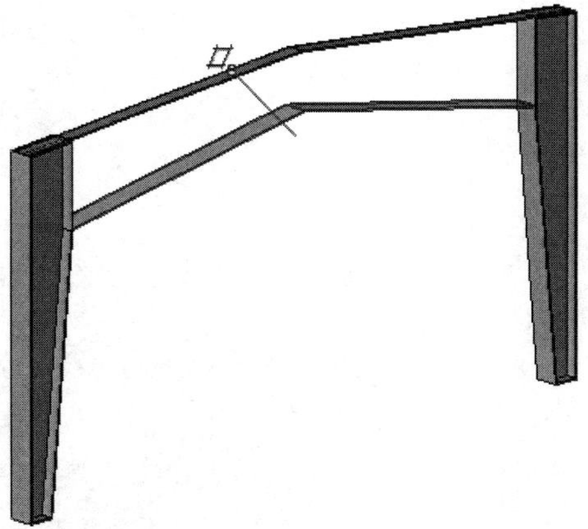

Figure 9–92

10. Return to the **Elevations: Front** view and add an extrusion using the interior of the previously created sweeps as shown in Figure 9–93. Set the *Extrusion End* to **3/8"** and *Extrusion Start* to (negative) **-3/8"**.

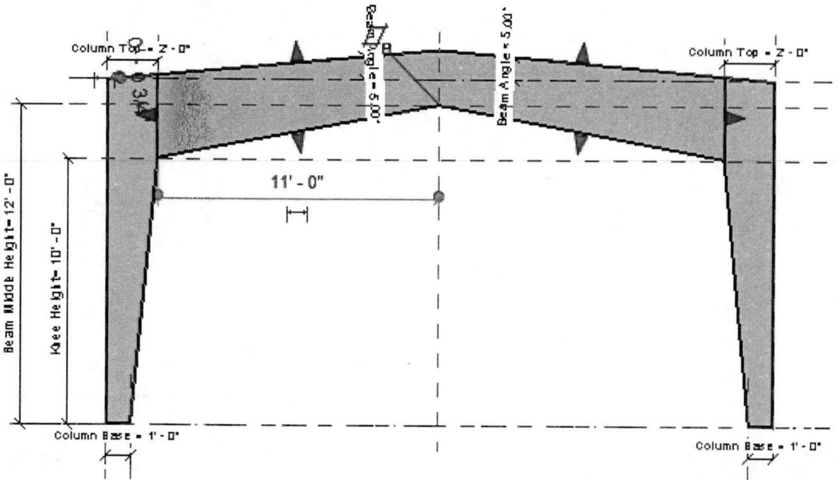

Figure 9–93

11. Open the 3D view and move it to one side. Open the Family Types dialog box and flex the geometry of the **Tapered Moment Frame** parameters as shown in Figure 9–94.

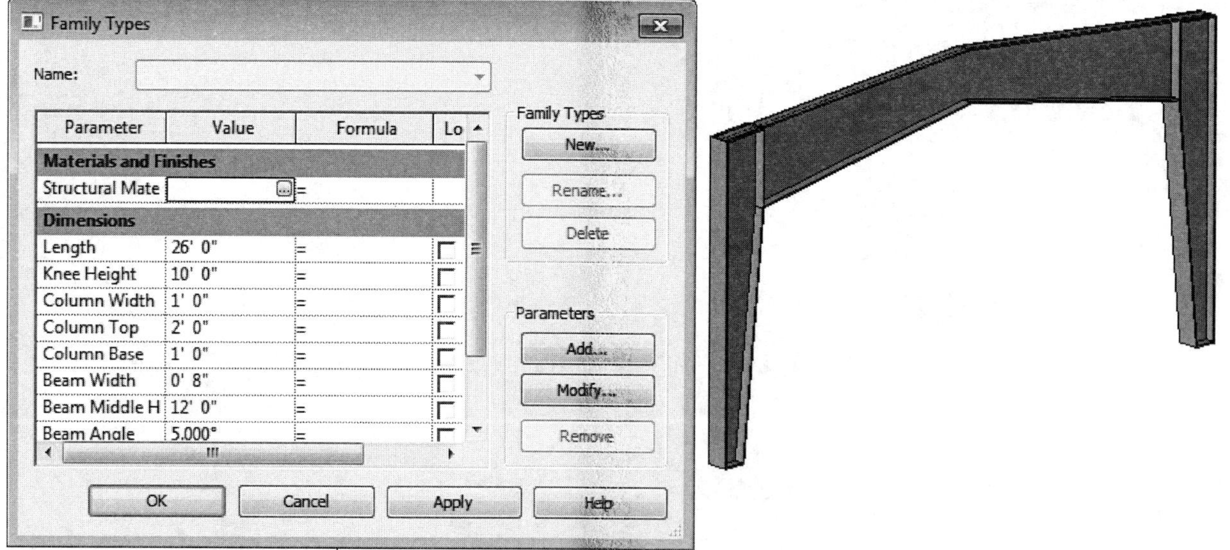

Figure 9–94

12. Save the family file.

13. Create a new structural project, load the family, and test it.

Chapter Review Questions

1. How do you change a standard dimension to a parameter-based dimension, as shown in Figure 9–95?

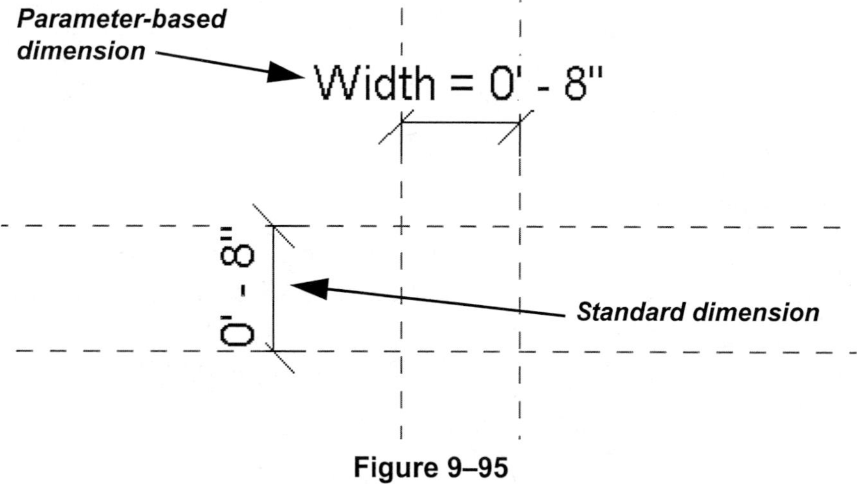

Figure 9–95

   a. In the Type Selector, select a parameter-based type.

   b. In Properties, change the *Value*.

   c. In the Options Bar, expand the Label options and select **<Add parameter...>**.

   d. In the Family Types dialog box, in the *Parameters* area click **Add**.

2. Why would you create an in-place family rather than a component family?

   a. An in-place family can be used several times in the current project only.

   b. An in-place family is used once in the current project for a specific location and size.

   c. A component family cannot be modified to fit the specific location.

   d. A component family cannot be joined to other elements in the project.

3. Which of the following Generic Model family templates would you use to create a slab depression?

   a. **Generic Model wall based.rft**

   b. **Generic Model face based.rft**

   c. **Generic Model line based.rft**

   d. **Generic Model floor based.rft**

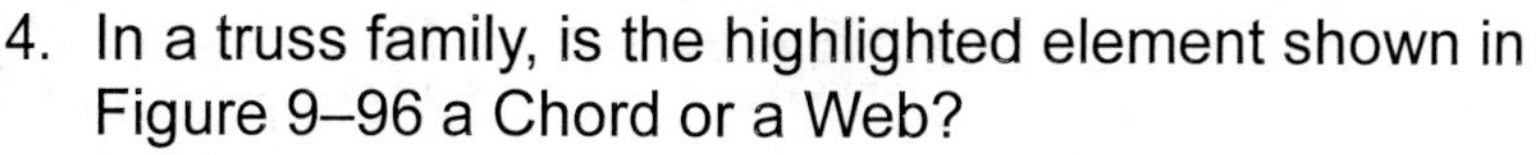

4. In a truss family, is the highlighted element shown in Figure 9–96 a Chord or a Web?

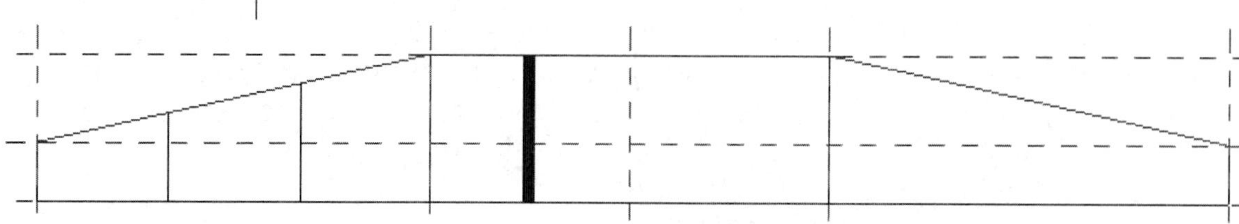

Figure 9–96

   a. Chord
   b. Web

5. When assigning material to a solid form you can select the **Associate Family Parameter** option. The option to select the material is grayed out, as shown in Figure 9–97. Where do you assign the material?

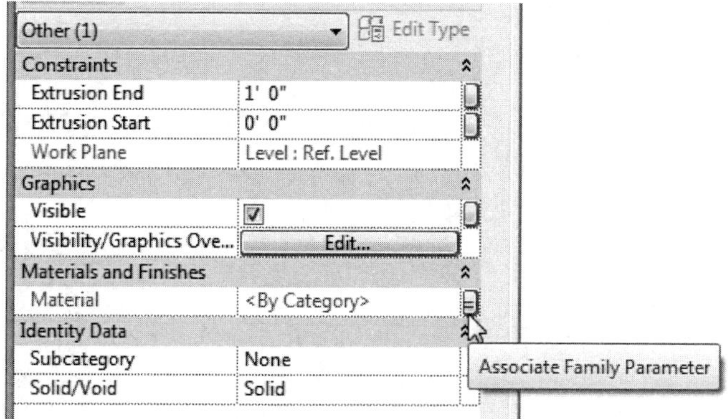

Figure 9–97

   a. In the Family Types dialog box, select the associated parameter and select the material.
   b. In the *Manage* tab>Settings panel, click (Materials) and select the material.
   c. In the *Modify* tab>Geometry panel, click (Paint).
   d. In the Family Category and Parameters dialog box, specify the material in the *Family Parameters* area.

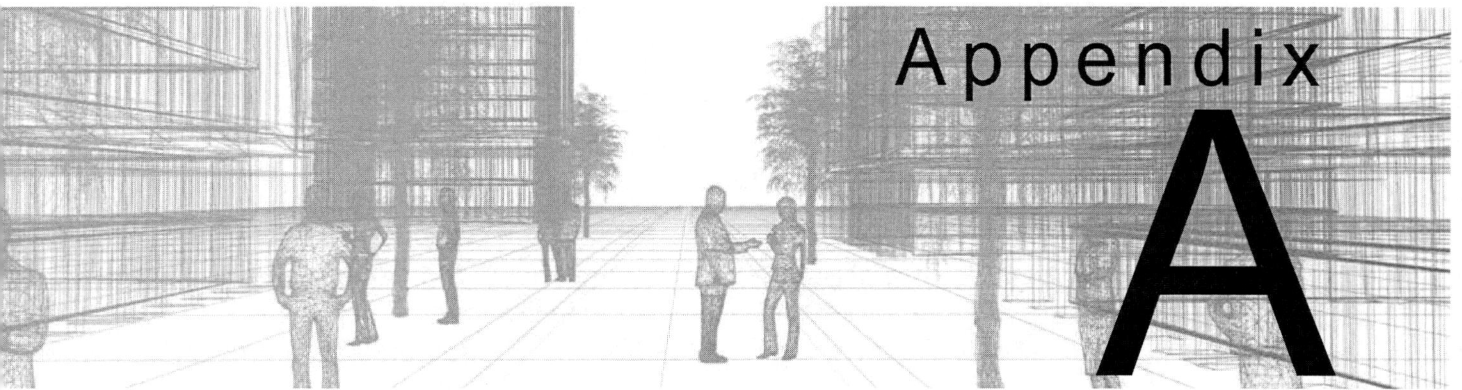

Additional Management Tools

There are many tools available in the Autodesk® Revit® software that you can use when creating and using templates and projects. This appendix provides details about several additional tools and commands that are related to those covered in this training guide.

Learning Objectives in this Appendix

- Set up units, snaps, temporary dimensions, and arrowheads.
- Create object styles for components that include line weight, color, pattern, and material.
- Create patterns used in the Filled Region command and for cut and surface patterns in materials.
- Create materials and material libraries that can be used in multiple projects.
- Review mechanical settings (for ducts and pipes) and electrical settings (for wiring, cable trays, and conduits).
- Review structural specific settings.
- Create sheet list schedules
- Modify the keyboard shortcuts and double-click function of the mouse.
- Create new Project Browser organization layouts for views and sheets.

A.1 General Settings

A variety of settings are available to help you customize the projects you create and keep everything consistent with your office's standards. Other settings can make the process of working in Autodesk Revit more productive.

Specifying Units

Even though you select Imperial or Metric units when you create the project template, you can set up the Project Units with specific formats and options. For example, if you are working on a Civil project where everything is set up in feet, you specify **Decimal Feet** as the Length Format. For an international metric project, you can specify the length units as Meters, Centimeters, or Millimeters.

How To: Set Up Project Units

1. In the *Manage* tab>Settings panel, click (Project Units) or type **UN**.
2. In the Project Units dialog box, in the *Format* column, click the button next to the unit type that you want to modify, as shown in Figure A–1. The related Format dialog box opens, as shown in Figure A–2.

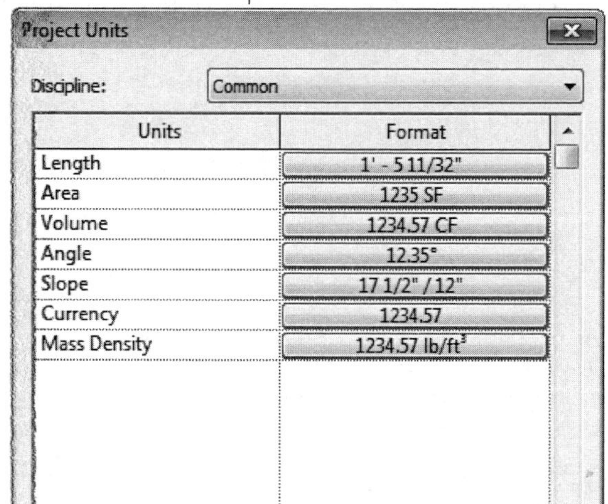

Figure A–1

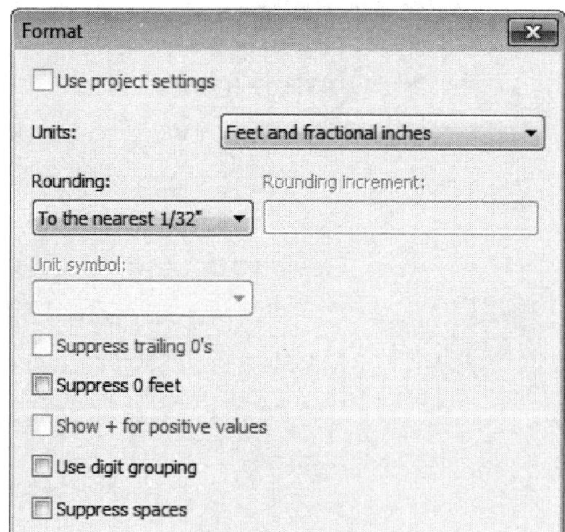

Figure A–2

3. Set the *Units*, *Rounding*, and other options as needed.
4. In the Project Units dialog box, you can also select the *Discipline* (**Common** by default) and change the Unit format for each discipline. The other options are **Structural**, **HVAC**, **Electrical**, **Piping**, and **Energy**. The options vary according to the discipline-specific software in use.
5. Click **OK** to close each dialog box.

Format Options

Each unit has specific formatting options. The option is grayed out if it is not applicable to that unit type.

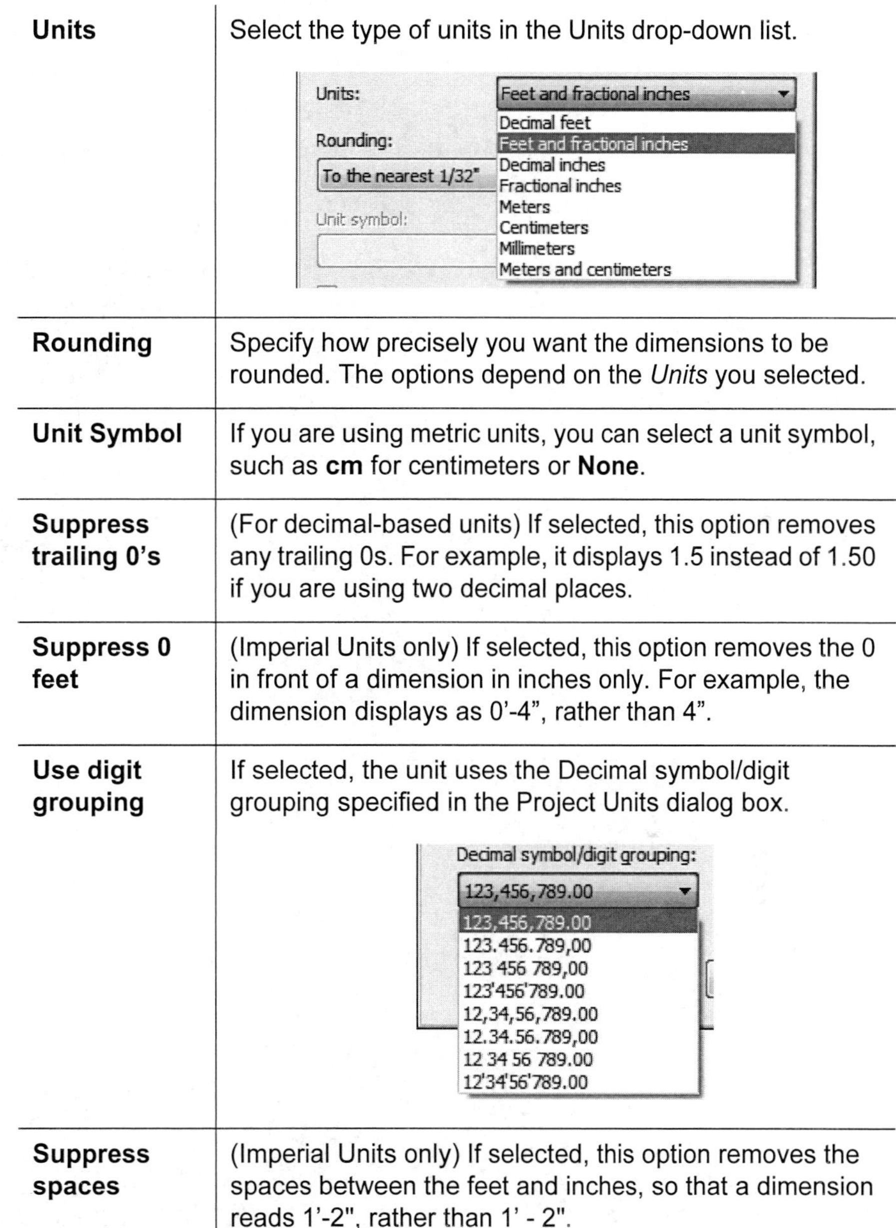

Units	Select the type of units in the Units drop-down list.
Rounding	Specify how precisely you want the dimensions to be rounded. The options depend on the *Units* you selected.
Unit Symbol	If you are using metric units, you can select a unit symbol, such as **cm** for centimeters or **None**.
Suppress trailing 0's	(For decimal-based units) If selected, this option removes any trailing 0s. For example, it displays 1.5 instead of 1.50 if you are using two decimal places.
Suppress 0 feet	(Imperial Units only) If selected, this option removes the 0 in front of a dimension in inches only. For example, the dimension displays as 0'-4", rather than 4".
Use digit grouping	If selected, the unit uses the Decimal symbol/digit grouping specified in the Project Units dialog box.
Suppress spaces	(Imperial Units only) If selected, this option removes the spaces between the feet and inches, so that a dimension reads 1'-2", rather than 1' - 2".

- The **Use project settings** and **Show + for positive values** options are grayed out when setting units for the project. This dialog box is used when creating dimension styles or specifying label formats when the options are available.

Snap Settings

The Snaps dialog box controls *Dimension Snaps,* which are the increments you see in temporary dimensions, and *Object Snaps*, which are the points on elements that you can select. In the *Manage* tab>Settings panel, click (Snaps) to open the dialog box, as shown in Figure A–3.

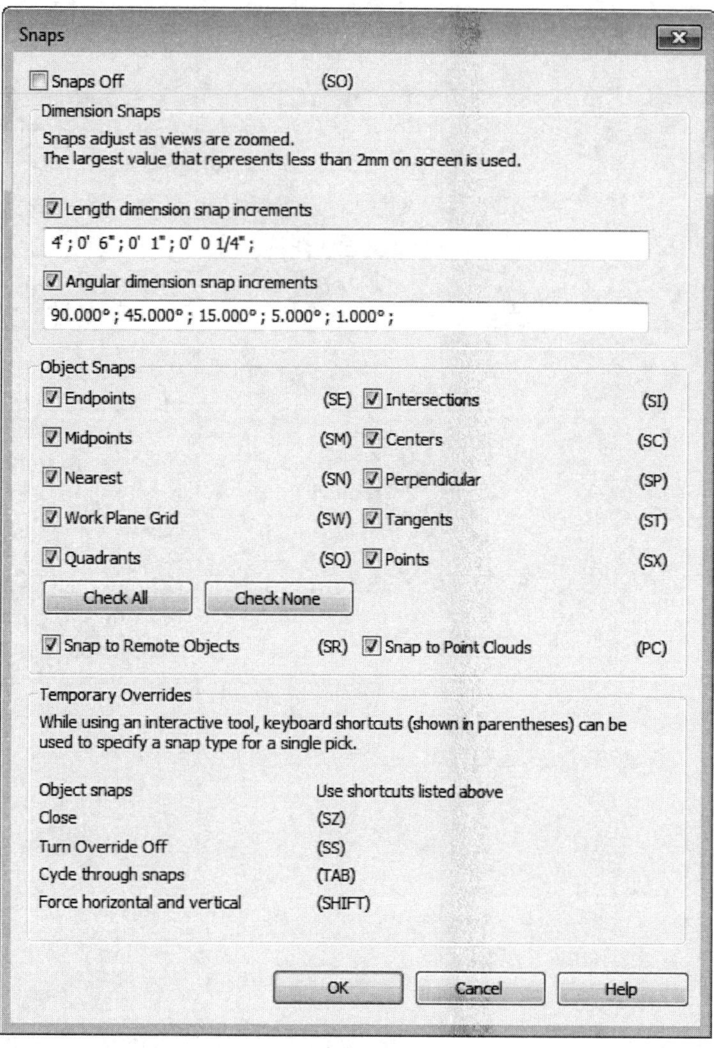

Figure A–3

Keyboard shortcuts, such as the ones listed here, are part of the User Interface and apply to all projects.

- Snap overrides are listed as keyboard shortcuts in parentheses, next to the corresponding snap. When a snap override is used, the cursor finds that specified snap type in your view until something is selected.
- Snap overrides can be used while inside a command by right-clicking and selecting **Snap Overrides** and then selecting the one that you want to use from the list.

Temporary Dimension Settings

Temporary dimensions display when you draw or edit building elements in the software. By default, they measure from the center lines of walls to the center lines of openings, as shown in Figure A–4.

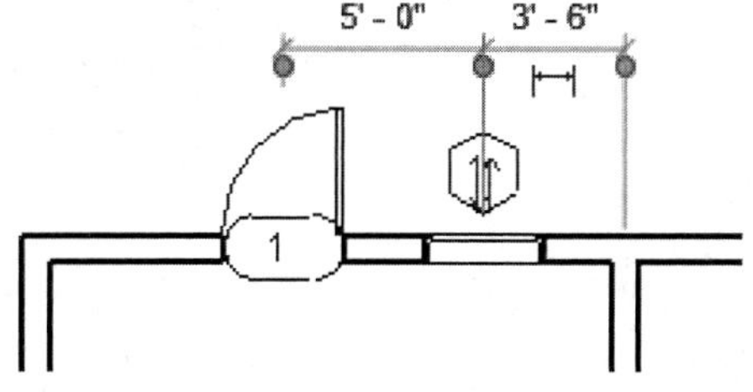

Figure A–4

- When you move a witness line to another element or part of an element, the location is remembered within the current session of program.

You can control where temporary dimensions are placed by default. In the *Manage* tab>Settings panel, expand (Additional Settings) and click (Temporary Dimensions) to open the Temporary Dimension Properties dialog box, as shown in Figure A–5.

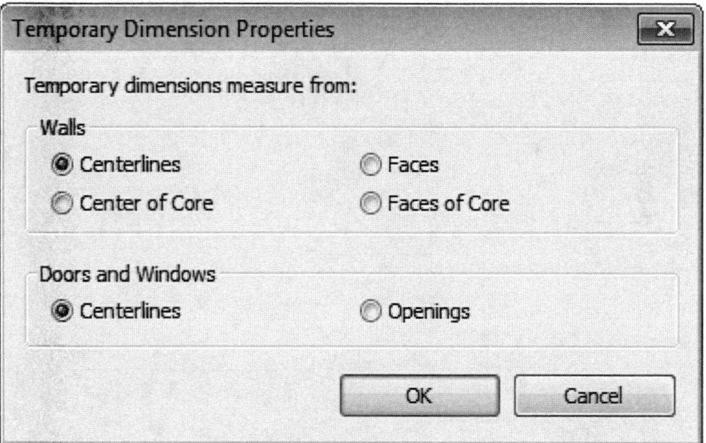

Figure A–5

- You can set up these properties in the project template file or modify them at any time. They do not affect existing elements in your project.
- You can control the size of the temporary dimension text in the Options dialog box in the *Graphics* area.

Setting Up Arrowheads

A variety of arrowhead types are supplied with the Autodesk Revit software, including open and filled arrow styles, tick marks, and dots. You can also create custom styles by duplicating an existing style and defining the parameters, such as the *Arrow Style* shown in Figure A–6. In the *Manage* tab>Settings panel, expand (Additional Settings) and click (Arrowheads) to open the dialog box.

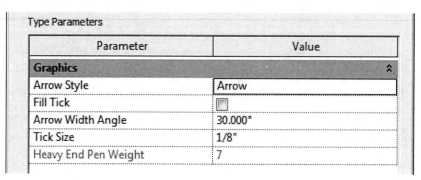

Type Parameters

Parameter	Value
Graphics	
Arrow Style	Arrow
Fill Tick	☐
Arrow Width Angle	30.000°
Tick Size	1/8"
Heavy End Pen Weight	7

Figure A–6

- Arrowheads are used by both text (with a leader) and dimensions.
- The MEP templates include loop leaders that can be set as *Tick Mark* for dimensions and *Leader Arrowhead* for text, as shown in Figure A–7. You can also create an Arrow Style using Loop in the Arrowhead settings.

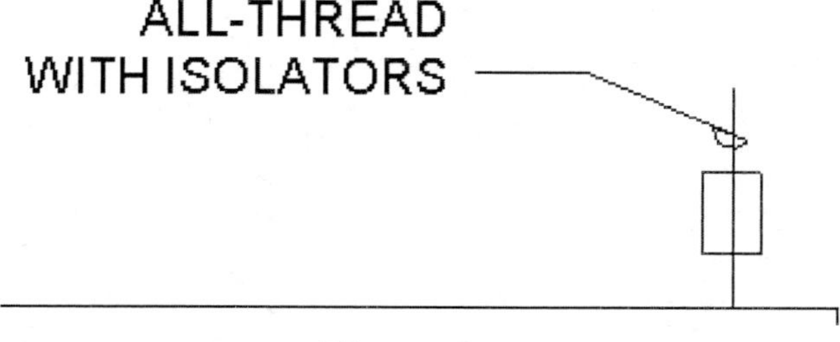

Figure A–7

A.2 Creating Object Styles

Every element in the Autodesk Revit software contains object styles that control how its components display in various views. For example, the *Doors* element includes the *Elevation Swing*, *Frame/Mullion*, *Glass*, *Hidden Lines*, *Opening*, *Panel*, and *Plan Swing* as components, as shown in Figure A–8. The *Elevation Swing* is set to a light line weight, displayed in black with a dashed line pattern. It does not have any material attached to it, but the *Glass* component does.

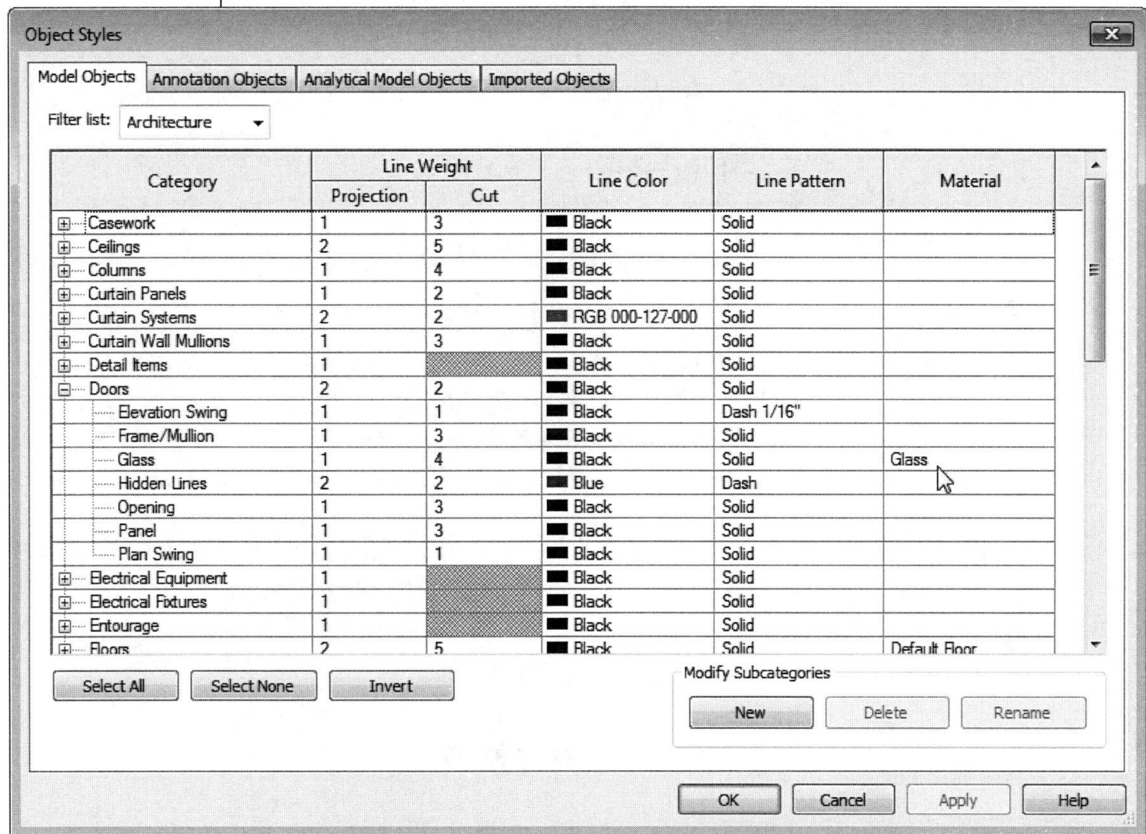

Figure A–8

- In the *Manage* tab>Settings panel, click (Object Styles) to open the Object Styles dialog box.
- To modify an object style, select the element to change and then select an option in the appropriate column in the drop-down list or dialog box.

Line Weights

Line weights for model elements can be set for a **Projection** (in elevations) or **Cut** (in section and plan). For example, a wall displays a heavy line weight while the door and window cuts are lighter, as shown in Figure A–9.

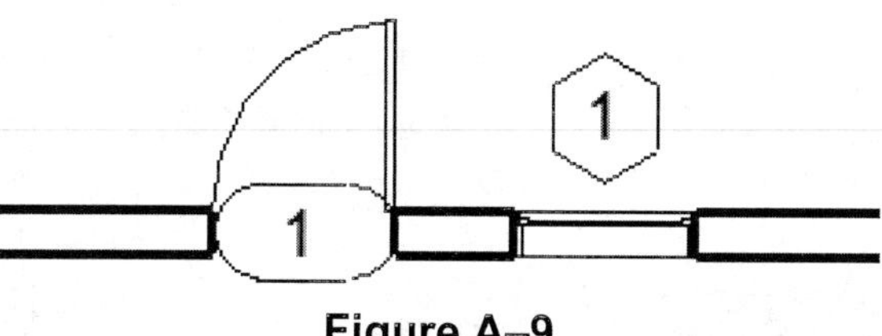

Figure A–9

- Annotation elements are only set for projection.

Hint: Setting Line Weights

In the *Manage* tab>Settings panel, expand (Additional Settings) and click (Line Weights) to open the Line Weights dialog box, as shown in Figure A–10. It contains three tabs: *Model Line Weights* (vary by scale), *Perspective Line Weights* (control elements like walls and windows in perspective views), and *Annotation Line Weights* (not dependent on the scale).

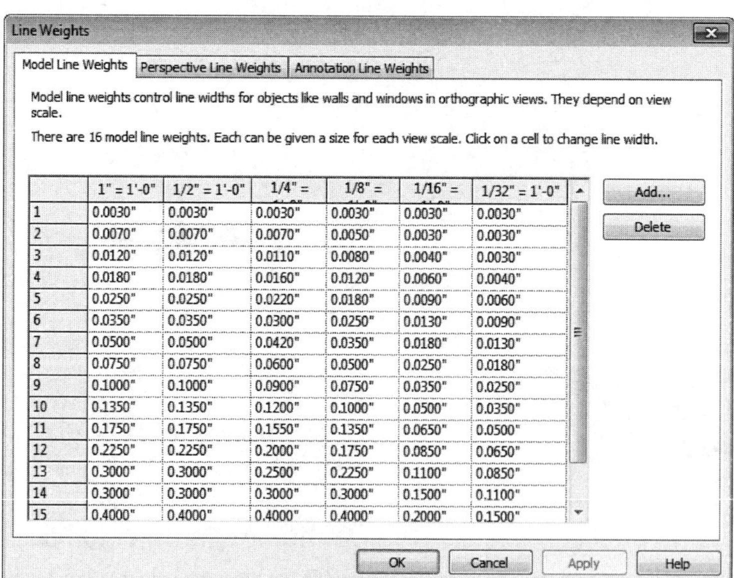

	1" = 1'-0"	1/2" = 1'-0"	1/4" =	1/8" =	1/16" =	1/32" = 1'-0"
1	0.0030"	0.0030"	0.0030"	0.0030"	0.0030"	0.0030"
2	0.0070"	0.0070"	0.0070"	0.0050"	0.0030"	0.0030"
3	0.0120"	0.0120"	0.0110"	0.0080"	0.0040"	0.0030"
4	0.0180"	0.0180"	0.0160"	0.0120"	0.0060"	0.0040"
5	0.0250"	0.0250"	0.0220"	0.0180"	0.0090"	0.0060"
6	0.0350"	0.0350"	0.0300"	0.0250"	0.0130"	0.0090"
7	0.0500"	0.0500"	0.0420"	0.0350"	0.0180"	0.0130"
8	0.0750"	0.0750"	0.0600"	0.0500"	0.0250"	0.0180"
9	0.1000"	0.1000"	0.0900"	0.0750"	0.0350"	0.0250"
10	0.1350"	0.1350"	0.1200"	0.1000"	0.0500"	0.0350"
11	0.1750"	0.1750"	0.1550"	0.1350"	0.0650"	0.0500"
12	0.2250"	0.2250"	0.2000"	0.1750"	0.0850"	0.0650"
13	0.3000"	0.3000"	0.2500"	0.2250"	0.1100"	0.0850"
14	0.3000"	0.3000"	0.3000"	0.3000"	0.1500"	0.1100"
15	0.4000"	0.4000"	0.4000"	0.4000"	0.2000"	0.1500"

Figure A–10

You can customize the line weights in each tab and add scales as needed for the model line weights.

Line Color

When you select the **Line Color** option in Object Styles or elsewhere, the Color dialog box opens, as shown in Figure A–11.

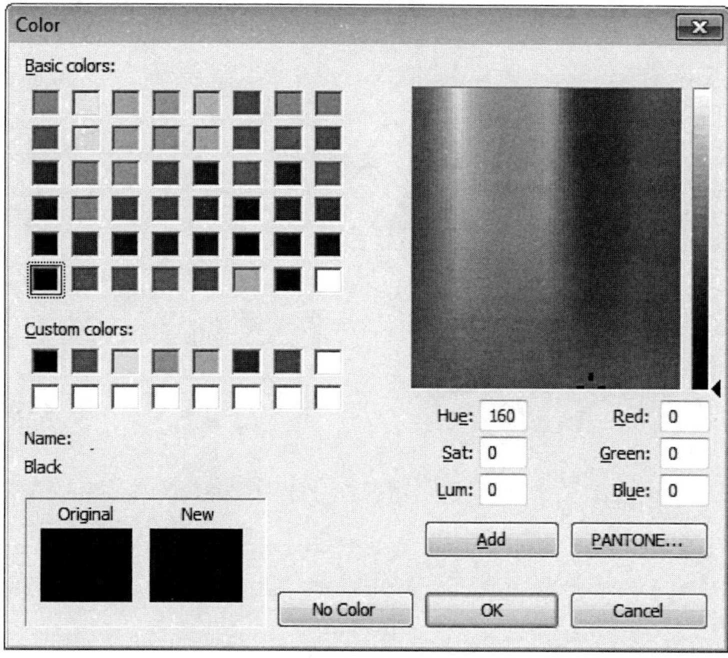

Figure A–11

You can select from thousands of colors, including options from the Pantone color system, as shown in Figure A–12. Typically, most elements in the Autodesk Revit software are black for printing purposes. When you shade and render, the elements take on the color of the assigned material. Therefore, these dialog boxes are opened when you create materials.

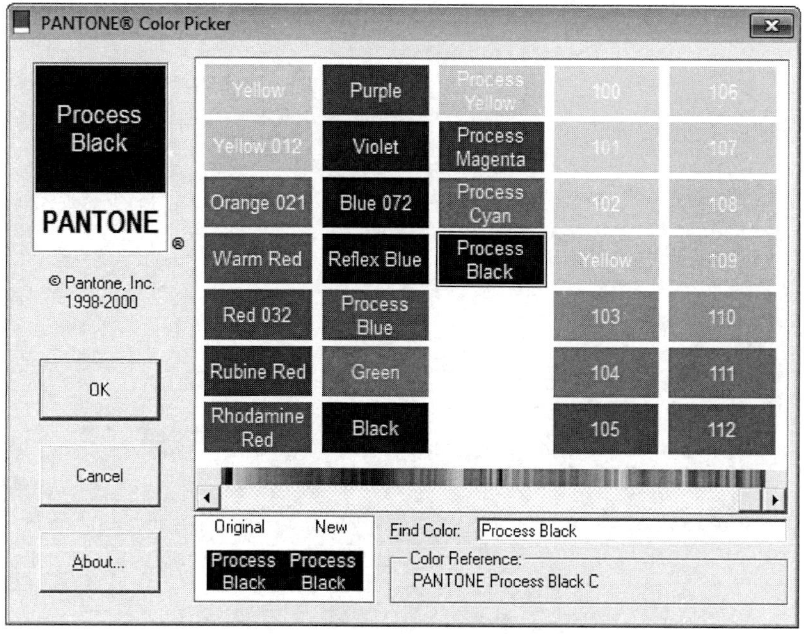

Figure A–12

Line Patterns

Line Patterns can be specified for various components in the Object Styles dialog box. In the example shown in Figure A–13, the line pattern **Dash** is used for the elevation door swing and the line pattern **Dash dot** is used for the callout bubble.

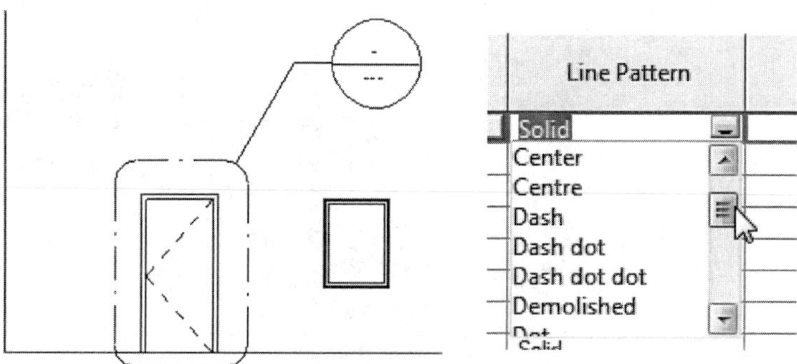

Figure A–13

How To: Create Line Patterns

A wide variety of patterns are supplied with the Autodesk Revit software and you can create your own with a series of dashes, spaces, and dots.

1. In the *Manage* tab>Settings panel, expand (Additional Settings) and click (Line Patterns) to open the Line Patterns dialog box.
2. In the Line Patterns dialog box, click **New**.
3. In the Name dialog box, type a name for the pattern and click **OK**.
4. In the Line Pattern Properties dialog box, add a list of dashes, spaces, and/or dots and specify their values, as shown in Figure A–14. *Space* is always the option after a dash or dot. You do not have to enter a value for dots.

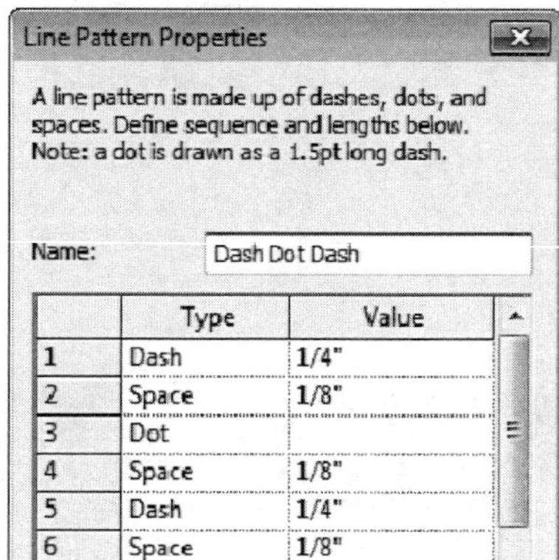

Figure A–14

Line Styles

Line styles are a specific type of object style used with the sketching tools. They are not part of the Object Styles dialog box, but are created in a similar way.

In the *Manage* tab>Settings panel, expand (Additional Settings) and click (Line Styles). In the Line Styles dialog box (shown in Figure A–15), specify the *Line Weight*, *Line Color*, and *Line Pattern* for each line style.

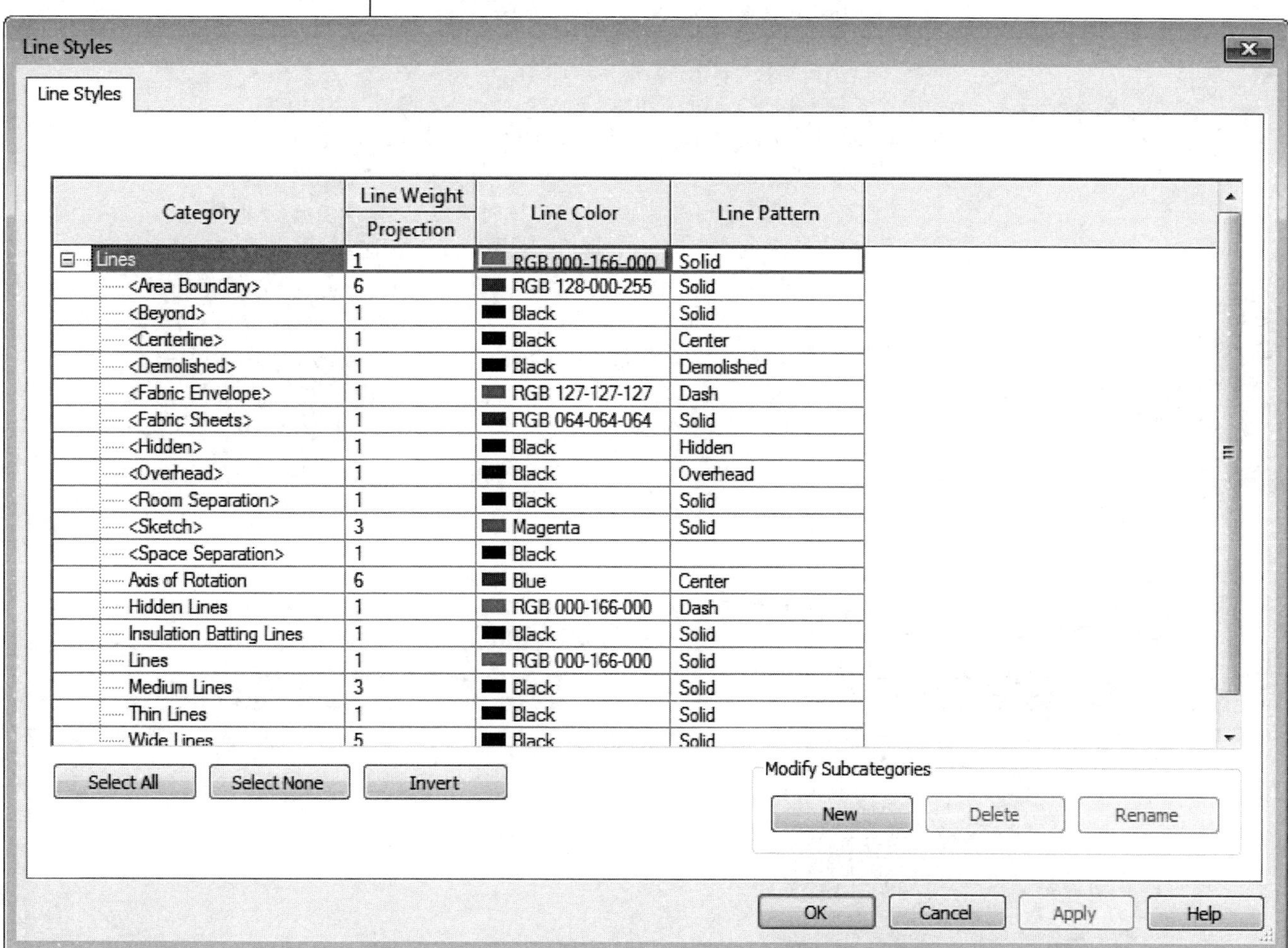

Figure A–15

A.3 Creating Fill Patterns

Fill patterns are used by the various **Filled Region** commands and material specifications. In the *Manage* tab>Settings panel, expand (Additional Settings) and click (Fill Patterns). Select from two types: **Drafting** and **Model**.

- Fill Patterns can also be created while you are working in the Visibility/Graphics dialog box. When you select a Fill Pattern override, the dialog box displays (Browse), which opens the Fill Patterns dialog box.

Drafting patterns are symbolic. The view scale controls the size of the pattern. Basic patterns include hatching and cross-hatching, as well as designs that display material (such as aluminum or concrete), as shown on the left in Figure A–16. These are used in details and in plan or section cuts.

Model patterns are full scale to the actual elements they represent. They do not change if the view scale changes. You can select brick, various sizes of tile, lines that are a specific distance apart, roof shakes, etc., as shown on the right in Figure A–16. These patterns are primarily used in elevation and plan views.

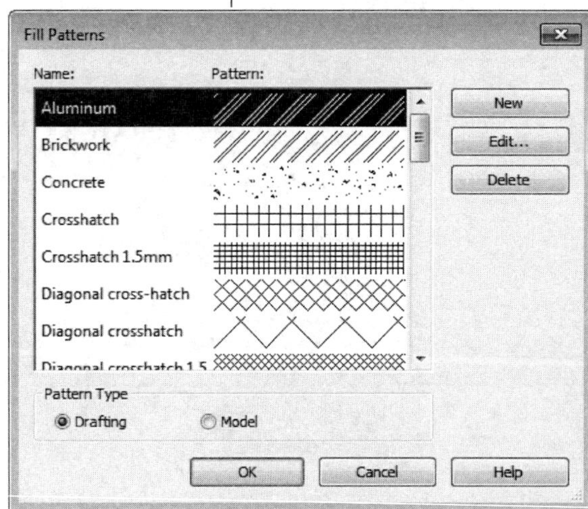

Drafting Fill Patterns

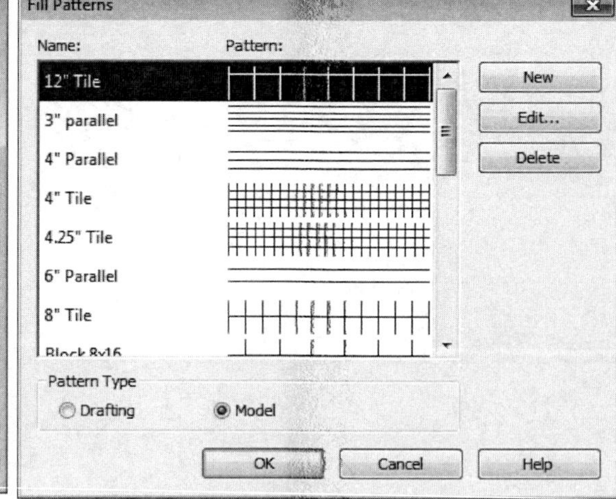

Model Fill Patterns

Figure A–16

How To: Create a New Simple Fill Pattern

1. In the *Manage* tab>Settings panel, expand (Additional Settings) and click (Fill Patterns).
2. In the New Pattern dialog box, select the type of pattern you want to create: **Drafting** or **Model**.
3. Click **New**.
4. In the New Pattern dialog box, specify the required **Orientation in Host Layers** option and select the *Simple* pattern style, as shown in Figure A–17.

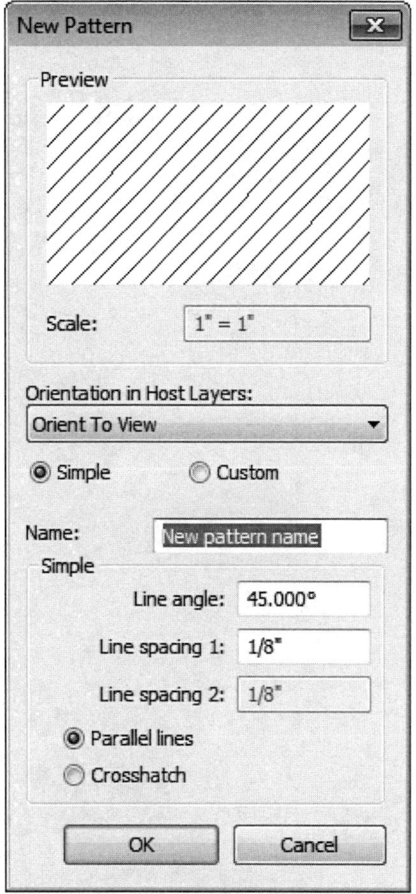

Figure A–17

5. Type a name for the new pattern.
6. Type values for the *Line angle* and *Line spacing*, and select the **Parallel lines** or **Crosshatch** option.
7. Click **OK** to finish. The pattern can now be used to create a filled region type or used in a material.

- Filled regions can have either **Opaque** or **Transparent** backgrounds. This is set up in the Type Properties of the filled region.

How To: Create a New Custom Fill Pattern

1. In the *Manage* tab>Settings panel, expand (Additional Settings) and click (Fill Patterns).
2. In the Fill Patterns dialog box, select the type of pattern you want to create: **Drafting** or **Model**.
3. Click **New**.
4. In the New Pattern dialog box, select the **Custom** option, as shown in Figure A–18.

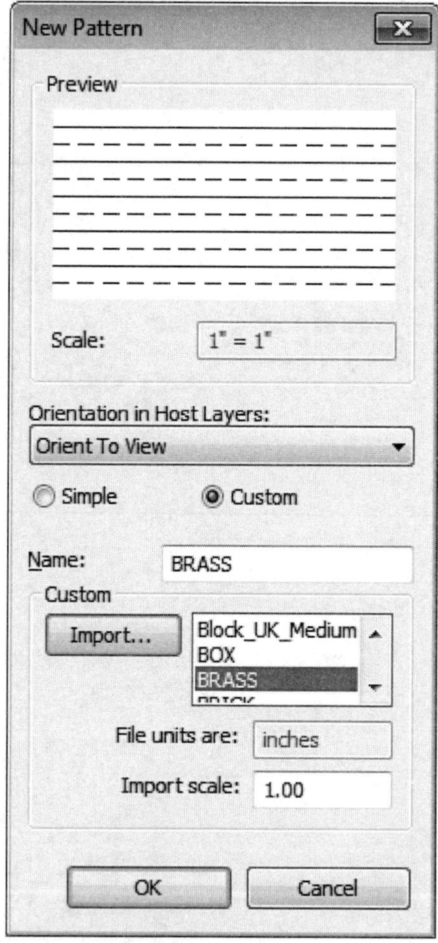

Figure A–18

5. Click **Import...** and select a PAT file. This file can be created specifically for the Autodesk Revit software or it can be based on an AutoCAD® PAT file.
6. Once the file is imported, a list of available patterns is displayed. Select the pattern. The name is automatically assigned. You can change the name after it has been selected.
7. Set the *Import Scale* as needed.
8. Click **OK** to finish.

Importing Patterns from AutoCAD

Pattern files (PAT) can be imported from the AutoCAD software, such as the Escher pattern shown in Figure A–19.

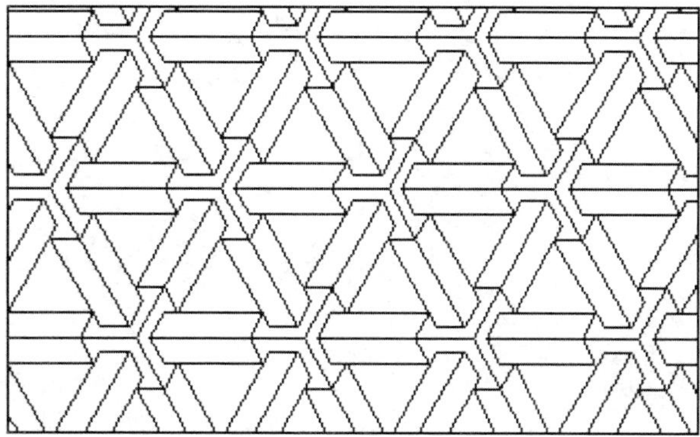

Figure A–19

One change must be made in the pattern file for it to be usable in the Autodesk Revit software. Open the PAT file and add the following line after the title of a custom hatch, as shown in Figure A–20.

```
*AR-HBONE, Standard brick herringbone pattern @ 45 degrees
;%TYPE=MODEL
45,        0,0,         4,4,                        12,-4
135,       2.828427125,2.828427125,   4,-4,         12,-4
```

Figure A–20

- A line with a semicolon in front it is ignored by the AutoCAD software, while the type is specified for the Autodesk Revit software.
- Any pattern in the **Acad.pat** file can be automatically brought in as a drafting pattern. However, you must add the string above to specify model patterns.

A.4 Creating Materials

When you shade or render Autodesk Revit models, their appearance depends on the materials that have been associated with the elements, as shown in a shaded view in Figure A–21. Materials also include information that can be used in material takeoff schedules and in energy and structural analysis. You can use materials supplied with the Autodesk Revit software or create custom materials.

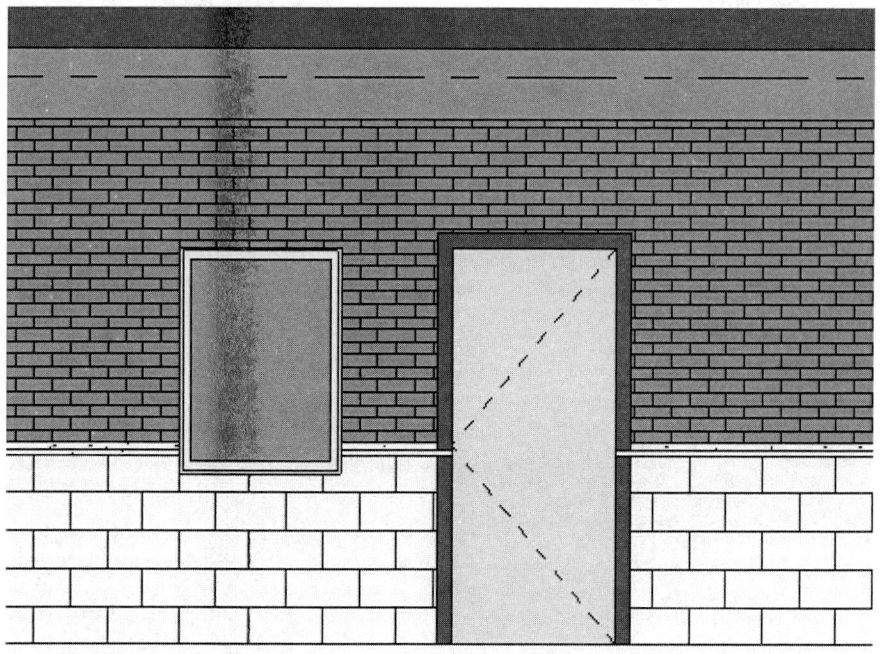

Figure A–21

- Materials can be assigned in several ways: using (Paint) on faces, setting it in the layers of system families (such as walls and floors), and in object styles. You also assign materials when you create other component families, such as doors and furniture.
- Materials consist of 2D graphic settings for shading, surface, and cut patterns and the 3D rendering appearance, as well as physical and thermal properties. You can also include identity data, such as manufacturer, model, and keynotes.

Material Libraries

Materials are stored in libraries, which can be shared between Autodesk Revit projects and the AutoCAD, Autodesk® Inventor®, Showcase and Autodesk® 3ds Max® software. Before creating custom materials it is recommended that you create a custom library. Materials can be copied between the current project and the library.

- You can also use Transfer Project Standards to bring in all the materials from another project into the current project.

How To: Create a Material Library

1. In the *Manage* tab>Settings panel, click (Materials).
2. At the bottom of the Material Browser, expand (Creates, opens and edits user-defined libraries) and select **Create New Library** as shown in Figure A–22.

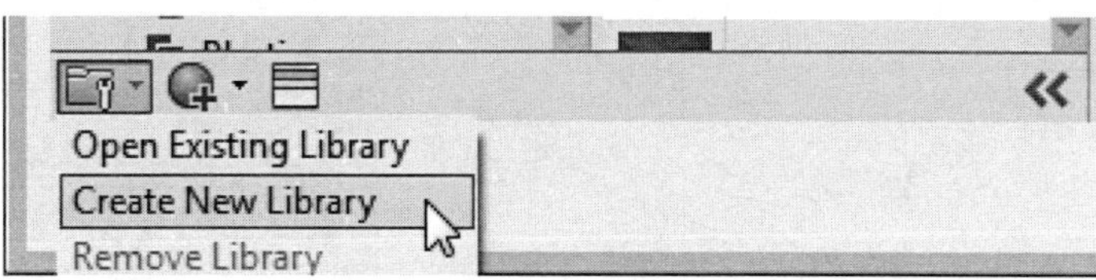

Figure A–22

3. In the Select File dialog box, navigate to the appropriate folder, type a *File name* and save the library. The Material Library displays in the Library list without any materials displayed.

4. Create new materials as needed in the current project. Then drag and drop both new and existing materials from the project materials into the new library as shown in Figure A–23.

Libraries that come with the Autodesk Revit software are locked. You can copy materials from them but cannot add or change materials in them.

Figure A–23

How To: Create a Material

1. In the *Manage* tab>Settings panel, click (Materials).
2. In the Material Editor, expand (Creates or duplicates a material), and select either **Create New Material** or **Duplicate Material** as shown in Figure A–24. If you are duplicating a material, select an existing material similar to what you want to use first.

You can also right-click on an existing material in the Material Browser and select ***Duplicate****.*

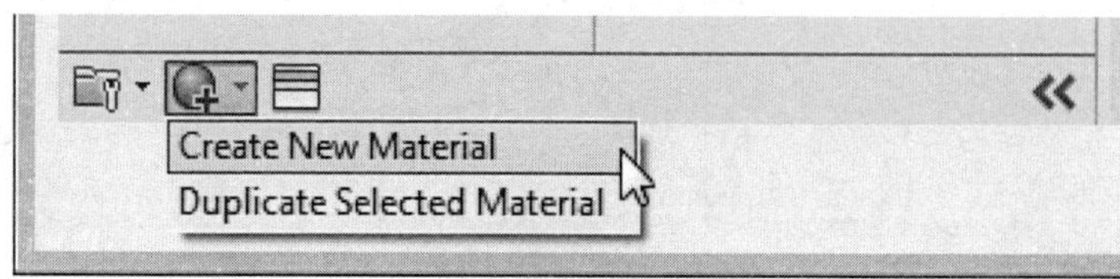

Figure A–24

3. A new material is added to the current project. Modify the name of the material or right-click on the Material name in the Material Browser and select **Rename**. Duplicated materials include all of the properties of the existing material as shown for a new Aluminum material in Figure A–25.

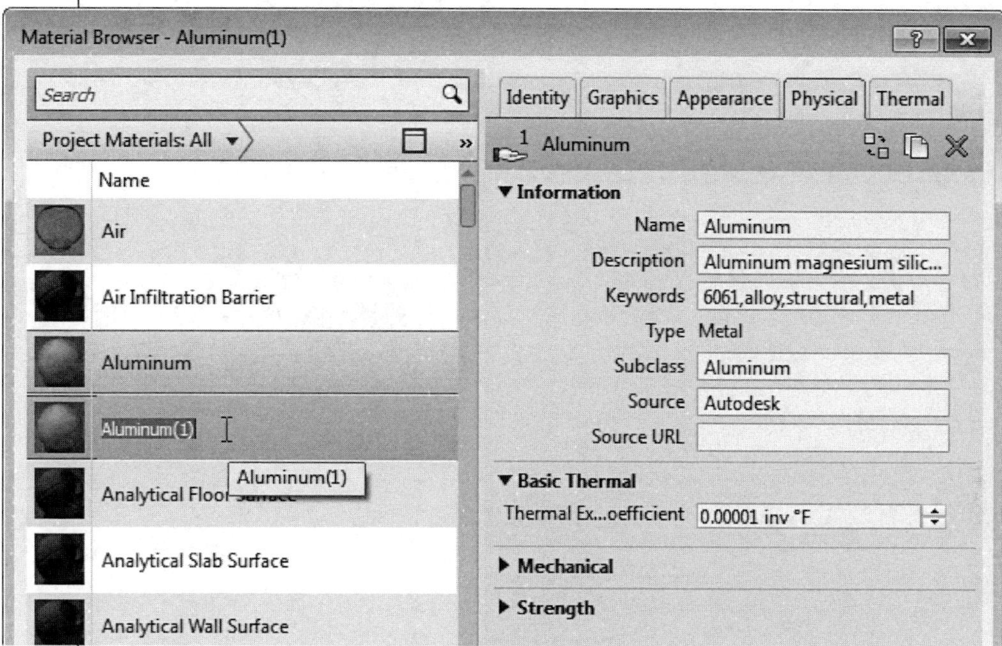

Figure A–25

- In the *Identity* tab, set up information, such as the related keynote, model, manufacturer, and other schedule-related items.
- In the *Graphics* tab, set up the *Shading*, *Surface Pattern*, and *Cut Pattern* as required.
- In the *Appearance* tab, set up the rendering design, including the properties of the elements to be rendered.
- In the *Physical* tab, specify the physical properties of the material including, **Basic Thermal**, **Mechanical**, and **Strength** properties, if applicable.
- In the *Thermal* tab, specify the advanced thermal physical properties of the material.

__Appearance__, __Physical__, and __Thermal__ parameters can be stored in Assets. Instead of making changes directly in these tabs, investigate the existing assets to determine whether you can swap one asset for another.

4. Click **Apply** and add additional materials as required.
5. Click **OK** when you are finished.

Identity Tab

In the *Identity* tab, you set up the *Descriptive Information*, *Product Information*, and *Revit Annotation Information* for the material, as shown in Figure A–26. This information can be used in material takeoff schedules.

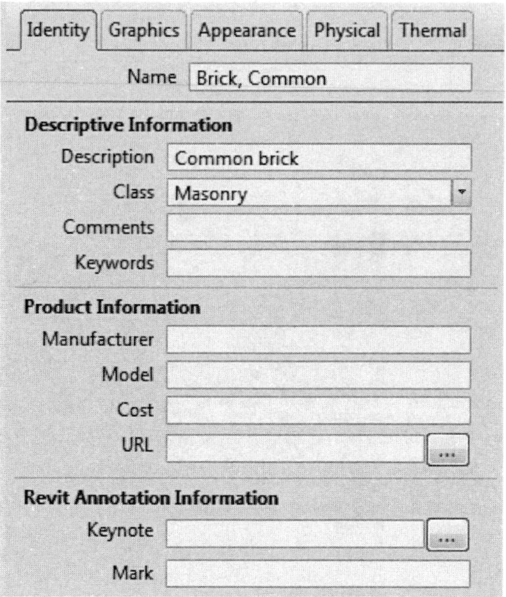

Figure A–26

Graphics Tab

In the *Graphics* tab (shown in Figure A–27), you set how a material displays in shaded or hidden line mode.

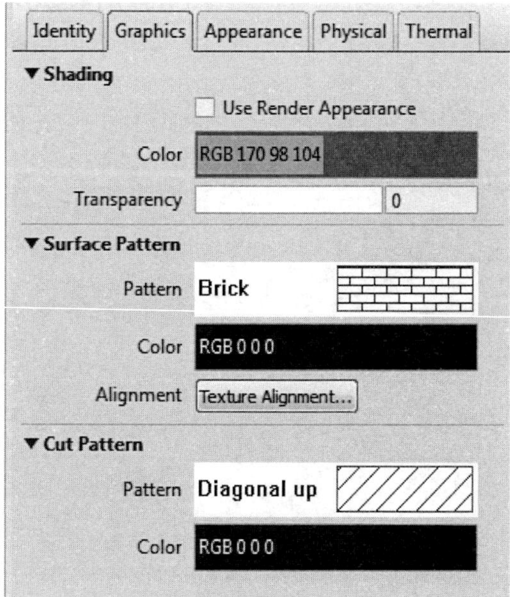

Figure A–27

- The *Shading* area controls how the material displays when an element is shaded. If you want the shaded view to resemble the rendered view, select the **Use Render Appearance** option. The color and other options are modified to match the Render Appearance selection. Alternatively, you can set up a color and transparency ratio. Only use this if the render appearance would be too dark or otherwise not suit the use.

- The *Surface Pattern* and *Cut Pattern* areas enable you to select a fill pattern and color to display on a surface or in a section cut. Surface patterns stay true to size. Cut patterns are drafting patterns, which change size according to the view scale.

Appearance Tab

In the *Appearance* tab, you can set different aspects of a rendered material, including **Information**, the type of material (in this example Masonry is selected as shown in Figure A–28) and other parameters depending on the type of material you are working with.

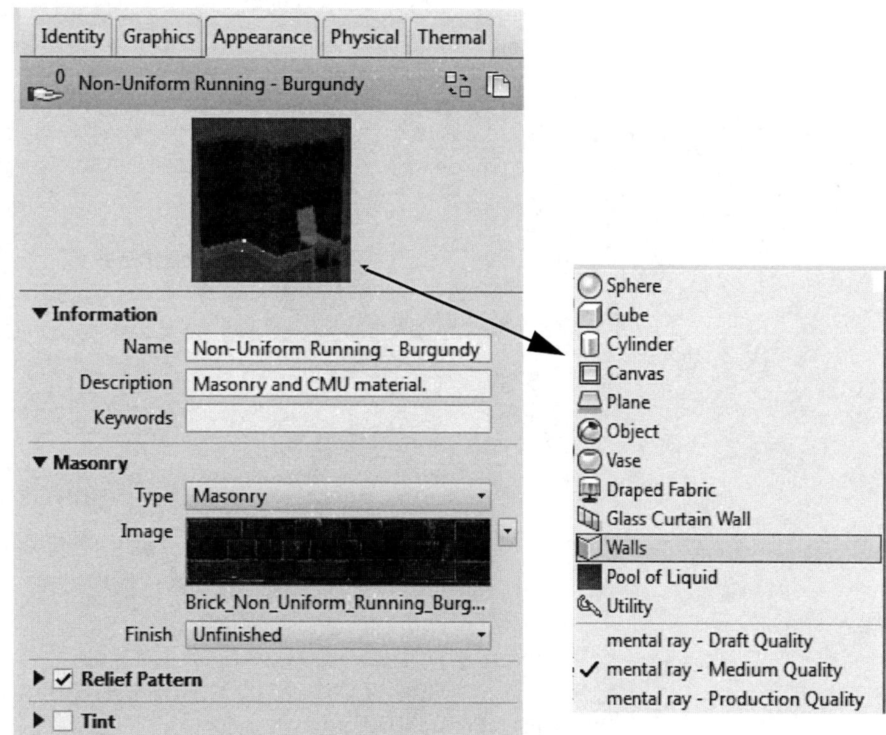

Figure A–28

- You can select how to display the preview of the material by expanding the list next to the graphic as shown in Figure A–28.

Physical Tab

In the *Physical* tab, you can specify information about the physical properties of the material. The options vary by material type, as shown in Figure A–29.

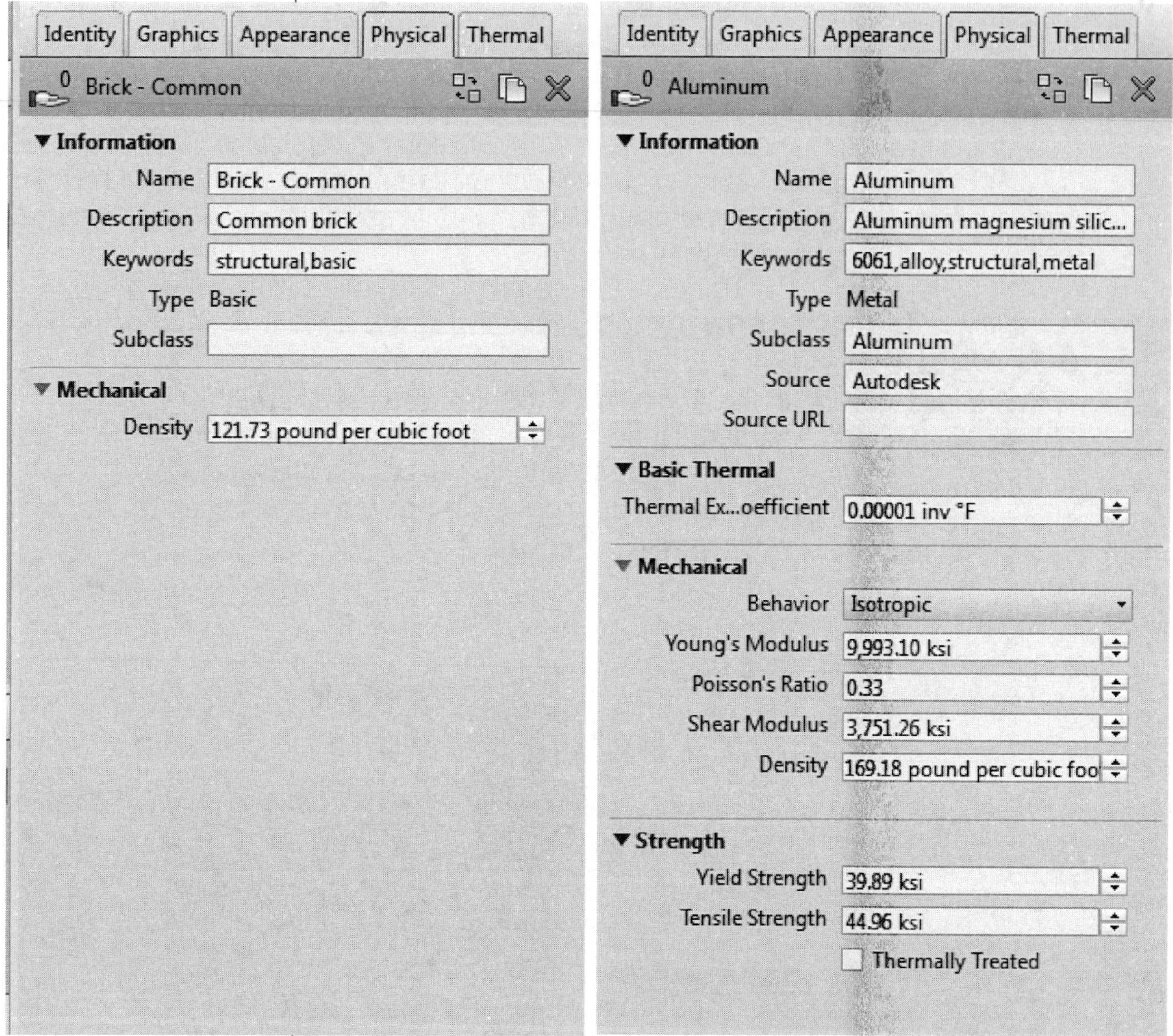

Figure A–29

- Some materials, such as **EPDM Membrane** or **Vapor Retarder**, do not have physical properties, but do have thermal properties.
- Analytical surface and Default materials do not have physical or thermal properties.

Working with Assets

Physical Assets contain preset parameters for materials. These assets are available in the *Appearance*, *Physical* and *Thermal* tabs. Autodesk provides an extensive library of assets that are divided by material type as shown in Figure A–30. You can modify and add assets to materials depending on the variables needed in the project.

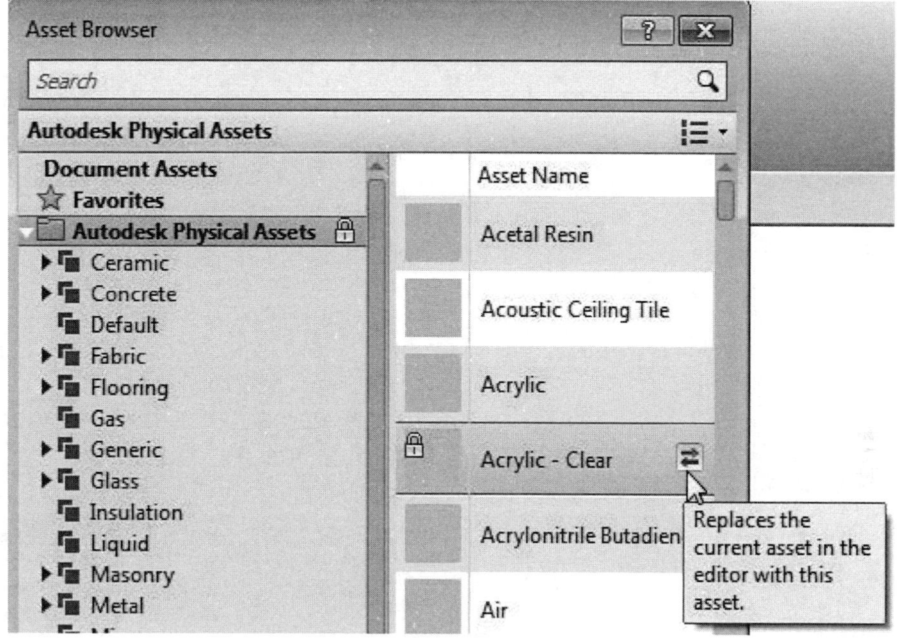

Figure A–30

How To: Replace an Asset for a Material

1. In the Material Browser, select the material that you want to modify.
2. In the *Appearance*, *Physical*, or *Thermal* tab, click (Replaces this asset).
3. In the Asset Browser, hover over the required asset and click (Replaces the current asset in the editor with this asset), as shown in Figure A–30.
4. Close the Asset Browser.

- The assets are filtered by type. Therefore, you need to modify the Appearance asset by clicking through the *Appearance* tab, etc.

- You can create new assets by duplicating an existing one and then modifying the parameters.

A.5 Settings for Mechanical and Electrical Projects

In addition to the standard settings common to all Autodesk Revit projects, the Autodesk Revit MEP software also has specific Mechanical and Electrical settings that can be used in a template or a project.

- In the *Manage* tab>Settings panel, expand (MEP Settings) and click (Mechanical Settings) or (Electrical Settings).
- Additional types of settings include Load Classifications, Demand Factors, and Building/Space Type Settings.
- You can also access the settings in the *Systems* tab by clicking on the related panel, as shown in Figure A–31.

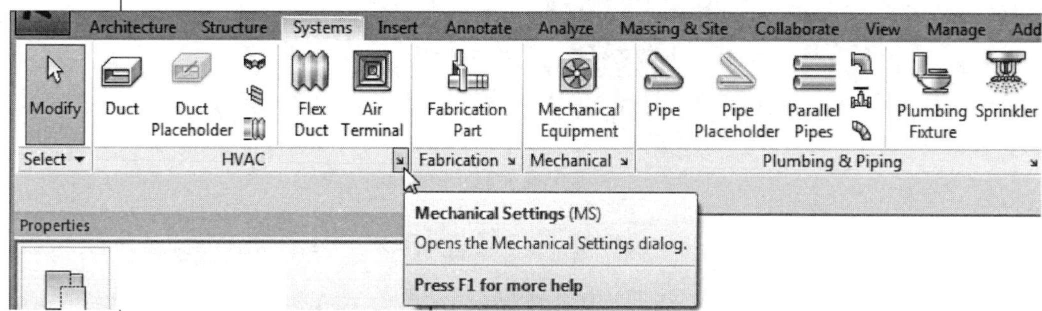

Figure A–31

Mechanical Settings

Mechanical settings enable you to preset many items relating to duct and pipe sizes, conversion settings, and other key defaults, as shown in Figure A–32.

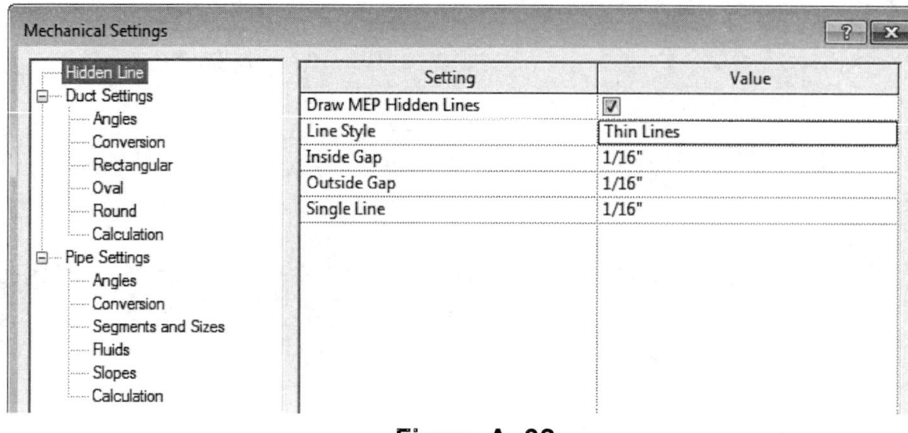

Figure A–32

Hidden Line

If the **Draw MEP Hidden Lines** option is selected, hidden lines are displayed to indicate pipes and ducts that are below other pipes and ducts, as shown in Figure A–33.

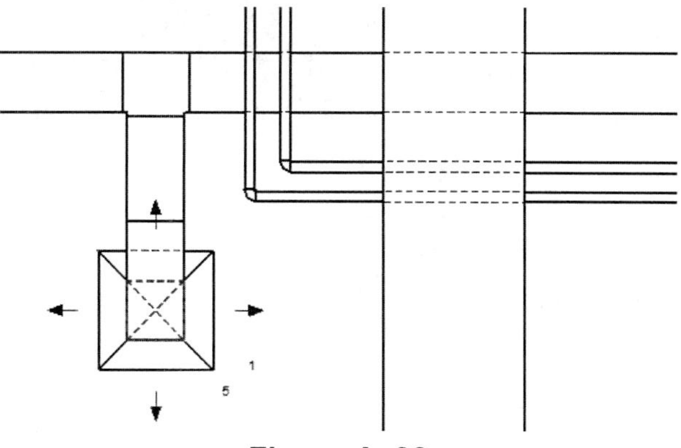

Figure A–33

The **Line Style**, **Inside Gap**, and **Outside Gap** options enable you to specify how the hidden lines are displayed.

Angles

In the *Angles* category you can specify the limits on the angles that can be used as ducts or pipes are being drawn, as shown in Figure A–34. This is most often used with piping.

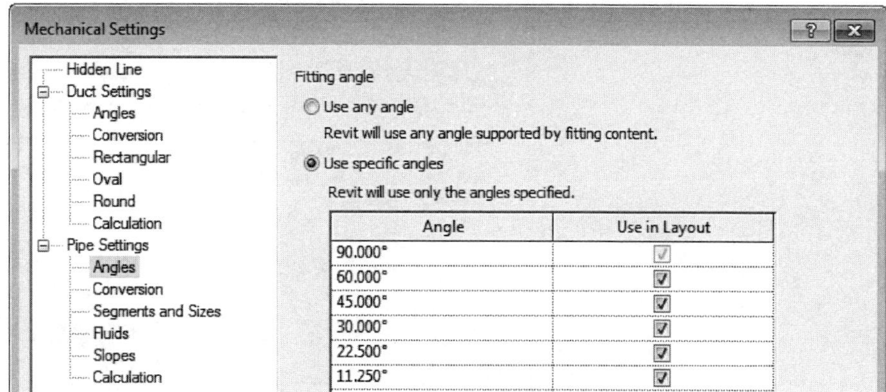

Figure A–34

- The **Duct Settings>Angles** options also include **Set an angle increment** for determining the angle values.

Duct Settings

For the **Single Line Fittings** option, the first two settings relate to how fittings are displayed in a single line view, as shown in Figure A–35. If the **Use Annot. Scale for Single Line Fittings** option is selected, the single line fittings are displayed at the same size on every sheet, regardless of the view scale. The **Duct Fitting Annotation Size** option enables you to specify the size at which the fittings should be displayed.

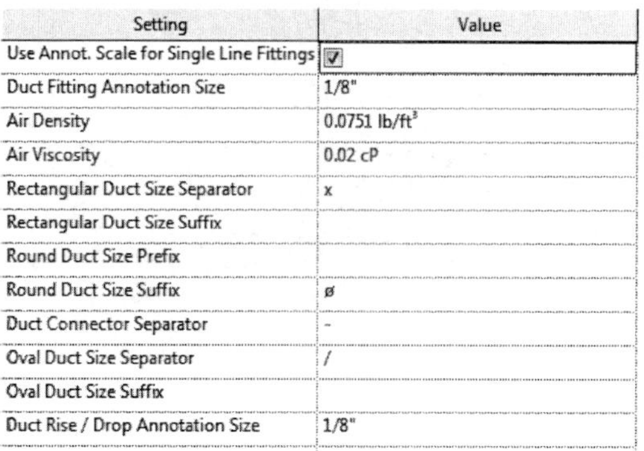

Setting	Value
Use Annot. Scale for Single Line Fittings	☑
Duct Fitting Annotation Size	1/8"
Air Density	0.0751 lb/ft³
Air Viscosity	0.02 cP
Rectangular Duct Size Separator	x
Rectangular Duct Size Suffix	
Round Duct Size Prefix	
Round Duct Size Suffix	ø
Duct Connector Separator	-
Oval Duct Size Separator	/
Oval Duct Size Suffix	
Duct Rise / Drop Annotation Size	1/8"

Figure A–35

The **Air Density** and **Air Viscosity** options are used when sizing the ducts. The remaining options set the separators and suffixes that are used for duct objects.

Conversion

The *Conversion* area is used to set the routing solutions for ductwork. The settings for the **Main** and **Branch** options can be set separately for each system type. You can specify the **Duct Type**, **Offset** from level, and **Flex Duct Type** settings (this is only applicable to branches), as shown in Figure A–36.

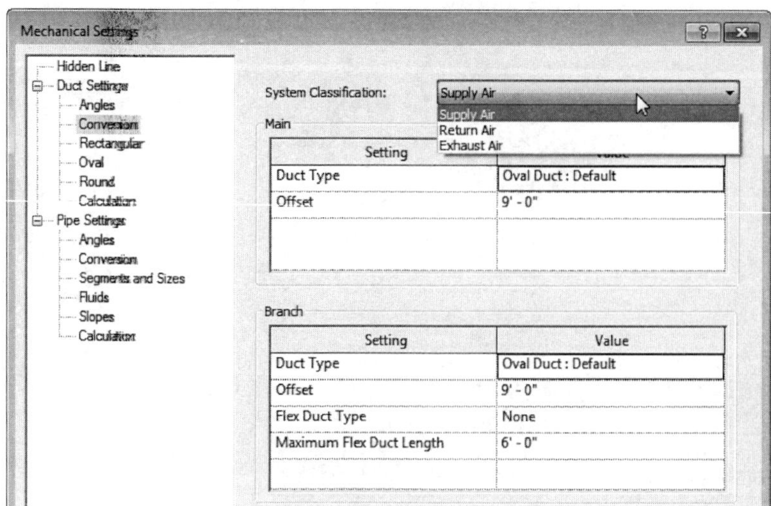

Figure A–36

Rectangular, Oval, and Round

The **Rectangular, Oval,** and **Round** options enable you to adjust the sizing tables. The *Size* column lists all of the sizes. You can click **New Size...** to add a size to the list and **Delete Size** to remove sizes from the list.

In addition, you can specify whether a size is displayed in the size lists (to be available when you place a new duct or select existing ducts) or if the size is used by the sizing routine (automatic sizing of duct branches and systems), as shown in Figure A–37.

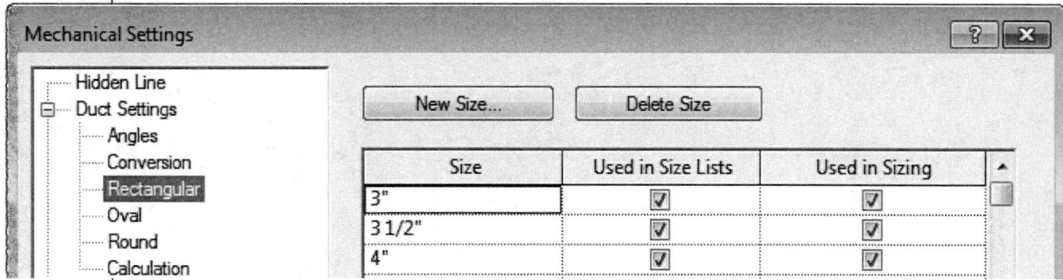

Figure A–37

- For example, if you do not want the sizing routine to use any odd numbered sizes, clear the **Used in Sizing** option for all odd-numbered sizes.

Pipe Settings

Most of the options in the *Pipe Settings* area are identical to or similar to those in the *Duct Settings* area.

Pipe Segments and Sizes

Pipe segment and size settings are more complex due to the many different materials and connections available. Each material has its own roughness, connection types, schedule/types, and sizes that correspond to each combination, as shown in Figure A–38.

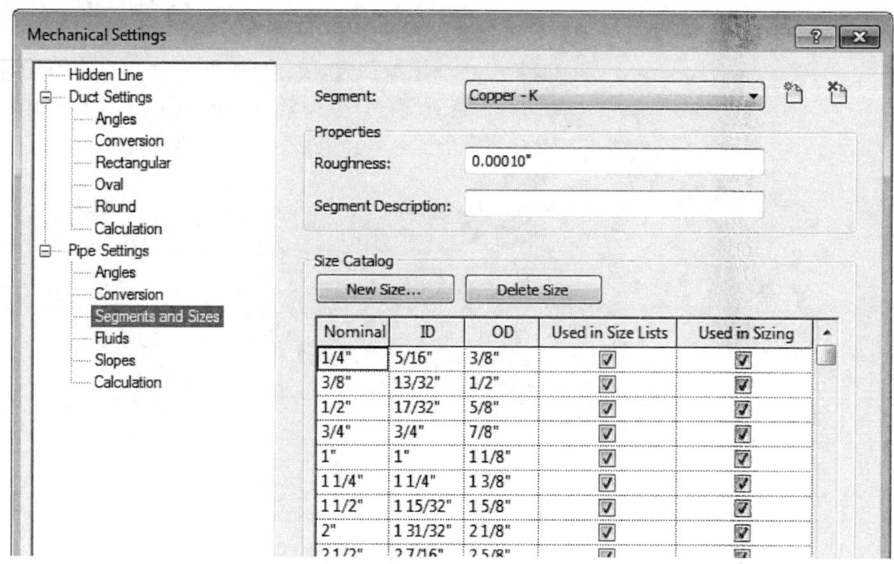

Figure A–38

The actual size lists are identical in function to those for ducts, except for the inclusion of the **ID** (inside diameter) and **OD** (outside diameter) parameters.

- Click (Create New Pipe Segment) to open the New Segment dialog box (as shown in Figure A–39), where you specify new Material, and/or Schedule/Type for the segment.

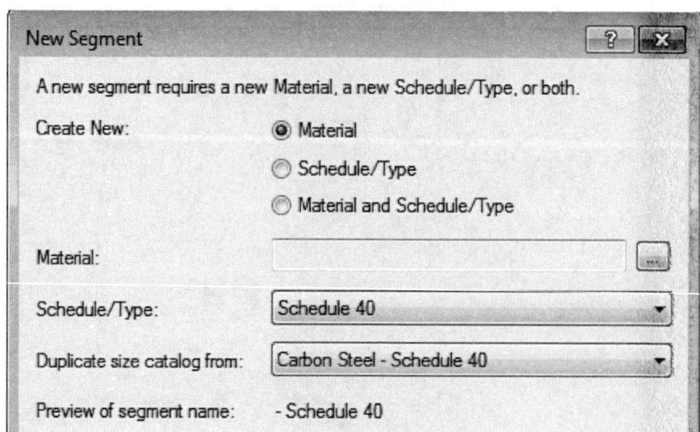

Figure A–39

- To delete a segment, click (Delete Pipe Segment).

Fluids

The *Fluids* area enables you to specify the viscosity and density of fluids at different temperatures, as shown in Figure A–40. You can add or delete fluid types, and for each type, add or delete temperatures.

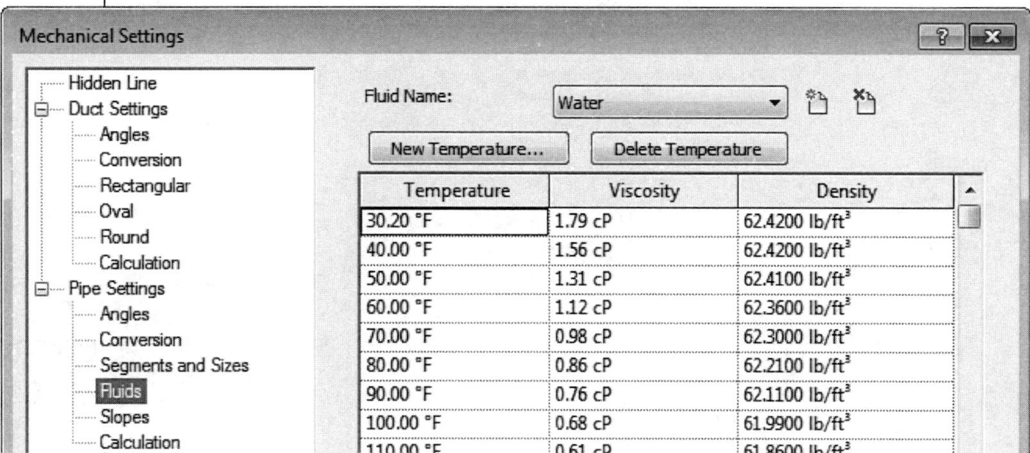

Figure A–40

Slopes

In the *Slopes* area, you can add or delete typical slopes used in a project, as shown in Figure A–41.

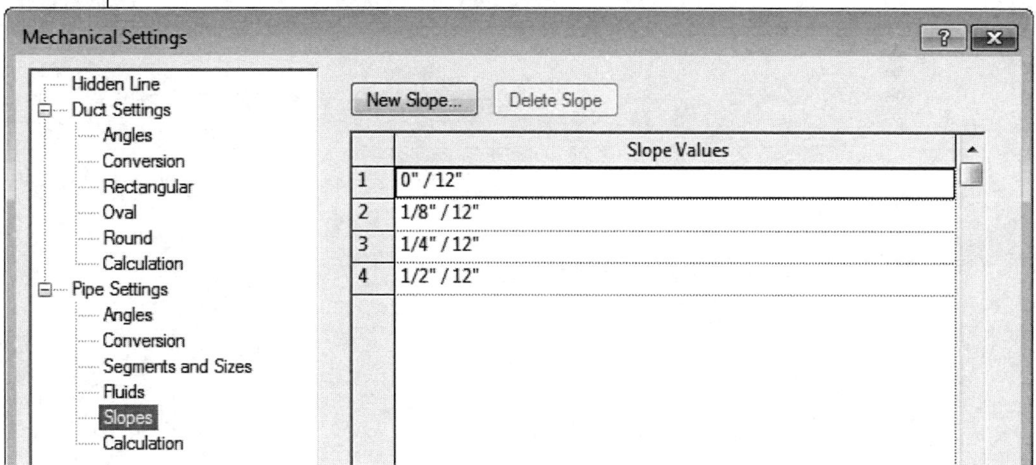

Figure A–41

Calculation

Both Duct and Pipe settings include methods for Pressure Drop that are specific to their needs, as shown for Pipe Settings in Figure A–42. For pipes, the calculation method for converting fixture units to flow for plumbing fixtures is also included.

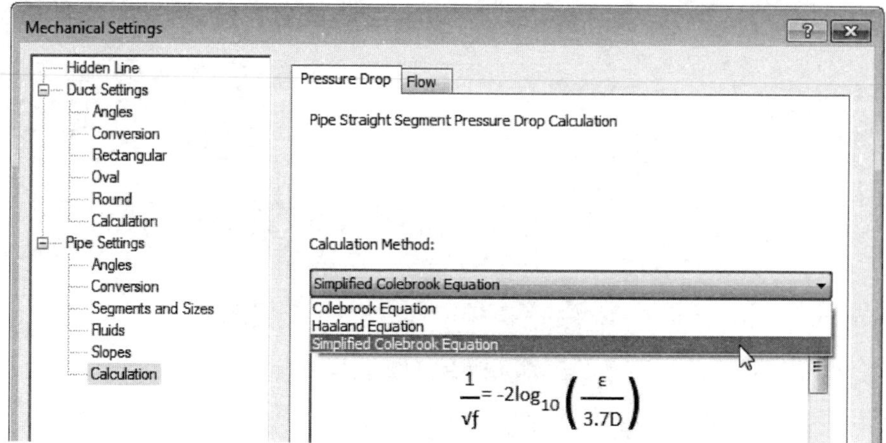

Figure A–42

Electrical Settings

In the Electrical Settings dialog box, you can preset many options relating to wiring, voltage, cable trays and conduits as well as Load Calculations and Panel Schedules, as shown in Figure A–43. Setting these up in the template for all of the commonly used electrical settings, saves time for individual projects and increases consistency.

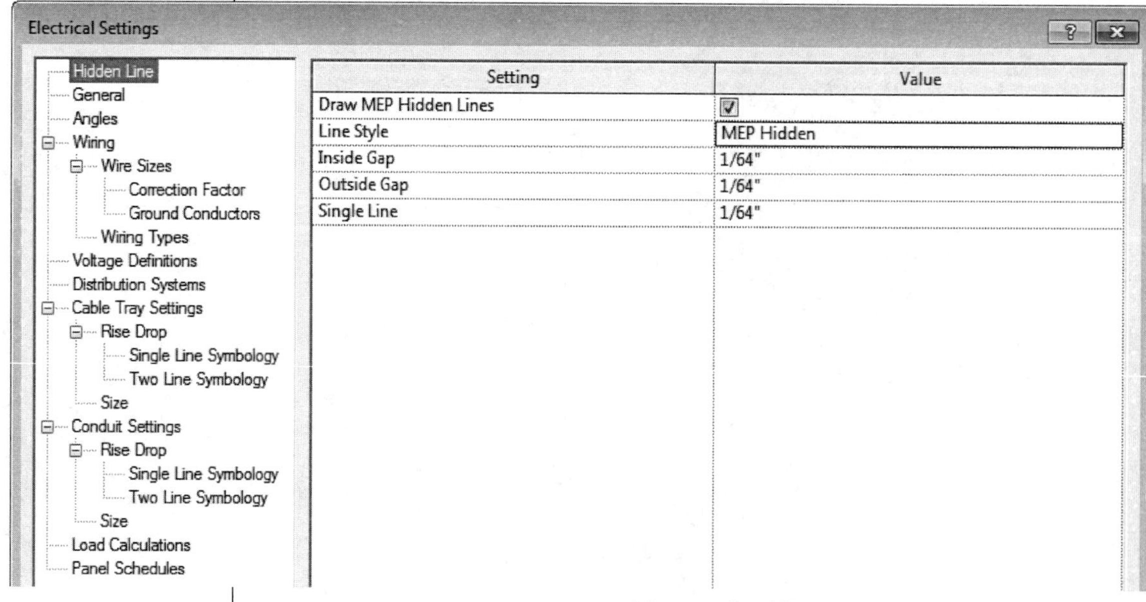

Figure A–43

General

The parameters in the *General* area affect the display of electrical information. The **Electrical Data Style** option sets how the power information is displayed in the **Electrical Data** parameter in the Instance Properties of the electrical component. The **Circuit Description** option sets the formatting of the circuit description.

Wiring

The *Wiring* area enables you to specify the ambient temperature for wiring and some annotation settings related to wiring. The **Tick Mark** families need to be preloaded into the template so that they can be selected in this area, as shown in Figure A–44.

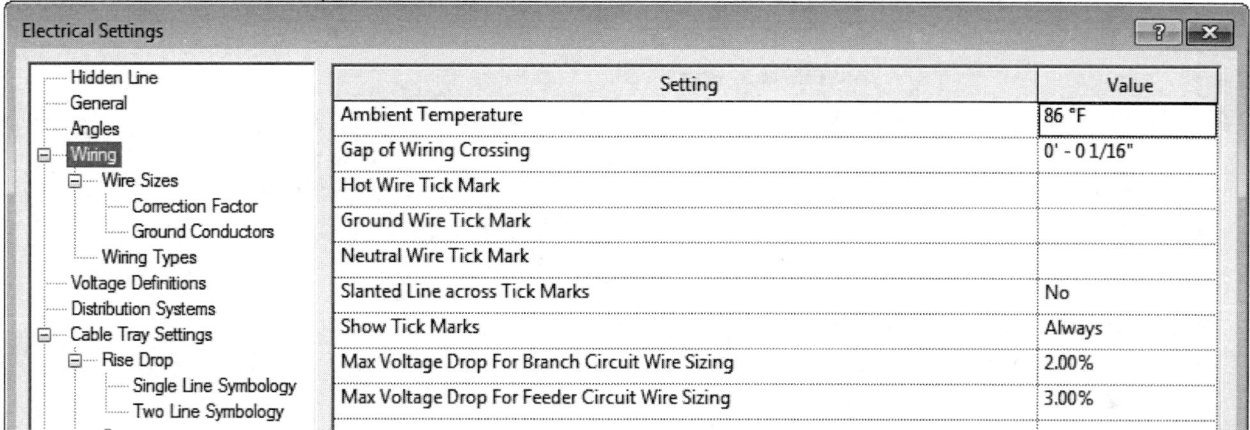

Figure A–44

Wire Sizes

The Wire Size settings enable you to create new wire materials, temperature ratings, and insulation types. You can then specify the ampacity, size, diameter, and if the size is used by the wire sizing tools.

- The functionality in the *Wire Sizes* area is similar to that in the *Pipe Sizes* and *Fluids* areas.

Correction Factor

The *Correction Factor* area enables you to specify the correction factors for different temperatures and for each wire type, as shown in Figure A–45.

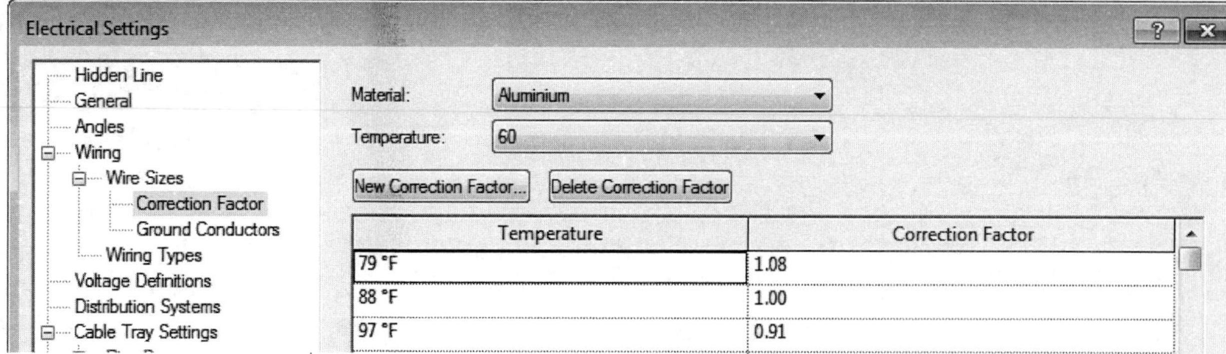

Figure A–45

Ground Conductors

The Autodesk Revit software sizes ground conductors according to their circuit rating. The *Ground Conductors* area enables you to customize the sizes it uses for each wire material and ampacity, as shown in Figure A–46.

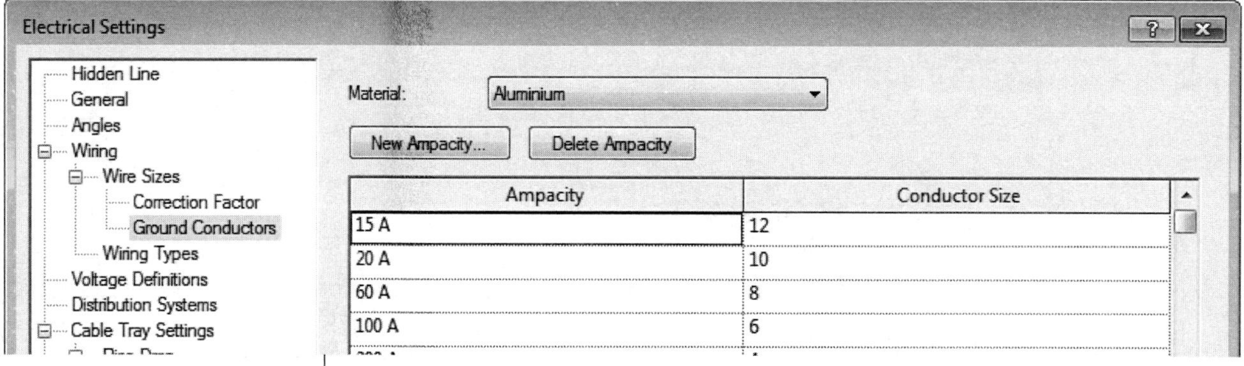

Figure A–46

Wiring Types

The *Wiring Types* area enables you to create wiring types by clicking **Add** and to delete them by clicking **Delete**. You can also specify their properties.

Voltage Definitions

In the *Voltage Definitions* area, you can specify the minimum and maximum voltages of devices that can be added to the distribution systems of a specified voltage. For example, the Autodesk Revit software permits devices ranging from 110V to 130V on a 120V distribution system, as shown in Figure A–47. You can also add or delete definitions.

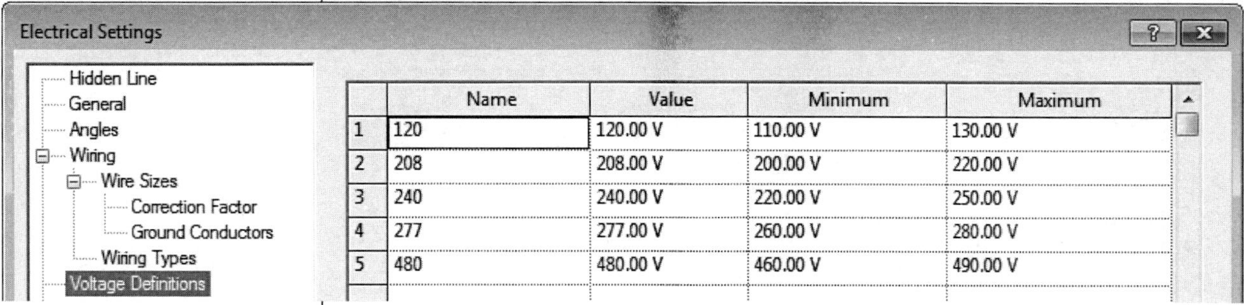

Figure A–47

Distribution Systems

The *Distribution Systems* area sets the distribution systems that are available in your project, as shown in Figure A–48. You can edit the names by selecting them. The options for *Phase* are preset and affect the options that are available in the *Configuration* and *Wires* columns and whether L-L Voltage is available. You can add more systems and delete existing systems if they are not assigned to devices in the current project.

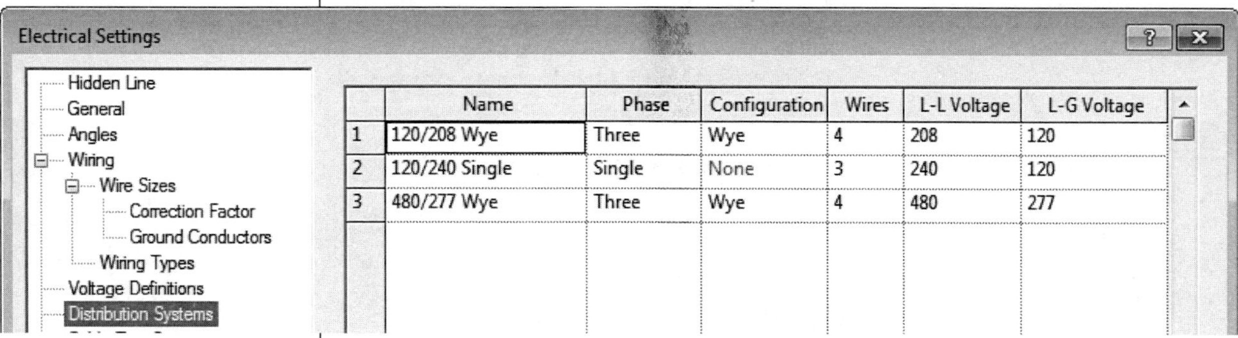

Figure A–48

Cable Tray and Conduit Settings

Cable Trays and Conduits settings include annotation, Rise Drop Symbology, and sizing, as shown in Figure A–49.

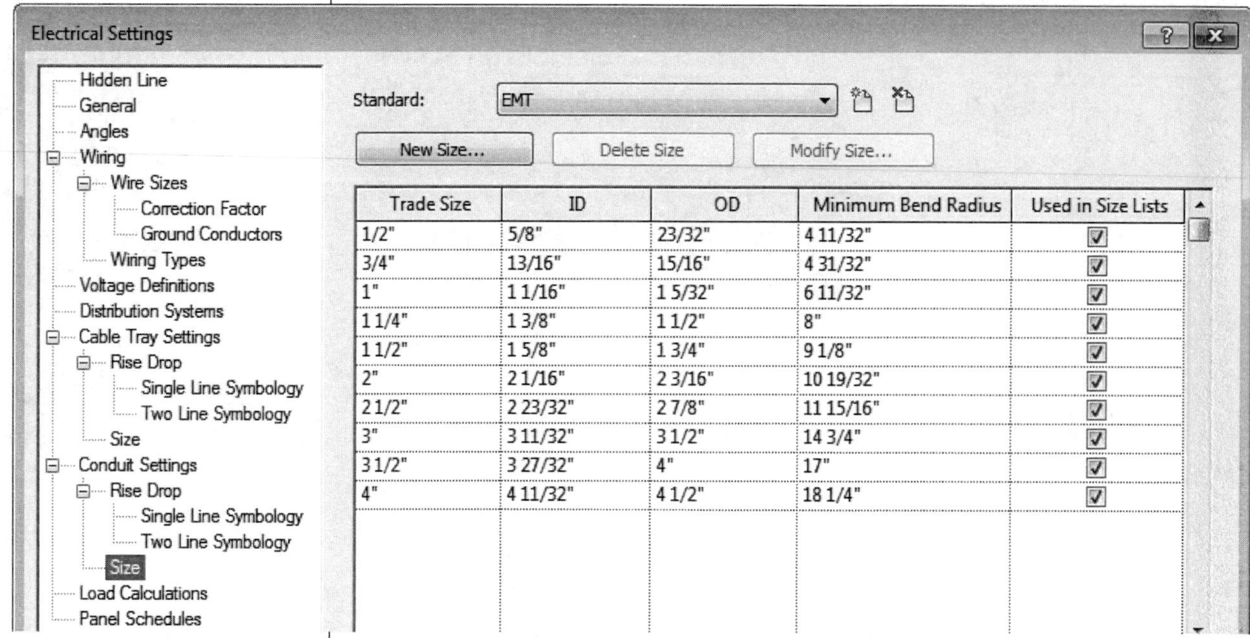

Figure A–49

Load Calculations and Panel Schedules

The *Load Calculations* area gives you access to additional dialog boxes where you can set Load Classifications and Demand Factors.

The *Panel Schedules* area enables you set up labels and other options for the default panel schedule.

A.6 Settings for Structural Projects

In addition to the standard settings common to all aspects of the Autodesk Revit software, there are structural settings that need to be customized.

- In the *Manage* tab>Settings panel, click (Structural Settings), or click the arrow in the *Structure* tab>Structure panel title.

Symbolic Representation

There are several tabs at the top of the Structural Settings dialog box, as shown in Figure A–50. The first tab, *Symbolic Representation Settings*, contains options that are mainly used for the graphical model and the common defaults.

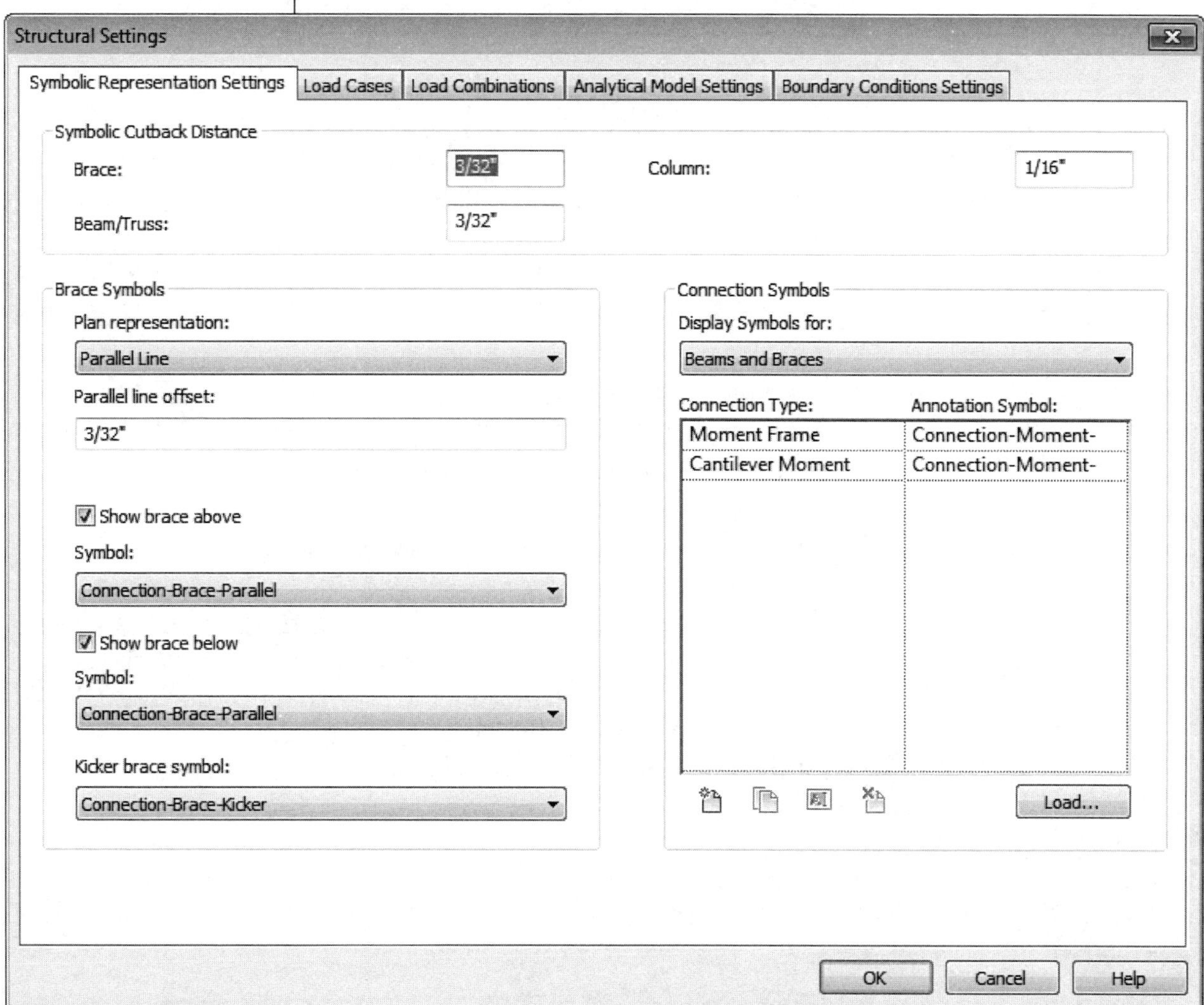

Figure A–50

The **Symbolic Cutback Distance** setting represents the distance that cuts a framing member back from a column or into another framing member. This cutback distance is symbolic and does not affect the 3D view. For example, in plan, beams do not extend into the column and there is a gap between them, as shown in Figure A–51. The same connection in 3D displays the beams back from the column, which is more consistent with a real-world situation.

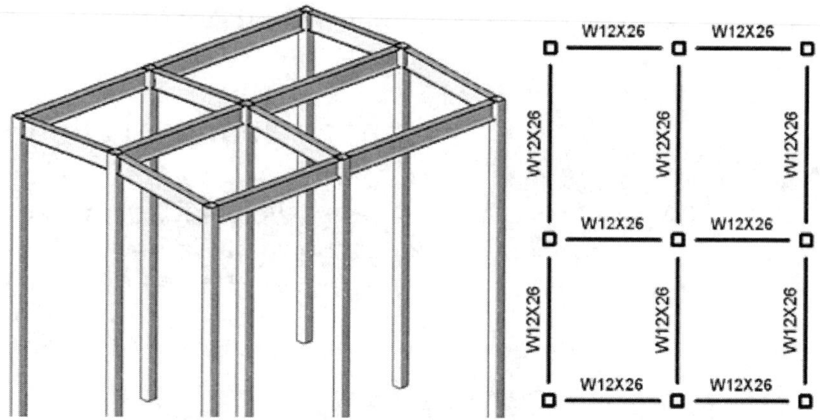

Figure A–51

The *Brace Symbols* and *Connection Symbols* areas enable you to specify the kinds of symbols that are shown for the braces in plan, and for frame and shear connections. For many of these items, only one symbol is loaded into the default template.

Load Cases and Load Combinations

Two tabs are related to setting up Load Cases and Load Combinations. These vary by project and by region and typically requires the input of a company's engineer.

Analytical Model Settings

The process of modeling a building simultaneously creates a digital model and its corresponding analytical model. As this is done, the software can check for consistency between these two models. The *Analytical Model Settings* tab contains settings for automatic checks, tolerances, and checks, as shown in Figure A–52.

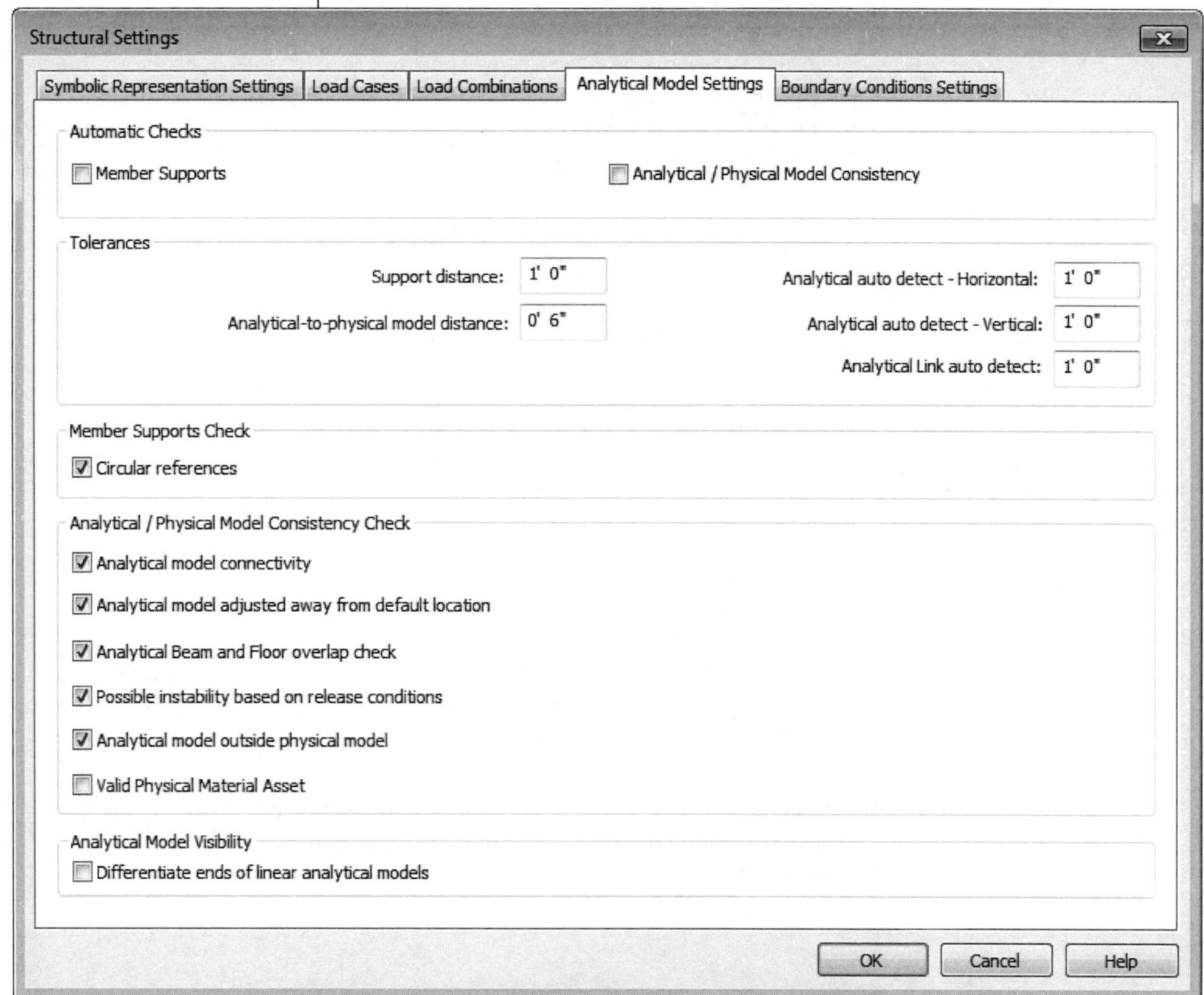

Figure A–52

It is a best practice not to enable **Automatic Checks** in the template, as in the early stages of design there might be many elements that are not supported, causing many warning boxes to open. Automatic Checks are very useful when most of the structure has been modeled and should be enabled at that time.

The values in the *Tolerances* area informs the kind of tolerance that is permitted for supports, differences between the analytical and physical distance, adjusting, and auto-detects. For example, when the **Analytical Model** parameters for a wall are set to auto-detect, the software adjusts the analytical lines, provided it can do so within the specified tolerances.

The *Analytical/Physical Model Consistency Check* area enables you to specify the kinds of consistencies to be checked for.

Boundary Condition Settings

In the last tab, you can select the family symbols used for boundary conditions, as shown in Figure A–53 (from left to right: **Fixed**, **Pinned**, **Roller**, and **User**).

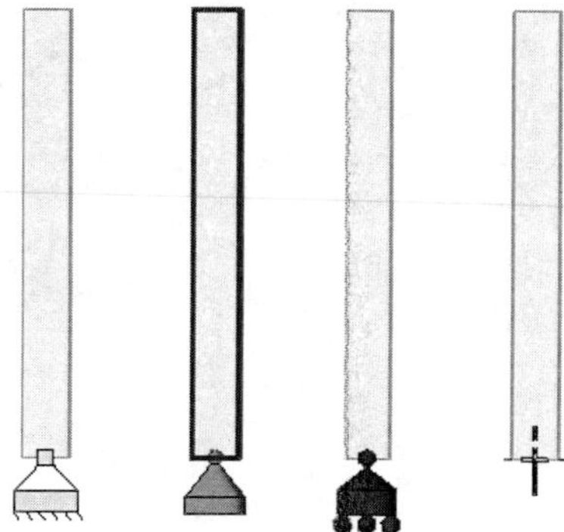

Figure A–53

The structural templates typically include four boundary condition families: **Fixed**, **Pinned**, **Roller**, and **Variable**. They are assigned to the corresponding boundary conditions, as shown in Figure A–54.

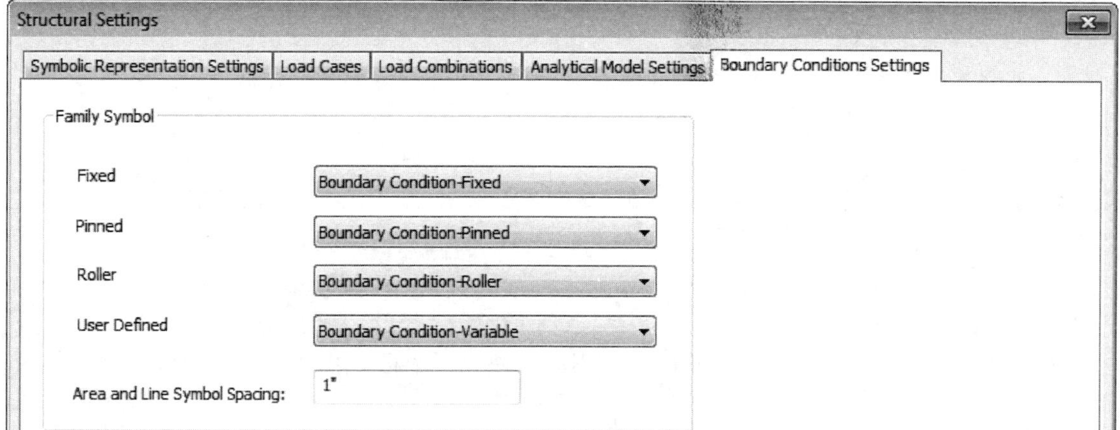

Figure A–54

- You can create new boundary condition families, load them into the template, and select them from the drop-down lists as needed.

A.7 Sheet List Schedules

Sheet List schedules are used to keep track of all of the sheets in a project and any list sheets that could be added. Sheet lists are often used by architects to display all of the sheets in a building project. This includes any consultant sheets that are not in the architectural project file, as they would only be in the Sheet List schedule and not created as actual sheets.

How To: Create a Sheet List Schedule

1. Open a project template file (or a project).
2. In the *View* tab>Create panel, expand (Schedules) and click (Sheet List).
3. In the Sheet List Properties dialog box, in the *Fields* tab, select the **Sheet Number** and **Sheet Name** and add them to the *Scheduled fields* list as shown in Figure A–55.

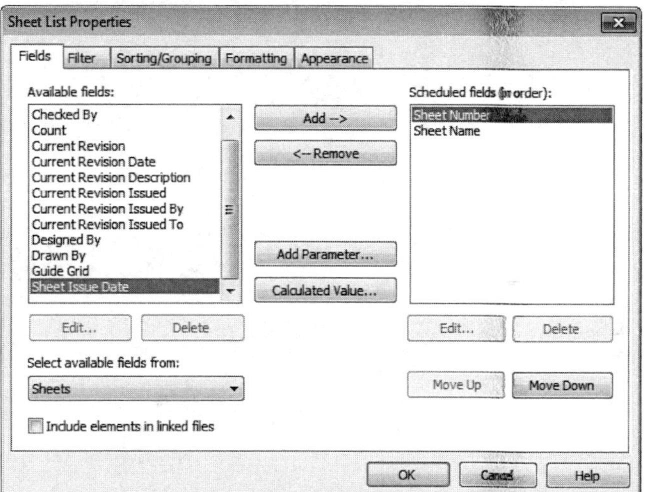

Figure A–55

4. Depending on the complexity of your sheet naming scheme, you can also modify items on the *Filter* and *Sorting/Grouping* tabs.
5. Click **OK** when you are finished.
6. You are placed in the schedule view with the two parameters displayed. Stretch out the columns as shown in Figure A–56.

If sheets are already in the project, they display here.

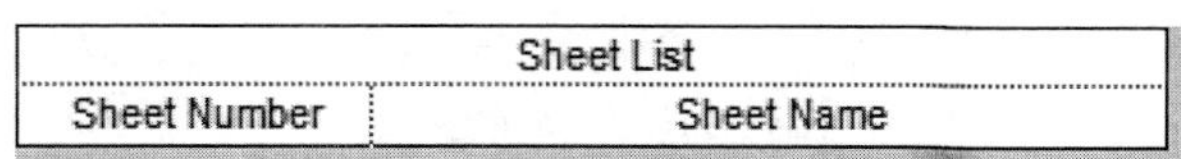

Figure A–56

7. In the *Modify Schedule/Quantities* tab>Rows panel, click (Insert Data Row).
8. A new row is added below the schedule names. If no sheets are in the project, it comes in automatically as **A101** and **Unnamed**.
9. Add as many rows as you have sheets. If you are using a numbering scheme such as A1xx for site plans, A2xx for Floor Plans, A3xx for Detail plans, etc., then you should rename the sheet number for the first row of a set before creating more rows so that they increment automatically.
10. Enter the name of each sheet. Once you have added a new name, it is available in the drop-down list, as shown in Figure A–57.

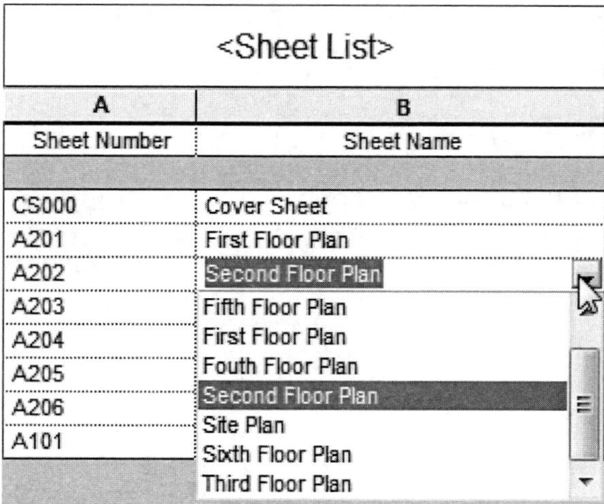

<Sheet List>
A
B
Sheet Number
Sheet Name
CS000
Cover Sheet
A201
First Floor Plan
A202
Second Floor Plan
A203
Fifth Floor Plan
A204
First Floor Plan
A205
Fouth Floor Plan
A206
Second Floor Plan
A101
Site Plan
Sixth Floor Plan
Third Floor Plan

Figure A–57

How To: Use Sheet List Tables

1. Create a new sheet. In the Sheet List Schedule view, in the *Modify Schedule/Quantities* tab>Create panel, click (New Sheet), You can also access the command on the *View* tab or right-click on *Sheets* in the Project Browser.
2. In the New Sheet dialog box, select the placeholder sheets you want to use, as shown in Figure A–58. To select more than one, hold down <Ctrl> or <Shift> as you select.

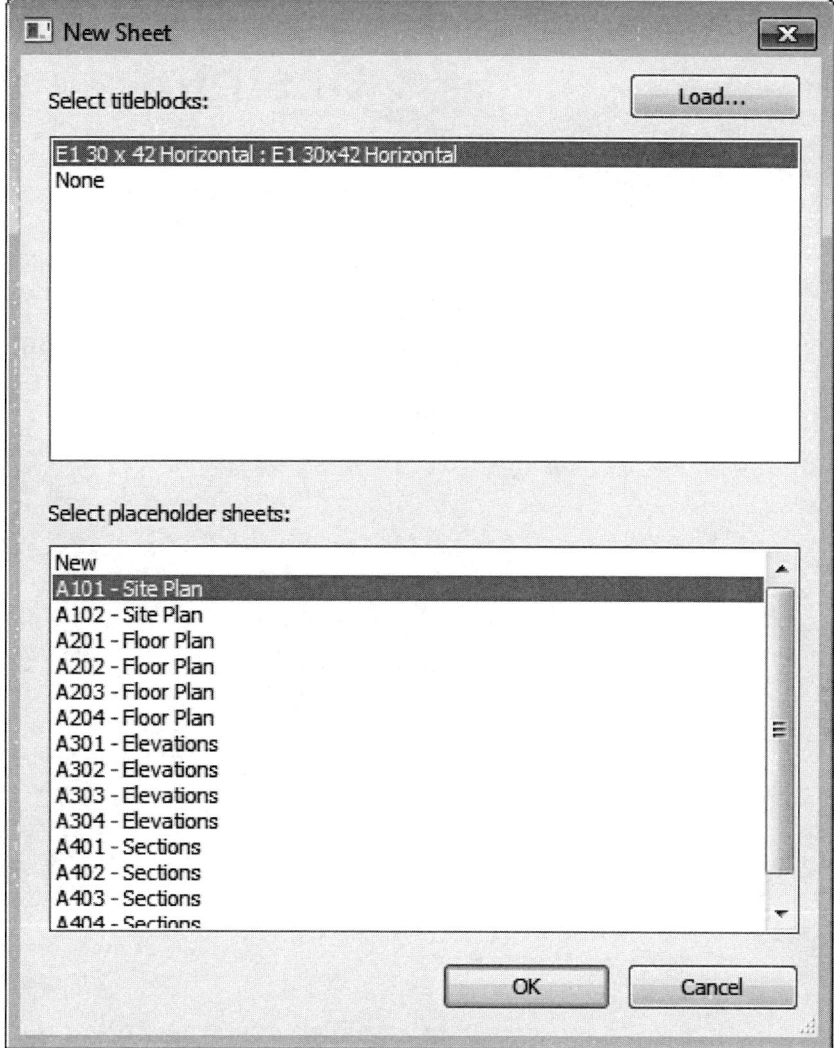

Figure A–58

3. Click **OK**. The new sheets are created in the project.

- This sheet list is available in all projects based on the project template where it was created.
- You can import the sheet list schedule into another project, but it does not import the associated sheets. That is because they are part of the project, not the schedule view.

A.8 Basic User Interface Customization

The Revit User Interface has a variety of features that can be customized. From moving Ribbon panels around by dragging and dropping them to a new location, to the activation and location of palettes, and items that display on the status bar. For example, if you do not use Design Options you can clean up the Status Bar by clearing **Status Bar - Design Options** as shown in Figure A–59.

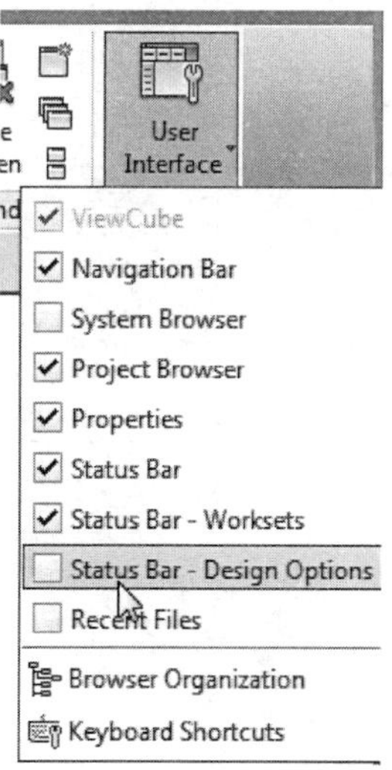

Figure A–59

You can also customize other parts of the Autodesk Revit software, including customizing keyboard shortcuts, specifying how the double-click functions work and setting up the Project Browser organization.

Customizing Shortcuts

When you type keyboard shortcuts, you do not need to press <Enter> or the <Spacebar>.

A helpful customization is to set up the keyboard shortcuts for frequently used commands. These are not saved by project, but on each computer. To display the shortcuts that are already available, hover the cursor over a tool, such as **Wall**, to display the tool tip that displays the associated shortcut, as shown in Figure A–60.

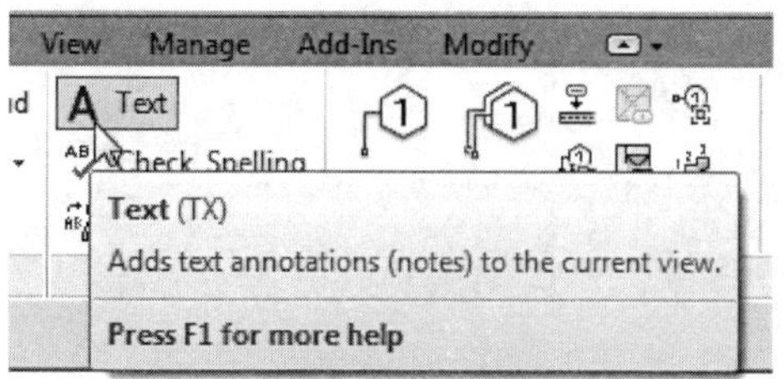

Figure A–60

How To: Customize the Keyboard Shortcuts

1. In the *View* tab>Windows panel, expand (User Interface), and click (Keyboard Shortcuts) or, in the Application Menu, click **Options**. In the Options dialog box, in the *User Interface* pane, in the *Configure* area, next to *Keyboard shortcuts:* click **Customize...**.
2. In the Keyboard Shortcuts dialog box, use the **Search** or **Filter** options to narrow the search, as shown in Figure A–61.

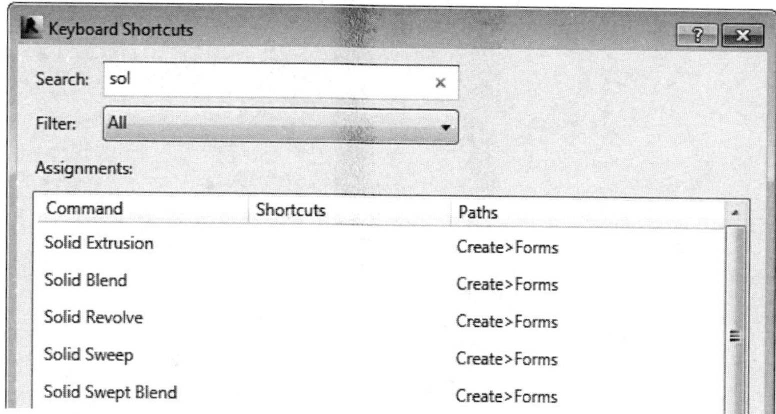

Figure A–61

3. Select the command that you want to add or modify.
4. In the Press new keys edit box, type the shortcut that you want to use (as shown in Figure A–62), and click **Assign**.

If a shortcut has already been assigned, you can still use it. Press the <Spacebar> to cycle through the options.

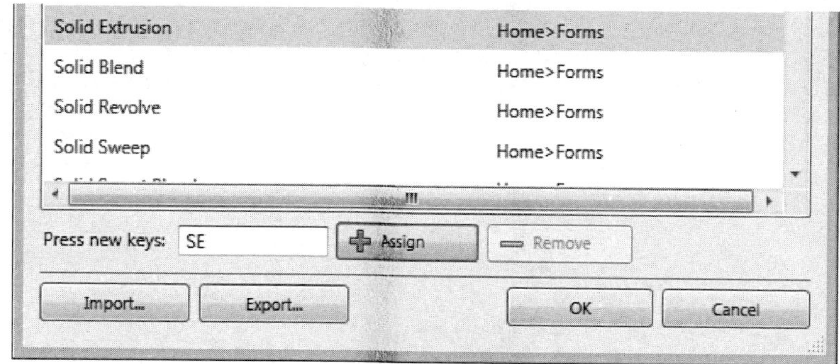

Figure A–62

5. Click **OK** when you have finished.

- To remove shortcuts, select the shortcut, and click **Remove**.
- You can import or export the shortcut file to be used in other stations of the Autodesk Revit software than the one on which they were created. When you export to XML, all of the commands are exported. You can then add the required shortcuts to the XML file and import them into your program.

Customizing Double-click Options

When you double-click on elements, you can specify what you want this to do. For example, by default, double-clicking on a family, such as a chair, column, or plumbing fixture, opens the element in the Family Editor. In most cases that is not something you want to do, editing the type would be better. You can change this in the Customize Double-click Settings dialog box as shown in Figure A–63.

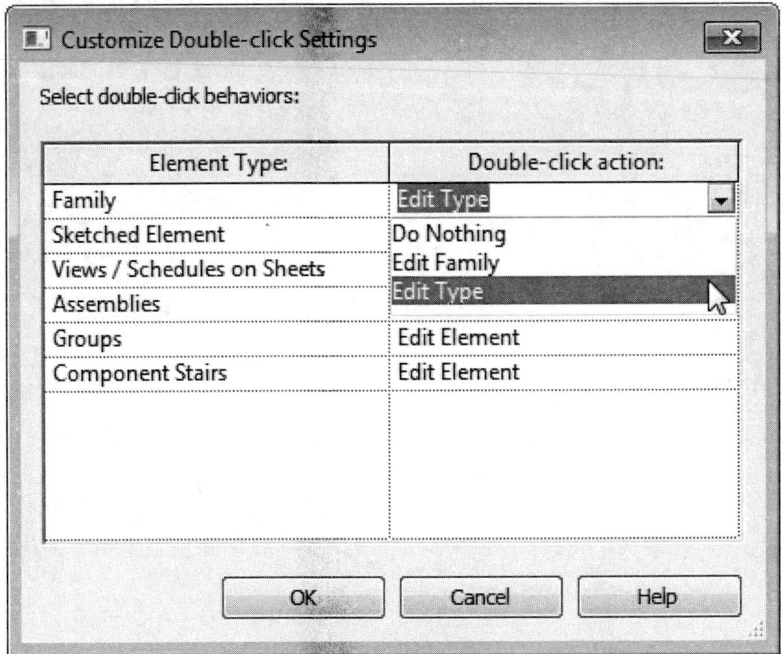

Figure A–63

How To: Customize Double-Click Options

1. In the Application Menu, click **Options**.
2. In the Options dialog box, in the *User Interface* pane, in the *Configure* area, next to *Double-click Options:* click **Customize...**.
3. Select the method that you want to use for each of the element types.

Customizing the Browser Organization

The name of the current organization type is in parentheses.

The Project Browser is the heart of the Autodesk Revit software. It is where you access all of the different views, sheets, family types, etc. When working with complex projects, it can help to organize the views and sheets so that it is most useful to the methods used in your firm. It is easy to switch between different organizations and you can create custom ones to suit your work.

- To switch between existing organization types, in the Project Browser, select the ***View*** node at the top of the palette, and in the Type Selector, select the type that you want to use as shown in Figure A–64.

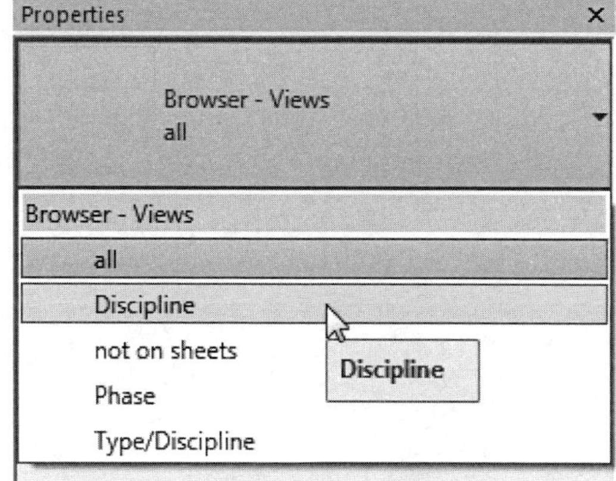

Figure A–64

- You can also switch sheet organizations by selecting the *Sheets* node and changing the type in the Type Selector.

How To: Customize the Browser Organization

1. In the *View* tab>Windows panel, expand (User Interface), and click (Browser Organization) or, in the Project Browser, right-click on the *Views* or *Sheets* node and select **Browser Organization**.
2. In the Browser Organization dialog box, select the *Views* or *Sheets* tab, and click **New**.
3. Name the new browser organization and click **OK**.
4. In the Browser Organization Properties dialog box, in the *Filtering* tab, specify the filtering rules as needed.

5. In the *Grouping and Sorting* tab, assign the various groups as needed. The example shown in Figure A–65 is by discipline, an organization that is used most often in the MEP templates.

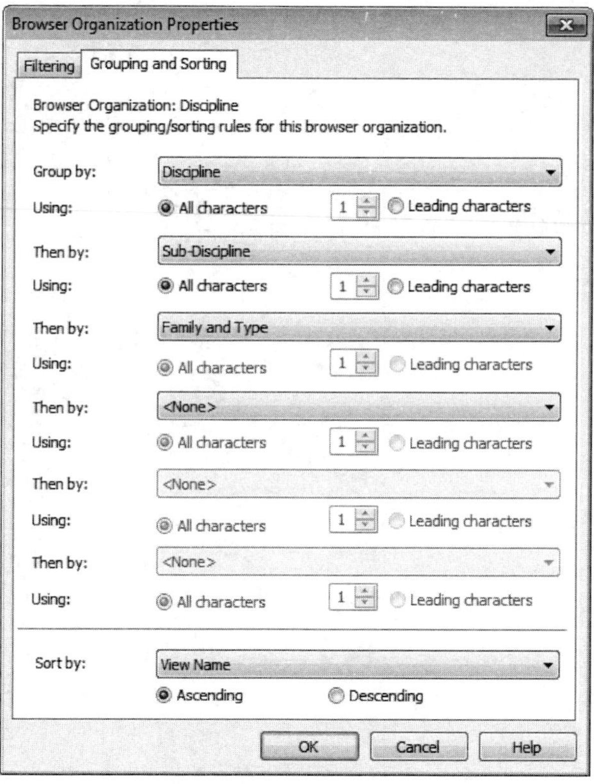

Figure A–65

6. Click **OK** to finish the property changes.
7. In the Browser Organization dialog box, you can continue to create the new organization and then select the current organization. Click **Apply** between sets if you are creating multiple organizations. Click **OK** when you have finished.

- In the Options dialog box, in the *General* tab, in the *View Options* area, you can set the *Default view discipline* that is used when you create new views, such as sections, callouts, and elevations, as shown in Figure A–66.

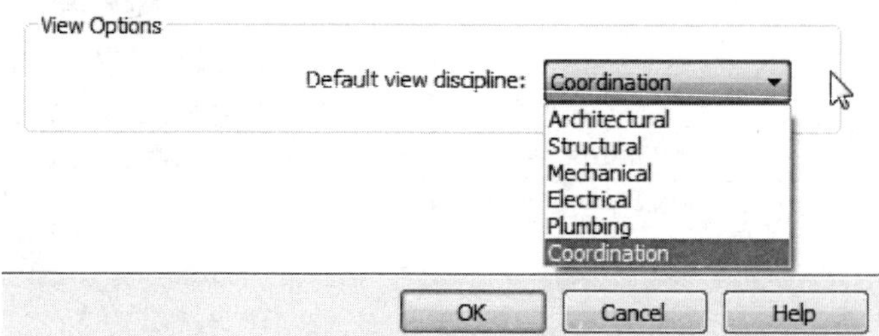

Figure A–66

- When you duplicate a view, it automatically assumes the discipline and other view properties of the original view.

Command Summary

Button	Command	Location
	Arrowheads	• **Ribbon:** *Manage* tab>Settings panel> expand Additional Settings
	Browser Organization	• **Ribbon:** *View* tab>Windows panel> expand User Interface • **Application Menu:** Options>*User Interface* tab
	Electrical Settings	• **Autodesk Revit MEP** • **Ribbon:** *Manage* tab>Settings panel expand MEP Settings
	Fill Patterns	• **Ribbon**: *Manage* tab>Settings panel> expand Additional Settings
	Keyboard Shortcuts	• **Ribbon:** *View* tab>Windows panel> expand User Interface • **Application Menu:** Options>*User Interface* tab
	Line Patterns	• ***Ribbon:*** *Manage* tab>Settings panel> expand Additional Settings
	Line Styles	• **Ribbon:** *Manage* tab>Settings panel> expand Additional Settings
	Line Weights	• **Ribbon:** *Manage* tab>Settings panel> expand Additional Settings
	Materials	• **Ribbon:** *Manage* tab>Settings panel
	Mechanical Settings	• **Autodesk Revit MEP** • **Ribbon:** *Manage* tab>Settings panel expand MEP Settings
	Object Styles	• **Ribbon:** *Manage* tab>Settings panel
Options	**Options**	• **Application Menu**
	Structural Settings	• **Autodesk Revit Structure** • **Ribbon:** *Manage* tab>Settings panel
	Temporary Dimensions	• **Ribbon:** *Manage* tab>Settings panel> expand Additional Settings

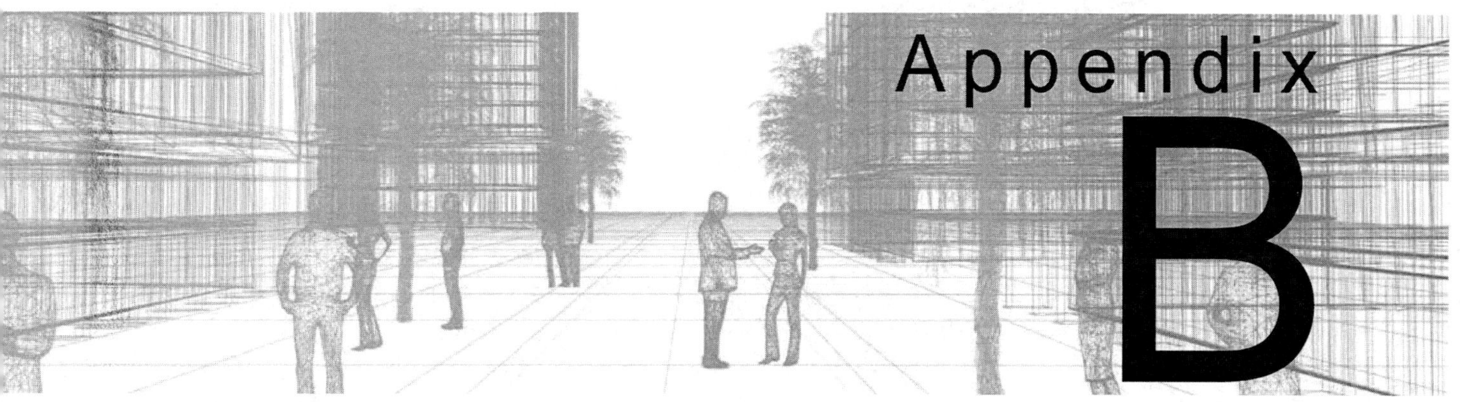

Autodesk Revit Architecture Certification Exam Objectives

The following table will help you to locate the exam objectives within the chapters of the Autodesk® Revit® 2016 training guides to help you prepare for the Autodesk Revit Architecture 2015 Certified Professional (Pro.) and User exams.

User	Pro.	Exam Objective	Training Guide	Chapter & Section(s)
User Interface				
✓		Change the view scale	• Revit Architecture Fundamentals	• 1.2
✓		Identify file types	• Revit Architecture Fundamentals	• 1.3
✓		Identify primary parts of the User Interface (UI)	• Revit Architecture Fundamentals	• 1.2
✓	✓	Use the Project Browser	• Revit Architecture Fundamentals	• 1.2
Collaboration				
	✓	Copy and monitor elements in a linked file	• Revit Collaboration Tools	• 2.3
	✓	Import DWG and image files	• Revit Architecture Fundamentals	• 3.4
			• Revit Collaboration Tools	• 3.1, 3.3
	✓	Use worksharing	• Revit Collaboration Tools	• 4.1, 4.3

User	Pro.	Exam Objective	Training Guide	Chapter & Section(s)
Documentation				
✓		Create a title sheet	• Revit Architecture Fundamentals	• 13.1
	✓	Create and modify filled regions	• Revit Architecture Fundamentals	• 16.3
	✓	Place detail components and repeating details	• Revit Architecture Fundamentals	• 16.2
	✓	Set the colors used in a color scheme legend	• Revit Architecture: Conceptual Design and Visualization	• 2.3
✓	✓	Tag elements (doors, windows, etc.) by category	• Revit Architecture Fundamentals	• 15.1
	✓	Use dimension strings	• Revit Architecture Fundamentals	• 14.1
	✓	Work with phases	• Revit Collaboration Tools	• 1.1
Elements				
	✓	Change elements within a curtain wall	• Revit Architecture Fundamentals	• 6.2, 6.3, 6.4
	✓	Create a new family type	• Revit Architecture Fundamentals	• 5.3
			• Revit BIM Management	• 4.5
	✓	Create a stacked wall	• Revit BIM Management	• 3.3
✓		Create and modify walls	• Revit Architecture Fundamentals	• 4.1, 4.2
	✓	Create compound walls	• Revit BIM Management	• 3.1
	✓	Differentiate system and component families	• Revit BIM Management	• 3.1 • 4.1
✓		Edit doors	• Revit Architecture Fundamentals	• 5.1
✓		Edit windows	• Revit Architecture Fundamentals	• 5.1
✓		Trim objects	• Revit Architecture Fundamentals	• 2.4
	✓	Work with family parameters	• Revit BIM Management	• 4.2
Families				
	✓	Assess review warnings in Revit	• Revit Architecture Fundamentals	• 12.1
	✓	Use Family creation procedures	• Revit BIM Management	• 4.1 to 4.7
✓	✓	Work with families	• Revit Architecture Fundamentals	• 8.1, 8.2

User	Pro.	Exam Objective	Training Guide	Chapter & Section(s)
Modeling				
✓		Add dimensions	• Revit Architecture Fundamentals	• 14.1
✓		Add model text to a plan	• Revit Architecture Fundamentals	• 14.2
	✓	Attach walls to a roof or ceiling	• Revit Architecture Fundamentals	• 11.2
	✓	Change a generic floor / ceiling / roof to a specific type	• Revit Architecture Fundamentals	• 9.1 • 10.1 • 11.2
	✓	Create a building pad	• Revit Architecture: Site and Structure	• 1.2
✓		Create a roof and modify roof properties	• Revit Architecture Fundamentals	• 11.2, 11.4
✓	✓	Create a stair with a landing	• Revit Architecture Fundamentals	• 12.1
	✓	Create elements such as floors, ceiling, or roofs	• Revit Architecture Fundamentals	• 9.1 • 10.1 • 11.2, 11.4
✓	✓	Define floor for a mass	• Revit Architecture: Conceptual Design and Visualization	• 1.8
	✓	Edit a model element's material	• Revit Architecture Fundamentals	• 5.3 • B.4
	✓	Edit room-aware families	• Revit BIM Management	• 4.4
	✓	Generate a toposurface	• Revit Architecture: Site and Structure	• 1.1
✓	✓	Model railings	• Revit Architecture Fundamentals	• 12.3
✓		Use grids	• Revit Architecture Fundamentals	• 3.2
Views				
	✓	Control visibility	• Revit Architecture Fundamentals	• 7.1
	✓	Create a duplicate view for a plan, section, elevation, drafting view, etc.	• Revit Architecture Fundamentals	• 7.2
✓		Create a schedule and add schedule tags	• Revit Architecture Fundamentals	• 15.1, 15.3 • B.10
	✓	Create and manage legends	• Revit Architecture Fundamentals	• 14.4
✓		Create section views	• Revit Architecture Fundamentals	• 7.4

User	Pro.	Exam Objective	Training Guide	Chapter & Section(s)
	✓	Define element properties in a schedule	• Revit Architecture Fundamentals	• 15.3
	✓	Manage view position on sheets	• Revit Architecture Fundamentals	• 13.2
	✓	Organize and sort items in a schedule	• Revit Architecture Fundamentals	• B.10
			• Revit BIM Management	• 2.2
✓	✓	Use levels	• Revit Architecture Fundamentals	• 3.1

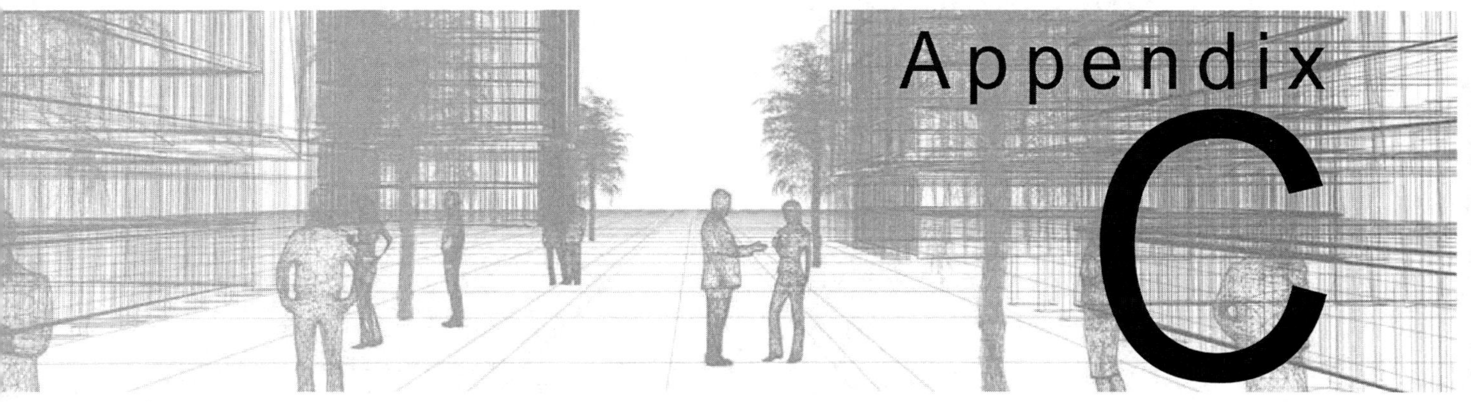

Autodesk Revit MEP Certification Exam Objectives

The following table will help you to locate the exam objectives within the chapters of the Autodesk® Revit® 2016 training guides to help you prepare for the Autodesk Revit MEP 2015 Certified Professional (Pro.) exam.

Pro.	Exam Objective	Training Guide	Chapter & Section(s)
Collaboration			
✓	Import AutoCAD files into Revit	• Revit MEP Fundamentals	• B.1
✓	Link Revit models	• Revit MEP Fundamentals	• 4.1
✓	Copy levels and setup monitoring	• Revit MEP Fundamentals	• 4.3
✓	Create floor plans	• Revit MEP Fundamentals	• 5.2
Documentation			
✓	Electrical: Tag components	• Revit MEP Fundamentals	• 14.1
✓	Mechanical: Tag ducts	• Revit MEP Fundamentals	• 14.1
✓	Plumbing: Tag items	• Revit MEP Fundamentals	• 14.1
✓	Create sheets	• Revit MEP Fundamentals	• 12.1
✓	Create schedules	• Revit MEP Fundamentals	• B.10
✓	Add and modify text	• Revit MEP Fundamentals	• 13.2
✓	Add and modify dimensions	• Revit MEP Fundamentals	• 13.1

Pro.	Exam Objective	Training Guide	Chapter & Section(s)
Modeling Electrical			
✓	Add and modify receptacles	• Revit MEP Fundamentals	• 11.2
✓	Add and modify panels	• Revit MEP Fundamentals	• 11.4
✓	Create and modify circuits	• Revit MEP Fundamentals	• 11.3
✓	Add and modify lighting fixtures	• Revit MEP Fundamentals	• 11.2
✓	Add and modify switches	• Revit MEP Fundamentals	• 11.2
✓	Create and modify lighting circuits	• Revit MEP Fundamentals	• 11.3
✓	Create and modify switching circuits	• Revit MEP Fundamentals	• 11.3
✓	Add and modify conduit	• Revit MEP Fundamentals	• 11.5
✓	Use cable trays	• Revit MEP Fundamentals	• 11.5
Modeling: Mechanical			
✓	Add and use equipment	• Revit MEP Fundamentals	• 8.1
✓	Add and modify air terminals	• Revit MEP Fundamentals	• 8.1
✓	Add and modify supply ducts	• Revit MEP Fundamentals	• 8.2, 8.3
✓	Add and modify return ducts	• Revit MEP Fundamentals	• 8.2, 8.3
✓	Add and modify duct accessories and fittings	• Revit MEP Fundamentals	• 8.3
✓	Work with heating and cooling zones	• Revit MEP Fundamentals	• 6.4
Modeling: Plumbing			
✓	Add and modify fixtures and supply piping	• Revit MEP Fundamentals	• 9.1, 9.2
✓	Add and modify sanitary piping	• Revit MEP Fundamentals	• 9.2
✓	Add equipment	• Revit MEP Fundamentals	• 9.1
✓	Create a plumbing system	• Revit MEP Fundamentals	• 10.1
✓	Add and modify pipe accessories	• Revit MEP Fundamentals	• 9.3
Views			
✓	View models	• Revit MEP Fundamentals	• 1.4
✓	Create a plumbing view	• Revit MEP Fundamentals	• 5.1
✓	Create detail views	• Revit MEP Fundamentals	• 15.1
✓	Create and label wiring plans	• Revit MEP Fundamentals	• 11.3 • 13.2

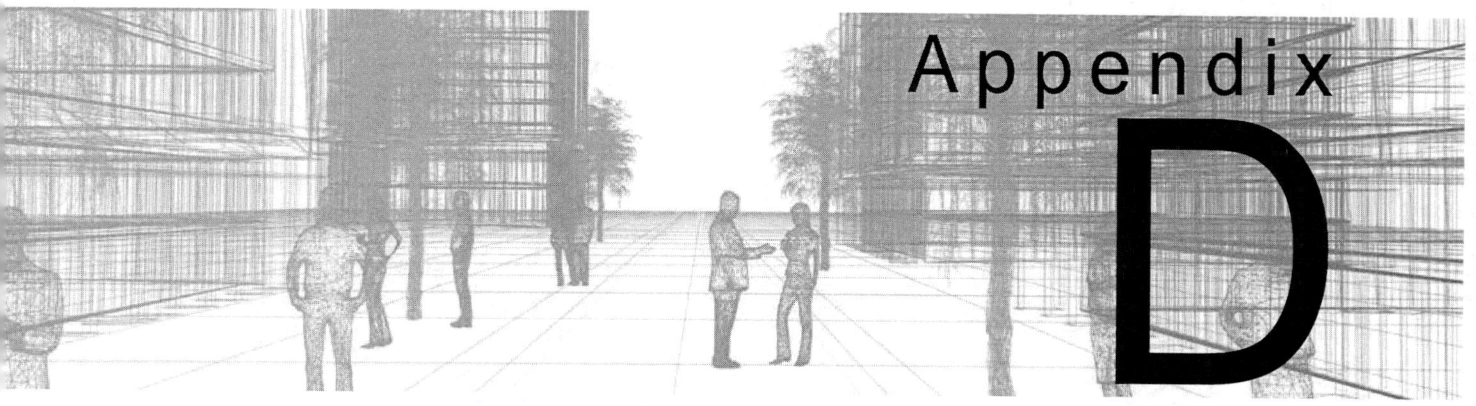

Autodesk Revit Structure Certification Exam Objectives

The following table will help you to locate the exam objectives within the chapters of the Autodesk® Revit® 2016 training guides to help you prepare for the Autodesk Revit Structure 2015 Certified Professional (Pro.) exam.

Pro.	Exam Objective	Training Guide	Chapter & Section(s)
Collaboration			
✓	Create and modify levels	• Revit Structure Fundamentals	• 3.3
✓	Create and modify structural grids	• Revit Structure Fundamentals	• 4.1
✓	Import AutoCAD files into Revit	• Revit Structure Fundamentals	• 3.1
		• Revit Collaboration Tools	• 3.1
✓	Link Revit models	• Revit Structure Fundamentals	• 3.2
		• Revit Collaboration Tools	• 2.1
✓	Control the visibility for linked objects	• Revit Collaboration Tools	• 2.2
Documentation			
✓	Using temporary dimensions	• Revit Structure Fundamentals	• 2.1
✓	Annotate beams	• Revit Structure Fundamentals	• 13.3
✓	Add and modify text annotations	• Revit Structure Fundamentals	• 13.2
✓	Add and use dimensions and dimension labels	• Revit Structure Fundamentals	• 13.1
✓	Use detail components	• Revit Structure Fundamentals	• 14.2

Pro.	Exam Objective	Training Guide	Chapter & Section(s)
✓	Create and modify column schedules	• Revit Structure Fundamentals	• 15.2
✓	Create and modify footing schedules	• Revit Structure Fundamentals	• 15.3 • B.8
		• Revit BIM Management	• 2.1 to 2.3
✓	Create and modify standard sheets	• Revit Structure Fundamentals	• 12.1, 12.2
Modeling			
✓	Place and modify structural columns	• Revit Structure Fundamentals	• 4.2
✓	Place and modify walls	• Revit Structure Fundamentals	• 5.1
✓	Create custom wall types	• Revit BIM Management	• 3.1
✓	Place footings	• Revit Structure Fundamentals	• 5.2, 5.4
✓	Create concrete slabs and/or foundations	• Revit Structure Fundamentals	• 8.1
✓	Create and modify stepped walls in foundations	• Revit Structure Fundamentals	• 5.2
✓	Place rebar	• Revit Structure Fundamentals	• 9.2
✓	Add beams	• Revit Structure Fundamentals	• 6.1
✓	Add beam systems	• Revit Structure Fundamentals	• 6.1
✓	Add joists	• Revit Structure Fundamentals	• 6.1
✓	Add cross bracing to joists	• Revit Structure Fundamentals	• 6.1
✓	Create and use trusses	• Revit Structure Fundamentals	• 6.3
✓	Create and modify floors	• Revit Structure Fundamentals	• 8.1
✓	Create and modify custom floors	• Revit BIM Management	• 3.1
✓	Create and modify sloped floors	• Revit Architecture Fundamentals	• 9.3
✓	Add floor openings for stairs	• Revit Structure Fundamentals	• 8.2
✓	Create and modify stairs	• Revit Architecture Fundamentals	• 12.1
✓	Create and modify ramps	• Revit Architecture Fundamentals	• 12.5
✓	Model and use roofs	• Revit Structure Fundamentals	• 8.1
		• Revit Architecture Fundamentals	• 11.2, 11.4

Pro.	Exam Objective	Training Guide	Chapter & Section(s)
Views			
✓	Create section views	• Revit Structure Fundamentals	• 7.4
✓	Create framing elevations	• Revit Structure Fundamentals	• 7.4
✓	Use callout views	• Revit Structure Fundamentals	• 7.3

A
Align Geometry **4-35**
Analytical Model Settings **A-36**
Annotation Symbols **6-18**
Arrowhead Types **A-6**
Assets **A-23**
  Appearance **A-23**
  Physical **A-23**
  Replacing Material Assets **A-23**
  Thermal **A-23**

B
Baluster Family **7-30**
Baluster Placement **7-24**
Blends **4-31**
Boundary Condition Settings **A-38**
Browser Organization **A-45**
Building Component Schedules **2-5**

C
Cable Tray Types **3-30**
Calculated Value Field **2-59**
Component Families **1-5**
Compound Ceilings **3-2**
Conduit Types **3-30**
Connectors **5-5**
  Families **5-4**
Coping **7-15**
Cornice **7-15**
Cut Geometry **3-22**

D
Detail Component **5-6**
Dimensions
  In Families **4-8**
  Labeling **4-9**
Doors **7-2**
Double-click Options **A-44**
Duct Types **3-25**

E
Electrical Settings **A-30**
Embedded Schedule **2-62**
Extrusions **4-30**

F
Families
  Add Controls **5-3**
  Component **5-6**
  Create **4-3**
  Create Types **4-57**
  File Locations **4-5**
  Formulas **4-12**
  In-place **6-2**
  Load into Project **4-59**
  Panel **7-30**
  Parametric Framework **4-6**
  Post **7-31**
  Rails **7-27**
Fascia **7-15**
Fill **A-12**
Filters
  Applying **1-42**
  Creating **1-41**
Flexing Geometry **4-10**
Floors
  Creating Types **3-2**
  Edit Assemblies **3-3**
Formulas **4-12**

I
Insert Views From File command **2-3**

K
Keyboard Shortcuts **A-43**

L
Labels
  Add to Dimension **4-9**
  Adding **1-20**
Lines
  Line Color **A-9**
  Model Lines **5-32**
  Patterns **A-10**
  Styles **A-11**
  Symbolic Lines **5-32**

Weights **A-8**
Load Cases **A-36**
Load Combinations **A-36**
Lock Geometry **4-35**

M
Masking Regions **5-33**
Material Takeoff Schedule **2-64**
Materials
Assets **A-23**
Assign **A-16**
Creating **A-18**
Libraries **A-17**
Properties **2-66**
Mechanical Settings **A-24**
Merging Regions **3-13**

O
Object Styles
Creating **A-7**
Subcategories **7-17**

P
Parameters **4-9**
Creating in Families **4-11**
Formulas **4-12**
Shared **6-24**
Parametric Framework **4-6**
Percentage Field **2-60**
Pipe Types **3-28**
Profiles **6-4**

R
Rail Profiles **7-27**
Rail Structure **7-23**
Railings **7-22**
Reference Planes **4-8**
Revision Schedules **1-22**
Revolves **4-32**
Roofs
Creating Types **3-2**
Editing Assemblies **3-3**
Fascia command **7-16**
Routing Preferences
Ducts **3-26**
Pipes **3-28**

S
Schedule Properties
Appearance Tab **2-10**
Fields Tab **2-6**
Filter Tab **2-7**
Formatting Tab **2-9**
Sorting/Grouping Tab **2-8**
Schedule/Quantities command **2-5**
Schedules
Appearance **2-35**
Building Component Schedule **2-5**
Columns **2-31**
Conditional Formating **2-61**
Creating **2-2**
Embedded **2-62**
Exporting **2-4**
Formulas **2-58**
Headers **2-33**
Importing **2-3**
Key Styles **2-50**
Parameter Values **2-30**
Rows **2-31**
Sheet List **A-39**
Titles **2-33**
Sheet List Schedule **A-39**
Tables **A-40**
Snap Settings **A-4**
Solid Forms **4-29**
Starting View **1-6**
Status Bar **A-42**
Sweeps **4-33**
Swept Blends **4-34**
Symbolic Representation **A-35**

T
Tags
Loading **1-14**
Types **1-12**
Templates
Create View Templates **1-38**
Set Default **1-7**
Sheets **1-5**
Views **1-5**
Temporary Dimensions **A-5**
Text Types **1-8**
Title Blocks **1-19**
Top Rail **7-25**

U
Units **A-2**

V
Vertically Stacked Wall **3-20**
Visibility Display Settings **5-31**
Visibility/Graphic Overrides **1-40**
Void Forms **4-29**

W
Wall Sweep command **7-15**
Walls
Edit Assembly **3-3**
Embedding **3-22**
Reveals **3-14**
Sweeps **3-14**
Vertically Compound **3-11**
Windows **7-2**